68th Annual Highway Geology Symposium (HGS 2017)

Better Highways Through Applied Geology

Marietta, Georgia, USA
1-4 May 2017

ISBN: 978-1-5108-5172-6

Printed from e-media with permission by:

Curran Associates, Inc.
57 Morehouse Lane
Red Hook, NY 12571

Some format issues inherent in the e-media version may also appear in this print version.

Printed by Curran Associates, Inc. (2017)

Highway Geology Symposium
C/O F.R. Glass
100 Wolfe Cove Rd
Asheville, NC 28804

Phone: (828) 252-2260
Fax: (828) 299-1273

frgeol@aol.com

Additional copies of this publication are available from:

Curran Associates, Inc.
57 Morehouse Lane
Red Hook, NY 12571 USA
Phone: 845-758-0400
Fax: 845-758-2633
Email: curran@proceedings.com
Web: www.proceedings.com

At-A-Glance Schedule of Events

Monday, May 1 – Thursday, May 4, 2017

Monday, May 1

11:00 AM – 5:00 PM
Highway Geology Symposium Registration Open in front of Meeting Room

1:00 PM – 4:00 PM
Transportation Research Board Technical Session:
"State-of-the-Art Practices in Subsurface Investigations"
Location: Cole Room (Downstairs)

5:00 PM – 8:30 PM
Highway Geology Symposium Exhibitor Area Open

6:30 PM – 8:30 PM
Ice Breaker Social—Sponsored by BGC Engineering
Location: Joe Mack Wilson (JMW) Foyer and Exhibitor Area

Tuesday, May 2

6:30 AM – 9:00 AM
Breakfast
Location: JMW Foyer

6:30 AM – 5:00 PM
Highway Geology Symposium Registration Open in JMW Foyer

8:00 AM – 5:00 PM
Highway Geology Symposium Exhibitor Area Open

7:30 AM – 8:30 AM
Welcome and Opening Remarks
Deana Sneyd, HGS Organizing Committee Chair
Dedication of Proceedings (Pete Ingraham)
Monica Flournoy, Georgia Department of Transportation State Materials and Testing Engineer
Dr. Tim Chowns, PG, Professor Emeritus, University of West Georgia
Location: JMW Ballroom

Highway Geology Symposium Guest Field Trip to Helen and Winery

8:30 AM – 4:30 PM
Field trip transportation sponsored by Maccaferri
Pick-up Location: Hotel Front Lobby

Tuesday, May 2 cont.

Technical Session 1 – Young Authors

Location: JMW Ballroom
Chris Ruppen, Moderator

Tuesday, May 2 cont.

Technical Session 2

Location: JMW Ballroom
Pete Ingraham, Moderator

2:20 PM – 2:50 PM
Afternoon Break—Sponsored by IDS Corporation
Location: Exhibitor Area

Technical Session 3

Location: JMW Ballroom
Bob Henthorne, Moderator

4:15 PM – 4:45 PM
Highway Geology Symposium Field Trip Preview
Presenters: Randy Kath
Location: JMW Ballroom

Enjoy the trip!

5:00 PM – 6:30 PM
HGS Steering Committee Meeting
Location: Cole Room (Downstairs)

Free evening to explore and dine in Marietta

Wednesday, May 3

6:00 AM – 7:00 AM
To-Go Continental Breakfast—Sponsored by Vulcan
Location: Hotel Front Lobby

Highway Geology Symposium Field Trip

6:45 AM – 7:15 AM
Load buses for Field Trip
Pick-up Location: Meet in Hotel Front Lobby

7:15 AM – 5:00 PM
Field Trip
Coffee and Pastries—Sponsored by Scarptec
Lunch—Sponsored by GeoBrugg
Afternoon Beverages—Sponsored by Golder Associates
(NO GLASS ALLOWED INSIDE BUSES)

5:30 PM – 6:30 PM
Highway Geology Symposium Social Hour—Sponsored by Access Limited Construction
Location: JMW Ballroom

Highway Geology Symposium Banquet Dinner

6:30 PM – 9:30 PM
Keynote Address by Dr. Scott Hippensteel:
Geoarchaeology and the Secrets of the Civil War Submarine "H.L. Hunley"
Location: JMW Ballroom

Thursday, May 4

6:30 AM – 9:00 AM
Breakfast
Location: JMW Foyer and Exhibitor Area

8:00 AM – 2:00 PM
Highway Geology Symposium Exhibitor Area Open
Exhibitors can break down after morning coffee break

Technical Session 4

Location: JMW Ballroom
Steve Sweeney, Moderator

9:30 AM – 10:00 AM
Morning Coffee Break—Sponsored by S&ME
Location: JMW Foyer and Exhibitor Area

Thursday, May 4 cont.

10:00 AM – 10:20 AM
345 **An Example of Risk-Based Geotechnical Asset Management**
Author(s): Scott Anderson, Mark Vessely, and Ty Ortiz

10:20 AM – 10:40 AM
364 **Montana's Rock Slope Asset Management Program (RAMP)**
Author(s): Darren Beckstrand, Aine Mines, Brent Black, Jeff Jackson, Bret Boundy, Scott Helm, David A. Stanley, and Paul D. Thompson

10:40 AM – 11:00 AM
384 **Foothills Parkway: Micropiles Support Closing the Missing Link**
Author(s): Anthony Sak

11:00 AM – 11:20 AM
399 **"Stub Pier" Stabilization Performance**
Author(s): Swaminathan Srinivasan and Jess A. Schroeder

11:20 AM – 11:40 AM
418 **Information Modeling Workflows for Using Geotechnical Data in Civil Engineering**
Author(s): Nicolas Loubier and Katie Aguilar

11:40 AM – 12:00 PM
426 **Road Widening in Creep-prone Bituminous Sandstone**
Author(s): Gresham D. Eckrich and Jerko Kocijan

12:00 PM
Closing Remarks and Adjournment

Travel safely; we'll see you next year in Maine

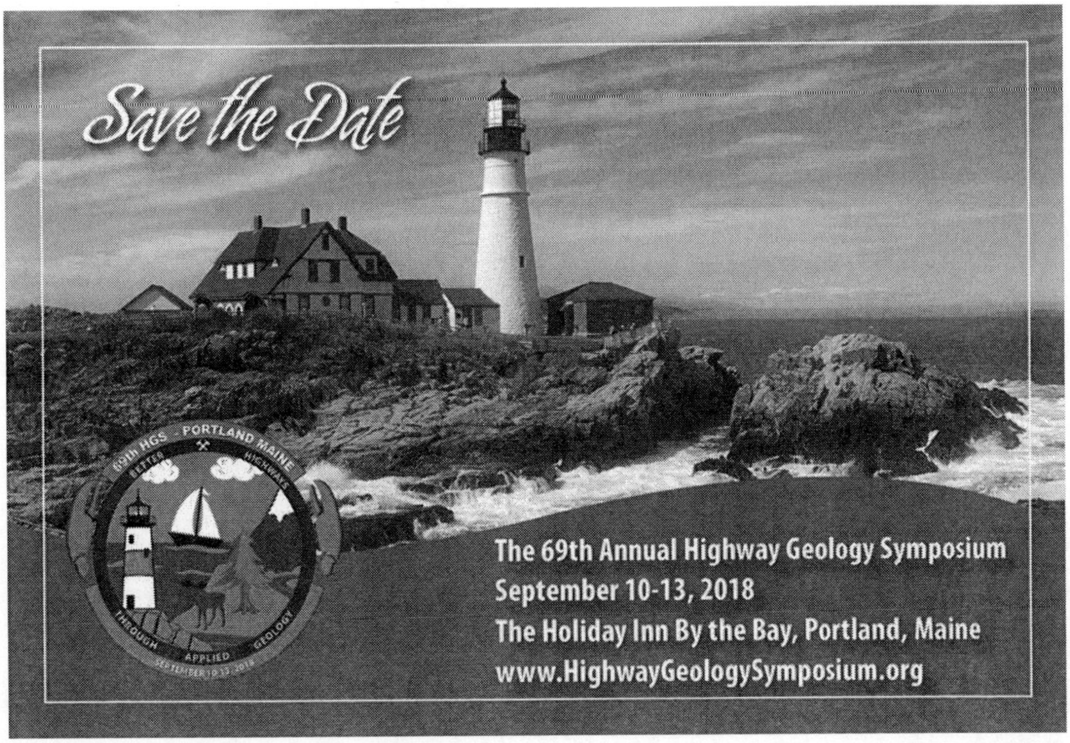

68TH HIGHWAY GEOLOGY SYMPOSIUM
MAY 1-4, 2017

Hilton Atlanta/Marietta Hotel and Conference Center

MARIETTA, GEORGIA

68th HGS – NORTH GEORGIA - 2017

BETTER HIGHWAYS THROUGH APPLIED GEOLOGY

Georgia

2017 Proceedings

BETTER HIGHWAYS THROUGH APPLIED GEOLOGY

68TH HIGHWAY GEOLOGY SYMPOSIUM

MAY 1-4, 2017

Dedication

The Proceedings of the 68th Highway Geology Symposium are dedicated to

Richard (Dick) Cross

Dick Cross (Rick to his family) was a fixture, along with his wife Linda, at HGS meetings for decades (from 1988 until 2015), often driving with Linda and their family to the HGS clear across the country in his van. Dick and Linda married in 1966, and he started his career with the New York State DOT in 1970. His early career with the DOT saw him working with geophysics in new transportation corridors and mapping/coring rock units supporting roadway and bridge construction all over the state. In 1988 he responded to a fatal rock strike on the NY State Thruway and was asked to become their Engineering Geologist. He developed the Thruway's rockfall hazard rating system, and was responsible for their rock slope assessment and maintenance program system-wide for over 900 slopes.

Dick retired from the NY Thruway in 1999 and went to work for Golder Associates in 2000, getting the opportunity to travel to Hawaii, California, Maine and Louisiana in support of FEMA projects, and he saw the completion of the largest Thruway contract for interchange reconfiguration at the Tappan Zee Bridge before finally retiring in August, 2012.

Dick hosted the 1991 HGS in Albany, helped Tom Eliassen host the 2003 HGS in Burlington, Vermont, and helped Mike Vierling and Steve Sweeney host the 2009 HGS in Buffalo. He received the Medallion award in 2009 for his many contributions serving on the Highway Geology Symposium Steering Committee for 22 straight years.

Dick always had time to help others, and was confident that things would always work out fine if folks just worked the problem. I will always remember him as he looks in this photo – the wry, confident smile, polo shirt and Woolrich shirt-jac that was worn year round – until it got too hot for the shirt-jac.

I would be remiss if I omitted that Dick was a man of God, and a very good engineering geologist. We lost Dick to cancer last year on July 4th. His faith helped him address his own mortality with the same grace and dignity with which he lived his life and career.

We dedicate these proceedings to his memory, and will miss him.

— Pete Ingraham

Dick Cross

68TH HIGHWAY GEOLOGY SYMPOSIUM
MAY 1-4, 2017

Hilton Atlanta/Marietta Hotel and Conference Center
MARIETTA, GEORGIA

68th HGS Local Organizing Committee

Deana Sneyd (Co-Chair)
Krystle Pelham (Co-Chair)
Pete Ingraham (Papers and Agenda)
John Pilipchuk (Website and Registration)
Shawnie Gruber (Logistics)
Randy Kath (Field Trip)
Ben Rivers (Field Trip Support)
Tararra Hill
Kathryn Haines
Randy Sullivan

Grateful Acknowledgments

HGS Steering Committee
HGS Local Organizing Committee
Brian Jones, City of Atlanta
Adam Bedell, Stantec
Dave Scarpato, Scarptec
Joe Stika, Buzzi Unicem USA
Rusty Simmons, US Army Corps of Engineers
Stanly Bearden, New Riverside Ochre Co.
Our Sponsors and Exhibitors
Golder Associates Inc.

At left: Rock dowel layout using rope access techniques on the Bellwood Quarry rock slope, as part of temporary rockfall mitigation efforts for tunneling associated with the City of Atlanta, Water Supply Program – Phase 1. *Photo courtesy Dave Scarpato, Scarptec, Inc.*

On the cover: A view of thea Bellwood Quarry project with the Atlanta skyline in the distance. The rectangular concrete pad at the bottom of the quarry is the tunnel boring machine (TBM) launch pad, and the construction going on the left side of the image is in support of the pumping stations. *Photo courtesy Atkinson-Technique J/V - Tunnel General Contractor.*

Hilton Atlanta/Marietta Floorplan

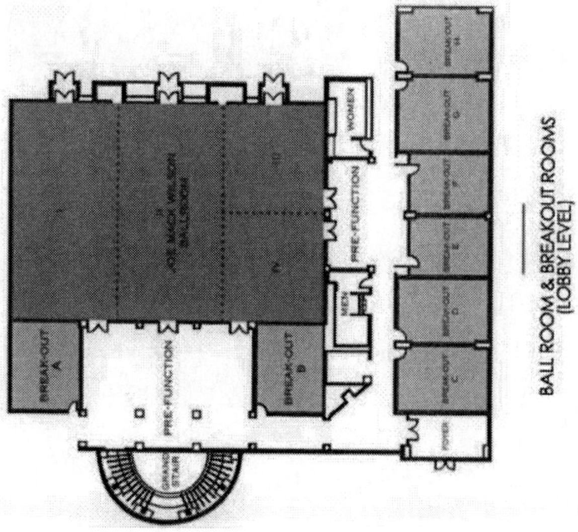

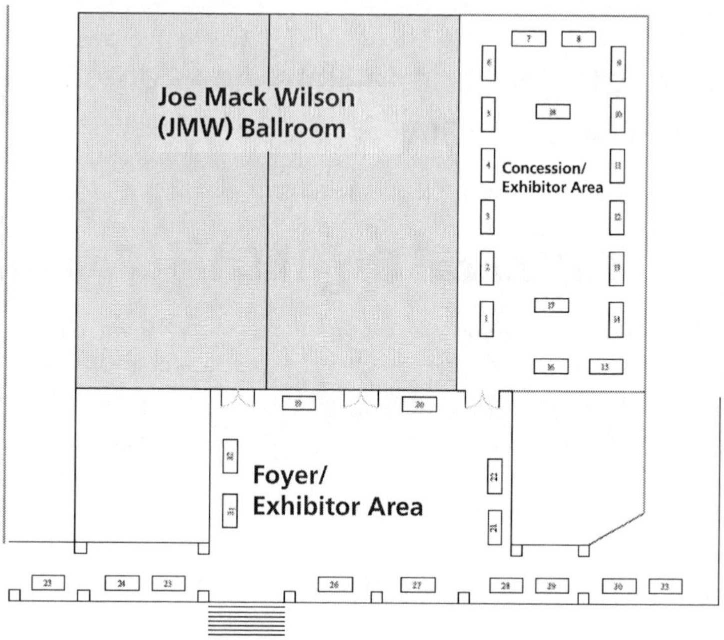

Cole Room located downstairs

Booth Locations in Exhibitor Area

Booth Number	Company	Booth Number	Company
1	Hayward Baker	17	Olson Engineering
2	Monotube	18	Maccaferri
3	SIMCO Drilling Equipment	19	S&ME
4	Vertek	20	Geobrugg
5	CME	21	BGC Engineering
6	Con-Tech	22	Access Limited
7	Vulcan	23	TenCate
8	RST Instruments	24	Golder Associates
9	GeoStabilization	25	EGSci
10	Atlas Pipe Piles	26	Ameritech Slope Constructors
11	GeoKon	27	IDS Corporation
12	Canary Systems	28	Pacific
13	KANE GeoTech	29	AMS, Inc.
14	Mabey Inc.	30	GEL Geophysics, LLC
15	HI-TECH	31	Scarptec
16	Williams	32	Stantec

Highway Geology Symposium: History, Organization, and Function

Inaugural Meeting

Established to foster a better understanding and closer cooperation between geologists and civil engineers in the highway industry, the Highway Geology Symposium (HGS) was organized and held its first meeting on March 14, 1950, in Richmond, Virginia. Attending the inaugural meeting were representatives from state highway departments (as referred to at that time) from Georgia, South Carolina, North Carolina, Virginia, Kentucky, West Virginia, Maryland, and Pennsylvania. In addition, a number of federal agencies and universities were represented. A total of nine technical papers were presented.

W.T. Parrott, an engineering geologist with the Virginia Department of Highways, chaired the first meeting. It was Mr. Parrott who originated the Highway Geology Symposium.

It was at the 1956 meeting that future HGS leader, A.C. Dodson, began his active role in participating in the Symposium. Mr. Dodson was the Chief Geologist for the North Carolina State Highway and Public Works Commission, which sponsored the 7th HGS meeting.

East and West

Since the initial meeting, 64 consecutive annual meetings have been held in 33 different states. Between 1950 and 1962, the meetings were east of the Mississippi River, with Virginia, West Virginia, Ohio, Maryland, North Carolina, Pennsylvania, Georgia, Florida, and Tennessee serving as host state. In 1962, the symposium moved west for the first time to Phoenix, Arizona, where the 13th annual HGS meeting was held. Since then, it has alternated, for the most part, back and forth from the east to the west.

The Annual Symposium has moved to different locations as listed on the next page.

Organization

Unlike most groups and organizations that meet on a regular basis, the Highway Geology Symposium has no central headquarters, no annual dues, and no formal membership requirements. The governing body of the Symposium is a steering committee composed of approximately 20 – 25 engineering geologists and geotechnical engineers from state and federal agencies, colleges and universities, as well as private service companies and consulting firms throughout the country. Steering committee members are elected for three-year terms, with their elections and re-elections being determined principally by their interests and participation in and contribution to the Symposium. The officers include a chairman, vice chairman, secretary, and treasurer, all of whom are elected for a two-year term. Officers, except for the treasurer, may only succeed themselves for one additional term.

A number of three-member standing committees conduct the affairs of the organization. The lack of rigid requirements, routing, and relatively relaxed overall functioning of the organization is what attracts many participants.

Meeting sites are chosen two to four years in advance and are selected by the Steering Committee following presentations made by representatives of potential host states. These presentations are usually made at the steering committee meeting, which is held during the Annual Symposium.

Upon selection, the state representative becomes the state chairman and a member pro-tem of the Steering Committee.

List of Highway Geology Symposium Meetings

No.	Year	HGS Location	No.	Year	HGS Location
1st	1950	Richmond, VA	2nd	1951	Richmond, VA
3rd	1952	Lexington, VA	4th	1953	Charleston, WV
5th	1954	Columbus, OH	6th	1955	Baltimore, MD
7th	1956	Raleigh, NC	8th	1957	State College, PA
9th	1958	Charlottesville, VA	10th	1959	Atlanta, GA
11th	1960	Tallahassee, FL	12th	1961	Knoxville, TN
13th	1962	Phoenix, AZ	14th	1963	College Station, TX
15th	1964	Rolla, MO	16th	1965	Lexington, KY
17th	1966	Ames, IA	18th	1967	Lafayette, IN
19th	1968	Morgantown, WV	20th	1969	Urbana, IL
21st	1970	Lawrence, KS	22nd	1971	Norman, OK
23rd	1972	Old Point Comfort, VA	24th	1973	Sheridan, WY
25th	1974	Raleigh, NC	26th	1975	Coeur d'Alene, ID
27th	1976	Orlando, FL	28th	1977	Rapid City, SD
29th	1978	Annapolis, MD	30th	1979	Portland, OR
31st	1980	Austin, TX	32nd	1981	Gatlinburg, TN
33rd	1982	Vail, CO	34th	1983	Stone Mountain, GA
35th	1984	San Jose, CA	36th	1985	Clarksville, TN
37th	1986	Helena, MT	38th	1987	Pittsburgh, PA
39th	1988	Park City, UT	40th	1989	Birmingham, AL
41st	1990	Albuquerque, NM	41st	1991	Albany, NY
43rd	1992	Fayetteville, AR	44rd	1993	Tampa, FL
45th	1994	Portland, OR	46th	1995	Charleston, WV
47th	1996	Cody, WY	48th	1997	Knoxville, TN
49th	1998	Prescott, AZ	50th	1999	Roanoke, VA
51st	2000	Seattle, WA	52nd	2001	Cumberland, MD
53rd	2002	San Luis Obispo, CA	54th	2003	Burlington, VT
55th	2004	Kansas City, MO	56th	2005	Wilmington, NC
57th	2006	Breckinridge, CO	58th	2007	Pocono Manor, PA
59th	2008	Santa Fe, NM	60th	2009	Buffalo, NY
61st	2010	Oklahoma City, OK	62nd	2011	Lexington, KY
63rd	2012	Redding, CA	64th	2013	North Conway, NH
65th	2014	Laramie, WY	66th	2015	Sturbridge, MA
67th	2016	Colorado Springs, CO	68th	2017	Marietta, GA
69th	2018	Portland, ME			

HGS History, Organization, and Function *cont.*

The symposia are generally scheduled for two and one-half days, with a day-and-a-half for technical papers plus a full day for the field trip. The Symposium usually begins on Wednesday morning. The field trip is usually Thursday, followed by the annual banquet that evening. The final technical session generally ends by noon on Friday. In recent years, this schedule has been modified to better accommodate climate conditions and tourism benefits.

The Field Trip

The field trip is the focus of the meeting. In most cases, the trips cover approximately 150 to 200 miles, provide for six to eight scheduled stops, and require about eight hours. Occasionally, cultural stops are scheduled around geological and geotechnical points of interests.

To cite a few examples: in Wyoming (1973), the group viewed landslides in the Big Horn Mountains; Florida's trip (1976) included a tour of Cape Canaveral and the NASA space installation; the Idaho and South Dakota trips dealt principally with mining activities; North Carolina provided stops at a quarry site, a dam construction site, and a nuclear generation site; in Maryland, the group visited the Chesapeake Bay hydraulic model and the Goddard Space Center. The Oregon trip included visits to the Columbia River Gorge and Mount Hood; the Central mine region was visited in Texas; and the Tennessee meeting in 1981 provided stops at several repaired landslide in Appalachia regions of East Tennessee.

In Utah (1988), the field trip visited sites in Provo Canyon and stopped at the famous Thistle Landslide, while in New Mexico, in 1990, the emphasis was on rockfall treatments in the Rio Grande River canyon and included a stop at the Brugg Wire Rope headquarters in Santa Fe.

Mount St, Helens was visited by the field trip in 1994 when the meeting was in Portland, Oregon, while in 1995 the West Virginia meeting took us to the New River Gorge Bridge that has a deck elevation of 876 feet above the water.

In Cody, Wyoming, the 1996 field trip visited the Chief Joseph Scenic Highway and the Beartooth Uplift in northwest Wyoming. In 1997, the meeting in Tennessee visited the newly constructed future I-26 highway in the Blue Ridge of East Tennessee. The Arizona meeting in 1998 visited the Oak Creek Canyon near Sedona and a mining ghost town at Jerrome, Arizona. The Virginia meeting in 1999 visited the "Smart Road" Project that was under construction. This was a joint research project of the Virginia Department of Transportation and Virginia Tech University. The Seattle Washington meeting in 2000 visited the Mount Rainier area. A stop during the Maryland meeting in 2001 was the Sideling Hill road cut for I-68 which displayed a tightly folded syncline in the Allegheny Mountains.

The California field trip in 2002 provided a field demonstration of the effectiveness of rock netting against rock falls along the Pacific Coast Highway. The Kansas City meeting in 2004 visited the Hunt Subtropolis, which is said to be the "world's largest underground business complex," created through the mining of limestone using the room and pillar method. The Rocky Point Quarry provided an opportunity to search for fossils at the North Carolina meeting in 2005. The group also visited the US-17 Wilmington Bypass Bridge, which was under construction. Among the stops at the Pennsylvania meeting, were the Hickory Run Boulder Field, the No. 9 Mine and Wash Shanty Museum, and the Lehigh Tunnel.

The New Mexico field trip in 2008 included stops at a soil nailed wall along US-285/84 north of Santa Fe, and a road cut through the Bandelier Tuff on highway 502 near Los Alamos, where rockfall mesh was used to protect against rockfall. The New York field trip in 2009 visited the Niagara Falls Gorge and the Devil's Hole Trail. The Oklahoma field trip in 2010 toured through the complex geology of the Arbuckle Mountains in the southern part of the state along with stops at Tucker's Tower and Turner Falls.

In the bluegrass region of Kentucky, the 2011 HGS field trip included stops at Camp Nelson which is the site of the oldest exposed rocks in Kentucky near the Lexington and Kentucky River Fault Zones. Additional stops at the Darby Dan Farm and the Woodford Reserve Distillery illustrated how the local geology has played such a large part in the success of breeding prized Thoroughbred horses and made Kentucky the "Birthplace of Bourbon."

In Redding, California, the 2012 field trip included stops at the Whiskeytown Lake, which is one in a series of lakes that provide water and power to northern California. Additional stops included Rocky Point, a roadway construction site containing Naturally Occurring Asbestos (NOA), and Oregon Mountain where the geology and high rainfall amounts have caused Hwy 299 to experience local and global instabilities since first constructed in 1920.

HGS History, Organization, and Function *cont.*

The 2013 field trip of New Hampshire highlighted the topography and geologic remnants left by the Pleistocene glaciations that fully retreated approximately 12,000 years ago. The field trip included stops at various overlooks of glacially-carved valleys and ranges; the Old Man of The Mountain Memorial Plaza, which is a tribute to the famous cantilevered rock mass in the Franconia Notch that collapsed on May 3, 2003; lacustrine deposits and features of the Glacial Lake Ammonoosuc; views of the Presidential Range; bridges damaged during Tropical Storm Irene in August 2011; and the Willey Slide, located in the Crawford Notch where all members of the Willey family homestead were buried by a landslide in 1826.

2014 presented a breathtaking tour of the geology and history of southeast Wyoming, ascending from the high plains surrounding Laramie at 7,000 feet to the Medicine Bow Mountains along the Snowy Range Scenic Byway. Visible along the way were a Precambrian shear zone, and glacial deposits and features. From the glacially carved Mirror Lake and the Snowy Range Ski Area, the path wound east to the Laramie Mountains and the Vedauwoo Recreational Area, a popular rock climbing and hiking area, before returning to Laramie.

In Sturbridge, MA, the 2015 field trip focused on the Connecticut Valley, a Mesozoic rift basin that signaled the breakup of Pangea, and the Berkshires, which represents the collision and amalgamation of an island arc system with the North American Laurentian margin.

The field trip in 2016 was an urban setting along the western edge of Colorado Springs and around Manitou Springs. Stops included the Pikeview Quarry, Garden of Gods Visitor Center, and several other locations where rockfall and debris flow mitigation, post-flooding highway embankment repair, and a nonconformity in the rock records that spans 1.3 billion years were observed.

Technical Sessions and Speakers

At the technical sessions, case histories and state-of-the-art papers are most common; with highly theoretical papers the exception. The papers presented at the technical sessions are published in the annual proceedings. Some of the more recent papers may be obtained from the Treasurer of the Symposium. Banquet speakers are also a highlight and have been varied through the years.

Member Recognition

Medallion Award. A Medallion Award was initiated in 1970 to honor those persons who have made significant contributions to the Highway Geology Symposium over many years. The award is a 3.5 inch medallion mounted on a walnut shield and appropriately inscribed. The award is presented during the banquet at the annual Symposium. The selection was and is currently made from the members of the national steering committee of the HGS.

Emeritus Members. A number of past members of the national steering committee have been granted Emeritus status. These individuals, usually retired, resigned from the HGS Steering Committee, or are deceased, have made significant contributions to the Highway Geology Symposium. Emeritus status is granted by the Steering Committee. A total of 34 persons have been granted Emeritus status. Fourteen are now deceased.

Dedications. Several Proceedings volumes have been dedicated to past HGS Steering Committee members or others who have made outstanding contributions to HGS. The 36th HGS Proceedings were dedicated to David L. Royster (1931 - 1985, Tennessee) at the Clarksville, Indiana meeting in 1985. In 1991, the Proceedings of the 42nd HGS held in Albany, New York were dedicated to Burrell S. Whitlow (1929 – 1990, Virginia). In 2013, the Proceedings of the 64th HGS held in North Conway, New Hampshire were dedicated to Earl Wright and Bill Lovell. The 2014 Proceedings of the 65th HGS held in Laramie, Wyoming were dedicated to Nicholas Michiel Priznar. The 2015 Proceedings of the 66th HGS were dedicated to Michael Hager, and the 67th HGS 2016 Proceedings were dedicated to Vern McGuffey. The 68th Proceedings are dedicated to Richard (Dick) Cross.

HGS Medallion Award Winners

Hugh Chase*	1970
Tom Parrott*	1970
Paul Price*	1970
K.B. Woods*	1971
R.J. Edmondson*	1972
C.S. Mullin*	1974
A.C. Dodson*	1975
Burrell Whitlow*	1978
Bill Sherman	1980
Virgil Burgat*	1981
Henry Mathis	1982
David Royster*	1982
Terry West	1983
Dave Bingham	1984
Vernon Bump	1986
C.W. "Bill" Lovell*	1989
Joseph A. Gutierrez	1990
Willard McCasland	1990
W.A. "Bill" Wisner	1991
David Mitchell	1993
Harry Moore	1996
Earl Wright*	1997
Russell Glass	1998
Harry Ludowise*	2000
Sam Thornton	2000
Bob Henthorne	2004
Mike Hager	2005
Joseph A. Fischer	2007
Ken Ashton	2008
A. David Martin	2008
Michael Vierling	2009
Dick Cross*	2009
John F. Szturo	2010
Christopher Ruppen	2012
Jeff Dean	2012
John Pilipchuk	2015
Peter Ingraham	2016

** Deceased*

Young Author Award Winners

2014 Simon Boone, "Performance of Flexible Debris Flow Barriers in a Narrow Canyon"
2015 Cory Rinehart, "High Quality H20: Utilizing Horizontal Drains for Landslide Stabilization"
2016 Todd Hansen, "Geologic Exploration for Ground Classification: Widening of the I-70 Veterans Memorial Tunnels"

Emeritus Members of the Steering Committee

RR. F. Baker*
John Baldwin
David Bingham
Vernon Bump
Virgil E. Burgat*
Robert G. Charboneau*
Hugh Chase*
Dick Cross*
A. C. Dodson*
Walter Fredericksen
Brandy Gilmore
Robert Goddard
Joseph Gutierrez
Mike Hager
Rich Humphries
Charles T. Janik
John Lemish
Bill Lovell*
George S. Meadors, Jr.*
Willard MaCasland
Henry Mathis
David Mitchell
Harry Moore
W. T. Parrot*
Nicholas Priznar*
Paul H. Price*
David L. Royster*
Bill Sherman
Willard L Sitz*
Mitchell Smith
Steve Sweeney
Sam Thornton
Berke Thompson*
Burrell Whitlow*
W. A. "Bill" Wisner
Earl Wright*
Ed J. Zeigler

** Deceased*

HGS National Steering Committee

HGS National Steering Committee *cont.*

Christopher A. Ruppen
Michael Baker Jr., Inc.
4301 Dutch Ridge Rd.
Beaver, PA 15009-9600
Phone: (724) 495-4079
Cell: (412) 848-2305
Fax: (724) 495-4017
Email: cruppen@mbakercorp.com

Stephen Senior
Ontario Min of Trans. (Retired)
11 Dewbourne Ave.
Richmond Hill, ON L4B 3G7
Canada
Phone: (416) 235-3734
Fax: (416) 235-4101
Email: sa.senior@rogers.com

Deana Sneyd
Petrologic Solutions, Inc.
3997 Oak Hill Road
Douglasville, GA 30135
Phone: (678) 313-4147
Email: dsneyd@gmail.com

Jim Stroud
Subhorizon Geologic Resources LLC
4541 Araby Ln.
East Bend, NC 27018
Cell: (336) 416-3656
Email: stroudjr@subhorizonresourc-es.com

Steven Sweeney
105 Albert Rd.
Delanson, NY 12053
Email: 2ssweeney@gmail.com

John F. Szturo
HNTB Corporation
715 Kirk Drive
Kansas City, MO 64105
Phone: (816) 527-2275 (Direct Line)
Cell: (913) 530-2579
Fax: (816) 472-5013
Email: jszturo@hntb.com

Michael P. Vierling
323 Boght Road
Watervliet, NY 12189-1106
Phone: (518) 233-1197
Email: rocdoc1956@gmail.com

Terry West
Earth and Atmospheric Science Dept.
Purdue University
West Lafayette, IN 47907-1297
Phone: (765) 494-3296
Fax: (765)496-1210
Email: trwest@purdue.edu

HGS Symposium Contact List

2009	New York	Mike Vierling		Rocdoc1959@gmail.com
2010	Oklahoma	Jeff Dean		jdean@odot.org
2011	Kentucky	Henry Mathis	859-455-8530	hmathis@iglou.com
2012	California	Bill Webster	916-277-1041	Bill_webster@dot.ca.gov
2013	New Hampshire	Krystle Pelham	603-271-1657	Krystle.Pelham@dot.nh.gov
2014	Wyoming	Jim Coffin	307-777-4205	Jim.coffin@wyo.go
2015	Massachusetts	Peter Ingraham	603-688-0880	pingraham@golder.com
2016	Colorado	Ty Ortiz	303-921-2634	Ty.ortiz@state.co.us
2017	Georgia	Deana Sneyd	678-313-4147	Dsneyd61@gmail.com

Opening Session Speakers

Monica Flournoy, State Materials Engineer, Georgia Department of Transportation

Monica L. Flournoy was named the State Materials Engineer for the Georgia Department of Transportation in July 2016. She has a Bachelor in Civil Engineering from Georgia Institute of Technology and is a registered Professional Engineer. She has been with the Department for 24 years and has varied experience in construction, construction claims analysis, project administration and management, construction bidding, and materials and testing.

Ms. Flournoy began her career with the Department as a Construction Project Engineer out of the Louisville Area Office, where she was responsible for administering construction contracts, including widening and reconstruction as well as bridge replacement projects. She oversaw the first soil-cement base construction project in the Tennille District. Prior to becoming the State Materials Engineer, she held the Administrator position for Office of Construction Bidding, where she oversaw the GDOT monthly Lettings and the Contractor Prequalification & Registration processes.

In addition to serving on GDOT's Estimating, Contractor Prequalification, and Pavement Design committees, among others, Ms. Flournoy has served on several state and national committees with the Georgia Partnership for Transportation Quality, Georgia Transportation partnership in Construction, and American Association of State Highway and Transportation Officials.

Dr. Tim Chowns, Professor Emeritus, University of West Georgia

Tim Chowns was born in London and graduated from the University of Leicester. After completing his PhD at Newcastle upon Tyne he served as a visiting assistant professor at the University of Georgia before moving to the University of West Georgia where he has taught for more than forty years. Although trained in sedimentology and stratigraphy, he has broad interests in geology and especially in the geology of Georgia.

He has published on the Pre-Cretaceous basement beneath the Coastal Plain and correlations with West Africa; the origin of sedimentary ironstones with special reference to the Birmingham ores; the formation of geodes by the silicification of evaporite nodules; and most recently, on drainage changes along the Georgia coast related to the breaching of inlets during the Holocene transgression.

Dr. Chowns has been a longstanding and enthusiastic member of the Georgia Geological Society, served as president, treasurer, and field trip leader over many years. Drawing on this experience he will provide us with a "Welcome to the Geology of Georgia."

Banquet Keynote Address

"Geoarchaeology and the Secrets of the Civil War Submarine H.L. Hunley"

Dr. Scott Hippensteel, University of North Carolina, Charlotte

Scott Hippensteel is an Associate Professor of Earth Sciences in the Department of Geography and Earth Sciences at the University of North Carolina at Charlotte. He joined UNC Charlotte after earning his PhD in Geology from the University of Delaware in 2000. His research and teaching interests involve using microfossils to solve environmental, historical, and geoarchaeological problems. His most recent publications have appeared in *Geosphere, GSA Today, Geoarchaeology, Journal of Coastal Research,* the Geological Society of America's Bulletin, as well as *The Chronicle of Higher Education.*

In 2004 he joined the research team investigating the sinking and later preservation of the Confederate Civil War submarine H.L. Hunley. This research is ongoing and represents a favorite lecture topic in his upper-level classes.

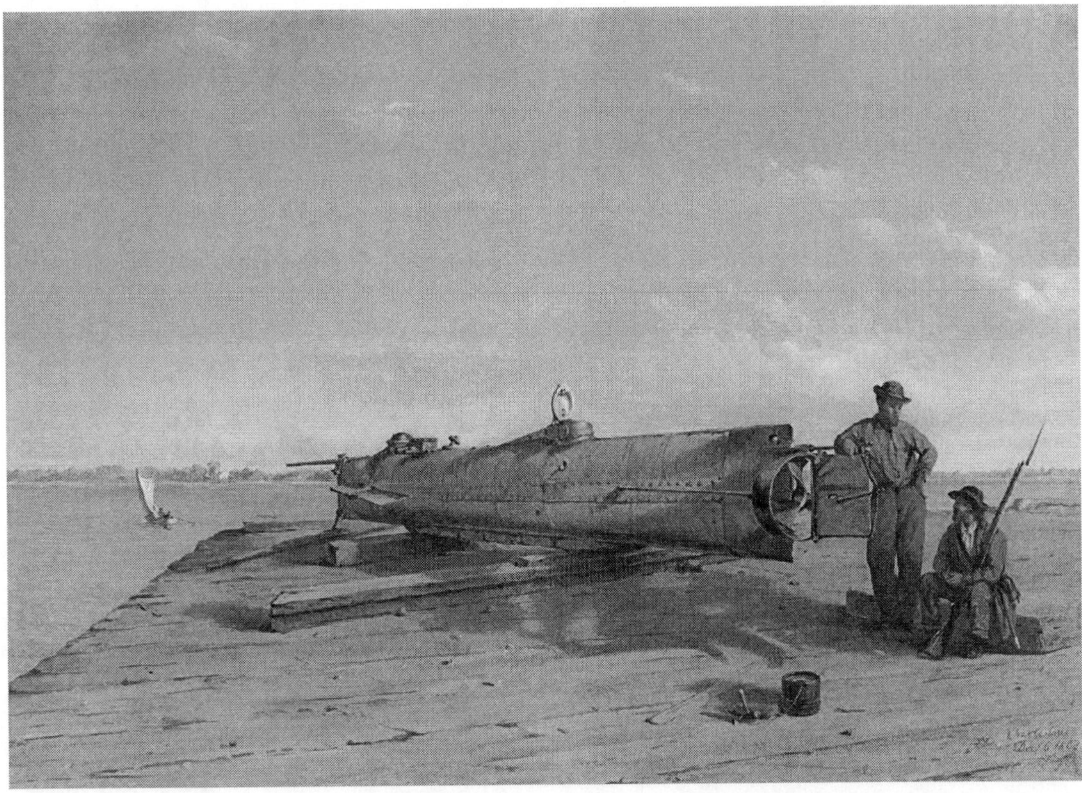

Submarine Torpedo Boat H.L. Hunley, Dec. 6, 1863 **by Conrad Wise Chapman (1864)**

Symposium Sponsors and Exhibitors

The following companies have graciously contributed toward the sponsorship of the Symposium. The HGS relies on sponsor contributions for refreshment breaks, field trip lunches, and other activities. We gratefully appreciate the contributions made by these generous sponsors.

Platinum Sponsor

GEOBRUGG®

BRUGG

Safety is our nature

Swiss company Geobrugg is the global leader in the supply of safety nets and meshes made of high-tensile steel wire. Many years of experience and intensive collaboration with universities and research institutes have made Geobrugg a reliable partner when it comes to protection and safety solutions.

A global network with branches and partners in over 50 countries ensures fast, thorough, and cost-effective solutions for customer requirements. With production facilities on four continents and more than 300 employees worldwide, Geobrugg combines short delivery times with local support for customers. We are partners, consultants, developers, and project managers for our customers.

Geobrugg North America, LLC
22 Centro Algodones
Algodones, NM 87001 USA

Phone: (505) 771-4080
Mobile: (505) 228-6425
geobrugg.com

Gold Sponsor

Ameritech Slope Constructors INC

Geotechnical Contractors

Ameritech Slope Constructors, Inc. is a multi-state licensed, specialty geotechnical construction firm located in Asheville, North Carolina. Our services include: manual rock scaling, high angle drilling, installation of rockfall barriers and rockfall drapes, as well as slope stabilization systems using soil nails and high strength mesh. Ameritech also installs rock bolts, cable anchors, rock dowels, and rock drains. Whether it is a rock face with loose debris or an unstable soil slope, we can install the system that is necessary to provide protection for people and property. The company is proud to offer a team of highly skilled professionals with over 100 years of combined experience in the rockfall and slope stabilization industry.

Ameritech Slope Constructors, Inc.
P.O. Box 2702
Asheville, NC 28802

Phone: (828) 633-6352
Fax: (828) 633-6353
ameritech.pro

Ameritech Slope Constructors INC

Geotechnical Contractors

- **Geological Solutions for Slope Stability**
- **Rockfalls**
- **Landslides**
- **Debris Flows**

Silver Sponsors

ACCESS LIMITED
CONSTRUCTION COMPANY
LIC #988368
San Luis Obispo

Located in San Luis Obispo, CA, Access Limited Construction is a General Contractor specializing in rockfall mitigation and slope stabilization systems, and is considered to be an industry leader in designing and installing rockfall protection, slope stabilization systems, and performing difficult access drilling throughout the United States.

Access Limited Construction Co.
225 Suburban Rd
San Luis Obispo, CA 93401

Phone: (805) 592-2230
info@accesslimitedconstruction.com
accesslimitedconstruction.com

HI-TECH
ROCKFALL
CONSTRUCTION

"The Rockfall Specialist"

HI-TECH Rockfall Construction, Inc., founded in 1996, is located in Forest Grove, Oregon, USA. HI-TECH is a General Contractor who specializes in rockfall mitigation and is considered to be the industry leader in designing and installing rockfall protection systems throughout the United States. HI-TECH constructs a vast array of rockfall mitigation systems in a variety of locations such as highways, railroads, dams, quarries, mines, construction sites, commercial and residential properties.

HI-TECH Rockfall Construction, Inc.
2328 Hawthorne St
Forest Grove, OR 97116

Phone: (503) 357-6508
hitechrockfall.com

Silver Sponsors

BGC Engineering Inc is a consulting firm providing specialist services in applied earth sciences with emphasis on geotechnical engineering (soil and rock mechanics), geoenvironmental engineering, permafrost engineering, hydrogeology, mining engineering and engineering geology. We have undertaken projects throughout Canada, United States, Mexico, Central and South America, Europe, Africa, Australia and Russia.

BGC Engineering Inc.
Suite 500, 980 Howe Street
Vancouver, BC
Canada V6Z 0C8

Phone: (604) 684-5900
Fax: (604) 684-5909
info@bgcengineering.ca
bgcengineering.ca

IDS GeoRadar is a world leader, designing and providing radar products for subsurface and surficial investigations. Various GPR and InSAR instruments are available.

IDS North America Inc.
14828 W. 6th Ave.
Ste. 12-B
Golden, CO 80401

Phone: (303) 232-3047 Ext. 121
Fax: (720) 519-1087
idsnabd@idscorporation.com
idscorporation.com/na

Bronze Sponsors

Scarptec, Inc. provides geological engineering and design solutions for challenging site conditions, with a laser-sharp focus on constructability. Whether it's a deep excavation or falling rock, ice, snow, and soil, Scarptec works collaboratively with our clients and other stakeholders to help mitigate the forces of Mother Nature.

Scarptec, Inc.
P.O. Box 326
Monument Beach, MA 02553

Contact: Dave Scarpato, P.E.
Phone: (603) 361-0397
dave@scarptec.com
scarptec.com

S&ME has a proven history of delivering engineering, environmental, and construction management solutions for transportation projects to include airports, rail/transit systems, roads, bridges and ports.

S&ME, Inc.
4350 River Green Parkway, Suite 200
Duluth, GA 30096

Contact: Jeffrey Doubrava
Phone: (770) -815-8343
jdoubrava@smeinc.com
smeinc.com

Bronze Sponsors

MACCAFERRI

"For nearly 140 years, Maccaferri has provided innovative solutions to the construction, geotechnical and mining industries. Renowned as the world leader in gabion retaining structures, we have diversified significantly over the past 20 years and now offer engineered solutions including reinforced soil structures, coastal protection, soil stabilisation, landfills, natural hazard mitigation and river control works.

Maccaferri, Inc.
10303 Governo Lane Blvd
Williamsport,MD 21795

Phone: (301) 233-6910
info@us.maccaferri.com
maccaferri-usa.com

Vulcan Materials Company

Vulcan Materials Company is the nation's largest producer of construction aggregates—primarily crushed stone, sand and gravel—and a major producer of aggregates-based construction materials, including asphalt and ready-mixed concrete. Our coast-to-coast footprint and strategic distribution network align with and serve the nation's growth centers.

Vulcan Materials Company
1200 Urban Center Drive
Birmingham, AL 35242
Contact: Tony Roberts

Cell: (678) 642-8997
Office: (205) 298-3000
info@vmcmail.com
robertst@vmcmail.com

HAYWARD BAKER
A KELLER COMPANY

Hayward Baker is North America's largest geotechnical contractor offering design-build services in grouting, ground improvement, earth retention, and structural support. Your site's challenging ground conditions have a Hayward Baker solution.

Hayward Baker
515 Nine North Court
Alpharetta, GA 30004

Phone: (770) 442-1801
Fax: (770) 442-8344
jrwolosick@HaywardBaker.com
haywardbaker.com

Exhibitors

Access Limited Construction Co.
225 Suburban Rd
San Luis Obispo, CA 93401
Phone: (805) 592-2230
accesslimitedconstruction.com

Ameritech Slope Constructors, Inc.
P.O. Box 2702
Asheville, NC 28802
Phone: (828) 633-6352
Fax: (828) 633-6353
ameritech.pro

AMS
105 Harrison Street
American Falls, ID 83211
Phone: (800) 635-7330
Fax: (208) 226-7280
ams@ams-samplers.com
ams-samplers.com

Atlas Pipe Piles
1855 E 122nd St
Chicago, IL 60633
(312) 262-1962
atlaspipepiles.com

BGC Engineering Inc.
Suite 500, 980 Howe Street
Vancouver, BC Canada V6Z 0C8
Phone: (604) 684-5900
Fax: (604) 684-5909
bgcengineering.ca

Canary Systems
5 Gould Road, PO Box 2155
New London, NH 03257
Phone: (603) 526-9800
canarysystems.com

Central Mine Equipment Company
4215 Rider Trail North
Earth City, MO 63045-1106
Phone: (800) 325-8827
info@cmeco.com
cmeco.com

Con-Tech Systems Ltd.
Delta BC, Brockville ON,
Charlotte NC, Tacoma WA
Phone: (604) 946-5571
Fax: (604) 946-5548
contechsystems.com

EGSci
1455 Lincoln Parkway #500
Atlanta, GA 30346
Phone: (770) 379-8590
egsci.com

GEL Geophysics, LLC
Charleston, SC
2040 Savage Road
Charleston, SC 29407
Phone: (843) 769-7379
gel.com/geophysics

Geobrugg North America, LLC
22 Centro Algodones
Algodones, NM 87001
Phone: (505) 771-4080
Mobile: (505) 228-6425
geobrugg.com

Geokon, Inc.
48 Spencer Street
Lebanon, NH 03766
Phone: (603) 448-1562
Fax: (603) 448-3216
geokon.com

Exhibitors

GeoStabilization International
P.O. Box 4709
Grand Junction, CO 81502
Phone: (970) 210-6170
Fax: (970) 245-7737
geostabilization.com

Golder Associates Inc.
3730 Chamblee Tucker Road
Atlanta, GA 30341
Phone: (770) 469-1893
golder.com

Hayward Baker
111575 Wadsworth Blvd.
Broomfield, CO 80020
Phone: (800) 864-4382
Fax: (303) 469-3581
haywardbaker.com

HI-TECH Rockfall Construction, Inc.
2328 Hawthorne St
Forest Grove, OR 97116
Phone: (503) 357-6508
hitechrockfall.com

IDS North America Inc.
14828 W. 6th Ave., Ste. 12-B
Golden, CO 80401
Phone: (303) 232-3047 Ext. 121
Fax: (720) 519-1087
idsnabd@idscorporation.com
idscorporation.com/na

KANE GeoTech
7400 Shoreline Drive, Suite 6
Stockton, CA 95219
Phone: (209) 472-1822
info@kanegeotech.com
kanegeotech.com

Mabey Inc.
6770 Dorsey Rd
Elkridge, MD 21075
Phone: (410) 567-1944
mabey.com

Maccaferri, Inc.
10303 Governo Lane Blvd
Williamsport, MD 21795
Phone: (301) 233-6910
maccaferri-usa.com

Monotube, LLC
1432 Maple Avenue NE
Canton, OH 44705
Phone: (330) 454-6111
monotube.com

Olson Engineering
12401 W. 49th Ave.
Wheat Ridge, CO 80033
Phone: (303) 423-1212
olsonengineering.com

Pacific Blasting & Demolition, Ltd
3183 Norland
Burnady, BC Canada V5B 3A9
Phone: (604) 291-1255
pacificblasting.com

Petrologic Solutions, Inc.
3997 Oak Hill Rd.
Douglasville, GA 30135
Phone: (678) 313-4147
dsneyd61@gmail.com

Exhibitors

RST Instruments
11545 Kingston Street
Maple Ridge, BC Canada V2X 0Z5
Phone: (604) 540-1100 (Local);
(303) 993-9230
Fax: 604-540-1005
info@rstinstruments.com
rstinstruments.com

S&ME
4350 River Green Parkway
Suite 200
Duluth, GA 30096
Phone: (770) 815-8343
smeinc.com

SIMCO Drilling Equipment
802 Furmas Dr
Osceola, IA 50213
Phone: (800) 338-9925
simocodrill.com

Scarptec, Inc.
PO Box 326
Monument Beach, MA 02553
Phone: 603-361-0397
scarptec.com

Stantec
2310 Parklake Drive NE, Suite 400
Atlanta, GA 30345-2915
Phone: (770) 493-0450
stantec.com

TenCate
PO Box 1955
Burlington, CT 06613
Phone: (860) 305-4441
tencate.com

VerTek - ARA
710 South Sycamore St.
Gardner, KS 66030
Phone: (800) 639-6315
vertekcpt.com

Vulcan Materials Company
1200 Urban Center Drive
Birmingham, AL 35242
Phone: (205) 298-3000
info@vmcmail.com
vulcanmaterials.com

Williams Form Engineering Corp.
251 Rooney Road
Golden, CO 80401
Phone: (303) 216-9300
Fax: (303) 216-9400
Joseph Erickson, Technical Sales Rep.
(303) 273-1005
williamsform.com

Flexible Rockfall Mitigation Design for Varying Site Conditions

Mallory A. Jones, GIT

KANE GeoTech, Inc.

7400 Shoreline Drive, Suite 6

Stockton, California 95219

(209) 472-1822

mallory.jones@kanegeotech.com

James Roth

HI-TECH Rockfall Construction, Inc.

2328 Hawthorne Street

Forest Grove, Oregon 97116

(503) 357-6508

Kevin Wiesman

Access Limited Construction

1102 Pike Lane

Oceano, California 93445

(805) 592-2230

Prepared for Submission to the 68th Highway Geology Symposium

Marietta, Georgia

ACKNOWLEDGEMENTS

The author(s) would like to thank the individuals/entities for their contributions in the work described:

Geobrugg North America

William F. Kane

Disclaimer

Statements and views presented in this paper are strictly those of the author(s), and do not necessarily reflect positions held by their affiliations, the Highway Geology Symposium (HGS), or others acknowledged above. The mention of trade names for commercial products does not imply the approval or endorsement by HGS.

ABSTRACT

Perry County, Pennsylvania is the location of the longest stone masonry arch railroad viaduct in the world. To provide sandstone blocks needed for bridge construction, quarries were developed near the bridge. A road was later constructed at the base of the quarry high-walls in the 1930s which later became U.S. Route 11, a heavily-traveled commuter highway.

Soon after highway construction, rockfall hazards to the highway developed. At the time of construction, rockfall mitigation from the highly fractured, sandstone highwall was not considered. In 2015, the Pennsylvania Department of Transportation (PENNDOT) retained a local geotechnical engineering company to perform the initial site investigation, analyses, and preliminary mitigation design.

Following the project bid award, the contractor was required to retain a consultant engineer specializing in rockfall mitigation. The engineer's responsibility was to finalize the mitigation design following clearing of the vegetation and scaling. The engineer then provided daily construction oversight which allowed design modifications to be made quickly and efficiently. The engineer and contractor could implement in-field changes as construction proceeded. Due to the highway closure, a strict deadline of forty days to complete the construction was put in place. Open and rapid communication between all parties allowed the project to be completed within time constraints. In addition, the formation of a team of the most qualified contractors and geohazard specialists available created the optimal environment for a successful project.

INTRODUCTION

Rockfall mitigation in the United States has become essential when constructing and improving roads and highways adjacent to rocky slopes or blasted high-walls. Due to the necessity of rockfall mitigation, the technology and infrastructure for rockfall protection has grown greatly in recent years. Improved options to mitigate different rockfall situations have allowed engineers and geologists to design and implement the most appropriate mitigation for each unique hazard present in a rockfall prone area. A combination of systems often provides the best mitigation for complicated rockfall hazards. By conducting a detailed site investigation and rockfall analyses, rockfall energies can be accurately modeled for specific sections of the site. The findings and results can pinpoint locations of higher rockfall energies for specific mitigation, thereby optimizing the overall mitigation with a combination of systems. In this way, the overall design maybe reduced in size and price, Figure 1.

1. Combined drapery mesh and rockfall barrier for rockfall mitigation

Mitigation options used in rockfall protection today include passive rockfall protection and active rockfall protection. Passive rockfall protection includes systems that allow rockfall to occur, but in a controlled manner. Examples of these systems include rockfall barriers, attenuators, draperies, ditches, and embankments. Active rockfall protection systems prevent rockfall from occurring. Active systems include pinned or tensioned mesh, rock bolts, and cable lashing. There are advantages and disadvantages with each system, but the distinct characteristics of the site conditions and the surrounding area determine the best option(s) for mitigating the hazards present.

BACKGROUND

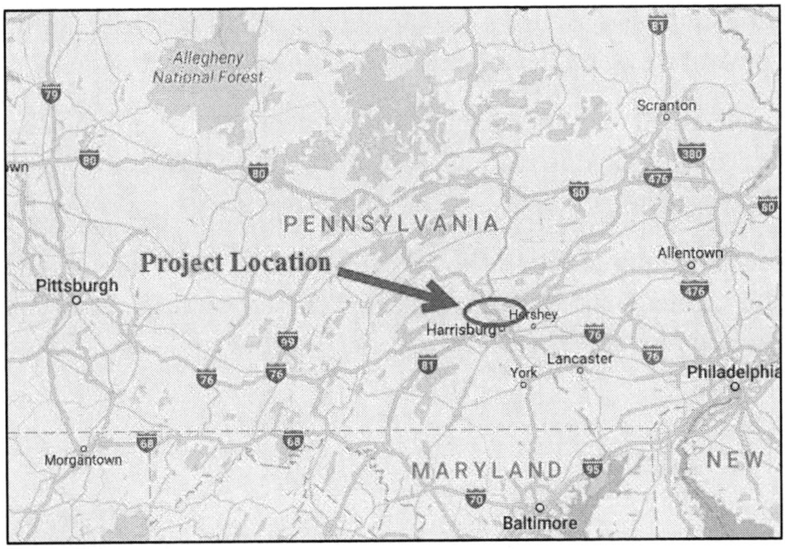

2. Project Location

The project site is in Perry County, Pennsylvania along U.S. Route 11. Geographically, Perry County is located in the State of Pennsylvania north of Harrisburg, along the Susquehanna River, Figure 2. Perry County is the location of the largest stone masonry arch railroad viaduct in the world. It is in still in use today. To construct the bridge, sandstone blocks were quarried from the adjacent river bluffs. Highwalls ranging in height from 300-ft to 1,100-ft high resulted from the blasting exposing interbedded layers of fractured sandstone and soft shale. Due to the differences in resistance to erosion, differential weathering occurs resulting in large overhangs of highly fractured sandstone throughout the area. Over time, these highwalls became highly vegetated with shrubs and trees varying in size reducing the visibility of the site's geologic exposures.

In the 1930's the interbedded sandstone-shale highwalls became the boundaries of U.S. State Route 11. During the initial construction of the road, rockfall was not considered resulting in numerous rockfall events impacting the road. These rockfall events were not only dangerous, they also created a great deal of maintenance work for the Pennsylvania Department of Transportation, (PENNDOT). Small catchment ditches were constructed along the toe of the highwalls to help limit the number of rocks entering the road, but have not been effective in providing protection to drivers, Figure 3. In 2015, PENNDOT decided to move forward with designing rockfall mitigation along U.S. State Route 11 in Perry County.

3. Catchment at the toe of the slope

The mitigation construction project was a design-build contract where preliminary plans, designed and engineered by a local firm, were utilized as a basis for design. The awarded contractor was responsible for obtaining an engineer experienced in rockfall mitigation to design and finalize the preliminary rockfall mitigation plan for two sections of highwalls. The final plan was submitted to PENNDOT for approval then drafted as working drawings for construction.

The retained rockfall engineer was to be present onsite daily during the construction to address any concerns and to work closely with PENNDOT and their contracted engineer if any in-field changes were necessary.

The site required full traffic control which included a total road closure for most the construction. This created major traffic delays and detours in the area. Because U.S. Route 11 is a heavily traveled road, the project had a strict deadline of 45-days for construction completion.

SITE GEOLOGY

The project site is part of the Valley and Ridge Physiographic Province of Pennsylvania. This area is located within the Appalachian Mountain Section and the Susquehanna Lowland Section of the Province. Compressional forces formed a series of synclines and anticlines throughout the area creating long, narrow ridges and valleys. The project area is located on the south limb of a regional fold known as the Cove Syncline.

The rock types found throughout the Perry County Project area are of the Bloomsburg Formation, (Boyer, 1984) and are mostly interbedded, fissile shale and fractured sandstone with a thick conglomerate bed at the southern-most section of the road cut. Due to the differential erosion, and fractured nature of the sandstone, over hangs dominate the slopes causing continual rockfall, Figure 4.

4. Fractured sandstone overhang; sandstone-shale contact

PENNDOT PRELIMINARY SITE INVESTIGATION AND MITIGATION DESIGN

Prior to the Project's bid, a local engineering company was retained to conduct a site investigation, rockfall analyses, and provide a preliminary mitigation design based on their findings.

The preliminary investigation was conducted before clearing and grubbing of the slope's thick vegetation. Conducting a rockfall investigation on a highly-vegetated slope is not ideal for design. Areas that may prove important may be overlooked if their location is covered with vegetation. Using visible outcrops, available reports detailing geology, and a surveyed topographic overview of the area, the retained engineering company could obtain general rockfall bounce heights and energies. These results were utilized in the design of the preliminary rockfall mitigation plan for the two highwalls within the project area.

5. Acid-bearing shale - sandstone contact

The bid plans outlined a general concept of installing a drapery mesh in areas producing lower energy rockfall. It was requested the drape be pinned along the base of the slope to help with limiting maintenance along the roadside. A large, low-deflection rockfall barrier was also specified for areas with higher energy rockfall. To mitigate smaller, overhanging areas of the slope, it was recommended to support the overhangs using re-enforced shotcrete. An additional area of concern was a thick, acid-bearing shale bed. This shale layer is extremely susceptible to erosion and underlies a more resistant sandstone layer creating large overhangs at the contact, Figure 5. The preliminary design called for this area to be further investigated and a recommendation for stabilization be provided to PENNDOT for approval.

These preliminary plans provided a general plan for mitigation, and was put out to bid to contractors for construction. The contractors were responsible for sub-contracting the finalization of the mitigation plan to an engineering consultant of their choice.

DESIGN FINALIZATION

Following the project's award to HI-TECH Rockfall Construction, KANE GeoTech was retained by HI-TECH to finalize the drapery and tensioned high-strength steel mesh designs, and submit construction drawings to be reviewed by PENNDOT and their consulting engineer. KANE GeoTech visited the site to verify the locations of the preliminary design for mitigation. However, heavy snowfall and vegetation made it difficult to verify the locations specified in the preliminary plans, Figure 6. KANE GeoTech used the available resources provided by PENNDOT to complete the final engineered design of the preliminary plans.

The final design consisted of anchor depths and locations for the previously recommended rockfall drapery. KANE GeoTech recommended that Geobrugg TECCO high-strength steel wire mesh be used for the drapery and that most of the localized re-enforced shotcrete support systems be replaced with tensioned Geobrugg Spider mesh systems to allow for a larger stabilization coverage area. Tensioned Spider mesh was also used in stabilizing large, sandstone blocks that were overhanging at the contact with the acid bearing shale seam.

6. Snow covered slope

Due to the highly-fractured nature of the slope, and the fractures' orientation, KANE GeoTech specified long rock dowels, approximately 20-ft, along the base of the slope. These bolts acted to create a buttress for the base of the slope and were also utilized as the maintenance pins for the bottom drapery rope requested by PENNDOT.

KANE GeoTech utilized the provided Geotechnical Report and the Post Tensioning Institute (PTI) Guidelines to calculate the anticipated loads and dowel depths for the TECCO drapery anchor design. The Geobrugg software, RUVOLUM, was used in the Spider sections' anchor design and layout.

The finalized design was reviewed by PENNDOT and their consultant engineer and the notice to proceed with construction was given to the contractor, HI-TECH Rockfall Construction, (HI-TECH).

IN-FIELD DESIGN CHANGES DURING CONSTRUCTION

Prior to the start of construction, the slope's heavy vegetation was cleared exposing a number of additional potentially hazardous areas. KANE GeoTech worked closely with HI-TECH, PENNDOT, and their contracted engineer to design and engineer additional protection in areas that were not concerning prior to vegetation clearing. The tensioned Spider mesh was easily utilized in several areas making field condition changes quick and efficient when needed. Revised final plans and drawings were submitted to PENNDOT for review and approved for construction, Figures 7 and 8.

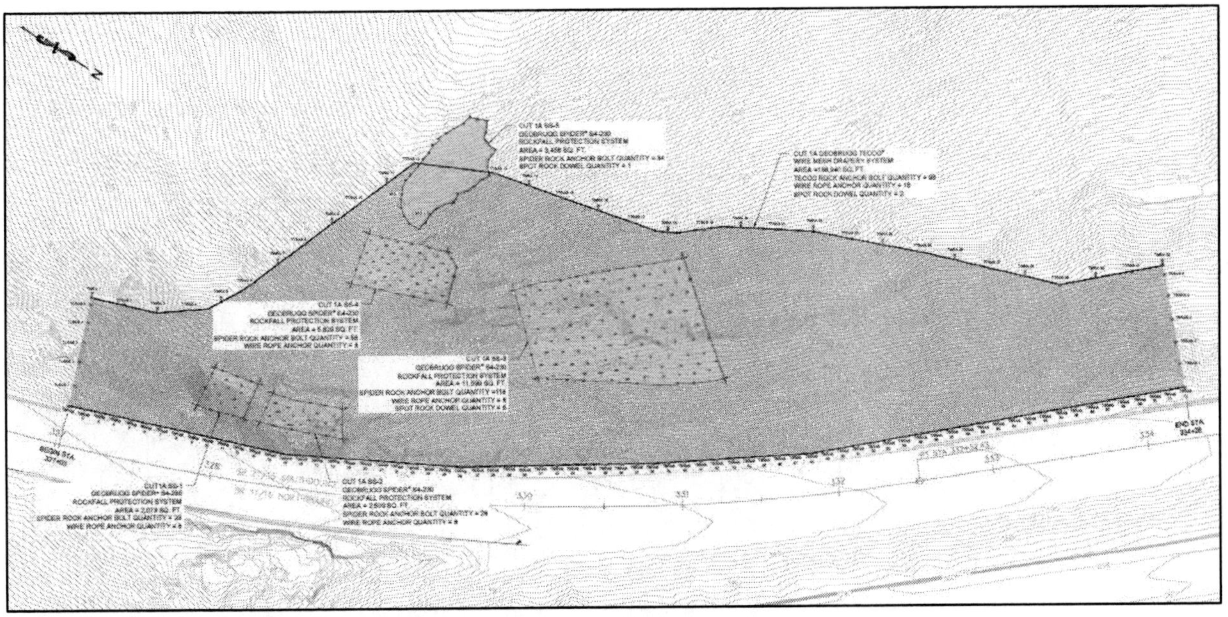

7. Final plan for Southern highwall using tensioned Spider mesh overlaid with TECCO drapery

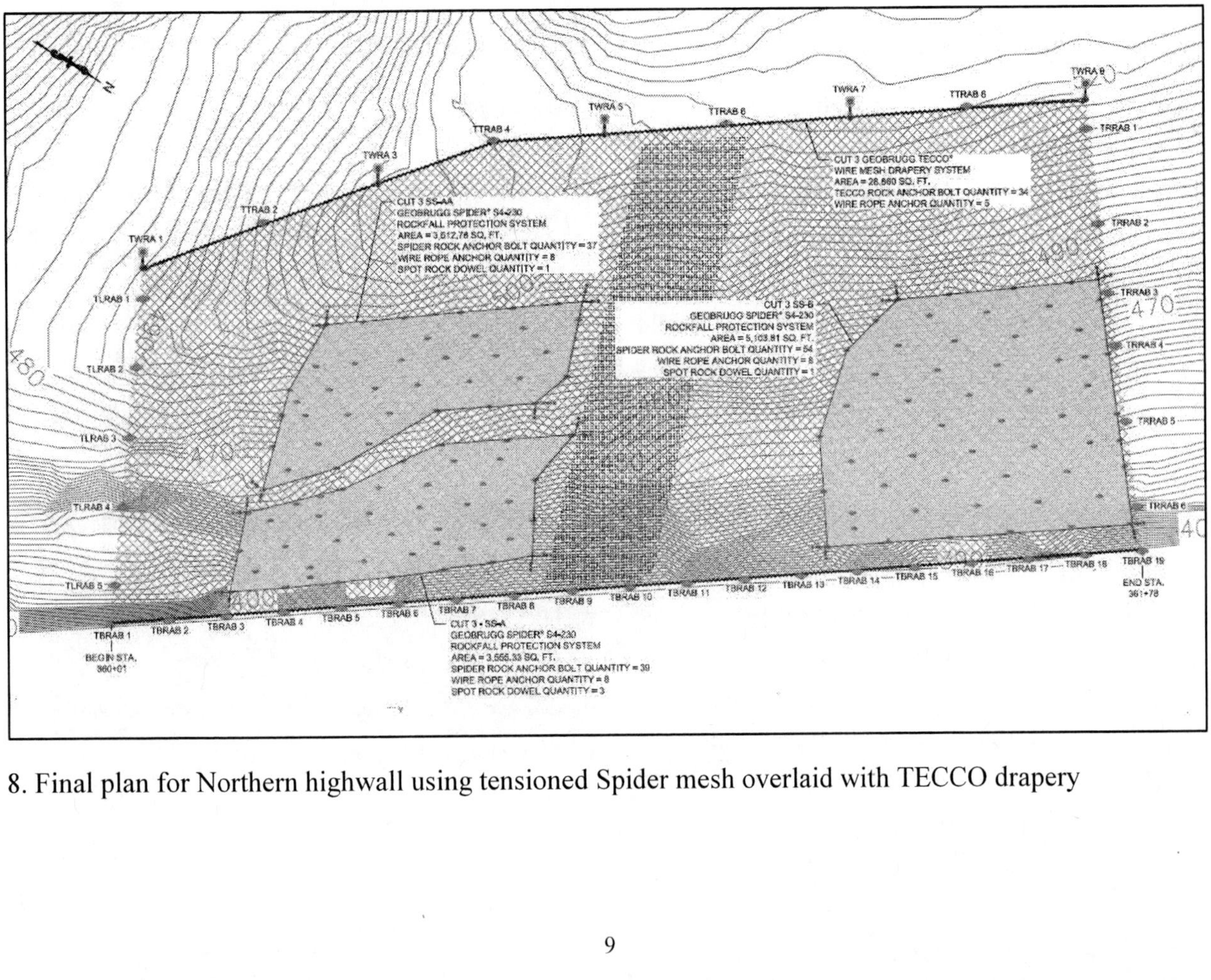

8. Final plan for Northern highwall using tensioned Spider mesh overlaid with TECCO drapery

CONSTRUCTION SEQUENCE

Due to the different components of the systems used to the rockfall mitigation systems, and the strict timeline for construction, a detailed plan for construction was created to maximize efficiency. Construction on the face of the slope was prohibited to prevent uncontrolled rockfall before road closure went into effect, however, drapery anchors were authorized to be drilled at the slope's crest prior to the road closure. This maximized the time for site preparation and reduced the amount of drilling during the allotted 45-days. The TECCO mesh was also cut and seemed into panels during this preparation time for quick and efficient installation.

9. Moveable rockfall drape used during scaling

Following the road closure, the slopes were completely cleared, grubbed, and scaled. As a part of the PENNDOT specifications, a temporary, moveable rockfall drape was designed and used during scaling to protect the road, workers, and the railroad located below U.S. Route 11, Figure 9.

Verification anchors were installed and tested to verify rock strengths in actual, worst-case field conditions, Figure 10. By conducting these tests, an in-field rock/grout bond strength was obtained so anchor depths could be drilled to their most appropriate depth.

Following clearing, grubbing, and scaling the tensioned Spider section anchor holes were laid out and drilled using two air rotary drills hung from cranes. An additional air rotary drill mounted to a Spider excavator subsequently drilled the bottom Spider section holes and 20-ft holes for the buttress anchors. Anchors were then installed and grouted.

10. Anchor verification test setup

The Spider mesh was hung and tensioned down at each location to conform to the slope using the manufacturer's anchor plates, Figure 11.

After the tensioned Spider mesh instrallation was complete, the TECCO drapery mesh was installed by a helicopter raising and laying the prepared panels on the crest of the slope, Figure 12. These panels were shackeled to the installed anchors and wire support rope at the crest of the slope. The TECCO panels were then laced together vertically and along the bottom of the mesh with wire rope. Finally, the bottom of the mesh was pinned along the toe where the buttress anchors were installed.

A final inspection of all connections and system components was conducted by KANE GeoTech following the construction completion. A letter of comformance was submitted to PENNDOT for review and was approved bringing the rockfall mitigation constrution on U.S. State Route 11 to a close.

11. Tensioned Spider mesh installation

12. TECCO drapery mesh lifted into place by helicopter

CONCLUSION

As a result of heavy rockfall, a need for mitigation along U.S. Route 11 in Perry County was inarguable. Because the hazards in this area are so complex, utilizing different rockfall mitigation approaches allowed for the best possible protection to be installed, Figure 13. By permanently stabilizing the large fractured areas with tensioned Spider mesh, the need for large barriers throughout the sectioned was eliminated. Also, by substituting the tensioned Spider mesh for the originally planned shot-crete, a larger area of the slope was able to be stabilized, thus increasing the rockfall protection.

The design-build approach taken by PENNDOT was successful. However, the number of in-field changes may have been avoided if the preliminary design was less specific. Once the slope was cleared prior to the final design, a more appropriate design for the site conditions was possible. Nevertheless, having an engineer's representative onsite daily allowed these in-field changes to be made quickly and efficiently.

13. Tensioned Spider mesh overlaid by TECCO drapery mesh

REFERENCES

Geobrugg AG (2016). *Dimensioning of the Slope Stabilization System TECCO® by the RUVOLUM Method.* CH-8590 Romanshorn, Switzerland. January 29, 2016.

Post-Tensioning Institute (PTI) (2014). Recommendations for Prestressed Rock and Soil Anchors DC35.1-14. 5th Ed., Post-Tensioning Institute, 38800 Country Club Farmington Hills, MI 48331.

Royer, Denise. (1984). "Geologic map of Perry County, Pennsylvania showing the locations of wells and springs". Pennsylvania Bureau of Topographic and Geologic Survey. Atlas W59, Plate 1. Harrisburg, Pennsylvania, 1984.

Results of tests to evaluate the tensile strength and the load bearing capacity of rockfall nets according to new ISO standards

Luca Gobbin
Maccaferri Canada Ltd.
400 Collier MacMillan Dr. Unit B
Cambridge, Ont, Canada, N1R 7H7
Telephone: 519-623-9990
lgobbin@maccaferri.ca

Giorgio Giacchetti
OFFICINE MACCAFERRI S.p.A.
Via Kennedy 10
40069 Zola Predosa
Ph: 01139051646000
giorgio.giacchetti_cons@maccaferri.com

Alberto Grimod
FRANCE MACCAFERRI
Valence, France (Fra)
alberto.grimod@maccaferri.fr

Ghislain Brunet
Maccaferri, Inc
10303 Governor Lane Blvd.,
Williamsport, MD 21795-3116
Ph: 301-223-6910
gbrunet@maccaferri-usa.com

Prepared for the 68th Highway Geology Symposium, April 29 to May 4 2017, Marietta, GA

Disclaimer

Statements and views presented in this paper are strictly those of the author, and do not necessarily reflect positions held by their affiliations, the Highway Geology Symposium (HGS), or others acknowledged above. The mention of trade names for commercial products does not imply the approval or endorsement by HGS.

Copyright Notice

ABSTRACT

Rockfall mitigation protections using meshes or ring and rope panels that can be divided in two main categories: secured and simple drapery systems. Secured drapery systems are composed by a rockfall net pinned to the slope with a pattern of nails. The goal of this solution is to stabilize the surficial portion of the rock slope (with the nails) and keep in place the unstable rock material that can move between the anchors (with the mesh). The characteristics used on the net design are: the tensile strength, the load bearing resistance, as well as, the punching deformation. Simple drapery systems are composed of a rockfall net fixed at the top of the slope with anchors and cables. The purpose of this solution is to drive the unstable rock-blocs to the bottom of the slope, by reducing their energy and velocity. The net is mainly designed using its own longitudinal tensile strength. In order to characterize the mechanical properties of the nets, laboratory tests must be carried on them. Since September 2016, the tensile strength and the load bearing capacity for rockfall ring nets and cable nets can be evaluated with new ISO standards: ISO 17745 and ISO 17746. ISO standards also introduce the concept of lifespan: based on the type of environment (ISO 9223) and on the type of coating, which can allow designers to estimate the design life of the rockfall netting. These standards give the designers the ability to compare different nets produced by the same manufacturing company or by other ones. The paper outlines some results of laboratory tests carried out according to these new standards.

Introduction

The experience demonstrates that among the cost effective solutions against the rockfall there are the draperies with steel meshes. Depending on several factors, like the size of instable blocks and slope morphology, the designer has to choose the most suitable intervention strategy of the two: secured drapery and simple drapery. The design process is based on the reliability of the calculation model and on the sensitivity of the designer to consider the most suitable mesh available. At this very last level of the design process, there are some basic questions to be addressed: what are the technical features of the mesh involved in the design process? Or, from another point of view, why the designer should choose a mesh instead of another one? And finally, how to compare the performances of the different steel meshes having so big differences in terms of fabric, mesh opening and constitutive material? The answers to these questions are not easy at all since they involve so many details of the design process.

The solution is given by the design principles which always analyze the performances in terms of resistance and deformation. Actually, for the draperies, resistance and deformation answer to the two main questions: "How much the mesh resists?", and "When is the maintenance needed?".

Outlook to the intervention strategies

Secured drapery systems (or pin drapery) consist of a combination of anchors and netting (see Fig. 1). They are very common solutions for the rockfall mitigation since they can improve the superficial rock face stability maintaining in place the debris/rock along the slope [1].

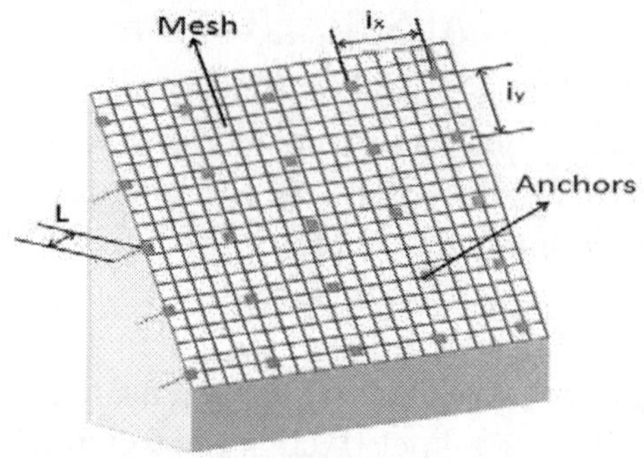

Figure 1 - Pin drapery system: the intervention is composed by a rockfall mesh and a pattern of anchors (L = length of the anchors; i_x and i_y = distance between the anchors, respectively horizontally and vertically).

The design of these systems could be very difficult because of the geomechanical input, most of the time very rough; several approaches are possible to design these systems; MacRO 1 (the software of Officine Maccaferri) runs two separate analyses: one for the stability mechanisms related to the anchors, and the other one for the mesh facing. MacRO 1 calculates and checks the minimum length and pattern for the anchors in order to improve the equilibrium condition of the rock face. The software also allows designing the most suitable mesh for the facing [2]. On site it can be easily noticed that under the weight of the debris, the mesh deflects and generates pocket of debris. The mesh cannot therefore be modeled as a beam which is able to transmit pressures uniformly distributed on a surface by means of the nails [1] [3]. In the case of a secured drapery the mesh can be thought as a membrane subjected to punch and tensile stress.

In the case of a simple drapery (or suspended drapery), the mesh is stretched by its self-weight, the accumulation of debris at the slope toe, and the load of the snow (if applicable). For this type of application, the mesh can be considered as a membrane subjected to a tensile stress (see for instance the calculation approach for MacRO 2 by Officine Maccaferri [4]).

Basic properties and tests

Considering the design process, the past experiences and the technical literature, the main properties to consider for the steel meshes are: the weight per unit area, the resistance and the related deformability under tensile and punch test.
Whereas there are no problems for the determination of the weight of the mesh, there are several concerns about the characterization of the tensile and punch resistance: how to determine the performance values; the validity of the tests procedures; physical issues; technological issues for the tests procedures.
These issues have driven to the Italian standard UNI 11437: 2012 [16] first, and then to the ISO 17745 [5] and ISO 17746 [6] in 2016. These ISO standards represent a milestone for the designers and market accordingly. For some of the above questions there is not a complete answer yet, so that some compromises have necessarily been introduced. Despite that, these standards are a big step forward for the following reasons:

- Performances: the tests are carried on with large samples. In this way the behavior of the mesh is significant: the effect of the single constitutive components (i.e. the wire resistance or the cables) becomes negligible and the performance of the entire mesh is closer to the reality. The size of the tested sample is relevant since, in the reality, the anchors are usually spaced more than 2.5 m.
- Validity of the tests: being the tests representative of the mesh behavior, they automatically allow the comparison of different type of meshes. For this reason, the punch test is carried on with a large pressure device to be able to push on any mesh opening. In the same way the tensile test can be carried on every type of mesh (i.e. the rings for the ring net).
- Technological details: the most relevant issue regards the punch test where the frame that restrains the mesh is squared but the pressure device has a circular footprint. These geometric differences give a non-homogeneous stress in the mesh and to some related issues in the results accordingly. However, this configuration allows fixing any type of fabric.
- Type of constraints: for the punch test, the main issue is how the sample is restrained on the frame. In the reality the meshes are usually restrained by four plates installed on the relative anchors, not only on four edges like in the test. Because of that, the best procedure apparently seems to be the punch test carried on with a single plate. Despite that, the test procedure described by the ISO gives some big advantages: (a) the results of the punch test are not affected by the anchor plate and the behavior of different meshes can be compared equally; (b) the use of homogeneous restrains gives clear results about the elastic properties of the mesh; (c) the feasibility of numerical models implementation (see Fig. 2); (d) it allows to find all the different properties for any type of mesh in any restrain condition by mean of numerical models (see Fig. 6).

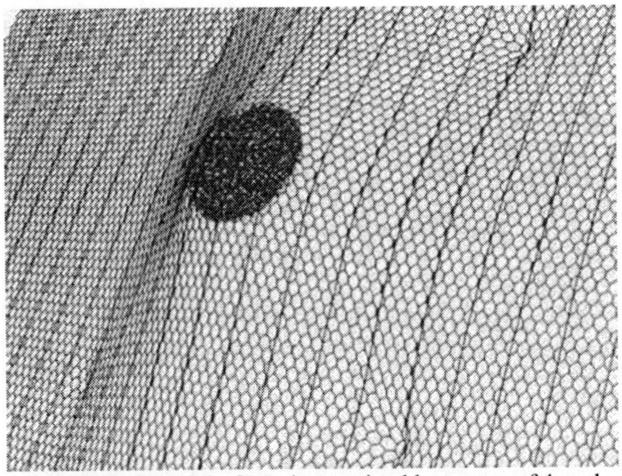

Figure 2 - Numerical model of Mesh restrained by means of 4 anchor plates.

The issue of these two new ISO fills this lack and makes everything more clear, scientific and standardized. As written above, and described also in the following chapters, the way the tests are carried on is basic to find reliable data since they can give values significantly different (Fig. 4). Because of these differences, the whole solution can fail or being not verified. Now the designer will be able to find, or to require, reliable values of strength, punch load and corrosion protection for all the types of netting tested according to common international standards.

Test procedures for net tensile resistance

The International Standard ISO 17745:2016 [5] and ISO 17746:2016 [6] describe the test procedure for determining the tensile strength (resistance and elongation) of steel wire ring panels and steel wire rope net panels and rolls.
This mechanical property is defined by testing a specimen connected to a metal frame equipped with load cells in order to acquire the load applied and the overall side reaction (longitudinal and transversal reactions). The sample has to be not less than 1,000 mm wide, with a minimum area of 1.0 m^2. It has to be fixed to the frame through lateral coupling devices, such as shackles or turnbuckles. The side coupling device is also free to slide along the longitudinal beams (see Fig. 3). The tensile strength reported at the end of the test is usually identified in kN/m.

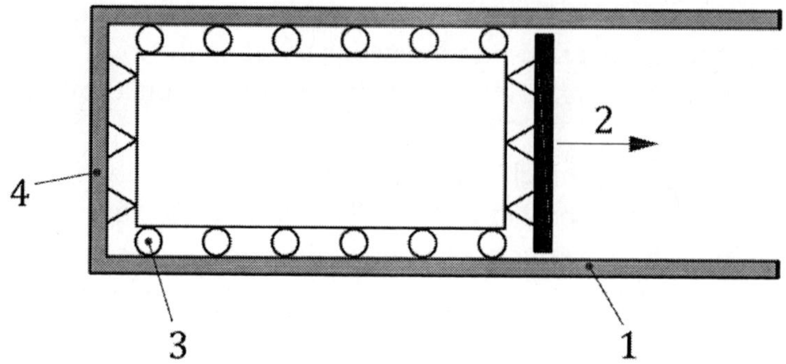

Figure 3 - Example of frame configuration for the tensile strength test. Legend: 1) Fixed frame; 2) Movable beam; 3) Lateral constraint; 4) Side connection device.

Maccaferri has tested its ring nets and HEA rope panels in order to get the CE mark and to implement the software MacRO 1 and MacRO 2 with these tested values.
The results of the tests are shown in the following tables (Table 1 and Table 2).

Nominal Tensile Strength for HEA Panels (ISO 17746)		
Nominal Mesh (mm)	**Panel Rope Diameter (mm)**	**Minimum Tensile Strength (kN/m)**
250 x 250	**8**	**200 ± 15 kN/m**
300 x 300	**8**	**165 ± 15 kN/m**
400 x400	**8**	**140 ± 15 kN/m**
300 x300	**10**	**250 ± 15 kN/m**
400 x400	**10**	**170 ± 15 kN/m**

Table 1 - Description of the Tensile Strength for Maccaferri's HEA Panels tested according to ISO 17746.

Nominal Tensile Strength for Ring Panels (ISO 17745)	
Type of Ring Panel	**Minimum Tensile Strength (kN/m)**
4PM7 (4 point of contacts, 7 loops)	**219 kN/m**
4PM9 (4 point of contacts, 9 loops)	**256 kN/m**
4PM12 (4 point of contacts, 12 loops)	**315 kN/m**

Table 2 - Description of the Tensile Strength for Maccaferri's Ring Panels tested according to ISO 17745.

Test procedures for net punch load capacity

The punch test is carried out on a sample having a size of 3.0 x 3.0 m ±20%, restrained into a large steel frame and loaded by mean of punching device with a diameter of 1.0 m (see Fig. 4).
The knowledge of the deformation is very important during the design of a secured drapery system, because of the following main reasons:

- When the deformation reaches the design limit, it means that the maintenance (cleaning) of the secured drapery is needed before that further displacements determine the mesh rupture. A simple visual monitoring let the owner plans the maintenance interventions.
- Too much deformed mesh implicates easy stripping on the anchors and lower durability of the intervention. The designer must be aware of this and he has foreseen the right mesh type accordingly.
- Since the meshes are largely deformable, the facing of the secured drapery could interfere with close infrastructures or vehicles.

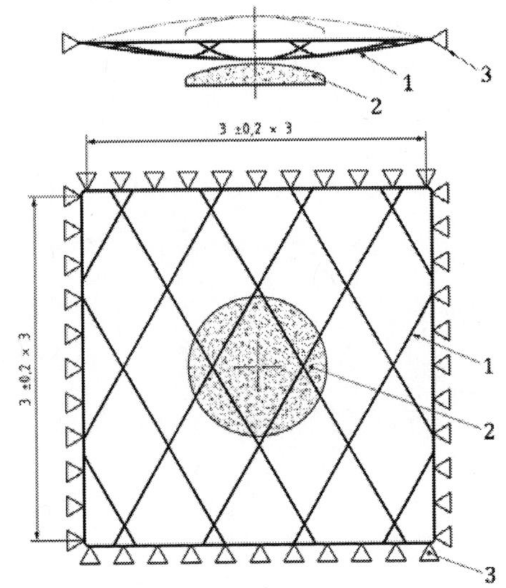

Figure 4 - Example of set up for punching test according to ISO 17745 and ISO 17746. Legend: 1) Tested mesh sample; 2) hemispherical shaped load sharing device (1.0 m in diameter); 3) Perimeter constraint between the mesh and the frame.

The comparison of the tests carried out with the University of Venice IUAV Lab [8], Turin Tech University [9] and CNR – Material labs [10] [11], shows that the resistance and deformation of the mesh under punch load changes substantially depending on the size of the samples (the scale effect) and on the configurations of the constraints (see Fig.6). This proves once again how fundamental is to carry on nets' testing according to a common standard in order to get reliable results. The general law of the scale effect is assumed in the following simplified form referred to the coordinate of the load-displacement diagrams (see Fig. 5):

$$x = x_0 \mu_x$$
$$y = y_0 \mu_y$$

Where:

(x, y) = generic coordinate of the scaled graphic
(x_0, y_0) = generic coordinate of the reference graphic
(μ_x, μ_y) = constants correlating the scaled to the reference graphic, which depends on the mesh type.

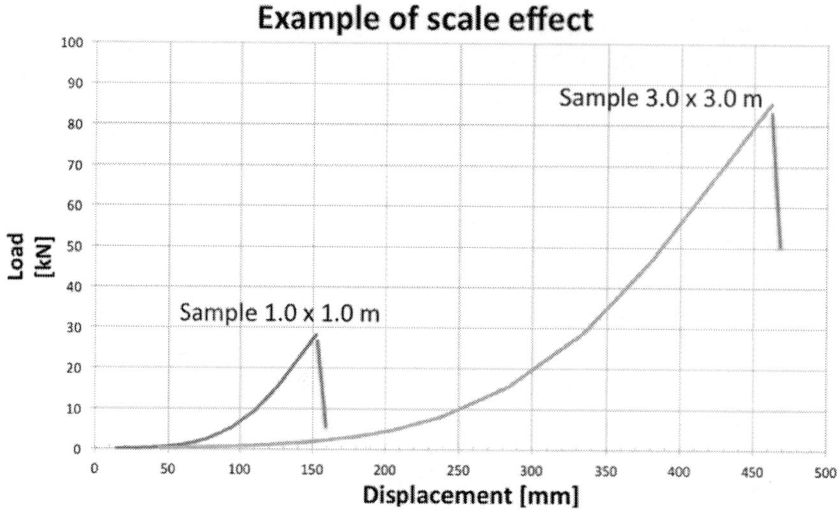

Figure 5 - Graph Displacement VS Load with the typical scale effect in the punch test.

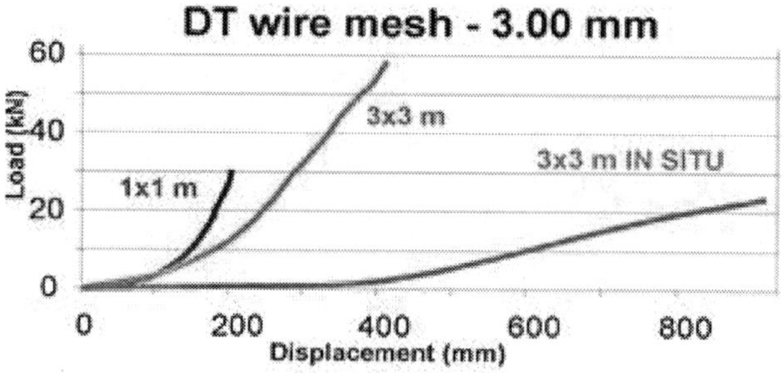

Figure 6 – Load-displacement curve for tests carried on with different configurations of constraints.

Maccaferri has tested its ring nets and HEA rope panels in order to get the CE mark and to implement the software MacRO 1 with the maximum bearing capacity as well as the maximum displacement.
The results of the tests are shown in the following tables (Table 3 and Table 4).

Punching Resistance for HEA Panels (ISO 17746)			
Nominal Mesh (mm)	Panel Rope Diameter (mm)	Minimum Ultimate Punch Load (kN/m)	Ultimate Punching Displacement (mm)
250 x 250	8	260 ± 15	240
300 x 300	8	250 ± 15	280
400 x400	8	200 ± 15	260
300 x300	10	400 ± 15	310
400 x400	10	300 ± 15	310

Table 3 - Description of the Punch Test resistance with relative displacement for Maccaferri's HEA Panels tested according to ISO 17746.

Punching Resistance for Ring Panels (ISO 17745)		
Type of Ring Panel	Minimum Ultimate Punch Load (kN/m)	Ultimate Punching Displacement (mm)
4PM7 (4 point of contacts, 7 loops)	**501**	**856**
4PM9 (4 point of contacts, 9 loops)	**578**	**833**
4PM12 (4 point of contacts, 12 loops)	**821**	**820**

Table 4 - Description of the Punch Test resistance with relative displacement for Maccaferri's Ring Panels tested according to ISO 17745.

Expected lifespan according to the chosen metallic coating

In addition to the mechanical characteristics (the tensile strength and bearing capacity), the designer must evaluate the type of mesh according to the jobsite environmental conditions too.

In 2013, the EN 10223-3 [15] introduced a new innovative concept: the expected lifespan for all the double twisted wire mesh products, such as Double twist net, Gabions, Reno mattresses, SteelGrid, etc. Thanks to this approach, it was possible to define the type of coating to apply to the metallic wire based on the corrosion-aggressiveness of the jobsite (ISO 9223) [12] and on the expected lifespan of the solution. Thus, it would be possible to design a double twist mesh product able to resist only 25 year in a low aggressive environment (i.e. using a Zinc-Class A wire coating), up to 120 years in a high aggressive environmental condition (i.e. using a Galfan + PVC or PA6 coating).

The two new ISO standards also define the ageing and corrosion resistance for rope and ring panels, introducing, as well as the EN 10223-3, the concept of lifespan for rockfall netting based on the results of the salt spray test (ISO 9227:2012) [13]. According to ISO 9227, the net samples shall not show more than 5% of DBR (Dark Brown Rusted) after they have been subjected to the neutral salt spray test described in the standard itself. This condition may emerge after an exposure period of 200, 500, 1000 or 2000 hours depending on the type of coating and, respectively, for:

- Zinc Class B (EN 10244-2 or EN 10264-2) [17] [18];
- Zinc Class A (EN 10244-2 or EN 10264-2);
- Zinc 95%+Aluminum 5% (or Galfan 95/5) Class B (EN 10244-2 or EN 10264-2);
- Zinc 95%+Aluminum 5% (or Galfan 95/5) Class A (EN 10244-2 or EN 10264-2);
- Zinc 90%+Aluminum 10% (or Galfan 90/10) Class B (EN 10244-2 or EN 10264-2);
- Zinc 90%+Aluminum 10% (or Galfan 90/10) Class A (EN 10244-2 or EN 10264-2).

These test results allow us to estimate the durability of the net based on the corrosivity conditions of the environment defined by ISO 9223.

In this way, it would be possible for the designer to choose the most suitable coating for any rockfall netting depending on the design life required for the intervention, as listed by the following table (Table 5).

Site environment level (in accordance with ISO 9223:2012, Table 4)	Coating	Class (ISO 7989-2)	Estimated working life of the product (year)
Low aggressive: (C2) Dry conditions. Temperate zone, atmospheric environment with low pollution, e.g. rural areas, small towns (over 100 m above sea level). Dry or cold zone, atmospheric environment with short time of wetness, e.g. deserts, sub-arctic areas.	**Zinc**	**A** **B**	**25** **10**
	Zn95%/Al5% alloy	**A** **B**	**50** **25**
	Advanced metallic coating	**A** **B**	**120** **50**
Medium aggressive: (C3) Dry conditions. Temperate zone, atmospheric environment with medium pollution or some effect of chlorides, e.g. urban areas, coastal areas with low deposition of chlorides, e.g. sub-tropical and tropical zone, atmosphere with low pollution.	**Zinc**	**A**	**10**
	Zn95%/Al5% alloy	**A** **B**	**25** **10**
	Advanced metallic coating	**A** **B**	**50** **25**
Medium aggressive: (C3) Wet conditions. Temperate zone, atmospheric environment with high pollution or substantial effect of chlorides, e.g. polluted urban areas, industrial areas, coastal areas, without spray of salt water, exposure to strong effect of de-icing salts, e.g. subtropical and tropical zone, atmosphere with medium pollution industrial areas, coastal areas, shelter positions at coastline.	**Zn95%/Al5% alloy**	**A**	**10**
	Advanced metallic coating	**A** **B**	**25** **10**
NOTE: Working life (product) - the period of time during which the performance of a product will be maintained at a level that enables a properly designed and executed works to fulfil the essential requirements (i.e. the essential characteristics of a product meet or exceed minimum acceptable values, without incurring major costs for repair or replacement). The working life of a product depends upon its inherent durability and normal installation and maintenance.			

Table 5 - Description of the environment of the installation site, coating wire rope requirement.

Conclusions

The design of ring nets and cable panels used for rockfall pin and simple drapery systems requires the complete knowledge of the nets' mechanical properties, as well as their expected lifespan. Since there are such large different types of meshes, the investigations must be focused on the tensile and punch resistance with tests procedures suitable for any net. The standards ISO 17745:2016 and ISO 17746:2016 are the solution to these problems.

These two standards set also the lifespan for rockfall netting based on the type of environment and mesh coating. The lifespan of the net has to be considered during the design phase together with the other mechanical characteristics mentioned before in order to match the expected design life for the intervention.

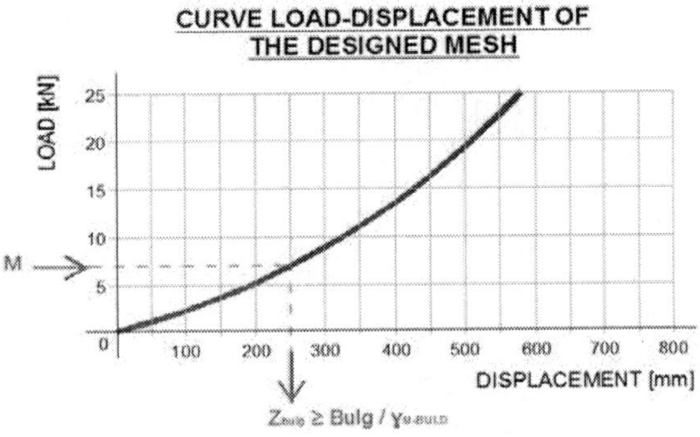

Figure 7 - Example of a curve load-displacement used for the design of the mesh at the Serviceability Limit State.

References

[1] Grimod A., Giacchetti G. (2013): A new design approach to design pin drapery systems. GeoMontreal 2013, 29th September – 3rd October 2013, Montreal (QC).

[2] Bertolo P. , Giacchetti G., (2008): An approach to the design of nets and nails for surficial rock slope revetment – Interdisciplinary Workshop on Rockfall Protection, June 23-25 2008, Morshach, Switzerland.

[3] Keith Turner A. and Schuster Robert L., Editors (2013) – Rockfall: Characterization and Control – TRB Transportation Research Board - The National Academies Keck Centre 500 Fifth Street, NW, Washington, DC 20001.

[4] Grimod A., Giacchetti G. (2013): New design software for rockfall simple drapery systems. 23rd World Mining Congress. 11-15 August 2013. Montreal QC, Canada.

[5] ISO 17745:2016 (2016): Steel wire ring net panels - Definitions and specifications.

[6] ISO 17746:2016 (2016): Steel wire rope net panels and rolls - Definitions and specifications.

[7] BS 8006-2:2011 (2011): British standard: Code of practice for strengthened/reinforced soils – Part 2: Soil nailing design.

[8] IUAV (2012): Internal report about the test series according to UNI 11437:2012.

[9] Bertolo P., Oggeri C., Peila D. (2009): Full scale testing of draped nets for rock fall protection Canadian Geotechnical Journal, No. 46 pp. 306-317.

[10] Bertolo P., Ferraiolo F., Giacchetti G., Oggeri C., Peila D., e Rossi B., (2007): Metodologia per prove in vera grandezza su sistemi di protezione corticale dei versanti – GEAM Geoingegneria Ambientale e mineraria, Anno XLIV, N. 2, Maggio-Agosto 2007.

[11] Bonati A., e Galimberti V., (2004): Valutazione sperimentale di sistemi di difesa attiva dalla caduta massi – In atti "Bonifica dei versanti rocciosi per la difesa del territorio" - Trento 2004, Peila D. Editor.

[12] ISO 9223:2012 (2012): Corrosion of metals and alloys - Corrosivity of atmospheres - Classification, determination and estimation.

[13] ISO 9227:2012 (2012): Corrosion tests in artificial atmospheres - Salt spray tests.

[14] ISO 7989-2:2007 (2007): Steel wire and wire products - Non-ferrous metallic coatings on steel wire - Part 2: Zinc or zinc-alloy coating.

[15] EN 10223-3:2013 (2013): Steel wire and wire products for fencing and netting. Part 3: Hexagonal steel wire mesh products for civil engineering purposes.
[16] UNI 11437:2012: Rockfall Protective Measures - Tests On Meshes For Slopes Coverage.
[17] EN 10244-2:2009: Steel wire and wire products. Non-ferrous metallic coatings on steel wire. Zinc or zinc alloy coatings.
[18] EN 10264-2:2012: Steel wire and wire products. Steel wire for ropes. Cold drawn non alloy steel wire for ropes for general applications.

Construction of Transportation Infrastructure in Weathered Volcanic Ash Soils

James M. Arthurs, Ph.D., P.E.
Federal Highway Administration – Central Federal Lands Highway Division
12300 West Dakota Avenue, Suite 210
Lakewood, Colorado 80228
720-963-3633
james.arthurs@dot.gov

Devin Dixon, E.I.T.
Federal Highway Administration – Central Federal Lands Highway Division
50 Makaala Street
Hilo, HI 96720
618-559-7678
devin.dixon@dot.gov

Khamis Y. Haramy
Federal Highway Administration – Central Federal Lands Highway Division
12300 West Dakota Avenue, Suite 210
Lakewood, Colorado 80228
720-963-3521
khamis.haramy@dot.gov

Prepared for the 68th Highway Geology Symposium, May, 2017

Acknowledgements

The authors wish to thank Andrew Coit (FHWA-CFL) for his help in describing construction methods on the project and his photos of construction on the Saddle Road West Side project.

Disclaimer

Statements and views presented in this paper are strictly those of the author(s), and do not necessarily reflect positions held by their affiliations, the Highway Geology Symposium (HGS), or others acknowledged above. The mention of trade names for commercial products does not imply the approval or endorsement by HGS.

Copyright

ABSTRACT

Roadway construction in volcanic materials is subject to numerous geologic and safety hazards, including lava tube collapse and encountering thick unstable soil deposits derived from weathering of volcanic ash. In saturated conditions, the weathered volcanic ash moisture contents commonly reach 250 percent. Saturated, thick weathered volcanic ash layers are frequently unconsolidated and prone to excessive settlement and strength loss when subjected to traffic loading. This material is frequently highly plastic and cannot support construction traffic without improvement. Conversely, dry volcanic ash deposits are nearly impossible to mix with water to meet the AASHTO compaction requirements, and highly mobile, causing dust plumes and hazardous conditions during construction.

The Central Federal Lands Division of the Federal Highway Administration recently completed the construction of Saddle Road, which is located in the saddle between the active Mauna Kea and Mauna Loa volcanos on the Island of Hawaii ("the Big Island"). The roadway was constructed through several phases and mostly traversed a new alignment within rugged volcanic terrain. The project encountered challenging conditions related to both dry and wet volcanic ash soils of varying thicknesses present in the subsurface along the alignment. On the west side, the climate is relatively dry; hence the volcanic ash soils in this region were subsequently dry with moisture contents near 25 percent. During construction the dry ash layers were scarified and inserted/mixed with crushed rock down to two feet from final subgrade elevation to reduce soil mobility and improve compaction. On the east side, however, the soils were saturated, highly plastic, soft, and compressible due to frequent, high-precipitation rates within this region. Moisture contents were commonly measured between 150 and 300 percent. Construction and long term-performance of the roadway prism and embankment fills over these soils was a major concern during the roadway design. To reduce long- and short-term potential settlements, the wet soft ash within the roadway subgrade and embankments was removed and replaced with a minimum 30-inch platform of geogrid reinforced crushed rock beneath the structural pavement section. This paper presents, lessons learned, effective construction practices, and remedial actions performed during construction of several miles of roadway on both wet and dry weathered volcanic ash material.

INTRODUCTION

Volcanic terrain presents potential for numerous geologic hazards to engineering works, in general, and roadway construction, in particular. In Hawaii, large voids (lava tubes) can frequently be present in the subsurface. These features can be difficult to detect and are fragile and prone to collapse due to load and vibrations generated by heavy construction equipment. Soils derived from volcanic ash also present difficulties for construction and long-term stability and maintenance of roadways. This paper presents background on volcanic ash soil composition and Hawaiian volcanism, as well as experiences gained through the development and construction of Saddle Road (also known as Daniel K. Inouye Memorial Highway) project on the Island of Hawaii (the "Big Island"). Geotechnical properties of the volcanic soils and implications for construction and maintenance concerns are discussed. Design and construction practices generally applicable to volcanic ash sites are presented, and best practices are recommended.

BACKGROUND

Volcanic Ash Characteristics

Pyroclastic is a general term used to describe the products of several explosive volcanic processes. Pyroclastic material is transported during an eruption in a "flow," "surge" or "fall." A pyroclastic *flow* is when a mixture of gas and melted volcanic rock flows downhill away from a volcano and stays close to the ground. A pyroclastic *surge* is a more rapid flow contains higher gas to rock ratio, allowing it to rise over ridges and hills. A pyroclastic *fall* is when material that has been ejected from the volcano into the atmosphere falls to the ground. Pyroclastic deposits are classified as ash, lapilli, and blocks based on their grain-size as shown in Table 1 and on their chemical and mineralogical composition. Volcanic ash is a fine-grained pyroclastic material originating from various types of eruption mechanisms, including Vulcanian, Plinian, and Surtseyan eruptions. Due to their formation process and small grain size, ash deposits are frequently widespread, while thickness of ash deposits varies significantly depending on proximity to the source vent, surface topography, and climatic factors such as wind and precipitation.

Table 1 – Pyroclastic Grain Sizes (1)

Clast Size	Name
> 64 mm (~2.5 in)	Block, bomb
2 – 64 mm (~0.1 in – 2.5 in)	Lapilli
< 2 mm	Ash

Ash deposits contains varying proportions of lithics (fragments of older rocks), crystals (new crystalline material formed from the pyroclastic source material), and glass (amorphous material formed from the pyroclastic source) (1). Typically, glass is the most common material (by volume) and lithics are the least common materials found in ash. Due to a combination of magma viscosity and entrained gasses, the glass phase of pyroclastic deposits are frequently vesicular, that is, full of air bubbles. As a result, the density of volcanic ash deposits is typically very low and porosity is very high (2) – (9). This is especially true of unconsolidated and lightly consolidated ash deposits.

The glass materials in ash deposits weather readily to form various types of amorphous materials and clay minerals depending on the original composition of the glass and the chemistry of the meteoric water infiltrating through the deposits (10) – (12). This in place weathering generates a highly porous deposit with a delicate microstructure that is susceptible to collapse and loss of strength (10), (12). Regardless of weathering characteristics, the high porosity of volcanic ash deposits allows for extremely high water contents. Moisture content by mass exceeding 100 percent is common, with values as high as 392 percent reported in the literature (10) – (12).

Atterberg limits testing for classification of volcanic ash soils are variable; however, most test results plot near Cassagrande's A-Line as shown in Figure 1. Minimally disturbed samples of volcanic ash frequently exhibit relatively high friction angles ranging between 30 and 45 degrees (10) – (12). Residual friction angles are in the range of 14 to 36 degrees (12). The reduction in residual friction angle is a consequence of the breakdown of a fragile microstructure (12).

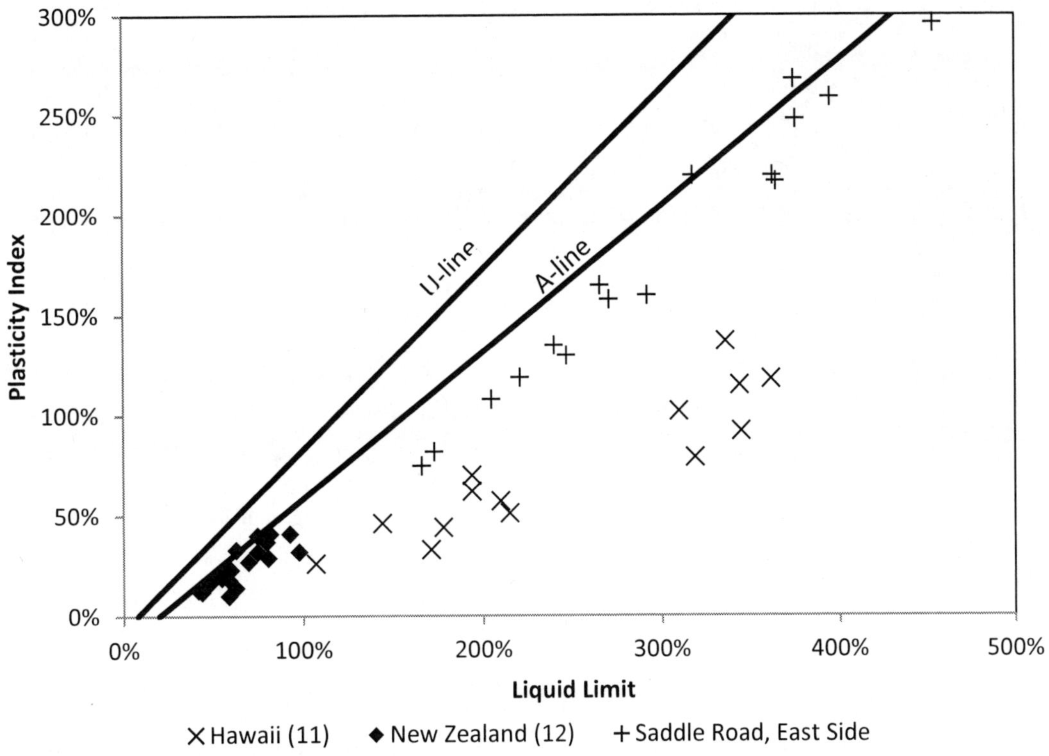

Figure 1 - Casagrande classification of select weathered volcanic ash samples from Hawaii and New Zealand.

Geology of the Island of Hawaii

The Island of Hawaii was formed by the coalescence of several major volcanoes which have erupted more or less continuously for more than one million years (13). The composition of

Hawaiian lavas is generally basaltic with a gradual increase in silica content noted during evolution of the volcanos (14).

The lava flow units, both a'a and pahoehoe, of these volcanoes frequently interfinger, creating a complex geologic history (Figure 2, (15) and (16)). The geology is further complicated by various air-fall pyroclastic units, sedimentary units (glacial, slope wash, eolian, etc.), and periods of weathering and soil formation. This has created a landscape with complex interfingering relationships between various types of soils and rocks.

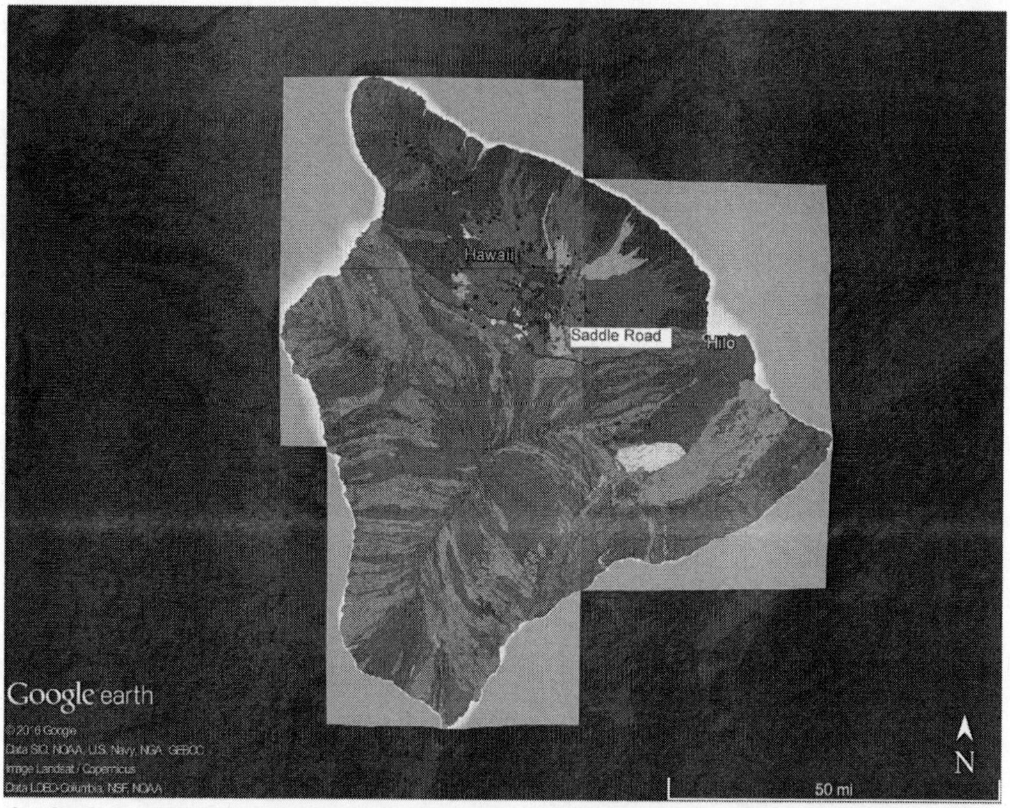

Figure 2 - Geology of the Island of Hawaii (15). Numerous flow and fall units from the five volcanos that comprise the island are evident, with complex geographic relationships.

Saddle Road Project

Project Development

Reconstruction of the cross island route, Saddle Road, on the "Big Island" was completed under several phases in a coordinated effort between the Hawaii Department of Transportation (HDOT) and Central Federal Lands Highway Division (CFLHD) of the Federal Highway Administration (FHWA). The project location is depicted in Figure 3.

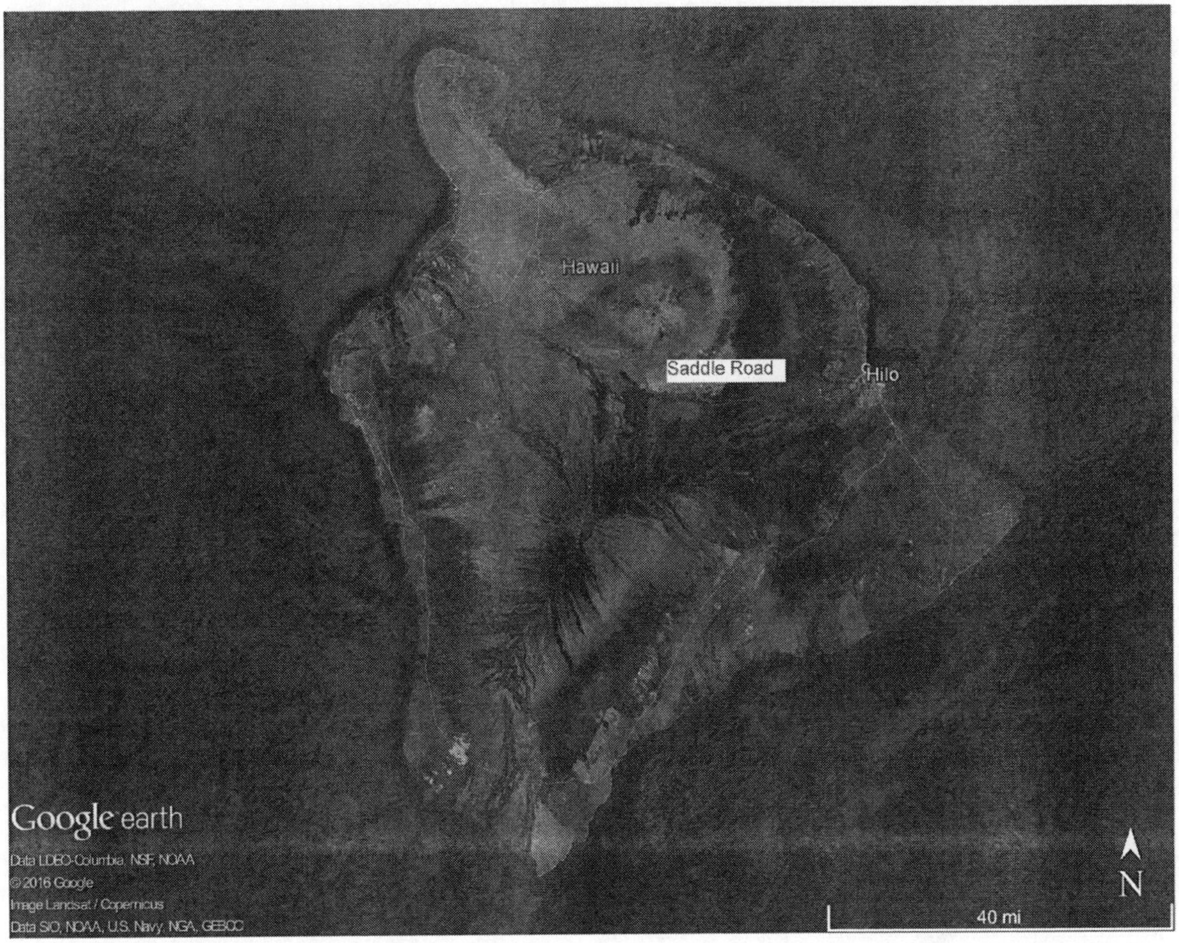

Figure 3 - Project location map (imagery from Google Earth, 2017)

The Saddle Road project was developed to improve transportation connections between the Kailua-Kona Coast and the City of Hilo. Improvements included segmental roadway realignment, vertical and horizontal geometry, public safety, and significant reduction in travel time across the island, as well as improvements in drainage structures to prevent flooding. The Saddle Road project included both reconstruction of existing roads and construction of new alignment in previously undeveloped areas.

Local Geology

Project geotechnical investigation, analysis and reporting were conducted by a local consultant for CFLHD (21). The Saddle Road subsurface investigations for the new alignment generally encountered lava flow deposits including pahoehoe and a'a, volcanic ash, and residual soils. The lava flow deposits varied from extremely hard and dense rock ("bluestone") to "clinker" deposits composed of gravel and cobble sized fragments of hardened lava. The volcanic ash composition includes pyroclastic fragments, phenocrysts, and minor lithic fragments.

On the west side , annual precipitation rates are relatively low, with measurements in Kailua-Kona recording approximately 32 inches (23) and weathering of volcanic materials progresses at a slow rate. This leaves much of the ash dry and relatively un-weathered. Conversely, the east side of the island receives a high rate of annual precipitation that results in high rates of weathering in the volcanic ash (24). Much of the original glassy material and some of the phenocrysts are altered to form various amorphous materials and clay minerals (11), (16), and (18).

FIELD CONDITIONS

The field conditions of volcanic ash on the east and west side of the Island of Hawaii contrast strongly. However, both present challenges for construction of roadway and engineered structures.

Dry Ash

Ash soils on the west side of the island tend to be loose, dry, granular, and mostly un-weathered. This material tends to classify as silty sand (SM) or sandy silt (ML) in accordance with the USCS with zero plasticity. The large proportion of fine grains and lack of plasticity leads to challenges working with the dry ash soils.

Table 2 - In place and Proctor density and moisture content of ash soils

Documented Studies	In Place Moist Density	In Place Moisture Content	Proctor Dry Density	Optimum Moisture
	(pcf)		*(pcf)*	
Wieczorek et al (11)	78.1 ± 5.8	228% ± 86%	-	-
Arthurs (12)	96.1 ± 5.6	72% ± 16%	-	-
West Side, Saddle Road (20)	58.7 ± 7.3	23% ± 9.7%	72.9 ± 11.8	39.9% ± 16.0%
East Side, Saddle Road (21)	72.2 ± 3.9	267% ± 116%	Results inconclusive, see Figure 4	

The dry volcanic ash does not readily absorb water and is therefore difficult to moisture condition and compact. Water added to the ash tends to drain off and is not readily incorporated into the soil. In addition, compaction curves for the volcanic ash are atypical, with very low maximum dry density and weak relationship between density and moisture content (Table 2, Figure 4). The low density, small particle size, and low moisture content of the ash soils cause them to be highly mobile in their natural state, leading to concerns about dust pollution that is difficult to suppress during construction activities (Figure 5, dusty photo). The ash also frequently coated construction equipment, causing visibility issues.

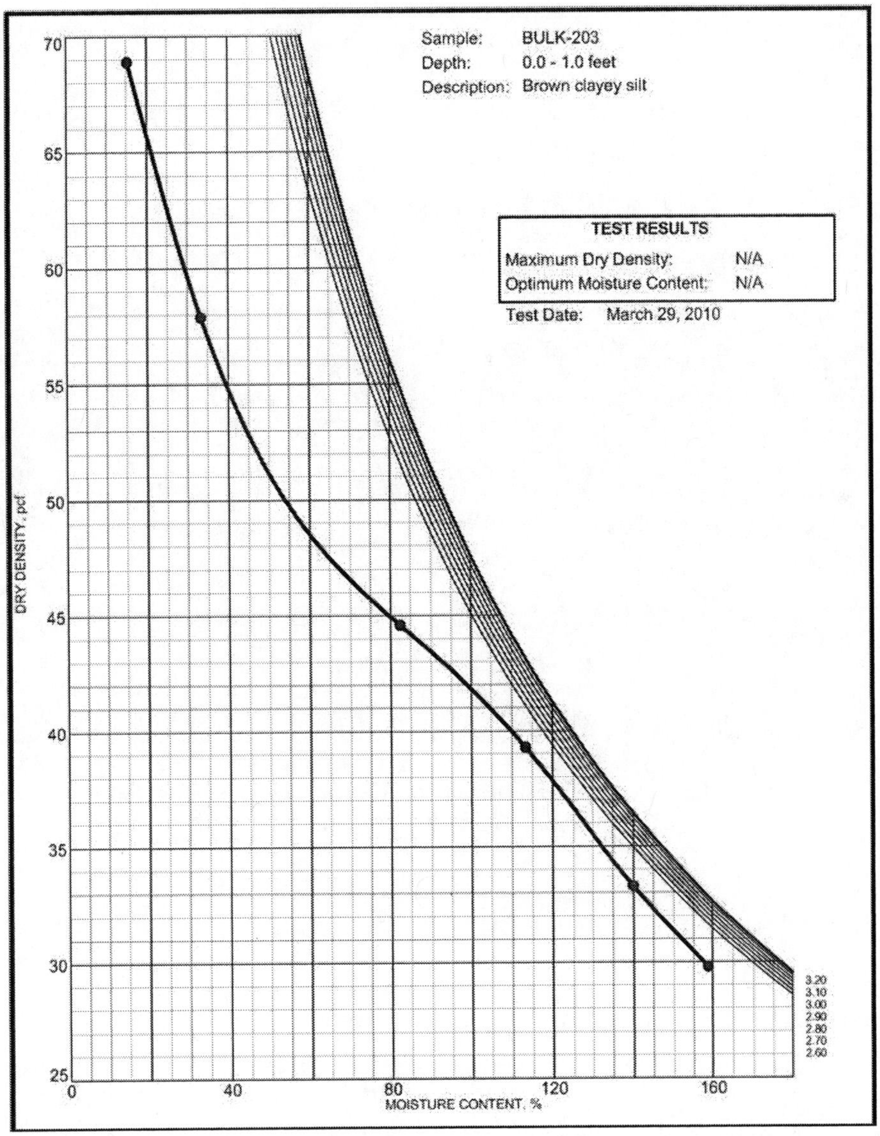

Figure 4 - Example Proctor test result from Saddle Road, East Side (21)

Figure 5 - Dusty conditions during construction work on the Saddle Road, West Side project.

Wet Ash

On the east side of the island, near Hilo, the average annual rainfall ranges from 100 inches near the coast to 300 inches at elevations between 2,000 and 3,000 feet above sea level (24). This high rate of precipitation increases the rate of weathering of geological materials, including volcanic ash. In addition, frequent rainfall tends to create saturated conditions in the ash within this region. The ash is highly plastic and often loses free water when remolded.

The saturated, weathered volcanic ash is typically very soft (Figure 6). Upon initial disturbance, the ash may be firm enough to support loads imposed by a passenger vehicle. However, subsequent disturbance leads to softening of the ash, making it difficult to traverse with vehicles with rubber tires. Careful excavation practices are required to prepare the subgrade to receive the embankment fill loads.

Figure 6 - Weathered ash material exposed during construction of the Saddle Road, East Side project. Note the degree of rutting and sheen visible.

For soft, wet, clayey material, long-term consolidation is a major concern. Based on one-dimensional consolidation testing, embankment subgrade settlements between 1 inch and 20 inches (average 7.3 inches) were predicted (Table 3).

Table 3 - Predicted and measured settlement for Saddle Road, East Side Project

Location	Approximate Ash Thickness	Proposed Embankment Height	Predicted 1-D Consolidation	Reported Settlement	Approximate Embankment Height (at time of survey)
Critical Section (survey monument)	*(ft)*	*(ft)*	*(in)*	*(in)*	*(ft)*
Sta. 1776+24	11	5	0.4	-	-
Sta. 1779+37	11	23	20.6	-	-
Sta. 1816+07 (Sta. 1815+00)	12.5	25	7.3	2.2 ± 0.4	9.5
Sta. 1821+04 (Sta. 1825+00)	9.5	25	6.2	1.0 ± 1.0	11.5
Sta. 1841+48	7	30	4.2	-	-
Sta. 1847+26 (Sta. 1849+75/1850+50)	7	30	5.1	0.4 ± 0.2	13

CONSTRUCTION METHODS

Compaction

To facilitate good subgrade compaction, as part of the special contract requirements for the Saddle Road project west side new alignment, the contractor was required to provide an "Ash Management Plan" that included detailed procedures to handle, stabilize, and compact ash materials prior to beginning construction. The project plans also included ash scarification to a minimum of 2-feet beneath the final roadway grade wetted and mix it with basaltic rock excavated from other areas within the project limits to improve bearing capacity and compaction (Figure 7). Theoretically, water contents between about 30 percent and 60 percent were necessary to achieve required compaction for the ash. With the addition of the rock, the moisture content requirements could be relaxed. Compaction efforts were typically controlled by method specifications due to difficulties in measuring in place density of the mixed rock and ash material.

Figure 7 - Mixing ash and rock to achieve compaction and stable embankment subgrade for Saddle Road, West Side project.

Dust Control

The contractor's "Ash Control Plan" also included provisions for dust control. Due to its fine particulate nature and dry condition, the ash soils are highly mobile. Regular application of water to areas of exposed ash was required.

Soft Subgrade

In contrast, saturated soft ash soils were cleared and stripped of heavy vegetation and organic topsoil, and then excavated a minimum of 30 inches below planned roadway subgrade elevations using tracked excavators and bulldozers. A minimum 30-inch thick section of reinforced crushed rock (minus 6-inch select topping) was then constructed on the exposed volcanic ash. A layer of geotextile was installed on top of the ash layer as a separator fabric to protect against fines migrating into the select topping, which was reinforced with two layers of biaxial geogrid. The geogrid was placed at the bottom of the section and at 18 inches above bottom (Figure 8 and Figure 9). Following placement of the last lift of select topping, the reinforced section was compacted with a steel drum roller. Depending on the finished grade, this could include subexcavation into the ash to provide the minimum reinforced section thickness. Where total embankment fill was greater than 4.5 feet, the geotextile and geogrid were omitted and rock fill (2-foot minus) was substituted for the select topping. Thicker initial lifts are common for the rock fill material in order to create a stable working platform for further construction.

Both the reinforced and unreinforced sections have performed well under heavy construction loading, including large haul trucks. Minor deflection has been noted when heavy loaded vehicles pass over some sections; however, cracking, has not been observed. Minor rutting has been noted in a few locations. Field observations have indicated that the 30-inch reinforced section has performed adequately in all cases. In areas of greater embankment thickness, the entirety of the fill is comprised of rock material. In these locations, fills greater than 60 inches in thickness have performed extremely well, while fills of thickness between 30-inch and 60-inch have exhibited various performance characteristics. For these intermediate fill heights, a number of factors may be affecting the fill stability, including thickness and properties of the weathered ash soil, size of rock in fill, and degree of compaction of fill.

Embankment Settlement and Stability

Long term performance of the embankments constructed on soft ash remains a concern. Settlement monitoring points were installed at the base of the embankment fills and have been monitored regularly during construction (Figure 10). Two settlement monitoring points were constructed at each location as summarized on Table 3. To date, no significant measurable movement has been recorded at the monitoring point locations. Monitoring will continue throughout construction as more fill is added and as traffic loads are imposed.

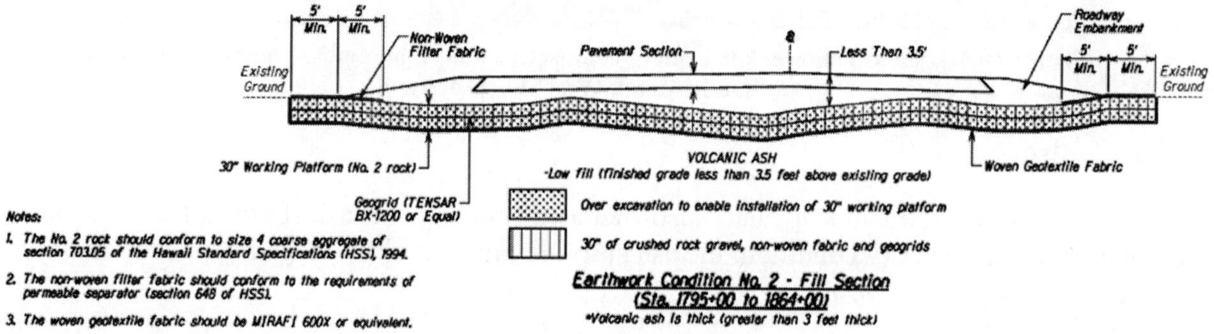

Figure 8 - Reinforced fill section used for Saddle Road, East Side project. This typical section was constructed in areas of relatively thick ash cover where only minor cut or fill was necessary to meet the final roadway grades.

Figure 9 - Installation of reinforced section including geotextile, two layers of geogrid, and crushed rock (select topping) material.

Figure 10 - Typical settlement monitoring point assembly. Assembly was installed near the base of embankment to measure settlement of subgrade ash.

SUMMARY AND CONCLUSIONS

Roadway construction in regions with deeply weathered volcanic ash presents specific challenges that are not common to other geologic settings. These challenges may vary considerably over a short distance depending on the local climate and precipitation rates, as evidenced by the construction of various segments along the new alignment within the Saddle Road project in Hawaii. Un-weathered, dry (near 25 percent moisture content), non-plastic volcanic ash does not readily accept moisture to meet required AASHTO compaction guidelines and also creates severe dust-control concerns during construction. Contrariwise, highly weathered, wet (near 250 percent moisture content) volcanic ash is soft, plastic, and sensitive to disturbance and remolding. In both cases, utilizing rock fills was instrumental in mitigating or improving compaction and load bearing characteristics that the ash materials present during roadway construction. These solutions are specifically effective in areas where rock excavation is planned within the construction project limits or within a close distance to the project. This is similar to construction practices in the Hilo area dating back to the mid-20th century (11). Although dry volcanic ash can generate a substantial amount of dust, control through ordinary dust palliative measures is possible. Although the roadway is performing well under heavy construction loads, the effects of long term settlement of the embankments is not clear and is currently under investigation through settlement measurements.

REFERENCES

(1) Winter, J.D. *An Introduction to Igneous and Metamorphic Petrology*. Prentice-Hall, Upper Saddle River, 2001. 699 pages.

(2) Wesley, L.D. "Shear strength properties of halloysite and allophane clays in Java, Indonesia." *Geotechnique*. Vol. 27, 1977, pp. 125-136.

(3) Guadagno, F.M. "Debris flows in the Campanian volcaniclastic soils." *Proceedings of the International Conference on Slope Stability: Slope Stability Engineering, Developments and Applications*. Thomas Telford, London, 1991, pp. 125-130.

(4) Tejakusuma, I.G. *Slope Movement and Faulting in Quaternary Deposits and Miocene Weak Rock at Beachlands, Auckland, New Zealand*. Unpublished M.Sc. Thesis, University of Auckland, Auckland, 1998.

(5) Bozzano, F., Gaeta, M., Mugnozza, G.S., and Valentini, G. "Numerical modelling of slopes surround ignimbrite plateaux." *Landslides in Research, Theory, and Practice*. Thomas Telford, London, 2000, pp 177-182.

(6) van den Bergen, J.L. *Geotechnical Properties, Stratigraphy, and Slope Movement at Ohuka Stream, Port Waikato*. Unpublished B.Sc. Honours Thesis, University of Auckland, Auckland, 2002

(7) Wesley, L.D. "Residual strength of clays and correlations using Atterberg limits." *Geotechnique*. Vol. 53, 2003, pp. 669-672.

(8) Calcaterra, D. and Antonio, S. "The January 10, 1997 Pozzano landslide, Sorrento Peninsula, Italy." *Engineering Geology*. Vol. 75, 2004, 181-200.

(9) Terribile, F., Basile, A., Mascellis, R.D., Iamarino, M., Magliulo, P., Pepe, S., and Vingiani, S. "Landslide processes and andosols: the case study of the Campania region, Italy." *Soils of Volcanic Regions in Europe*. Springer, Berlin, 2007, pp. 545-563.

(10) Moon, W.G., Lowe, D.J., Cunningham, M.J., Wyatt, J.B., de Lange, W.P., Churchman, G.J., Mörz, T., Kreiter, S., Kluger, M.O., and Jorat, M.E. "Sensitive pyroclastic-derived halloysitic soils in northern New Zealand: interplay of microstructure, minerals, and geomechanics." *Volcanic Rocks and Soils. Proceedings of the International Workshop on Volcanic Rocks and Soils, Lacco Ameno, Ischia Island, Italy*. Taylor & Francis, London, 2015, pp. 3-21.

(11) Wieczorek, G.F., Jibson, R.W., Wilson, R.C., and Buchanan-Banks, J.M. *Geotechnical Properties of Ash Deposits Near Hilo, Hawaii*. U.S. Geologic Survey Open-File Report 82-279. United States Government Printing Office, Washington, 1982. 20 pages.

(12) Arthurs, J.M. *The Nature of Sensitivity in Rhyolitic Pyroclastic Soils from New Zealand*. Unpublished Ph.D. Thesis, University of Auckland, Auckland, 2010.

(13) Lockwood, J.P. and Lipman, P.W. "Holocene Eruptive History of Mauna Loa Volcano." *Volcanism in Hawaii*. U.S. Geological Survey Professional Paper 1350. United States Government Printing Office, Washington, 1987, pp. 509-535.

(14) Wolfe, E.W., Wise, W.S., and Dalrymple, G.B. *The Geology and Petrology of Mauna Kea Volcano, Hawaii – A Study of Postshield Volcanism*. U.S. Geological Survey Professional Paper 1557. United States Government Printing Office, Washington, 1997, 129 pages.

(15) Wolfe, E.W. and Morris, J. "Geologic Map of the Island of Hawaii". Miscellaneous Geologic Investigations Map I-2524-A. U.S. Geological Survey, Washington, 1996. Scale 1:100,000.

(16) Peterson, D.W. and Moore, R.B. "Geologic history and evolution of geologic concepts, Island of Hawaii." *Volcanism in Hawaii.* U.S. Geological Survey Professional Paper 1350. United States Government Printing Office, Washington, 1987, pp. 149-189

(17) Hirashima, K.B. "Highway experience with thixotropic volcanic clay." *Highway Research Board Proceedings*. Vol. 28, 1949, pp. 481-496.

(18) Hirashima, K.B. "Highway construction problems involving plastic volcanic ash." *Highway Research Board Bulletin*. Vol. 44, 1951, pp. 1-10.

(19) Federal Highway Administration. *Continuing the FHWA and HDOT Peer to Peer Partnership*. Federal Highway Administration, https://flh.fhwa.dot.gov/projects/hi/hidot/. Accessed February 21, 2017.

(20) Hawaii Department of Transportation. *Eleven Highway Bridge Projects to be Expedited with Innovative Federal/State Partnership.* Hawaii Department of Transportation, http://hidot.hawaii.gov/highways/eleven-highway-bridge-projects-to-be-expedited-with-innovative-federalstate-partnership/, July 15, 2014. Accessed February 21, 2017.

(21) Geolabs, Inc. "Geotechnical Data Report, Hawaii State Route 200 (Saddle Road), Milepost 11 to Milepost 5.6 (Saddle Road East Side), Island of Hawaii." Report No. W.O. 4983-30&40. Geolabs, Inc., July 2, 2015.

(22) Geolabs, Inc. "Geotechnical Engineering Report, Hawaii State Route 200 (Saddle Road), PTA-1 Alignment (Milepost 28 to Milepost 42), Island of Hawaii." Report No. W.O. 3411-20. Geolabs, Inc., October 10, 2003.

(23) U.S. Climate Data. *Climate Kailua – Hawaii.* U.S. Climate Data, http://www.usclimatedata.com/climate/kailua/hawaii/united-states/ushi0132. Accessed on March 1, 2017.

(24) National Weather Service Weather Forecast Office. *Hilo, Hawai'i first order observing station information.* National Weather Service Weather Forecast Office, http://www.prh.noaa.gov/hnl/climate/phto_clim.php, November 24, 2016. Accessed on February 22, 2017.

Application and Cost Analysis of TDA for Slope Stability

Matthew Kraus

Missouri Department of Transportation

1617 Missouri Boulevard Jefferson City, MO 65102

573-526-4320

Matthew.Kraus@modot.mo.gov

Prepared for the 68th Highway Geology Symposium, May, 2017

Acknowledgements

The author would like to give his appreciation and thanks for the individuals and entities for their invaluable contributions in the work described:

Kevin McLain Geotechnical Section Director Missouri Department of Transportation

Geology Laboratory Missouri Department of Transportation

Dr. Dana Humphrey University of Maine

Terry Gary Tetra Tech

ABC Tire, LLC

Disclaimer

Statements and views presented in this paper are strictly those of the author, and do not necessarily reflect positions held by their affiliations, the Highway Geology Symposium (HGS), or others acknowledged above. The mention of trade names for commercial products does not imply the approval or endorsement by HGS.

Copyright Notice

ABSTRACT

Tire Shreds used as Tire Derived Aggregate (TDA) have been noted and used in civil engineering projects as a green lightweight fill for the core of embankment fills and backfill material for retaining walls. The TDA possesses unique engineering properties of being lightweight, allowing drainage, and having cohesive abilities. The TDA can also have an added bonus of cost savings to the project budget. The Missouri Department of Transportation's (MoDOT) goal is to effectively use TDA for slope remediation, and employ key factors to successfully applying TDA as an alternative to traditional slope remediation. These key factors include site evaluation, proper design, practicality, and a local source of tire shreds.

The project site is a slide area located just south of Platte City, MO on the west embankment of southbound I-435. The slide is 240 ft long and with a slope height of around 34 ft. A concrete drainage ditch along the base of the original embankment is cracked, saturating the toe of the soil causing the embankment composed of lean clay to slide along a shale layer. Instead of using a traditional rock wedge repair, a two TDA layered application was evaluated using researched data. The data was applied to a slope analysis, resulting in an above required satisfactory factor of safety. The proximity of the site to several tire shredding operations made the cost of TDA considerably less than local rock fill. Thus, saving the project $160,000-$200,000 and using 3,300-3,800 tons of tire shreds (approximately 330,000-380,00 tires) and eliminating 6,100-7,700 tons of rock fill. The layered design and placement of the TDA is similar to placing rock fill lifts so no special equipment or techniques are needed that would add extra cost to the project.

MoDOT did elect to go with a traditional rock wedge repair and not the TDA solution. This case study along with construction and environmental guidelines can hopefully be used as a blueprint for future site evaluations, and possible solutions to resolve concerns MoDOT had with the TDA solution.

\

Introduction

According to the latest data provided by the Rubber Manufacturers Association (RMA) in 2015, 4038.8 thousand tons of tires were generated and 87.9% were utilized or went to market (*1*). This left 487.5 thousand tons of tires left to be baled or to be put in landfills, which in 2015, all but 36.3 thousand tons of unutilized tires were of managed this way. The extra 36.3 thousand tons is waste with nowhere to go. Additionally from 2007-2015 the United States has generated more scrap tires than utilized or went to market adding to the extra 36.3 thousand tons generated every year with the trend looking to continue.

Tire Derived Aggregate (TDA) used in civil engineering applications is one way to close the gap between generated and used tire shreds. TDA are tire shreds that are compacted and used like aggregate. Tire shreds by definition are waste tires cut in sizes of 2-12 inch pieces. Tire chips are any sizes less than 2 inches. Unless the tires are old, the shreds will contain steel belts. The sharpness of the knives cutting the tires will determine how much wire is embedded in the rubber. Duller knives tend to produce more free wire (un-embedded wire) than sharper ones. For engineering applications, the less free wire there is in TDA the better. Figures 1 and 2 show the difference between +30% free wire and minus 30% free wire.

Figure 1 - +30% Free Wire

Figure 2 – Minus 30% Free Wire

Rubber fines, dirt, and debris can be seen in Figures 1 and 2 and are a byproduct of the shredding process and should be kept to an extreme minimum for engineering applications (2). The sizes shown in Figures 1 and 2 was proposed for use in MoDOT's project site consist of 3-6 inch shreds and is considered Type A TDA by ASTM D6270-08 (*4*). Table 1 shows typical engineering values and properties of tire shreds compared to typical rock fill (*2*)(*3*)(*4*)(*5*).

Table 1 Typical Engineering Values		
	Rock Fill	TDA
Unit Weight (pcf)	125-135	40-60
Phi Angle (°)	28-30	19-25
Cohesion (psf)	0	160-240

The unique property of TDA is that it can be considered an aggregate and is drainable, and when compacted does achieve cohesion unlike typical rock fill slopes. These values achieved for TDA are obtained using the same or modified versions of current ASTM Standards (3)(4). Unfortunately, the one drawback of the larger shred sizes are that modifications need to be made to the testing apparatus's or required testing equipment needs to be acquired that may not be readily found or is available which is discussed later.

TDA has been successfully used to build embankments, roadways, remediate landslides, used in ground improvement, and even as backfill for retaining walls removing potential tires from going to landfills and reducing projects cost *(2)(6)(7)*. Also, due to Missouri's geography rock is not readily available in our southeastern and northern regions, making the traditional rock wedge solution very expensive in those areas. TDA provides us with an alternative and a significant way to reduce the cost of these repairs. This alternate solution is by no means a fix all, but rather a tool that could be used in MoDOT's "tool belt" of slide remediation. The cost savings and friendly environmental impact are the two main reasons along with its established track record is why MoDOT considered using TDA as an alternative solution for slide remediation.

Site Evaluation

The project site is a slide area located just south of Platte City, Missouri on the west embankment of southbound I-435. The slide is approximately 240 ft long and with a slope height of around 34 ft. A concrete drainage ditch is located at the base of the embankment. This became cracked allowing saturation of the toe of the embankment triggering the lean clay to slide along an underlying shale layer. What made this site an ideal place to use TDA was the lower stable undisturbed in place soil, a large area of right away to work within, and a large sized slide where the cost savings of using TDA could be realized. The proposed site can be seen in Figure 3 with the slide area outlined in red.

Figure 3 – Aerial Map of Project Site

After drilling and sampling the top and toe of the slide and analyzing the results the slide was modeled for failure. Solutions were drawn up for both a Rock Wedge and TDA solution. The TDA solution consists of two different versions based on adding a compacted soil layer along the critical failure surface or not. The overall design for both solutions calls for a 1.5:1 benched cutback slope starting at the edge of the roadway shoulder. The top of the slope remediation will be 10ft wide with a 2:1 finishing slope. At the toe, rock fill was to be placed 3ft below and above the toe. Alternating layers of TDA (maximum of 10ft high) with a 1ft layer of rock fill was then to be constructed. This was repeated up the slope until the top is capped off with a 1ft layer of rock fill and a 3ft layer of compacted fill. The exception being on the second solution placing a compacted soil layer between the bottom layer of rock and first layer of TDA The top lift TDA was planned to be only be 5ft thick to allow enough room for the soil cap. Each layer of TDA was wrapped on all sides by separation geotextile. This is to aid in compaction and to limit the amount of oxygen the shreds come in contact with, thus, limiting the chance for instantaneous combustion. Additionally covering the finished slope made with the borrowed fill material from the excavation of the benched slope limits the possibility of combustion. American Association of Tranportation Officials (AASHTO) code states that all slide remediation recommended must have a minimum Factor of Safety of 1.3. Figures 4 and 5 below are the slope analysis of the two TDA solutions of with and without the lower soil layer respectively.

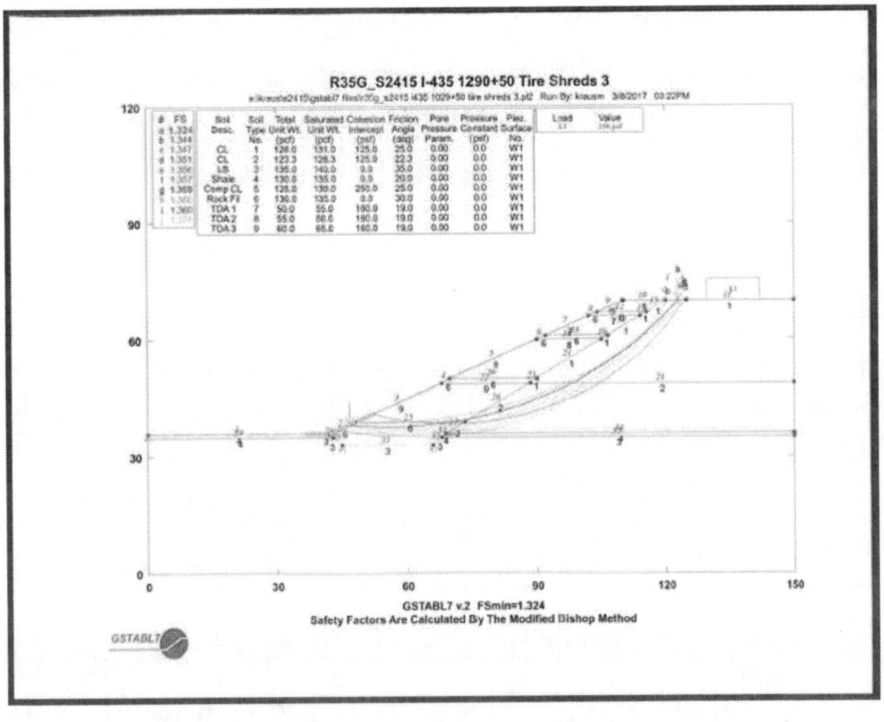

Figure 4 - TDA Slope Analysis Without Lower Soil Layer

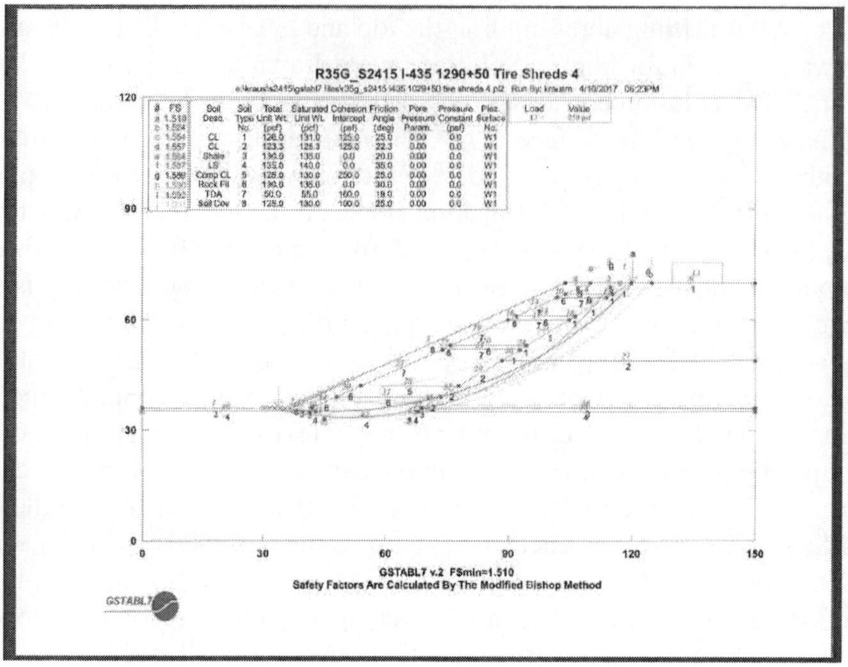

Figure 5 – TDA Slope Analysis With Lower Soil Layer

As shown above the two solutions both meet the 1.3 Factor of Safety requirements while only assuming very low cohesion of the undisturbed soil and using conservative TDA properties.

Cost Analysis

The unit prices used in the cost analysis came from the average cost for the item on all projects in the Kansas City District in 2016 and the cost for the tires is an average from several local suppliers in Kansas City. Table 2 shows the breakdown of the cost analysis and the significant savings the TDA can provide.

Table 2 Cost Analysis of Slide Remediations

	Rock Fill			TDA w/o Soil Layer		TDA w/ Soil Layer	
Item	Quantity	Unit Price	Total	Quantity	Total	Quantity	Total
Furnishing Rock Fill (cyd)	5951	$26.83	$159,665.33	1540	$41,318.20	860	$23,073.80
Placing Rock Fill (cyd)	5951	$22.36	$133,064.36	1540	$34,434.40	860	$19,229.60
Furnishing TDA (ton)	0	$10.00	$0.00	3822	$38,220.00	3363	$33,630.00
Geotextile (syd)	0	$3.41	$0.00	5204	$17,745.64	5204	$17,745.64
Total			$292,729.69		$131,718.24		$93,679.04
Savings					$161,011.45		$199,050.65

One hundred passenger tires is equivalent to 1 ton of tire shreds, so 330,000 to 380,000 tires would be needed helping to close the gap between scrap tires generated and utilized every year.

Constructability Concerns

As mentioned before testing these larger tire shreds for their engineering properties are more difficult than traditional testing and require uncommon size or specifications to run the test making testing and quality control more expensive. Additionally, the use of TDA means now MoDOT is responsible for the tire shreds for perpetuity which causes natural initial concerns if a problem arises and the TDA need to be disposed of, which would become an additional cost to consider. Currently the Missouri Department of Natural Resources (DNR) do not seem concerned in using TDA in this application. The DNR only require a scope of the project before construction and a form designating where and how many tires were used after construction is complete.

When compacting the shreds, there is no current method to measuring the compaction and ensuring it meets with design standards. Best practice is 4 to 6 static passes of a 10 ton roller. The shreds will compress as more weight is added on top of them, so final compaction and compacted unit weight won't be achieved until final construction is finished (*2*). This is a design factor considered in calculating the extra thickness per layer needed to compensate for the compression and slight settlement the TDA will go under during construction. The main reason the TDA solutions were rejected by MoDOT was the potential difficulty and lack of experience of performing material and QC tests for the TDA.

The Future

Again it is important to remember this is not a "fix-all" solution. Each site needs to be evaluated with a variety of potential solutions and the best solution then chosen. Sites that make for TDA as a viable remediation option have adequate right of way space for construction, and the close proximity of a permitted tire shredder and haulers (dependent on state laws). Mobile shredders can also be brought on site to shred if required, though extra paperwork and care needs to be completed in those events The State of Missouri tire allows shreds from across state lines, but the shredder and hauler must obtain Missouri DNR specific permits.

Even though in this case TDA was not used, further research needs to be done to find cost effective ways for testing larger shreds, but also provide quality control and quality assurance testing during construction, specifically for shear testing. For this current problem MoDOT is investigating and looking for solutions so the next time a TDA suitable slide occurs, TDA will be a more viable option for remediation so MoDOT can play more of a part in reducing scrap tire waste.

REFERENCES

ASTM International. (2012). "D6270-12 Standard Practice for Use of Scrap Tires in Civil Engineering Applications." *ASTM International*, D6270-08 (Reapproved 2012),.

Federal Highway Administration (FHWA). (1997). "User Guidelines for Waste and Byproduct Materials in Pavement Construction." *Federal Highway Administration Research and Technology*, 148,.

Graham, M. A., Kjartanson, B. H., and Lohnes, R. A. (2001). "Shear Strength of Large-Siz Shredded Scrap Tires." *TRB: Journal of the TRB*, 1755, 168-172.

Humphrey, D. N. (1998). "Highway Application Of Tire Shreds." New England Transportation Consortium (NETC)., 210 Colchester Avenue Burlington, VT, 05405

Humphrey, D. N. (2005). "Tire Derived Aggregate-*A New Road Building Material* Barriers and Opportunities." *Recycled Materials in Road and Airfield Pavement*

Humphrey, D. N. (2009). "Civil Engineering Applications of Tire Derived Aggregate." Resource Recovery Fund Board, Halifax, Nova Scotia.

Humphrey, D. N. (2011). "Civil Engineering Applications Using Tire Derived Aggregate (TDA)." CalRecyle, <http://www.calrecycle.ca.gov/publications/Documents/tires%5C20110 8.pdf> (Nov. 30, 2016).

Humphrey, D. N., and Katz, L. E. (2001) "Tire Shreds Placed Below The Water Table." *TRB: Journal of the TRB*, 1755, 18-24.

Humphrey, D. N., Dunn Jr., P. A., and Merfeld, P. S. (2000). "Tire Shreds Save Money for Maine." *TR NEWS*, 206, 42-44.

Mills, B., and McGinn, J. (2010). "Design, Construction, and Performance of a Highway Embankment Failure Repaired with Tire-Derived Aggregate." *TRB: Journal of the TRB*, 2170, 90-99.

Rubber Manufacturers Association (2016). "2015 U.S. Scrap Tire Management Summary." *Rubber manufacturers association*, <https://rma.org/sites/default/files/RMA_scraptire_summ_2015_0.pdf> (Nov. 30, 2016).

State of Ohio Department of Transportation (ODOT). (2005). "Embankment Construction using Recycled Materials." *Supplemental Specification 871*, 871.01-871.05.

Design-Build Success:

A Multi-Use Recreational Trail with Minimal Subsurface Investigation

Sara E. Hansen, P.E.
Yeh and Associates, Inc.
2000 Clay Street, Suite 200
Denver, CO 80211
(303)-781-9590
SHansen@yeh-eng.com

Prepared for the 68th Highway Geology Symposium, May, 2017

Acknowledgements

The author would like to thank the following individuals and entities for their valuable contributions:

Matt Brenkle – CEI Constructors, Inc.
Scot Grossman and Scott Timmons – Jefferson County Open Space
Martha Tableman – Clear Creek County Open Space
Muller Engineering Company

Disclaimer

Statements and views presented in this paper are strictly those of the author(s), and do not necessarily reflect positions held by their affiliations, the Highway Geology Symposium (HGS), or others acknowledged above. The mention of trade names for commercial products does not imply the approval or endorsement by HGS.

Copyright Notice

ABSTRACT

Ideally, a complete and thorough subsurface geotechnical investigation is conducted as part of the design process for a large transportation project, particularly for a virgin alignment. This paper exhibits a multi-use recreational trail project in the Front Range of Colorado for which design and successful construction of several thousand linear feet of retaining walls, as well as temporary ground anchor shoring and drilled shaft bridge foundations, were completed with minimal subsurface investigation activities prior to pioneering work, all of which was located on the highway side of Clear Creek. Factors which contributed to this success include the Design/Build project delivery method, substantial past experience with the soils and rocks in the project area, structure types chosen, ample flexibility of the contractor and client regarding final appearance, readily available materials, and involvement of the geotechnical engineer throughout the 4-year project timeline (and still going). As the construction crew encountered conditions which differed from those anticipated, walls and temporary shoring were added or removed, with quick responses and approvals from the design team and client. Many unexpected conditions during construction resulted in cost increases; others resulted in a cost decrease, or were cost-neutral. However, it is unlikely that all of these conditions could have been known ahead of time, even with a detailed geotechnical subsurface investigation including the opposite side of the creek, which would have required drilling equipment placed by helicopter, and added several months to the design schedule. Even if the team had chosen to spend up to $200,000 on geotechnical subsurface work prior to construction, the costs of some of the above changes would not have been any less. An additional benefit to this approach was the flexibility to revise the alignment several times throughout the lengthy design process, including moving the trail to the opposite side of the creek, to meet hydraulic or environmental permitting requirements, without excessively “wasting” geotechnical information.

INTRODUCTION

The Peaks to Plains Trail is a long-term planning initiative to create a 65-mile paved recreational trail between the confluence of the South Platte River and Clear Creek, in Adams County, Colorado, to the Continental Divide in Clear Creek County. Portions of the trail have been constructed and used by the public for many years; however, from Golden, Colorado, westwards up 14-mile-long Clear Creek Canyon, the only access is by US Highway 6 (owned and maintained by the Colorado Department of Transportation, CDOT); bicycles are not legally permitted on the highway due to four 2-lane tunnels with no shoulders, and many other areas with very little shoulder or clear zone. This canyon has many exciting opportunities for recreation, including hiking, whitewater rafting/kayaking, fishing, rock climbing, mountain biking, panning for gold and silver, and appreciating the scenery and the history of the railroad and stagecoach roads which were in use until the early 20th century.

In 2012, Clear Creek County Open Space (CCCOS) and Jefferson County Open Space (JCOS) jointly applied for and won funding from Great Outdoors Colorado (GoCo) to design and construct a 4-mile portion of the Clear Creek Canyon Segment of the Peaks to Plains Trail (see Figure 1 and Figure 2). A 10-foot wide concrete surface trail was proposed. This is a very challenging section, with steep slopes, hard metamorphic and igneous bedrock, high creek flows in the spring runoff season, varying bedrock depths, rockfall, landslides, and a need to preserve water quality and sensitive environmental areas. The project was advertised as a Design/Build project and awarded based on design costs, team qualifications, and project approach. The work was awarded by the co-owners (both Counties) in early 2013 to CEI Constructors, Inc., Muller Engineering Company, and Yeh and Associates, Inc., along with several other sub-consultants.

The Counties had each previously conducted feasibility studies for the trail, including determination of a preliminary trail alignment. This alignment underwent significant revisions once the design phase was underway, after several limitations with the preliminary alignment were encountered. These included a significant historic landslide, rockfall hazards, environmental and water quality concerns, a threatened species with habitat in the flat areas near the creek, very steep slopes, and geometric constraints on the longitudinal grade and curve radii. Because of all of these limitations, and the many other unknowns, the design team found that a more careful, measured approach to the geotechnical investigation was more successful, rather than rushing out to drill borings along the first alignment.

Figure 1 – General project location: Clear Creek and Jefferson Counties, Colorado

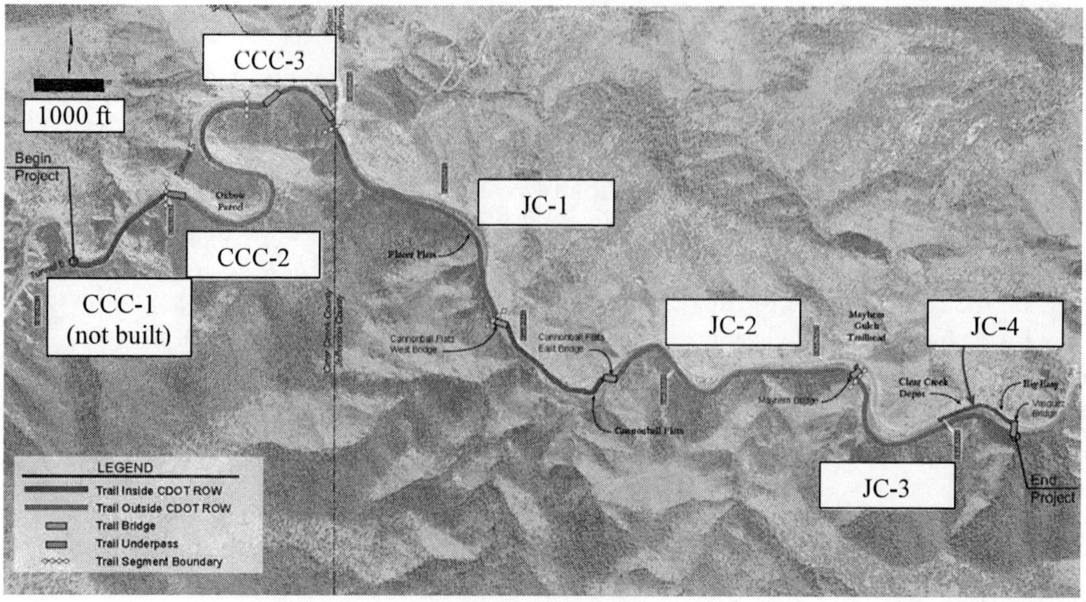

Figure 2 – Project extents and construction segments (CC: Clear Creek County, JC: Jefferson County)

GEOLOGIC SETTING

The USGS 7.5-minute geologic quadrangle maps for this area are the Squaw Pass Quadrangle (Sheridan and Marsh, 1976) and the Evergreen quadrangle (Sheridan, Reed and Bryant, 1972); these are shown below in Figure 3. The bedrock within this segment of the trail is described as Precambrian-aged metamorphic and igneous rocks that form the base of the Front Range uplift, and is known locally as the Idaho Springs Formation. These include primarily feldspar-rich gneiss, in some areas interlayered with hornblende gneiss, amphibolite, and other gneisses, and with intrusions of Precambrian-aged white, pink, or light gray hard to very hard granitic pegmatites. The gneiss bedrock encountered within the project area was moderately to slightly weathered. These similar rock units vary in color, including light- to dark gray, tan, pinkish-gray, and/or greenish-gray, and are generally hard to very hard. Bedrock weathering, extent of fracturing or foliation, and most other bedrock conditions vary throughout the project area. A geologically inactive fault zone crosses the highway at the east end of the project area, and may include highly fractured or brecciated rocks at depth. Cuttings from the drilled shaft excavation appeared to be weakened and altered bedrock, possibly due to the mapped fault.

Surficial deposits include artificial fill, alluvium, and colluvium of varying thicknesses overlying the bedrock. Recent alluvial deposits occur in the modern floodplain of Clear Creek. The alluvium is normally consolidated and composed of cohesionless silt- to boulder-size, moderate to poorly graded sediment deposits. Colluvial deposits derived from the bedrock are present at the base of and on the sides of some slopes, and are generally similar to the alluvium. Artificial fill soils are generally reworked on-site soils which were placed for the historic railroad bed, and also include large piles of cobbles and gravel which were the result of past large-scale placer mining activities.

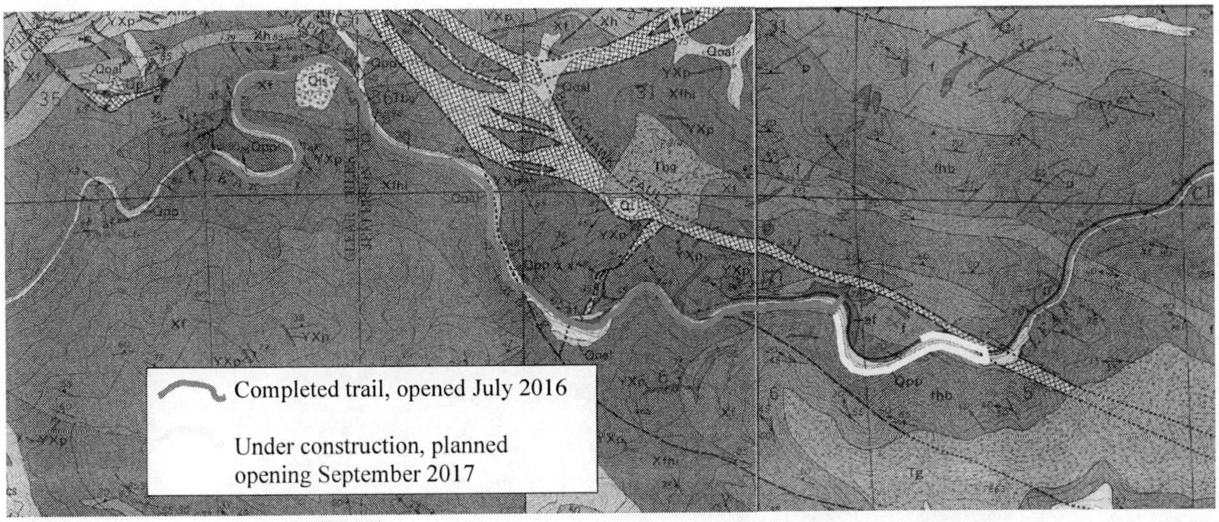

Symbols from Geologic Map of the Squaw Pass Quadrangle (Sheridan and Marsh, 1976)

Squaw Pass		Evergreen
Qpp	Post-Piney Creek Alluvium (Upper Holocene)	Qpp
Qls	Landslide Deposits (Upper Holocene)	
YXp	Pegmatite (Precambrian X and Y)	p
Xf	Feldspar-rich Gneiss (Precambrian X)	f
Xfhi	Feldspar-rich Gneiss interlayered with Hornblende Gneiss, Amphibolite, and other Gneisses (Precambrian X)	fhb

Symbols from Geologic Map of the Evergreen Quadrangle (Sheridan, Reed, and Marsh, 1972)

Figure 3 – Combined Geologic Map of the Subject Site. (Green and yellow lines indicate approximate trail location)

Geotechnical borings which were drilled in the shoulder of the highway indicated that the surficial deposits ranged from 0 to over 40 feet thick, with an average thickness of approximately 20 feet. However, excavations during construction revealed that the depth of surficial deposits appears to decrease further up the slopes, although somewhat unpredictably.

SITE CONDITIONS

The existing site conditions for the trail include steep slopes (generally 35 to 45 degrees from horizontal) and a close proximity to Clear Creek and US 6. The project area consists of undeveloped public and private land, with no apparent building structures nearby. The slopes surrounding the trail alignment are vegetated with large conifer trees, shrubs, and grasses. Numerous bedrock outcrops are also visible on the slopes. In some sections there is a small flat bench for most of the year between the water's edge and the toe of the slope; this is a riparian/wetland area in many places. In other sections, higher benches appear to be relic terrace deposits of the creek, and historic railroad grades, walls and a bridge abutment are also present in a few places. Several highway bridges also carry US 6 across the creek within the project limits.

Figure 4 – Example existing conditions before trail construction: (L) steep slopes at proposed trail elevation, and (R) riparian bench next to creek found in some areas.

Groundwater flow in the native alluvial soils and bedrock is influenced primarily by Clear Creek, surface water infiltration, and fracture flow within the bedrock. Due to the topography and geology of the site, it is likely that groundwater levels vary both throughout the year, and from year to year. Very little seeping groundwater has been encountered from soils or rock during construction of the trail.

A large landslide, which has caused closures of the highway in the past, is present near the confluence of Clear Creek and North Clear Creek. This is known as the Forks Landslide or the Junction Landslide. It is an unstable rock and colluvial slope, with movement occurring along bedrock foliation planes which dip to the north towards the highway. Movement has historically occurred as rock and debris slides, and has been observed here since the 1940s or 1950s. While this landslide threatens the highway and associated structures, and was included as No. 3 on a list of 46 "Critical Landslides of Colorado" by the Colorado Geologic Survey

(Rogers, 2005), observed movement of this slide has diminished following reconfiguration of the US6/SH119 intersection in the 1990s.

TRAIL ALIGNMENT SELECTION

From the beginning of the project, the Design/Build team was aware that the Owners had several major constraints on where the trail alignment could be, including:

- Maximum trail grades of 5% (creek gradient is 3% on average through this section), with minimal sections up to 8% according to 2012 AASHTO Guide for Development of Bicycle Facilities.
- Minimum curve radii and sight distances to accommodate cyclists.
- Maximize length of trail located on the opposite side of the creek from the highway to enhance the user experience; where this was not possible, trail needed to be as safe as possible.
- Reduce environmental impacts to vegetation and wildlife habitat, and avoid removing mature trees where feasible.
- Avoid the Forks Landslide and other geohazards (mainly rockfall hazards), or mitigate where feasible.
- Aesthetically integrated with the surrounding natural environment.
- Schedule: both Counties were required to spend the GoCo grant monies by June 2015; additional funds from County budgets did not have strict timing restrictions.

During the design process, several additional issues arose which created further limitations on the trail alignment location:

- Discovery of a threatened species, the Preble's Meadow Jumping Mouse, with habitat in the flat riparian zones immediately adjacent to the creek.
- Old survey information which did not accurately reflect the topography in many locations (+/- 3 feet vertically in some places)
- Limiting encroachment into the floodway as per Section 404 of the EPA; most sections of the trail were able to be permitted under a Nationwide Permit Individual Permit, but one section did require an Individual Permit.
- Hydraulics: final trail grades above at least the 10-year design flood event (major structures higher), but need to pass under two existing highway bridges with adequate clearance.
- Other areas of Colorado experienced severe flooding in the fall of 2013, leaving many engineers more sensitive to potential scour impacts.

Due to these and other additional constraints, the trail alignment selection process was iterative and took several months for each trail segment, and several "alignment walks" with the primary members of the design/build team. In addition, the geohazards and geotechnical design issues were somewhat secondary to other concerns; since the geology of the area is fairly predictable (yet challenging), the design was approached with the mindset that geological problems generally could be resolved more easily than other problems.

The trail alignment was split into sections based on where it traversed CDOT Right-of-Way, where bridges were planned, and other geographic features. This resulted in three trail segments within Clear Creek County (CCC-1, CCC-2 and CCC-3), and five trail segments within Jefferson County (JC-1, JC-2 West, JC-2 East, JC-3, and JC-4). The scope of work for the project included design only for segment CCC-1; this was later eliminated due to budget overruns in CCC-3. JC-2 was originally one segment, but was later extended to the west, had an additional bridge added, and the segment was split into a section inside CDOT ROW (JC-2 West), and a section outside CDOT ROW (JC-2 East). Approval and permitting requirements were different inside and outside the highway ROW.

In general, the design, permitting, pioneering and construction work has progressed from the west towards the east in both Counties' trail sections simultaneously. CCC-2 and JC-1 were designed and constructed first, followed by CCC-3 and JC-2 East, then JC-2 West and JC-3. CCC-2, CCC-3, JC-1, and JC-2 West & East were all opened to the public in July 2016. JC-3 and JC-4 are still currently under construction at the time of this writing, and schedule to be complete this summer (August 2017).

PRELIMINARY GEOTECHNICAL INVESTIGATION

11 geotechnical borings were drilled in October 2013, after the trail alignment had been essentially set for both trail segments in Clear Creek County, but only one segment in Jefferson County. At this point in the design process, it was known that many thousands of feet of retaining structures would be needed, and wall types were also selected. The two primary structure types which the Counties chose to incorporate were rockeries up to 12 feet high and large-block-faced MSE walls where larger or more vertical structures were needed. No trail pioneering or temporary creek crossings had yet been built, so locations on the opposite side of the creek from the highway were not accessible by standard truck- or track-mounted drilling equipment. While drilling contractors with lightweight equipment that can be placed by helicopter do exist, and we have used them on past projects, the budget and scope proposed for this project did not include this type of geotechnical investigation.

This decision was not made lightly; there was some potential for budget impacts if some significant change from assumed conditions were later discovered. However, Yeh and Associates personnel have many years of experience on other projects in Clear Creek Canyon, and with the terrain and the rock and soil types involved. The cost involved in completing a thorough geotechnical investigation with a helicopter-access drill rig could have easily reached $200,000, while the financial risk to the project of a differing geologic condition was projected to be equal to this estimated cost.

The primary geologic factor affecting the construction budget was the depth to bedrock; if additional retaining walls were needed in certain areas, or additional rock excavation in other areas, it would primarily be because bedrock was shallower or deeper than anticipated. Areas where rock excavation was anticipated were fairly obvious (see Figure 5). To reduce the potential for wall design to be affected by bedrock being too deep, in most places the walls were designed assuming that rock was deep enough to not contribute to bearing resistance or global/overall stability.

Figure 5 – An area where rock cuts were obviously required. (JC-2 East)

In October of 2013, four deep borings were drilled at the accessible (highway-side) potential bridge locations, four deep borings at the large concrete box culvert (CBC) trail underpass structure in CCC-3, and three shallow borings at proposed parking lot locations. Each of the deep borings at the underpass encountered bedrock between 37 and 43.5 feet below the highway grade; at the bridge locations in Jefferson County, which were spread out over a few miles, each of the deep bridge borings encountered bedrock between 6 and 50 feet below the highway grade. As a result of these borings, and the findings on bridge scour from the hydraulics report, shallow foundations for the pedestrian bridges were precluded.

Table 1 – Summary of Preliminary Geotechnical Investigation

Boring Type	Number	Location	Depth to Bedrock
Deep	4	Potential Bridge Abutment Locations	6 to 50 ft
Deep	4	CBC Underpass structure (CCC-3)	37 to 43.5 ft
Shallow	3	Proposed parking lot locations	Not encountered (max boring depth 10 ft)

GEOHAZARD EVALUATION STUDY

As part of the preliminary design work, we conducted a Geohazard Evaluation Study; the primary geohazard for this project area is rockfall onto the trail. This work required accessing the opposite side of the creek in some areas by wading across the creek. Rock outcrops, talus fields, eroding colluvial slopes, and proposed rock cut areas were observed and photographed. In conjunction with the Counties, a process was created for determining where mitigation should be constructed, taking into account the severity of the hazard, the exposure to the public, sight distance along the trail, the steepness of the slope, the access for constructing mitigation, maintenance requirements, and the estimated costs. Three different levels of hazard reduction, with corresponding mitigation measures, were presented to the Counties, and mitigation options for these rockfall areas were chosen.

Part of this study was ensuring that the Counties and their respective legal counsel were aware of the potential liability from constructing a public-access trail in such a geologically active area. There were clear signs of recent rockfall activity in many places, and a few small to moderate events have occurred in the 4 years since the project began, including some rocks which have fallen onto the completed trail surface and caused chips or cracks in the concrete. Various measures including suspended rockfall mesh, anchored cable net, and rock scaling, were implemented. In one area, where an outcrop is overhanging the trail, in order to scale some large and hazardous loose rocks a series of overhead electric poles and lines had to be relocated. However, in many areas the only measures the County chose to incorporate were appropriately placed warning signs.

ALIGNMENT REVISIONS AND ADDITIONAL GEOTECHNICAL INVESTIGATION

By the summer of 2014, some trail pioneering and permitted temporary bridge crossings had begun to allow access to the opposite side of the creek. At the same time, challenges meeting the other design criteria and constraints discussed above had resulted in alignment changes, especially in segment JC-2. In this trail segment, one bridge was moved 1000 feet west in order to avoid a very steep slope with exposed rock structurally oriented with a dip-slope (Figure 6), another bridge was added, and a section of the trail was switched from being on the highway side of the creek to the opposite side. This led to the division between JC-2 West and JC-2 East, and another consequence was that several of the borings which had been drilled in the fall of 2013 were no longer located near a structure. Other alignment changes in segment CCC-2, JC-1, and JC-3 had occurred which were more fine-tuning and did not have a large effect geotechnically.

Since construction equipment had already mobilized to the site for pioneering a construction bench (which was anticipated to take several months in total), but the design of bridge foundations and abutments was not yet complete, five test pits were excavated using a backhoe, at each abutment of two of the bridges in trail segment JC-2, and at one abutment of the bridge at the end of JC-2/beginning of JC-3. These were known as the Cannonball West bridge, the Cannonball East bridge, and the Mayhem Gulch bridge, respectively.

Figure 6 – A steep dip-slope in JC-2 which was avoided by changing the proposed location of a bridge.

These test pits were up to 14.4 feet deep, and bedrock was encountered in three of them, at depths between 8 and 13 feet below ground (the test pits were dug as close as possible to creek elevation, which was low in August 2014). In addition, a downhole hammer drill, which was being used to install rockfall mesh anchors in another area of the project, was mounted to a small skid-steer and used to drill one boring without sampling at the north abutment of the Mayhem Gulch bridge. This boring encountered consistently slow-drilling, hard rock at approximately 25 feet, which was assumed to be the hard metamorphic bedrock found and mapped elsewhere on the project.

A third round of geotechnical investigations in JC-3 and JC-4 was completed in the fall of 2015; at this time, the design and permitting of JC-3 was at an approximate 50% level, and the design and permitting of JC-4 was still in the preliminary stages. This investigation included a seismic refraction study within JC-3, one geotechnical boring at a proposed pedestrian bridge abutment, three borings for pavement design of a new relocated section of US Highway 6, and non-sampling probe holes drilled with a rock anchor construction (downhole hammer) drill on the cut slope side of the existing highway at nine locations. These non-sampling holes were drilled both vertically and at an angle at each location, in an attempt to discover where the bedrock surface was located and where retaining structures would be required on the uphill side of the highway. Other areas of the uphill/cut side had exposed bedrock and would require rock excavation.

The seismic refraction study was helpful to identify the likely depth to bedrock below two large piles of cobbles and gravel which had been dredged from the creek during historic placer mining activities. These piles made it very difficult to discern what the topography below them was, and whether bedrock was likely to be deep or shallow in these locations. One seismic refraction line was also located at the proposed bridge abutment on the opposite side of the creek from the highway in JC-3. See Figure 7 below for all locations.

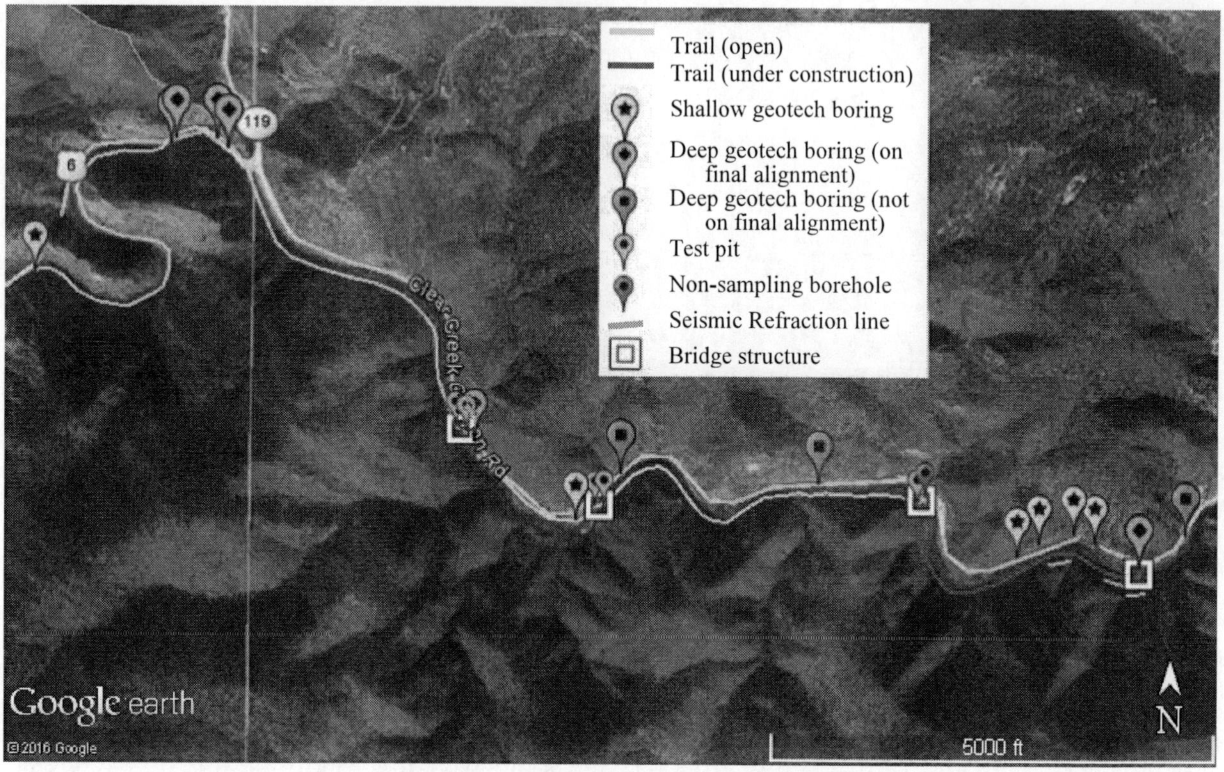

Figure 7 – Locations of all subsurface geotechnical work. (Not all non-sampling borings are shown for clarity)

In summary, the geotechnical information used for design of structures on the 4-mile long trail project came from 15 geotechnical borings, 10 non-sampling borings, five test pits, and 5 seismic refraction lines (See Table 2). Many of the subsurface explorations encountered bedrock, but three of the borings were drilled too early in the design process, and the information from these was not ultimately used in design of any structures.

Table 2 – Summary of All Geotechnical Investigation Work			
Boring Type	Number	Location	Depth to Bedrock
Deep	3	Potential bridge abutment locations (not on final alignment)	12 to 50 ft
Deep	2	Bridge abutment locations (on final alignment)	20 to 34 ft
Deep	4	CBC underpass structure (CCC-3)	37 to 43.5 ft
Shallow	6	Proposed parking lot locations	Not encountered (max boring depth 11 ft)
Non-sampling	1	Bridge abutment location (on final alignment)	25 ft
Non-sampling	9	Soil nail wall and rock cut on uphill side of highway	2 to 19 ft
Seismic Refraction Lines	4	JC-3 - Piles of cobble-sized material (dredged from the creek)	8 to 35 ft
Seismic Refraction Line	1	Bridge abutment location (on final alignment)	30 to 35 ft

FINDINGS DURING CONTRUCTION

As is the usual occurrence, there were many instances during construction when the subsurface conditions did not match what was anticipated. Many of these involved the bedrock surface being either deeper than expected, or shallower than expected. Some of the highlights included:

- An area of fine sandy soil just below a granitic bedrock outcrop, where it appeared an eddy in the creek had previously caused deposition of these soils on an inside curve of the creek. This was managed with the addition/extension of an uphill rockery wall.
- Immediately adjacent to this fine sandy soil were two large bedrock protrusions which were not excavated on the cut side, and instead incorporated into the rockery wall.
- A significant quantity of temporary shoring, over what was estimated, had to be added in segments JC-2 West and JC-2 East, because of deeper bedrock than anticipated and slope disturbances made during pioneering.
- At least three significant rockfall events: two occurred during construction, and one which happened after a portion of the trail was open to the public. Two of these events caused

some damage to the concrete surface of the trail, while one was in an area where pioneering had not yet taken place.

- A small "sinkhole" was discovered during installation of a wooden fence in a pullout off the highway; the finished concrete trail was within 3 feet, and was somewhat undermined, yet was not damaged. This void appears to have been either left over from original construction of the embankment, or due to erosion and loss from flood events over the years.

Figure 8 – A collapsed void was encountered during installation of a fence.

Figure 9 – Bedrock outcrops incorporated into a rockery in JC-2 East.

- Bedrock encountered during excavation of the lower lift of a ground nail wall, where soil was anticipated. Several ground nails were eliminated as a result.

- In one area, it was known that bedrock lay directly under the trail subgrade surface, and the construction plan was to expose this rock, place base course soil, and concrete pavement. However upon excavation, the bedrock surface formed a small cliff which would be under the trail, then sloped down towards the creek. To deal with this, the D/B team designed a hybrid structure with a portion of cantilevered concrete trail deck anchored into the rock at the back, and a portion of cast-in-place concrete wall, also anchored into the rock through the wall footing.

Many of these unexpected conditions during construction resulted in cost increases; others resulted in a cost decrease, or were cost-neutral. However, it is unlikely that all of these conditions could have been known ahead of time, even with a detailed geotechnical subsurface investigation including the opposite side of the creek. Even if the team had chosen to spend up to $200,000 on geotechnical subsurface work prior to construction, the costs of some of the above changes would not have been any less.

The temporary shoring in particular resulted in spending much of the construction contingency funds which had been included in pricing by the contractor. While some temporary shoring was unavoidable due to the terrain and the choice of wall types by the Owners, much of the increase could have been avoided with a different pioneering approach, which likely would have impacted the creek more, taken more time to construct, and required import of additional temporary embankment materials. In the end, this may not have cost the project any less, in money or in time.

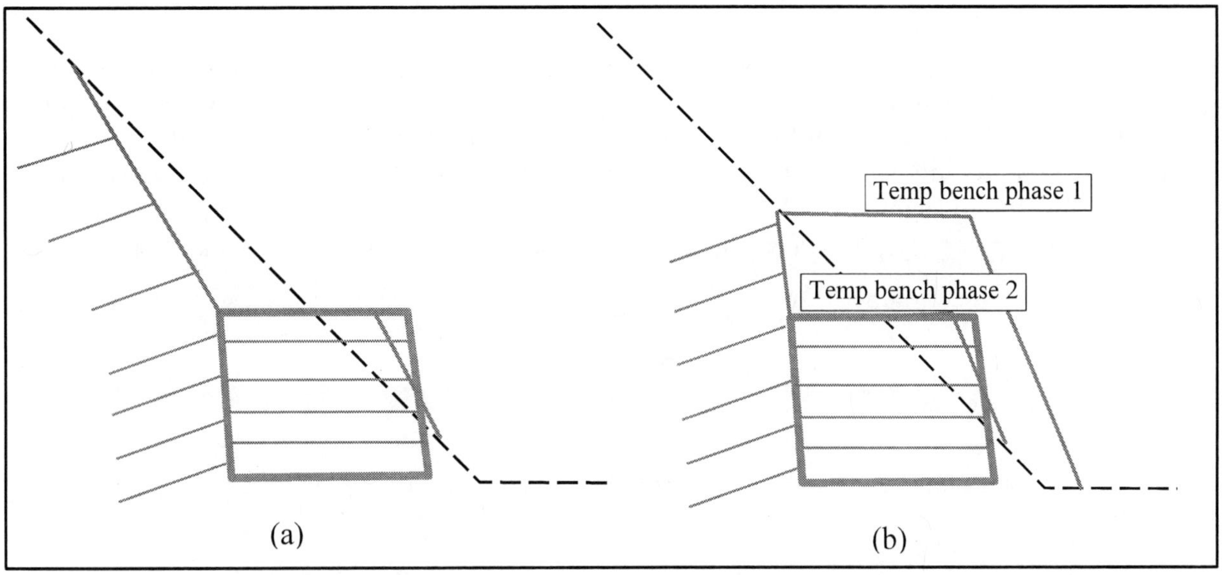

Figure 10 – (a) Pioneering approach used for the project, requiring greater height of temporary shoring. (b) A different approach may have required less temporary shoring, but with other consequences.

LESSONS LEARNED: FACTORS IN SUCCESS

The project was delivered as Design/Build, however in practice the delivery was more similar to a CM/GC. The contractor had money set aside for construction contingencies, based on a calculated risk of overrunning certain bid items. The engineering design team, including all subconsultants, was involved throughout the design phase, including submission of 90% and 100% design level plans. The Owners and the design team were flexible in their expectations and approach. These factors all contributed to the successful ability of the Design/Build team to deal with unforeseen, yet not altogether surprising conditions during construction, and to do so without having a complete picture of the subsurface geotechnical conditions before construction started. The following additional factors led to this success:

- Yeh and Associates' collective substantial past experience with the soils and bedrock types in the project area
- Ample flexibility of the client and contractor regarding final appearance
- Structure types chosen: Large-block (wet-cast) faced MSE walls, rockeries, rock embankments, and deep (drilled caisson) bridge foundations.
- Readily available materials, including a large nearby quarry/aggregate plant which is mining rocks of the same type as those exposed on the project
- Quick responses of the design team to the challenges presented, to make revisions as required.
- The nature of the project: a recreational trail rather than a critical highway corridor.

The author recognizes that these circumstances are unique, and that for most projects this approach to geotechnical subsurface investigation is not appropriate. If any of the above factors were not present, this success would not be as likely. For future projects in the vicinity, especially the continuing work to connect the Clear Creek Segment of the Peaks to Plains Trail to its next closest trail segments in Idaho Springs and Golden (approximately 6 and 10 miles west and east, respectively), however, the lessons learned in this project can be readily applied.

REFERENCES:

1. Rogers, W.P. *Critical Landslides in Colorado: A Year 2002 Review and Priority List*. Open-File Report 03-16. Colorado Geological Survey, 2005.

2. Sheridan, D.M. and Marsh, S.P. *Geology Map of the Squaw Pass Quadrangle, Clear Creek, Jefferson, and Gilpin Counties, Colorado*. Map GQ-1337. U.S. Geological Survey, 1976.

Geosynthetically Confined Soil Walls for Roadway Reclamation

Walter Turnbull, EIT
GeoStabilization International
543 31 Road
Grand Junction, CO 81504
(970)-210-6170
walter@gsi.us

Prepared for the 68th Highway Geology Symposium, July, 2017

Acknowledgements

For example: The author(s) would like to thank the individuals/entities for their contributions in the work described:

Matthew Reihl, P.E. – GeoStabilization International
Reid Bailey, P.E., – GeoStabilization International
Roger Teesdale – GeoStabilization International

Disclaimer

Statements and views presented in this paper are strictly those of the author(s), and do not necessarily reflect positions held by their affiliations, the Highway Geology Symposium (HGS), or others acknowledged above. The mention of trade names for commercial products does not imply the approval or endorsement by HGS.

ABSTRACT

Geosynthetically confined soil (GCS) walls have been readily adapted to fit along roadways to achieve required grade when situations such as slope failure or extreme erosion occur. There are several important aspects of GCS walls that make them a far superior fix compared to other traditional methods including cost, material availability, equipment availability, geometry flexibility, simplicity of design, speed of construction, and minimal required development length. An interesting example of GCS road reclamation is the 10th Division Road failure in Fort Benning, GA. Heavy rains experienced in the final days of December 2015 activated a slope failure on 10th Division Rd. just north of Russ Pond. Approximately 400 linear feet of roadway from the centerline and the slope on the northern, downhill side of the road sloughed off towards Upatoi Creek. The road had recently been constructed by placing fill atop native soils from the Eutaw Formation consisting of Coastal Plain sedimentary rocks and soils including unconsolidated fine-detrital clays and course-detrital sands (USGS).

Geostabilization International (GSI) was contracted to install a design/build slope stabilization system that included construcing the roadway platform up to grade. By the time GSI mobilized to the site, the entire roadway had subsided and ground elevation was approximately sixteen feet below where it once laid on the outboard shoulder. The sliding mass below the road surface was secured in place and material was excavated from the shoulder thereby unloading of the driving mass. A GCS wall was built to bring the roadway platform back up to the required elevation. The ability to shape the GCS wall to the shape of the excavation allowed adaptable engineering as only the material above the slide plane was removed, and excavation ceased once stable material was encountered below the slide plane. This case study will be discussed at length in addition to discussing how GCS walls are a great tool engineers need to keep in the top of their toolboxes.

INTRODUCTION

Geosynthetically Confined Soil (GCS) walls are an underutilized system that would likely gain significant use if more engineers understood its benefits. GCS walls have been used for retaining walls, box culverts, foundations, rockfall barriers, avalanche and debris flow barriers, and much more. The Federal Highway Administration (FHWA) has even published guidelines to build bridge abutments out of these structures (*1*).

The technology is not new; however, the flexibility of the system and the plethora of applications for which it can be used have just started to be realized in the past two decades. The purpose of this paper is to describe just one way in which these systems are used – roadway reclamation. Situations such as slope failure, extreme erosion, or the need for increased shoulder are instances in which a GCS wall can be readily adapted to bring a surface up to required grade.

A brief discussion on theory, components, and construction of Geosynthetically Confined Soil walls will be presented along with several advantages over other traditional roadway build-up methods. Also included is a case study on a large GCS wall road reclamation project in Chattahoochee County, Georgia in which a GCS wall was used to rebuild a site to grade after a landslide caused a significant grade change.

GEOSYNTHETICALLY CONFINED SOIL WALLS

Geosynthetically confined soil® (GCS®) walls, also known as geosynthetically reinforced soil (GRS) walls, consist of densely spaced layers of granular fill and geosynthetic reinforcement. This combination creates a GeoMonolith, a superior product in compared to its individual components (*2*). Much like the properties of individual components of concrete do not discern how a concrete mix will behave, the strength and behavior of a GCS wall is not distinguished by the properties of its individual components.

The tight spacing of the geosynthetic reinforcement confines the particles such that in order to fail in shear the failure plane must go through the particles instead of around them. This degree of confinement also limits the need for structural facing elements, for the confined soil exhibits no lateral pressure. The capacity of a GCS wall is dependent on the spacing of the reinforcement, thetype of soil used within the reinforced mass, and the tensile strength of the reinforcement. GCS walls act as a flexible mass and have the following characteristics:

1. The spacing of the geosynthetic is small – typically 12 inches or less, typically
2. The geosynthetic is frictionally connected to the facing elements
3. The facing elements are not mechanically connected and do not act as a single rigid element

GCS walls are very different from Mechanically Stabilized Earth (MSE) walls, though both systems are often inappropriately grouped together. Describing the differences is beyond the scope of this paper; however, Figure 1 highlights the main differences and more information on this topic can be found at www.gcswall.com. In general, MSE walls overall stability is dependent on the tieback analogy while GCS walls create a monolith, providing an internally stable mass.

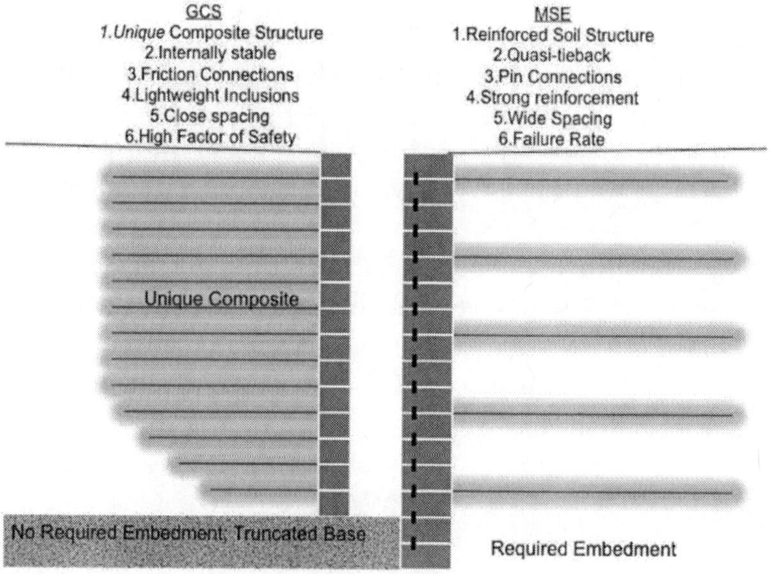

Figure 1: Differences between GCS walls and MSE walls (*3*)

GCS Wall Components

The three components needed to build a GCS wall are select fill, facing elements, and geosynthetic reinforcement. The equipment required to assemble the wall is either owned by most contractors or DOTs or can be easily rented from a local rental company. Material and equipment availability are one of the reasons this system is such a good resource. Large or custom equipment doesn't have to be brought to the sight, and custom fabrication is not needed.

Select Fill

The select fill inside a GCS wall is the structural component of the system and therefore arguably the most important. The fill needs to be free draining and angular. Most material suppliers across the U.S. carry suitable fill materials. Crusher run, 57 stone, road base, and many other names are applied to materials that will be sufficient. Most GCS walls for road reclamation receive relatively small loading as compared to a GCS wall used for a large bridge abutment. Some reference manuals, such as FHWA GRS-IBS, will specify specific gradations.

Geosynthetic Reinforcement

The mechanism by which GCS walls gain their strength is by confining the soils. The geosynthetic reinforcement restrains the soil in a compacted state, preventing dilation, yet only a small amount of load is transferred to the reinforcement. A geosynthetic reinforcement of sufficient strength can typically be purchased at a local material supply store. Figure 2 shows a large stack of Jersey barriers placed atop a GCS structure that was built with bed sheets for reinforcement. Note the facing blocks were removed and the inclusions still confine the soil.

Figure 2: Bedsheets as confining inclusions (*3*)

Concrete Masonry Units

Facing elements serve as a form when constructing a GCS wall and keep soil particles outside of the confining area from raveling out of the wall. They serve minimal structural purpose; therefore, whatever facing element is on sale that day can be used to construct a GCS wall. Facing elements should meet the durability requirements required to suit local climate, and be resistant to chemicals used in regular or seasonal road maintenance practices.

Required Equipment

Most contractors and DOTs either own all equipment required to build a GCS wall or have access to it via a local rental company. The equipment includes:

- Compactor – hand operated plate compacter sufficient

- Excavator or equivalent – all the material could be moved by hand but typically done with a piece of equipment

- Broom – sweep off blocks prior to placing next lift

GCS Wall Construction

One aspect of the elegance of GCS walls is their ease of construction, for a skilled work force is not required. Using common tools and materials that are easily procured simply follow the construction sequence below as depicted in Figures 3-8.

1. Excavate to required elevation and compact existing soil
2. Construct foundation (leveling pad/reinforced soil foundation)
3. Place the first row of blocks
4. Place select fill
5. Compact fill
6. Lay out geosynthetic reinforcement
7. Repeat steps 3-6 until required elevation is reached

Figure 3 (left): Site excavated and ready for GCS wall foundation
Figure 4 (right): Installation of a reinforced soil foundation

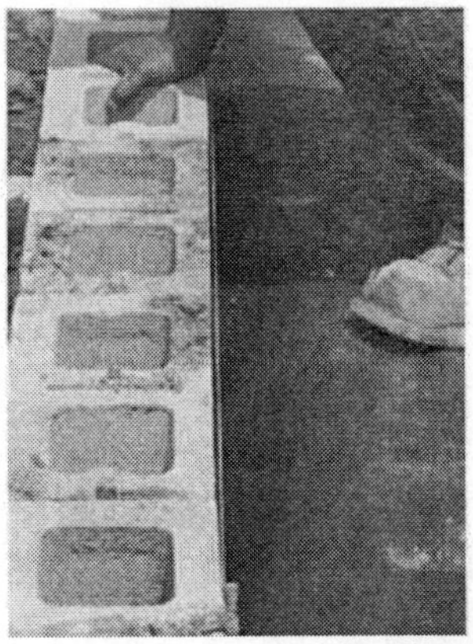

Figure 5 (left): Placing a row of CMUs
Figure 6 (right): Placing select fill

Figure 7 (left): Compacting fill with vibratory plate compactor
Figure 8 (right): Laying out geosynthetic reinforcement

GCS Walls – Advantage and Innovation

There are multiple reasons why GCS walls are a better alternative to reclaim lost roadway compared to other traditional methods including cost, speed of construction, material and equipment availability, simplicity, and flexibility in design.

Removing and replacing completely or building a MSE wall can close roads down completely for extended periods of time. Since GCS walls cannot be failed internally, fabric width is a function of global stability (*2*). The required footprint is small, on the order of 0.3 base to height, so roads can generally remain open while the wall is being built.

Pile and lagging walls require specialized equipment and labor and are very costly compared to GCS walls. Sheet pile walls can only be used in certain soils, and depending on the height or loading can often require costly tie-backs.

GCS WALL CASE STUDY: 10TH DIVISION RD

10th Division Road in Chattahoochee County, Georgia was originally built by hauling in a significant amount of sandy clay fill, on the order of ten plus feet in some locations. The fill was placed atop native soils from the Eutaw Formation consisting of Coastal Plain sedimentary rocks. The native soils where generally unconsolidated fine-detrital clays and course-detrital sands (*4*).

Heavy rains experienced in the final days of December 2015 activated a slope failure on 10th Division Rd. just north of Russ Pond. Approximately 400 linear feet of roadway from the centerline and the slope on the northern, downhill side of the road sloughed off towards Upatoi Creek. Engineering from the U.S. Army Core of Engineers (USACE) performed a site visit on January 5, 2016 and observed cracking not only in the roadway but also on the slope face below the road. Figures 9 and 10 are photos of the road and embankment failure during the initial USACE visit.

Figure 9 (left): Slope failure undermining roadway platform
Figure 10 (right): Slope failure below roadway platform

It is difficult to discern what instigated movement: the undermining of the toe at Upatoi Creek or the rain saturatingthe soils to the point in which the embankment could no longer support itself. The area of the hill in which the project is located drains from the south to the north where the creek is located just to the north of the roadway. The road embankment was built up such that a large area above the road forms a basin allowing for ponding. A small culvert located on one end of this basin would not have adequate capacity to drain this large area during intense rain events. It is likely a perched water condition developed causing an increase in hydrostatic pressure coupled with the lower effective stress which started the slope movement (ref***).

Figures 11-14 show the meandering of Upatoi Creek from 1993 to the current position of the creek captured in satellite imagery from May 2016 where an oxbow lake has formed. It can be discerned that soils at the toe of the slope are exceedingly soft, loose, and saturated.

Figure 11: Project location 1993

Figure 12: Project location 2003

Figure 13: Project location 2014

Figure 14: Project location May 2016

Initial Site Visit

The first attempt by others at reconstructing the slope was solely with grading. Erosion control was set in place, vegetation was removed, and traditional earthwork slope construction commenced. Material was removed and stockpiled then placed back in lifts and mechanically compacted. Slope reconstruction efforts can be seen in Figures 15 and 16 below. During slope reconstruction in April 2016 tension cracks began to form and it became evident that the slope had begun moving again.

Figure 15 (left): From above – site visit during grading operations
Figure 16 (right): From below – site visit during grading operations

Geostabilization International was contacted in April 2016 to perform a site visit and propose a solution. Despite being more than ten feet below required road elevation, stress cracks were forming during construction of the slope as shown in Figures 17 and 18. It was evident at this point that more robust structure would be required to regain the lost shoulder width given the geometry of the slope.

Figure 17 (left): Stress cracking during slope reconstruction
Figure 18 (right): Stress cracking during slope reconstruction

Repair Plan

Due to the size of the unstable slope below the road and the depth of the sliding mass, the proposed design encompassed both the roadway platform and the road shoulder. The slope on the shoulder of the road is acting as a driving force for the landslide and had to be removed. The existing road platform had completely subsided and had to be built back up to required grade.

Engineering Design

Borings taken on site, inclinometer data, and information gleaned from site visits were inputs in the slope stability software used to develop a solution. A limit equilibrium back analysis was performed. The conceived design included significant excavation to unload the slope and rebuilding of the road and shoulder with a GCS wall system.

Proposed Solution

The proposed solution occurred in multiple phases. First, the unstable material was safely and permanently removed. The slope was unloaded in a top down construction sequence which provided a stable platform for the reinforced soil foundation. The flexibility of the system allowed the crew to field fit the foundation along the stable contour of the excavation. Being able to form the foundation reduces excavation and the amount of material required to build the GCS wall. Construction of a vertical wall on stable material after removing unstable fill provides the opportunity to reduce the slope angle to a more stable condition.

A GCS wall was built up to the required grade. The inherent flexibility of GCS is ideal for the bit of settlement that will occur. The system can also easily be shaped to any geometry allowing for field fitting of the wall. Surveyors marked where and how high they needed the wall to be, and the GCS wall was built out and up to accommodate. The minimal required base to height for GCS walls also allowed for minimal excavation. An example cross-section below, Figure 19, shows how a truncated design could be fit into the current slope geometry to which the existing slope had settled.

The flexibility in the design of the system and the ability to field fit was very important at this project. The slide was still active; therefore, a precise wall geometry was not attainable. Not requiring precise stationing and elevations allowed the excavation contractor to optimize their excavating efforts and remove undesirable soils when they were encountered. Unit pricing for the GCS wall ensured the crew field fitting this project were not bound by numbers on a page but could really use their experience and expertise to build what was required for a successful project.

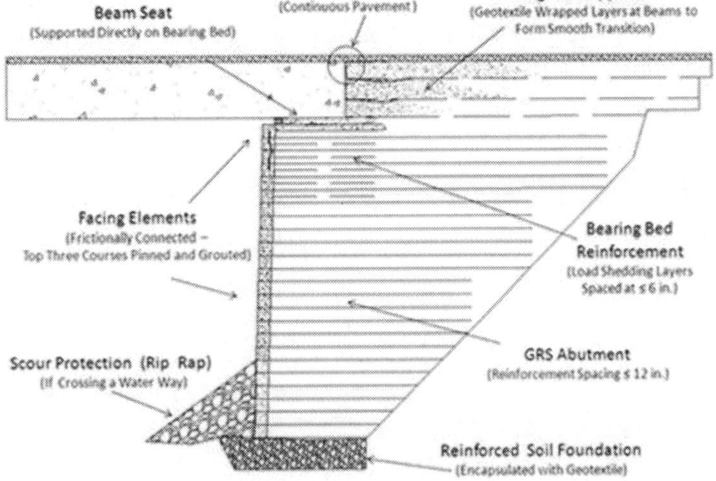

Figure 19: Example truncated GCS wall (*1*)

Construction

Submittals and paperwork were approved in September 2016, permitting GeoStabilization International was able to begin work on the 10th Division Road landslide. When work started, the original road surface was sixteen feet below original grade (Figure 20). This once relatively heavily trafficked road had been closed for eight months, leaving drivers with a time consuming detour.

The first step was to determine the horizontal alignment of the future wall so that the batter of the wall could be planned leaving a safe distance for shoulder width, and a fence to be installed on top of the GCS Structure. The disheveled slope was graded smooth and surveyors marked locations as shown in Figure 21.

Figure 20 (left): Sight conditions at commencement of work
Figure 21 (right): Surveyed line mark top of wall

Top-down excavation commenced with one lift being completed at a time. Up to five feet of additional material was excavated at a time, Figures 23 shows the walls being constructed. Note the flat ledge at the top of the wall that will serve as the foundation for the GCS wall. Figure 22 also shows water at the front of the wall that was apparent the entire project and will likely persist indefinitely.

Figure 22: Slope prepared to begin construction of GCS wall

While excavating the fourth lift in the center portion of the wall, there was evidence of the Eutaw Formation. Highly weathered sedimentary rocks, clay, and course detrital sands all littered with fossilized sea shells were found in the excavation. A photo of the observed stratigraphy is shown below.

Figure 23: Eutaw Formation observed during construction

Once construction of the GCS wall commenced the wall went up rapidly as workers laid blocks, placed and compacted fill, placed geosynthetic reinforcement, and repeated. Figures 24 and 25 show the project from either ends during GCS wall construction.

Figure 24 (left): Construction of GCS wall facing east
Figure 25 (right): Construction of GCS wall facing west

Finished Project

The composite wall was completed in November 2016 with final grading and paving following shortly after. The road is currently open to the public who drive across the project without realizing they are right above what was once, recently, a large landslide

Figure 28: Completed project looking east

Figure 29: Completed project looking east

Figure 30: Completed project looking west

Figure 31: Completed project looking west

Figure 32: Finished roadway platform

CONCLUSION

Geosynthetically confined soil walls are a great tool to use to reclaim lost roadway. There are multiple reasons why GCS walls are a better alternative to reclaim lost roadway compared to other traditional methods including cost, speed of construction, material and equipment availability, simplicity, and flexibility in design. They are not the right tool for every job; however, it is a great tool for engineers to keep in the top of their toolboxes.

REFERENCES

1. Adams, M., Nicks, J., Stabile, T., Wu, J., Schlatter, W., Hartmann, J. (2011) "Geosynthetic Reinforced Soil Integrated Bridge System, Interim Implementation Guide." Federal Highway Administration, Office of Infrastructure Research and Development, Washington, D.C.

2. Barrett, R. *GCS® GeoMonolith Theory, Design, and Construction.* gcswall.com, July 2010. Accessed March 16, 2016.

3. Barrett, R. *GCS® Wall vs MSE Wall.* gcswall.com. Accessed March 16, 2016.

4. Stephenson, L. W. and Veatch, O. *Geology of the Coastal Plain of Georgia* Bulletin No. 26, U.S. Geological Survey, 1911

Emergency Landslide Stabilization and Roadway Repair
Otsego County, NY

Eric List, EIT
GeoStabilization International
543 31 Road
Grand Junction, CO 81504
(970) 210-6170
eric.list@gsi.us

Prepared for the 68th Highway Geology Symposium, May, 2017

Acknowledgements

The author wishes to thank Nate Beard, P.E. and Perry Kairis, P.E. for their support and resources provided.

Disclaimer

Statements and views presented in this paper are strictly those of the author(s), and do not necessarily reflect positions held by their affiliations, the Highway Geology Symposium (HGS), or others acknowledged above. The mention of trade names for commercial products does not imply the approval or endorsement by HGS.

Copyright

ABSTRACT

Due to heavy rains, the slope below a portion of County Route 31 in Otsego County, NY dropped several feet. The entire outboard lane and shoulder had disappeared, falling over 20 feet down the slope and closing the roadway. Tension cracks extended into the opposite travel lane, and the guiderail, still in place, bridged the 75-ft long crevasse with posts now hanging far above the new ground surface.

The landslide was observed to be rotational and occurred in glacial till containing a range of material from silt and clay to cobbles and large amounts of water. In areas upslope from the landslide, bedrock was present on the slopes or covered by only a thin layer of soil. Within days of the slide occurring, GeoStabilization International (GSI), in coordination with Cobleskill Stone Products, Inc., installed a design-build-warranty permanent stabilization system that included a combination of advanced geohazard mitigation technologies such as geosynthetic reinforced soil systems and soil nails.

A soil nail wall faced with shotcrete was installed along the existing scarp to stabilize the rest of the slope and prevent successive failures or further sloughing of the soil. Deep foundation elements were installed at the base of the soil nail wall to increase the stability of the slope and provide a suitable foundation for a reinforced soil fill wall that was then built up to regain the roadway width lost in the slide. PVC drains were installed to control water within the slope and guide the drainage appropriately to a creek below.

Through cooperation with the client, Otsego County Department of Highways, Forestry and Parks, GSI was able to respond quickly to the problem. This case study discusses the geologic and hydrologic conditions that contributed to the failure as well as the innovative design-build construction used to repair the emergency landslide and reconstruct the roadway to its original state utilizing advanced techniques that have proven to be economical, robust, and quick to safely construct.

INTRODUCTION

The slope underlying a segment of County Route 31 in Otsego County, NY, failed catastrophically and dropped over 20 feet after heavy precipitation in the area. The landslide extended back to the center line of the road and the outboard shoulder and lane had fallen away. For seventy five linear feet, the guiderail hung suspended across the gap where the shoulder had once been. Tension cracks formed in the opposite lane, signaling the potential of further progressive failures farther into the road.

Figure 1 (left): CR 31 Site condition upon initial site visit. The pavement in the foreground would later fall into the gap created by the landslide.
Figure 2 (right): Landslide as seen from below. Note guiderail posts near top of photo.

The slide closed CR 31 and caused access issues to a local summer camp. In order to access the camp, travelers from Cooperstown to the south had to detour to a smaller nearby road until it intersected with CR 31 a few miles to the north, and then south to the camp. The detour road was not meant for the increased volume of traffic and passed by many homes and personal properties.

The client, Otsego County Department of Highways, Forestry and Parks, contacted GeoStabilization International (GSI) to visit the site of the landslide and propose a solution to stabilize and rebuild the roadway.

Geologic Setting

The site is located approximately five miles north of Cooperstown, NY, along County Route 31. CR 31 travels along the east side of Otsego Lake and is the main route for travel in the area. Other sections of CR 31 had experienced slope stability problems in the past that had since been remedied. At the site of the landslide, CR 31 was only a couple hundred feet from the shores of Otsego Lake.

The site is located in central New York in an area that was sculpted by glaciers. Otsego Lake, like the Finger Lakes to the west, was formed by the retreat of continental glaciers at the end of the last glaciation period around 10,000 years ago (*1*). This period of glaciation left much of the region covered in glacial till of varying thickness. Due to its origin, this till is composed of materials varying in texture, size, sorting, and compaction. According to the surficial geologic map of Otsego County, this till varies in size from silt to boulders, and ranges in thickness from as little as one meter to 50 meters (*2*). Because it is related to particle size and compaction, this material has highly variable permeability as well.

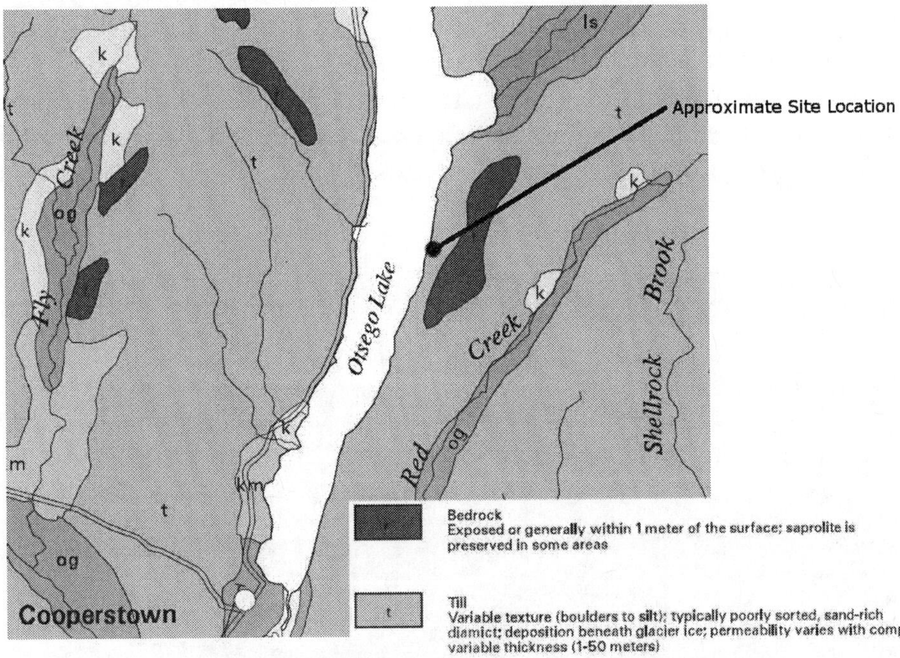

Figure 3: Regional surficial geology map (*2*)

The variability of this type of soil presents issues in determining its behavior. Its composition is heterogeneous and difficult to classify without site-specific information.

The till is underlain by bedrock of varying formations. The Marcellus formation, comprised of shale and limestone, surrounds most of Otsego Lake, with shale and siltstone of the Panther Mountain formation are present higher up on slopes around the lake (*3*). Bedrock depth varies greatly in the area due to differences in till deposition. The site is located near an area of shallow bedrock, as shown in Figure 3.

EMERGENCY RESPONSE

Site Visit and Characterization

Within twenty-four hours of the landslide occurring, a GSI representative visited the site and created a preliminary design to stabilize the landslide and rebuild the roadway. During the site visit, measurements and surveys were taken to determine the dimensions of the slide and develop a cross section through the worst area. Several photos were taken showing the extents of the landslide as well as the soil and groundwater conditions.

During this visit, the soil was observed to be a mixture of sand and silt, with gravel and cobbles. Rock outcrops with thin, horizontal bedding features were present on the slope across the road, indicating relatively shallow bedrock beneath the road. In addition, rock outcrops were exposed in the sides of the landslide scarp. Near the center of the slide, the scarp exposed only glacial till materials.

Beneath the toe of the slide and debris was a small stream flowing on bedrock. The slide mass had blocked the upper portion of the stream and the debris was saturated, with pools of standing water present among downed trees, fence posts, chunks of pavement, and soil. Water flowed out from below the slide debris into a stream that continued down the slope. On the north side of the scarp, a small metal pipe segment was exposed. A few feet north of the slide, a corrugated plastic culvert extended out from the slope, although no water was leaving the pipe or culvert during the site visit. Much of the water was traveling from the opposite side of the road along the soil/rock boundary.

Figure 4: Ponded water was present in the debris at the base of the slide. The existing stream entered the debris from the right of this photograph and continued downslope behind the tree stumps.

Past slope instability was a problem at this location, as evidenced by several layers of asphalt pavement that had adhered together. These layers, combined in an overall pavement thickness of more than a foot, were exposed in the landslide crown.

The slide was observed to be rotational, following a circular failure surface. The primary trigger of the slide was loss of shear strength due to saturation following a rainstorm. Presumably, water that infiltrated into the till layer percolated down until it reached the bedrock layer, where it gathered. The water within the slope increased the weight of the soil and decreased the shear strength of the material at the interface between the till and bedrock. The failure did not occur along the surface of the bedrock, but was observed to be near to this boundary. Water was observed trickling out of the base of the scarp into the slide debris.

Engineering

Due to the urgent need to reopen CR 31, engineering a repair was performed with little data. The information available was primarily limited to what was observed during the site visit. Soil borings, inclinometers, piezometers and laboratory testing were not performed. The soil was characterized based on visual observation during the site visit as was the location of water within the slope.

The slope failure was modeled using RocScience Slide limit equilibrium software. Measurements taken during the site visit were compiled into a cross section. An assumed bedrock surface was included in the model from the locations of rock outcrops observed at the site. A water surface was also included based on the areas where water was exiting the slope near the toe of the slide. Preliminary properties for the upper till soil layer and lower bedrock layer were assumed and then refined based on the output of the model. The goal of this back analysis was to show a failure surface approximating the actual slide with a Factor of Safety of close to 1.0.

Properties for the bedrock were determined from average values for similar rock types. Based on its appearance and the regional geology, the rock was determined to be limestone of varying degrees of weathering. Site observations indicated that the failure occurred entirely above the surface of bedrock. The joint sets present in rock outcrops supported this observation in that none of the joint sets dipped at an angle or orientation close to the slide surface. The bedrock was modeled with properties resembling that of a soft sedimentary rock.

Figure 5 (left): Rock outcrops on the south flank of the landslide
Figure 6 (right): Rock outcrops to the north of the landslide, outside of the slide extents

In order to repair the road damage, two major elements were needed in the design: a way to stabilize the scarp, and a way to rebuild the lost roadway width. A combined system utilizing soil nails and a reinforced fill wall founded on deep foundation elements was selected to solve both of these issues. Additionally, horizontal PVC drains were designed at the base of the wall to drain water from the slope and discharge it properly to the creek below.

A soil nail wall was proposed to stabilize the oversteepened scarp face below the roadway due to the ease and safety of installation: the upper portion of the scarp could be stabilized working from the roadway platform. For lower areas, work could safely progress beneath the stabilized slope. The soil nail wall provided both temporary stability while the roadway was rebuilt and long-term stability. The design called for soil nails up to 30-ft in length, in order to have embedment into bedrock.

With several outcrops of rock around this site, a relatively shallow bedrock surface was assumed, and the soil nails were designed to terminate in rock. The soil nail system was analyzed for global stability using the soil and rock parameters determined by the back analysis. A sensitivity analysis was performed by looking at the effect that increasing bedrock depth had on the global Factor of Safety. The final design, including wall height, nail length, and nail spacing, were determined by this sensitivity analysis. The wall was designed to terminate below the slide material, and nails lengths and spacing were designed to provide a final global Factor of Safety greater than 1.5.

Deep foundation elements were included at the base of the soil nail wall to provide a suitable foundation for the fill wall as well as for slope stability. Uncompacted slide material, or even the unfailed till beneath it, did not have the bearing capacity upon which to found the fill wall.. In addition, the lateral and axial capacities of the foundation elements contributed to the stability of the slope.

The foundation elements were analyzed using LPile to determine their lateral capacity and using a t-z analysis to determine axial capacity. The results of these analyses were incorporated into the Slide model. While the soil nail wall was designed to stabilize the existing scarp, the foundation provided stability for the slope below the fill wall and acted to transfer much of the weight of the wall to the bedrock.

As part of the client's requests, the reconstructed roadway was to be widened to include 10-ft of width outside of the roadway. This allowed for a setback between the guiderail and edge of the fill wall; in addition, a 6-ft tall chain link fence was to be installed behind the wall face for safety.

The review process was expedited to allow construction to begin as quickly as possible. GeoStabilization included a five-year warranty on the performance of the system, relieving the risk the client otherwise would have taken on by a quick review.

Construction

Within days of the initial site visit, upon approval, GSI mobilized to the site and began stabilizing the existing scarp. In cooperation with Cobleskill Stone Products, Inc., construction began with the soil nail wall to prevent further movement, sloughing, or collapsing of the soils underneath the road. The soil nails were drilled in rows along the scarp and subsequently each row was encased in a reinforced shotcrete facing. Because construction progressed in lifts from the top of the slide to the base, the slope behind each lift was stabilized before work began on the next. The first three rows were drilled with equipment on the road, but due to reach limitations, subsequent rows were drilled from below. Cobleskill Stone Products provided an access ramp so that equipment could reach the lower areas of the design.

Figure 8: Soil nail wall in progress

Constant communication between the construction Superintendent and the engineering team allowed the design to be modified in real time based on drilling and excavation observations. The original design maintained a degree of flexibility so that changes made in the field were possible. The stability model was refined from drill logs taken during drilling of each row of soil nails. As seen in Figure 8, nails were not trimmed behind the shotcrete facing for ease of construction, but were buried within the wall backfill.

The fill wall design also allowed for some flexibility in construction. Along the outer areas of the slide, rock outcrops protruded from the slope. Varying the wall shape to work around rock outcroppings was easily accommodated: the facing blocks were saw cut to shape the face of the wall around the rock outcrops, the reinforcement layers were cut to shape, and the backfill was placed around the rocks as needed. During construction, especially as the wall grew in height, a fall protection system was put in place so that workers could continue to build the wall safely.

Figure 9: Facing block placement around rock outcrop

After completion of the wall, Cobleskill Stone Products installed asphalt pavement in the section of roadway that was lost in the slide and to nearly the face of the wall, to replace the shoulder section and provide a walkway along the slope through the repair. In the end, the constructed wall measured 75 LF in length and 26 ft in height. The full construction of the stabilization system, including soil nail wall, foundation, and fill wall, took two weeks.

CONCLUSION

After the landslide occurred along CR 31, taking out one lane of the highway, a rapid reaction from Otsego County Department of Highways, Forestry and Parks allowed for GSI to quickly mitigate the impact of the site. Within days of the slide, GSI had visited the site, engineered a solution, and began construction. The design-build nature of the construction provided constant communication between the engineers and field personnel. Changing conditions at the site were incorporated into the design in real time and the changes relayed back to the operators on site. Since the construction, this site has not experienced any slope stability or construction-related issues and is performing as designed.

Figure 10: Completed repair

During construction, County Route 31 was closed to traffic for a limited time following the slide. During this time, the scarp was stabilized, the roadway area rebuilt and repaved, and guiderail and fence installed. The road was reopened in the time that an average bid project would have only been awarded. With a five-year warranty included as part of the project, the risk of future problems at the site arising from the swift pace of events was shifted away from the client.

The landslide mitigation technologies employed on this project have been proven to be cost-competitive, robust, and quick to construct safely. Numerous landslides of varying magnitudes have been repaired with these technologies under similar emergency circumstances. Repairs such as this have shown to be faster to install and very competitive with other current technologies. This innovative design-build-warranty construction reduces the time to stabilize the site and the risk to the clients, while providing a long-term economical and safe repair.

REFERENCES

1. School of Natural and Mathematical Sciences. *About Otsego Lake*. SUNY Oneonta. http://www.oneonta.edu/academics/biofld/otsegolake.asp. Accessed March 9, 2017.

2. *Surficial Geology Map Otsego County, New York*. Natural Resources Conservation Service. https://www.nrcs.usda.gov/Internet/FSE_MANUSCRIPTS/new_york/NY077/0/maps/surficial_geology.pdf. Accessed March 10, 2017.

3. *Mineral Resources On-Line Spatial Data*. United States Geologic Survey, February 28, 2017. Accessed March 10, 2017.

Seismic refraction tomography for post-flooding roadway reconstruction design

Miriam Moller
Olson Engineering
12401 W 49th Ave.
Wheat Ridge, CO 80033
(303)-423-1212
mimi.moller@olsonengineering.com

Todd Schlittenhart
Yeh and Associates
2000 Clay St.
Denver, CO 80211
(303)-781-9590
tschlittenhart@yeh-eng.com

Prepared for the 68th Highway Geology Symposium, May, 2017

Disclaimer

Statements and views presented in this paper are strictly those of the authors, and do not necessarily reflect positions held by their affiliations, the Highway Geology Symposium (HGS), or others acknowledged above. The mention of trade names for commercial products does not imply the approval or endorsement by HGS.

Copyright Notice

ABSTRACT

In 2013 a massive flood washed through Big Thompson Canyon, in the Rocky Mountains of northern Colorado. Big Thompson Canyon is a very steep, minimally vegetated canyon, created by the 78 mile long Big Thompson River. During the historic 2013 flood large portions of US Highway 34 were either damaged or destroyed by washout, particularly along roadway segments not founded on bedrock. Temporary repairs were implemented; however, this being the second destructive flood in the canyon since 1976, the Colorado Department of Transportation decided to consider permanent, large-scale changes to the highway to improve the resiliency of the roadway during future flood events. Resiliency improvements include regrading riverside embankments, rerouting the highway, and adding scour resistant elements beneath the roadway. Two geophysical investigations were conducted using Seismic Refraction Tomography to aid engineers with construction and design of the planned improvements. Objectives of the first investigation were to map depth to competent rock and determining rippability for construction purposes, which were imaged in areas within the canyon that were inaccessible to a drill rig. The objective of the second investigation was to map depth to competent rock for design purposes. The geophysical results have, and will continue, to contribute to the design of a permanent roadway better able to withstand future flood events.

INTRODUCTION

Built in 1903 to 1904, US Highway 34 through Big Thompson Canyon is a two lane road which serves as the main connection between Loveland, CO and the national treasure known as Estes Park, CO; both the city and the U.S. National Park. The current 18-mile long, paved highway follows much of the original wagon track through the canyon. Additionally, there are two small towns within the canyon which are only accessible via US34.

Rock within the canyon primarily consists of a Precambrian metamorphic complex, overlain by Quaternary fluvial deposits, colluvium and talus. Big Thompson Canyon itself was formed by the 78-mile long Big Thompson River, which has an average annual flow of about 73 (*1*) cubic feet per second. The river formed the canyon by following the path of least resistance, generally following the east/west trending Big Thompson Canyon fault. The canyon consists of varying topography. Much of it is comprised of steep, near vertical walls and very narrow canyon floor (Figure 1 and Figure 2). The two towns are located where the canyon opens up and is significantly wider. Vegetation along the steep walls is minimal, while in wider portions trees and shrubs can be found.

Flash floods occur regularly within the mountainous canyons of Colorado. Due to the picturesque settings of these canyons many people choose to live and visit often; the Big Thompson River and canyon are no exception. Big Thompson Canyon is home to many people, with visitors in the tens of thousands during the summer months. During the height of summer travel, on July 31st, 1976, up to 14 inches of rain fell within 4 hours. The flash flood that ensued caused over 40 million dollars in damages. 418 houses were destroyed, 138 were damaged, and most tragically 145 lives were lost. Today, this flood is still considered the largest natural disaster in Colorado history.

Figure 1. Photograph of the damage after the 1976 flood.

After the 1976 flood US34 was repaired and, for the most part, rebuilt in the same position it was, just with better drainage control and soldier-pile walls to mitigate erosion. A 2-mile section of the highway known as The Narrows was raised above the 100-year flood level through a series of four retaining walls and a bridge. On September 12th, 2013 the Big Thompson Canyon was hit by another significant flood (Figures 2 and 3). The flood has been referred to as the 500-year flood by many. Fortunately the damages were not nearly as substantial, but lives were still tragically lost. Immediately following the 2013 flood 30 million dollars in emergency repairs were made to the road. However, the Colorado Department of Transportation (CDOT) determined some major redesigns to the road were needed to improve the resiliency of the highway and avoid disasters like this in the future.

Figure 2. Photograph of "The Narrows" after the September 2013 flood.

Figure 3. Photograph of highway destruction after 2013 flood.

Two of the major design concepts to improve highway resiliency were to relocate sections of the roadway onto competent rock (Figure 4), or add scour resistant elements under the roadway along sections particularly vulnerable to high-energy erosion (Figure 5). Relocating the roadway will ideally prevent any major damage in future floods, as the roadway will be moved above 100-year flood level line and onto a scour resistant source of metamorphic rock. In locations where it is not feasible to shift the roadway onto bedrock, scour resistant elements will help prevent damage and destruction in future floods. In the 2013 flood there were areas where both lanes of the roadway were compromised (Figures 2 and 3), limiting access to people trapped within the canyon to only a helicopter. The scour resistant element concept is intended to maintain at least one lane of highway from being compromised during another flood.

Figure 4. Schematic of roadway relocation design.

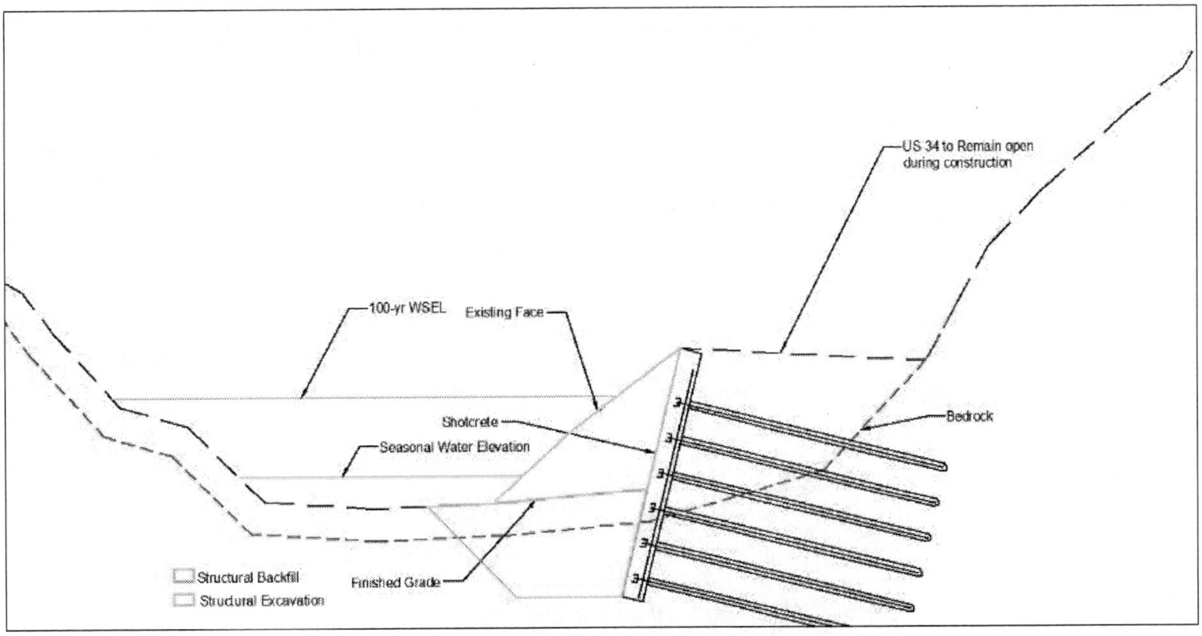

Figure 5. Schematic of potential scour resistant element design.

GEOPHYSICAL INVESTIGATION

To aid the proposed highway construction and re-design, two geophysical investigations were conducted. The first investigation took place in spring 2016, with the objective to determine rippability and map depth to competent rock for construction purposes. During the fall of 2016 the second investigation was conducted, with the objective of mapping depth to competent rock for design purposes. Objectives from both investigations were met using Seismic Refraction Tomography (SRT) to obtain compressional wave (P-wave) velocity data. A total of 10,975 line feet of seismic data were collected during the two investigations, both on- and off-road.

Seismic Refraction Tomography Method

Seismic refraction is based on the fundamental principles of Snell's law, which describes how waves propagate across a medium, in this case soil and rock layers, at varying velocities (*2*). A seismic energy source is created at a known location and time, referred to as time zero, and the energy from this propagates into the ground and away from this source. The energy propagates until it reaches a faster layer or interface, such as bedrock, referred to as a refractor. The energy propagates along this interface at the velocity of the deeper and faster layer, while simultaneously energy is emitted off this refractor back to the surface. On the surface a receiver records this energy and time for the refracted wave to propagate from the source to said receiver. This information is recorded at multiple locations along the receiver array. Using source and receiver geometry with observation times, the depth and velocity of the refractor layer can be calculated.

In the presence of a gradually increasing velocity gradient with depth, as often seen in soils with increasing overburden pressure or, in this case, decreasing weathering of rock, the wave-front path basically '*curves*' back to the surface. This can be viewed as an infinite number of layers that behave as refractors described above.

Two set ups were used to collect the P-wave velocity data. One utilized a receiver spacing of five feet, with seismic source positions every 15 feet along the array. For this tight spacing either 24 or 48 receivers, known as geophones, were used. With this set up the geophones were planted into the ground with three inch spikes. The second set up also had a geophone spacing of five feet, and this time with a source position every 30 feet. Only 24 geophones were used at a time with this set up. With this set up the geophones were mounted on weights to couple the geophone to the ground, utilizing a system known as a 'landstreamer' (Figure 6). In both cases the seismic source was a sledge hammer impacting a plate on the ground.

Figure 6. Photograph of the seismic survey deploying the 'landstreamer'.

There are two major steps in processing SRT data; first arrival picking and data inversion. The first arrival picking consists of picking the time where the first arrival of wave energy is observed at a geophone location (Figure 7). In the example below, the seismic source was located 15 feet off one end of the receiver array. There is a signal, or trace, for each geophone. The red crosses indicate the picked time for that particular trace.

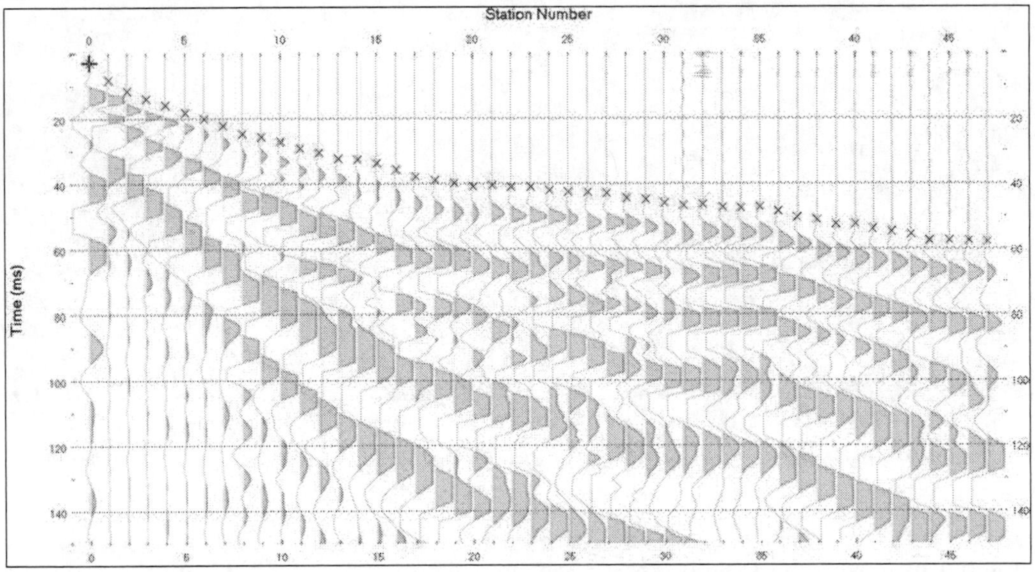

Figure 7. Example of first arrival picking, indicated by red crosses, from a seismic line collected during the first investigation in the Big Thompson Canyon.

After picking is completed, a two-dimensional (2D) P-wave velocity model is generated which best fits the first arrival picks. This is done by iteratively modifying a P-wave velocity grid model until the misfit between the modeled travel times and real travel time is minimized.

Site Specifics

As mentioned above, the objective of the spring 2016 geophysical investigation was to aid in construction of rerouting the highway. Rerouting was chosen for areas which sustained major damage or destruction in the 2013 flood. Several seismic lines were also collected for the construction of a temporary haul road, to be used for access to dump, process, and crush blasted waste rock and store construction materials. A total of 3,875 line feet of data collect were collected along 16 separate seismic lines. The topography was extreme and 15 of the 16 lines were located in areas not easily accessible by a drill rig prior to construction. Figure 8 is from one of these area, informally known as the Horseshoe Bend.

Figure 8. Photograph from Horseshoe Bend, where several seismic lines were collected. Drill rig access to this area was difficult, and expensive, prior to construction efforts.

This 0.4-mile stretch of the highway was completely destroyed in both the 1976 and 2013 flood events. Because of the high river energy through this curve, the preferred design alternative is to bypass the curve with 2 bridge crossings and cutting through the rock on the inside of the curve (Figure 9). Drill rig access to this area, and similar areas for this investigation, required difficult access drilling equipment and helicopters to mobilize the equipment across the river and up the steep terrain. Due to the high costs associated with helicopter access drilling, a geophysical investigation was preferred by CDOT. The SRT equipment easily mobilized across the river and up the steep terrain, making the geophysical investigation a budget-friendly option in the geotechnical design. Additionally, the geophysical investigation required minimal time to acquire and analyze subsurface data across lengths of the roadway.

Figure 9. Computer generated image of roadway relocation around Horseshoe Bend.

During the 2016 fall investigation, 7,100 line feet of seismic data were collected on or next to the roadway (Figure 6). Recall, the purpose of this survey was to aid in the design of scour resistant elements under the roadway. Results of this investigation were used in addition to over 130 boreholes, drilled at selected locations along the 18-mile stretch of highway, to create detailed profiles of the subsurface. That is, the seismic results are used to fill in the gaps between boreholes. One stretch of roadway, The Narrows, required detailed information on bedrock depth from the canyon wall to the far side of the highway (Figure 2), as rock depth is highly variable in this area. To map the lateral change under the road at this location, two parallel lines were collected, one down the middle of the east bound lane and one down the middle of the west bound lane (see Figure 16 for line location map). The bedrock profile from the parallel seismic lines and the boreholes helped generate a 3-D bedrock surface and cross-sections (Figure 10). The generated bedrock surface provided a reasonable estimation of the highly variable bedrock through The Narrows. This estimated surface can be used to determine the preferred scour resistant design option and establish quantities for construction. Collecting data in this manner not only provided more comprehensive information on lateral rock depth changes, it was also more cost effective than increasing the drilling effort in the area, which requires significant roadway closures, traffic controls, and monetary expenses (compared to seismic investigations).

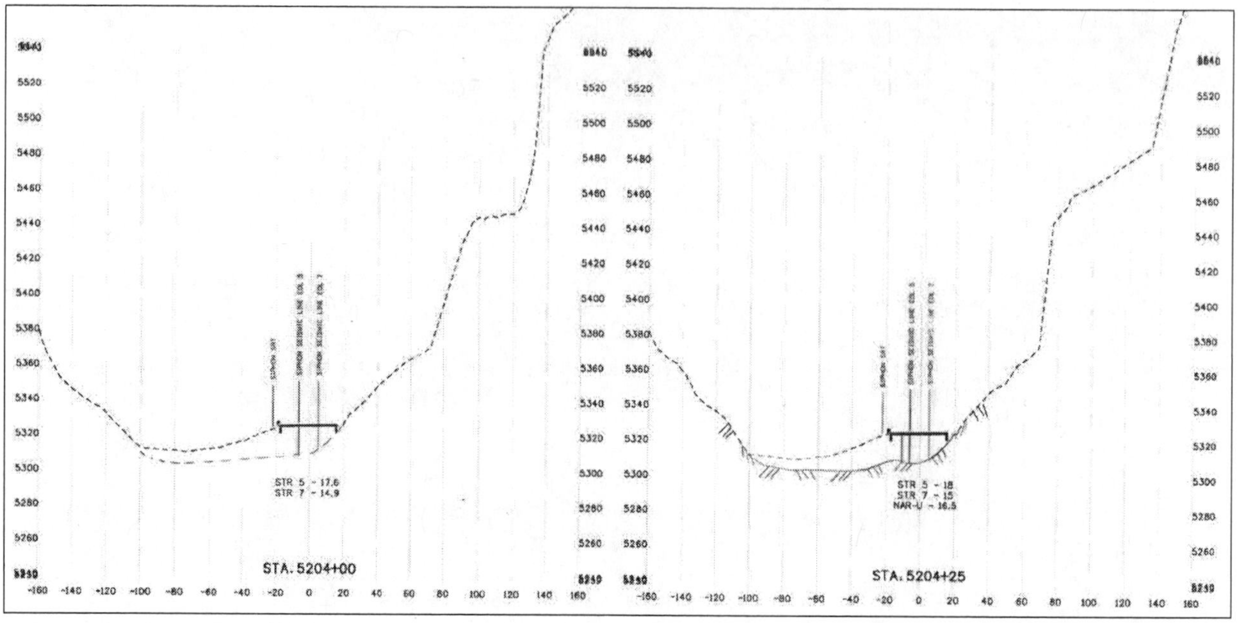

Figure 10. Cross sections created with results from drilling (blue) and geophysical (red) investigations in The Narrows.

SURVEY RESULTS

Figures 11 through 16 present P-wave velocity results and line location maps from the two seismic surveys. For each of the results, the horizontal and vertical dimensions are shown in feet. On the color scale, cool colors (e.g., blue) represent lower velocity values and warm colors (e.g., red) represent higher velocity values. Ground surface topography was incorporated when applicable, and a simplified representation of proximal borehole logs were overlain on the profiles. The interpretation of the 2D P-wave profiles were based on a velocity gradient analysis, and correlation to borehole data where available. Where interpretation of the seismic results lack borehole data, it is imperative to understand that the velocity data presented are modeled through an iterative process based on the travel time of refracted waves. Thus, these color velocity plots do not objectively represent intrinsic properties of the geology and structure beneath the seismic lines. The velocity gradient analysis is based on a rapid change in seismic velocity over a short depth range, which are visualized as abrupt color changes. Velocity gradients are typically indicative of a transition from soft to hard layers or materials including the degree of weathering in rock, though not necessarily indicative of the actual geologic interface. Refraction tomography will always produce a gradient at velocity transitions or a layer interface, no matter how sharp the geologic interface or boundary is physically.

Figures 11 through 13 present the velocity results from the first investigation in the canyon. As mentioned before, the objective was to map depth to competent rock and determine rippability. Annotated on the profiles are two contour lines chosen to represent two layer boundaries, between rippable, marginally rippable, and non-rippable according to the Caterpillar Handbook of Rippability for a D10R bulldozer (*3*). The other two contour lines represent the interpreted top of weathered rock and unweathered/competent rock, at 4,000 feet per second (ft/sec) and 8,000 ft/sec, respectively. These contour lines were chosen based on velocity gradients and rock outcrops observed near the seismic lines. Due to the metamorphic bedrock complex in this area, use of a single contour line for the interpreted top of weathered and unweathered / competent rock boundary would not reflect the actual velocity of those layers. Those velocities are probably closer to 6,000 ft/sec for weathered rock and 10,000 ft/sec for unweathered / competent bedrock. Of note is Line 13 in Figure 12. The low velocity contour zone observed in the profile represents a localized shear zone in the metamorphic rock mass. Published geologic maps do not identify the localized shear zone. As part of the geologic investigation, a surficial mapping exercise of the area suspected localized faulting, but the extent of the faulting was unknown. Final design of the rock cut slope was adjusted based on confirmation of the existence of the shear zone through the geophysical investigation. Additionally, the geophysical investigation helped define the extents of the shear zone. The design slope grade was made less steep to improve slope stability and reduce the impacts of rockfall.

Results from the second investigation are presented in Figures 14 through 16. As mapping depth to competent rock was the only objective, these profiles are annotated with only two velocity contour lines, those chosen in the first investigation, and proximal borehole data. The first velocity contour at approximately 4,000 feet per second generally correlated with the transition to denser or more compacted overburden, as indicated by the majority of the borehole logs. However, there are several instances where this 'interface' correlated with groundwater encountered during drilling. The second velocity contour at approximately 8,000 feet per second correlates well with the transition to unweathered or competent bedrock encountered in the boreholes. Several boreholes indicate bedrock being encountered shallower below the ground surface than what the P-wave velocity results indicate. This is interpreted to relate to areas where the RQD is low, likely caused by a high density of fractures and high degree of weathering, which would reduce P-wave velocity. Construction has commenced at several areas where the seismic data were acquired. In one such area, the "Omega line" (Figure 15), the 2D seismic profile proved accurate to what was encountered in the field.

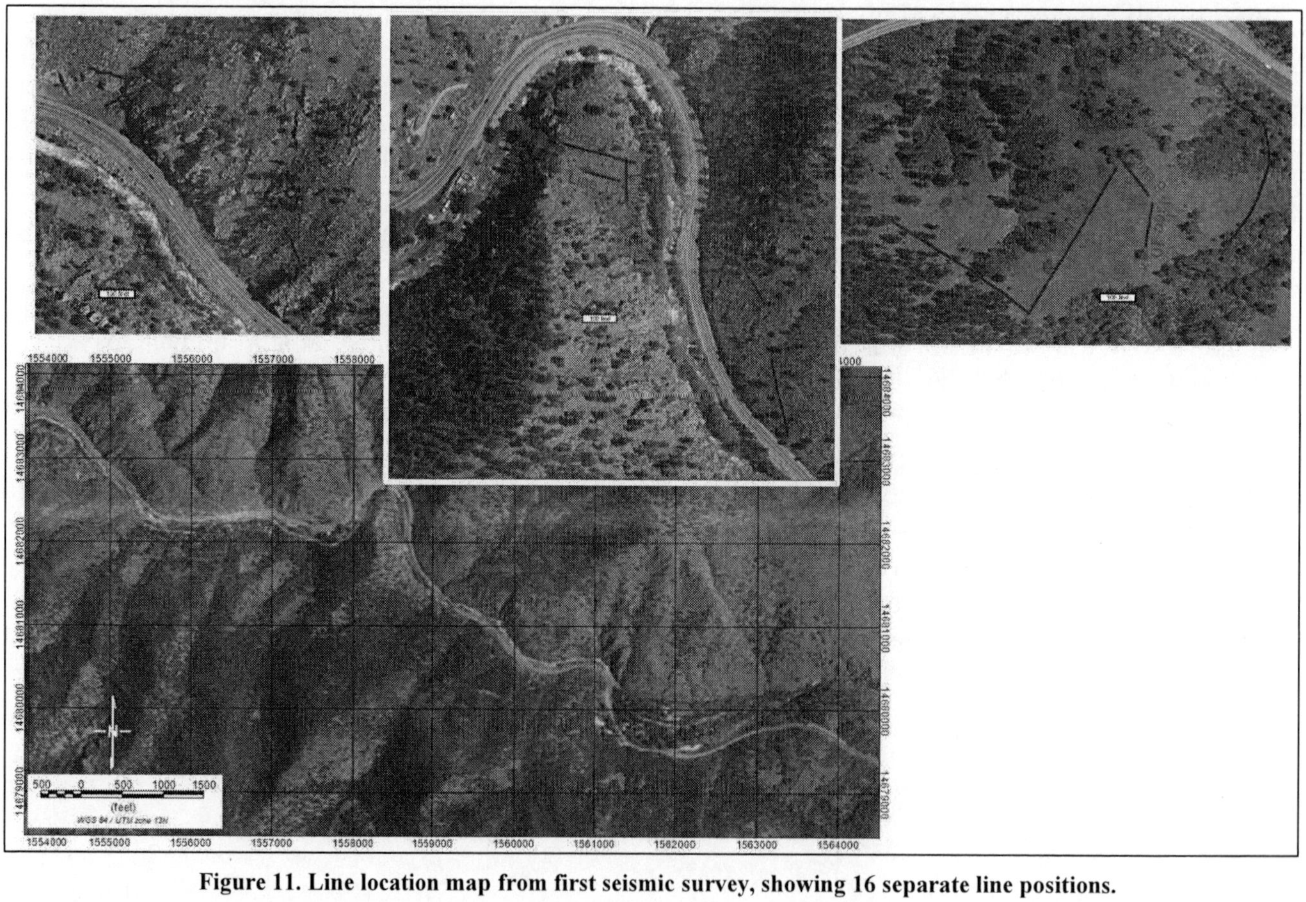

Figure 11. Line location map from first seismic survey, showing 16 separate line positions.

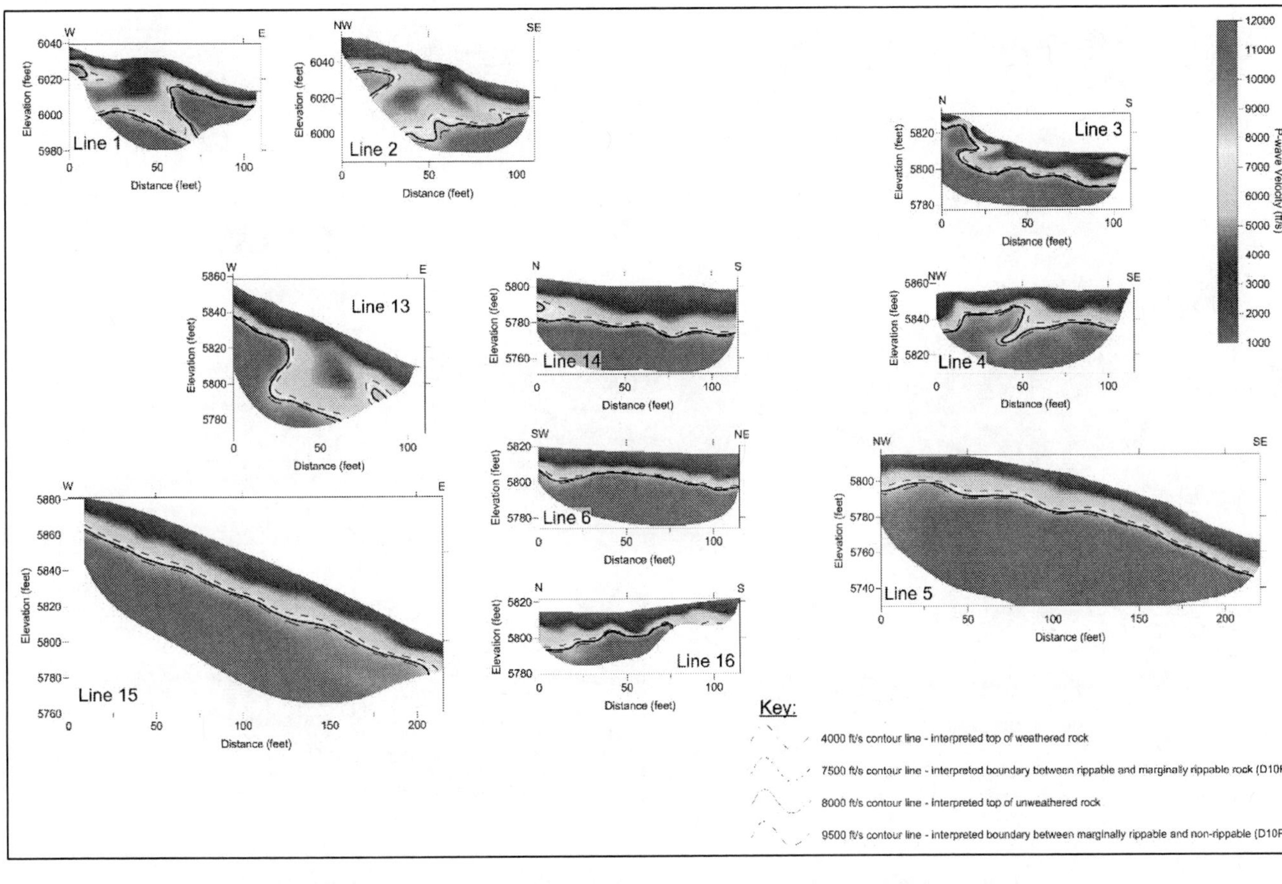

Figure 12. SRT Results from first seismic survey; in the west and central area of investigation.

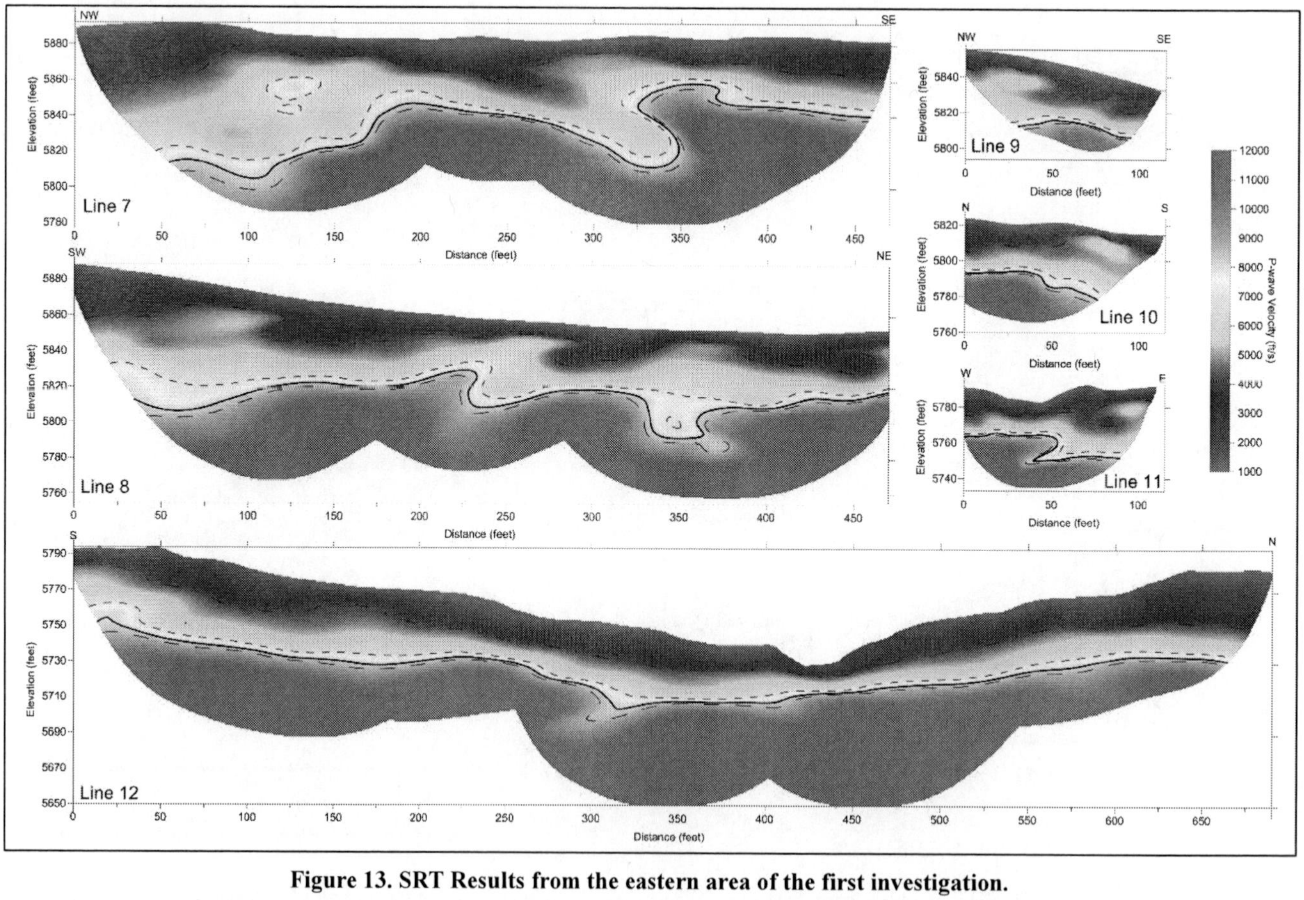

Figure 13. SRT Results from the eastern area of the first investigation.

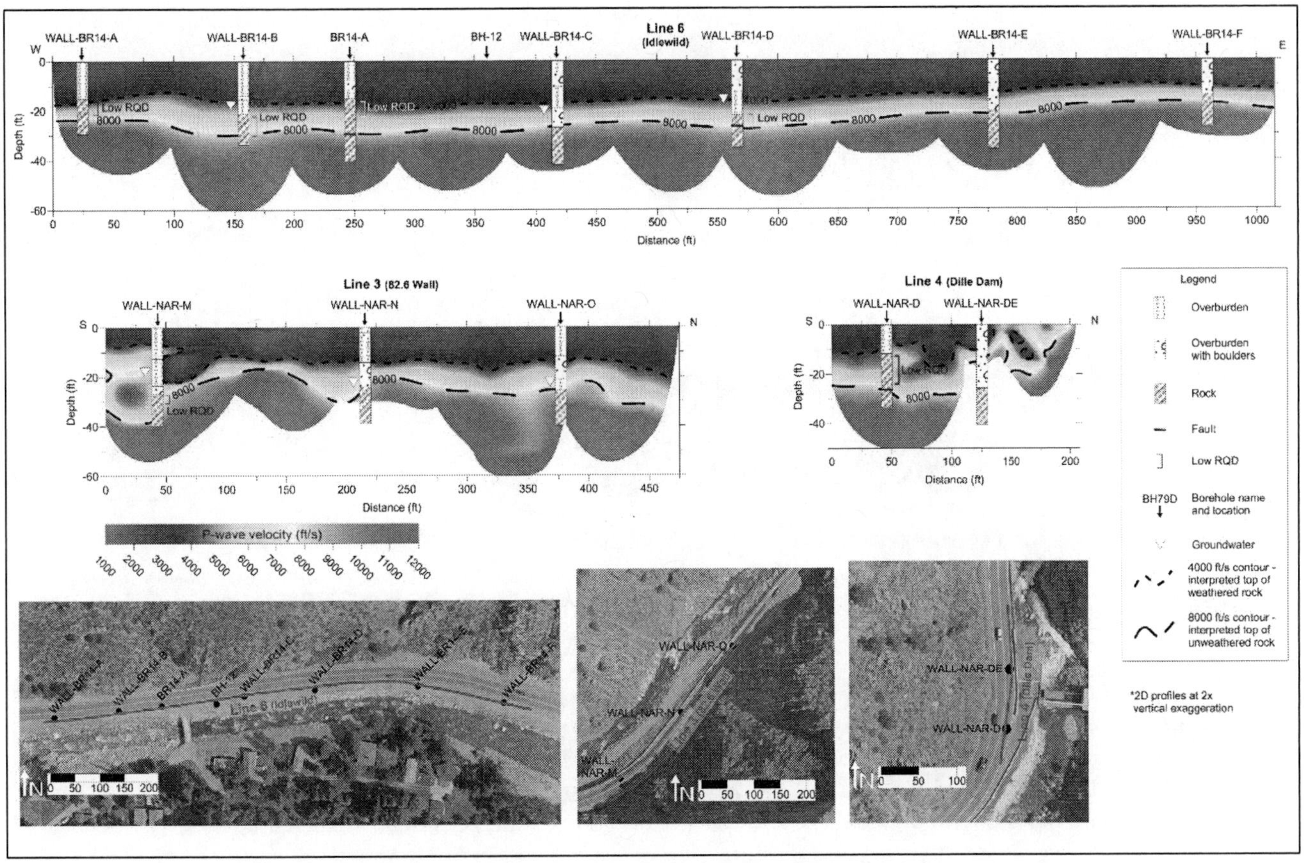

Figure 14. SRT Results from the second seismic investigation; Lines 6, 3, and 4.

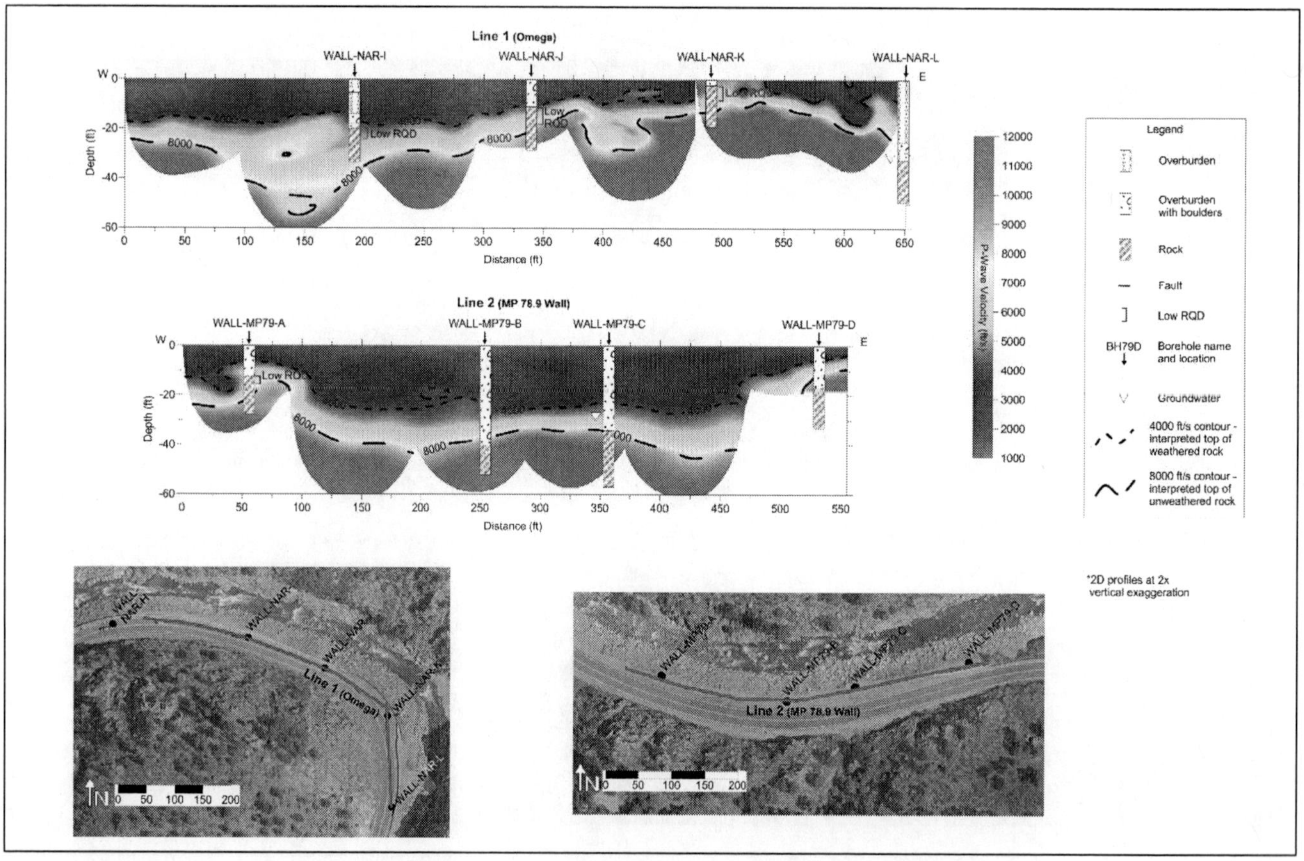

Figure 15. SRT Results from the second investigation; Lines 1 and 2.

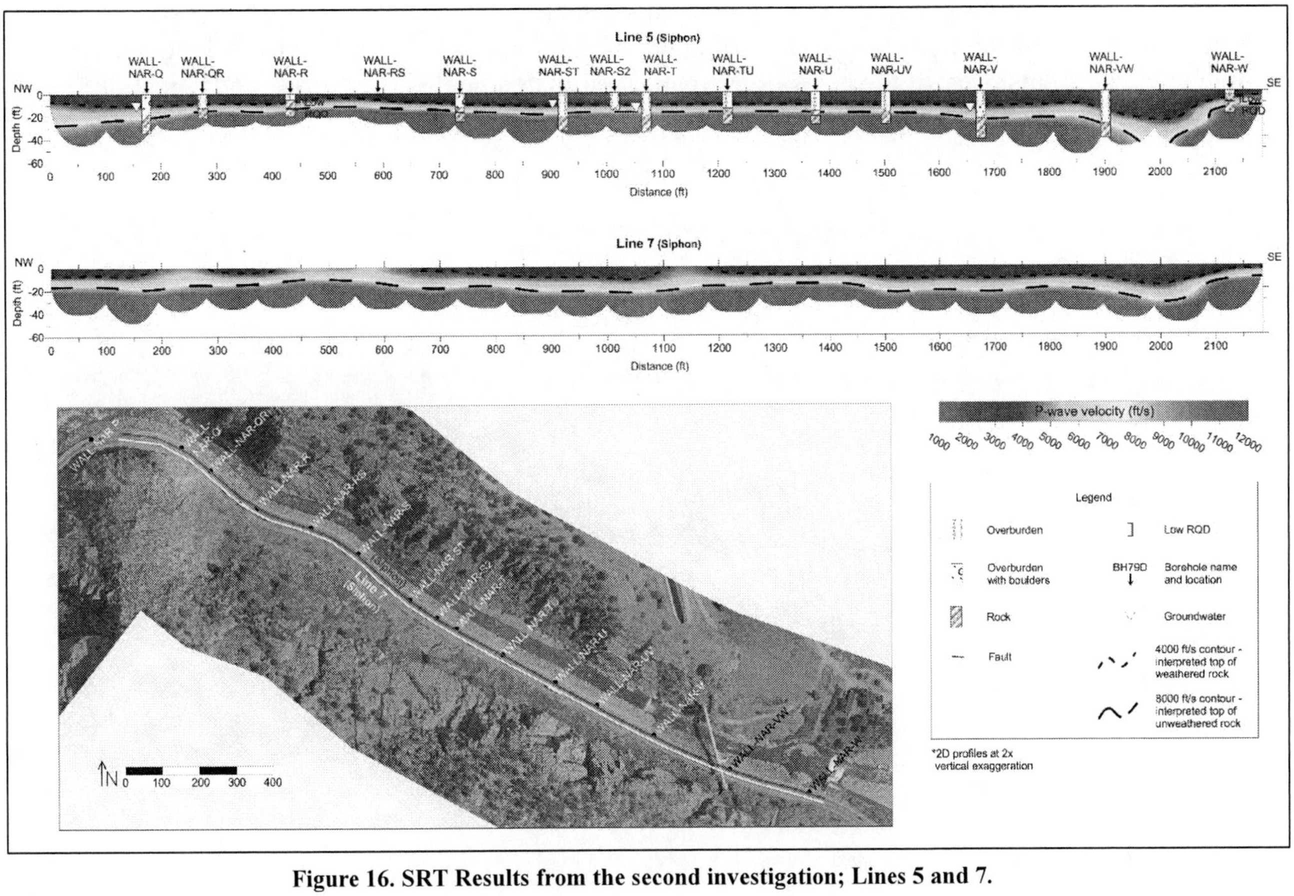

Figure 16. SRT Results from the second investigation; Lines 5 and 7.

SUMMARY

Large-scale flooding in Big Thompson Canyon has had massive impacts over numerous years, and several roadway improvements were implemented to lessen these impacts during future floods events. Approximately 11,000 line feet of seismic data were acquired over the course of two geophysical investigations within Big Thompson Canyon to aid in construction and design of these highway improvements. The objectives of the investigations were to determine rippability and map depth to competent rock. While the purpose of each investigation was different, the Seismic Refraction Tomography method was utilized for both surveys and final results presented to the client were tailored to meet the individual needs of each investigation. The use of geophysics was invaluable in rugged terrain and areas not easily accessible to a drill rig, and also for providing detailed information regarding rock depth between boreholes. For a majority of the areas investigated, estimated depths to bedrock and rippability are accurate to what is being encountered during construction.

REFERENCES

1. United States Geological Survey. *Surface Water data for USA: USGS Surface Water Annual Statistics.* U.S. Department of the Interior. waterdata.usgs.gov/nwis/annual?. Accessed March 1, 2017.

2. Mussett, Alan E., Kahn, M. Aftab. *Looking Into the Earth.* Cambridge University Press, Cambridge, United Kingdom, 2000.

3. Caterpillar. *Handbook of Ripping.* Twelfth Edition. Caterpillar, Peoria, Illinois, February 2000.

Marina Fire Rockfall Protection: A rapid, practical response

Simon P. Boone, GIT

Yeh and Associates, Inc.

391 Front Street, Suite D

Grover Beach, California 93433

(805) 481-9590

sboone@yeh-eng.com

John D. Duffy, PG, CEG

Yeh and Associates, Inc.

391 Front Street, Suite D

Grover Beach, California 93433

(805) 481-9590

jduffy@yeh-eng.com

Prepared for Submission to the 68th Highway Geology Symposium

Marietta, Georgia

ACKNOWLEDGEMENTS

California Department of Transportation

-Joe Blommer

Access Limited Construction

-Brian McNeal, Kevin Weisman, and Danni Jonsson

Geobrugg North America

-Tim Shevlin and Saleh Feidi

ABSTRACT

Marina Fire Rockfall Protection: A rapid, practical response

Prepared for Submission to the 68th Highway Geology Symposium

Marietta, Georgia

Simon P. Boone, GIT
Yeh and Associates, Inc.
391 Front Street, Suite D
Grover Beach, California 93449
(805) 481-9590
sboone@yeh-eng.com

John D. Duffy, CEG
Yeh and Associates, Inc.
391 Front Street, Suite D
Grover Beach, California 93449
(805) 481-9590
jduffy@yeh-eng.com

The Marina Fire (July 2016) resulted in significant increase in rockfall risk to a portion of US 395 in Mono County, California. This corridor is the main north-south transportation artery on the east side of the Sierra Nevada providing access to Yosemite National Park, Mammoth ski area and Southern California. The roadway is cut into a slope comprised of colluvium and interspersed tufa from relict shorelines of Mono Lake.

Due to rockfall events impacting the roadway immediately post-fire, safely keeping the road open to traffic was a priority for Caltrans. A rapid response was executed to assess the hazard and implement a mitigation strategy for this segment of roadway. Working in conjunction with Caltrans, Yeh engineering geologists evaluated the rockfall trajectories impacting the roadway and provided recommendations for installing a temporary flexible rockfall fence.

Two Geobrugg flexible rockfall fences, 2,000 linear feet of GBE 500AR (3m height) and 1,500 linear feet of model GBE 500AR (4m height) were recommended. A significant constraint was the presence of a critical fiber optic utility at the base of the slope, which was installed via horizontal directional drilling. Due to the uncertainty of the precise location of this utility, excavation and drilling to install post support system and ground anchors for the post support were a primary concern.

Design modifications were made to accommodate the site conditions. Of particular note was the design of the post support system. The fence posts did not have the typical concrete foundation type support systems but instead were supported globally by a wire rope support system and locally with number 10 threaded bar, 18 inches in depth. This design did not encroach on the buried utility and facilitated rapid construction, enabling the installation to be completed within a matter of weeks.

INTRODUCTION

Rockfall is a hazard commonly impacting transportation corridors in mountainous terrain. It can originate from natural or cut slopes and the risk of rockfall can be exacerbated by natural events such as intense rainfall and wildfire. In the western United States, dry season wildfire events adjacent to mountain highways can result in an elevated rockfall risk immediately, during, and post-fire. Often, highway managers approach the reality of wildfire-elevated rockfall risk by mitigating the hazard through engineered solutions. Methods that are commonly used include protection systems like wire mesh systems to cover the slopes and flexible fences to prevent rockfall from entering the traveled way.

Advantages of flexible rockfall fences include economic feasibility, expedited analysis, and rapid installation. In use in California since the late 1980s *(1)*, this type of mitigation is now common throughout the world. Comprised of steel posts, wire mesh, and wire rope components, flexible rockfall fences can be manufactured, delivered and installed under an expedited schedule to provide rockfall protection to a site within a timeframe of a few weeks. Of particular interest to geoprofessionals working in transportation corridors with traffic safety and utility constraints, flexible rockfall fence design can be optimized to accommodate site conditions, Figure 1.

Figure 1. Flexible rockfall fence posts installed on US Highway 395, Mono County, California.

Although flexible rockfall fences are now commonplace, a recurrent issue for installation contractors and owners is the post support requirements that are often recommended by designers. Frequently, large excavated foundations are specified that call for a significant quantity of steel reinforced concrete. This translates into a significant portion of the overall cost and a longer construction timeframe for the installation of flexible rockfall fences as compared to other methods of rockfall fence post support.

This paper presents a case study in Mono County, California where temporary, flexible rockfall fences were installed to provide rockfall protection to US 395, Figure 2. A unique aspect of this project was the uncertain location of a critical utility and design modifications that allowed the rapid installation of the fence posts while minimizing risk to the utility.

Figure 2. Project Location on US 395 near Lee Vining, California.

BACKGROUND

US 395 traverses the western margin of Mono Lake and in the project area is cut into slopes comprised of colluvium, alluvium, and interspersed tufa from relict shorelines. Tufa is a precipitated calcium carbonate formation that can be found in proximity to many lakes of the eastern Sierra Nevada. Interestingly, it can precipitate as "tufa towers", which can be seen in nearshore environments around the lake. Often formed by calcium laden spring water percolating into the alkaline (carbonate) waters of endorheic systems like Mono Lake, these formations are then exposed when lake levels drop, Figure 3.

Figure 5. Tufa towers at Mono Lake. Paoha Island, a volcanic cone, is in the background.

Figure 4. The Marina Fire burning on the slopes above US 395 on June 26, 2016.

The Marina Fire, Figure 4, burned 654 acres of steep slopes between its ignition on June 24, 2016 and its containment on July 7, 2016 *(2)*.The fire was named for its proximity to the nearby old marina on Mono Lake. Prior to the fire, Caltrans had programmed a phased rockfall mitigation project to address rockfall originating from six cut slopes along US 395 from post mile 53.2 to 53.7. Phase 1 of that project had been completed in 2015, with an anchored wire mesh system consisting of cable net with double twist wire mesh backing installed on several cut slopes within the Caltrans right of way where US 395 traverses around Mono Lake. Subsequent to the Marina Fire, this section of roadway experienced rockfall events from source areas outside the Caltrans right of way, and resulted in a need for rapid mitigation of the increased rockfall risk. Phase 2 was temporarily suspended after the fire, until the post fire rockfall risk was mitigated.

ROCKFALL INVESTIGATION

At the request of Caltrans, Yeh engineering geologists performed a field investigation on July 19 and 20, 2016 to evaluate the rockfall conditions at the site and to provide recommendations for temporary rockfall mitigation. Yeh personnel walked the alignment within the project limits to document the site conditions, measure the catchment width, slope angles, and interview construction and maintenance personnel to obtain verbal accounts of rockfall activity post-fire. At the time of the field investigation, Caltrans was in the process of placing concrete K-rails along the edge-of-traveled way on the southbound lane of the highway as an immediate response to rockfall events at the site.

Yeh engineering geologists observed rockfall in the catchment area, Figure 5, and evidence of rockfall impacts to the K-rail along the fog line and to the pavement in the traveled way. Caltrans maintenance personnel reported that a rockfall event had occurred that spalled concrete from the top of the K-rail and crossed into the highway.

Figure 5. Largest rockfall observed in catchment area during field investigation on July 19, 2016.

ANALYSIS AND DESIGN

Based upon the field investigation and meetings with Caltrans personnel to understand the project constraints, a temporary, flexible rockfall fence was recommended to mitigate the increased risk of rockfall for a 3,500 linear feet (lf) portion of the highway. The temporary mitigation was requested to have a service life of up to 5 years and was required to have no impacts to a critical fiber optic utility line that was installed via horizontal directional drilling along the southbound shoulder of the highway. Yeh engineering geologists evaluated rockfall trajectories at the site to provide recommendations for the flexible rockfall fence energy rating and height.

The potential energy equation can be used to calculate the maximum possible energy that could result from a design rockfall event.

$$P.E. = mgh$$

$$where\ m = mass\ of\ rock, g = gravitational\ acceleration\ (32.17\ ft\ (9.80\ m) per\ S^2),$$

$$and\ h = vertical\ height\ of\ rock$$

As a part of the analysis, the natural slopes above the highway were evaluated for rockfall potential. Given the slope heights and geometries, the maximum vertical height for potential rockfall to impact the roadway was estimated at 200 feet. The vertical height was used to calculate the maximum potential energy that could result from rockfall originating at this height and impacting a flexible rockfall fence along the highway.

The largest boulder observed by Yeh engineering geologists in the catchment area within the project limits was a blocky-shaped granodiorite boulder with dimensions 2-ft x 2-ft x 1-ft, Figure 5. The design rock used in this analysis had dimensions 3-ft x 3-ft x 1-ft and was based on rocks observed in the rockfall source areas at the site. The design rock was igneous (granodiorite) and was assumed to have a unit weight of 160 pcf. The weight of the design rock was estimated to be 1,500 lbs. Utilizing the potential energy equation, the maximum potential energy that could be expected from the design rockfall event is approximately 150 ft-tons (408 kJ).

Given this maximum potential energy, a 185 ft-tons (500 kJ) rockfall fence was recommended. Although the maximum potential energy was calculated to be 150 ft-tons, the maximum kinetic energy that could impact the fence in the design rockfall event would be much less due to energy losses as the rock travels from the source area towards the catchment *(3)*.

The slope inclinations along the alignment were measured at 31 degrees or flatter. On slopes flatter than 38 degrees, the rockfall trajectory will be a rolling motion on the ground *(4)*. To provide a design height recommendation for the rockfall fence, the Rockfall Catchment Area Design Guideline (RCADG) *(5)* and the Ritchie Criteria *(6)* were used.

By measuring the heights, slope angles, and catchment area along the project alignment, and using these parameters as input to the RCADG and Ritchie Criteria, a percent retention can be developed for rockfall originating from a particular slope configuration. Using these methods and a 95% retention criteria, fence heights were recommended for the project area. The southern portion of the project (2000 lf) had overall flatter slopes and wider catchment and a 10 ft (3m) fence height was recommended. The slopes on the northern end (1500 lf) were steeper and required a 13 ft (4m) height to fulfill the retention criteria.

Due to the emergency nature of the project, and the presence of the poorly located underground fiber optic utility, it was necessary to avoid excavating large foundations for the rockfall fence posts. By recommending a commercially available rockfall fence, the manufacturing timeframe could be minimized. Working with the contractor, the Geobrugg GBE series rockfall fence was selected for the project, and recommended design modifications were coordinated with the manufacturer to optimize the functionality and feasibility of installation at the site.

The GBE A model rockfall fence post design is a hinged post and baseplate configuration supported by upslope and lateral wire ropes, while the GBE AR model rockfall fence posts consist of a braced post design welded to a baseplate, without support ropes. For this project, the braced post design was modified, for global support, to accommodate a post base support wire rope, Figure 6, and upslope and lateral support wire ropes. The post base wire rope is attached to the base of the post and attached to a ground anchor installed at the base of the slope. The intent is to prevent the post from kicking out upon impact. The upslope and lateral wire rope supports were installed in typical fashion. For local stability, threaded bars were grouted 18 inches into the ground. The post baseplate was placed on the bar and secured with a nut. While not commonplace today, this type of rockfall fence post support has been in use in California since the early 1990s.

Figure 6. GBE 500AR temporary rockfall fence with upslope and lateral support wire ropes and threaded bar post support.

CONSTRUCTION

Caltrans directed the contractor that was performing the phased anchored mesh installation to install the temporary fence system. Since they were already on site, layout of the temporary, flexible rockfall fence was conducted on July 20, 2016 and consisted of 30 ft. post spacing, with upslope and lateral anchors marked out per the manufacturer's recommendations to facilitate rapid drilling and installation.

Geobrugg provided upslope and lateral support rope ground anchor loads required for the 500 kJ-rockfall flexible fence system. These loads were used to estimate anchor embedment depths to provide the required pull out resistance. The contractor proposed to drill 4-inch diameter, 6-feet deep holes for the ground anchors. In lieu of testing production anchors, the contractor installed sacrificial anchors along the project alignment and pulled to failure, to verify that the bond strength and pullout capacities were sufficient to withstand the loads that could be transferred to the anchors in a rockfall event.

The construction sequence involved drilling holes for the wire rope anchors and post support threadbar, grouting the anchors and threadbar in place, testing the sacrificial anchors to verify pull out strength, and assembly of the fence components. Using air rotary drilling methods with an excavator mounted drill, the anchor holes were rapidly advanced into the subgrade. Delivery of the wire rope anchors began the first week, and the first posts were fabricated and delivered to the site within a week of ordering. Due to the temporary nature of the mitigation and the urgency of having the system in place, corrosion protection was not required, and greatly expedited the delivery of the posts. It was reported that 57 posts were installed and over 700 linear feet of mesh was hung by August 5, 2016.

Yeh personnel visited the site on August 10, 2016 and the fence was nearly completed, Figure 7. Final completion of the fence was delayed until later in the fall to facilitate the completion of the temporally suspended Phase 2 of the anchored mesh system.

Figure 7. Completed rockfall fence looking north along US 395

PERFORMANCE

After the fences were installed, Phase 2 of the programmed anchored mesh project began and was completed by the end of November 2016. No significant rockfall events were reported until March 12, 2017 when the 3m GBE 500AR fence experienced an impact in the early morning hours. The event resulted in the rock being retained by the fence, but the impact elongated the

TECCO high tensile steel mesh enough that the force of the impact struck the adjacent concrete K-rail, resulting in a portion of the concrete spalling off into the traveled way, Figure 8.

Figure 8. Rockfall event occurring on March 12, 2017 retained behind 3m Geobrugg GBE500AR flexible rockfall fence. (Photo credit: Joe Blommer, Caltrans)

The source area of the rockfall was approximately 100 ft vertical height on a slope above the 10ft (3m) fence. Based on the damage to the K-rail and the deformed Tecco mesh, it appears that the rock rolled across the catchment with a significant rotational and translational velocity, struck the K-rail and was retained by the fence mesh. Rocks observed in the catchment behind the impacted fence were 3ft x 3ft x 2ft, 2ft x 2ft x 1.5ft, and 1.5ft x 1.5ft x 1ft. K-rail debris was 6 inches and smaller and were reported by the California Highway Patrol to be scattered across the traveled way. The impact did not engage the friction brakes, indicating that the impact was below the upper energy capacity of the fence. While the Tecco mesh permanently deformed, it did not require any repair to maintain its effectiveness.

SUMMARY AND CONCLUSIONS

By utilizing simple and effective field methods, in conjunction with published reference documents, a rapid rockfall assessment can be performed, allowing mitigation to be designed and implemented under an expedited timeframe. Once a mitigation solution was chosen, working with the manufacturer permitted design modifications to be made which accommodated the project constraints. The rapid mobilization of the contractor and the expedient delivery of fence components allowed the installation to be completed in short order. While likely not feasible for all installations, a practical, innovative post support system functioned effectively for this project, allowing the system to be installed rapidly in response to the elevated rockfall risk.

The system has performed as anticipated, retaining rockfall events up to the design event effectively. Due to the optimization of the catchment area and the need to prevent errant vehicles from impacting the fence, the K-rail remained in place and within the elongation zone of the mesh after the installation of the fence. While preventing traffic from striking the fence, the K-rail has the potential to be impacted by rockfall events and spall off into the traveled way. This occurrence was anticipated at the design phase, and measures to mitigate this undesirable result were recommended, including hay bales and timber lagging between the fence mesh and the K-rail.

REFERENCES

1. Duffy, J, D., Field Tests and Evaluation of Rockfall Restraining Nets, Proceedings of the 7th International Congress on Rock Mechanics, Aachen, Germany, Balkema, Rotterdam, 1991.
2. InciWeb (2016). "Marina Fire". https://inciweb.nwcg.gov/incident/4809/
3. Smith, D.D., Duffy, J.D., Field Tests and Evaluation of Rockfall Restraining Nets, 1989, California Department of Transportation Research Report No. CA/TL - 90/05.
4. Turner A.K., Schuster, R.L., 2013. Rockfall Characterization and Control, Transportation Research Board, National Academy of Sciences, Washington, D.C.
5. Pierson, L. A., C. F. Gullixson, and R. G. Chassie, 2001, Rockfall Catchment Area Design Guide. Final Report SPR-3(032), Research Group, Oregon Department of Transportation, Salem, 77 pp.
6. Ritchie, A.M., 1963. The evaluation of rockfall and its control. Highway Research Record, Number 17, Highway Research Board of the Division of Engineering and Industrial Research, National Academy of Sciences, Washington, D.C.

Low deflection Kevlar® reinforced gabion rockfall protection embankments.

Dr Dave Cheer
Technical Developments Manager, Linear Composites Ltd
Vale Mills, Oakworth, Keighley,
West Yorkshire, UK, BD22 0EB
+44 (0) 1535 643363
dcheer@linearcomposites.com

Adrian Koe
Associate Director, Ground Engineering, AECOM
Royal Court, Basil Close,
Chesterfield, UK, S41 7SL
+44 (0) 1246 244511
adrian.koe@aecom.com

Prepared for the 68th Highway Geology Symposium, May, 2017

Acknowledgements

For example: The author(s) would like to thank the individuals/entities for their contributions in the work described:

Name 1 – Affiliation
Name 2 – Affiliation
Name 3 – affiliation

Disclaimer

Statements and views presented in this paper are strictly those of the author(s), and do not necessarily reflect positions held by their affiliations, the Highway Geology Symposium (HGS), or others acknowledged above. The mention of trade names for commercial products does not imply the approval or endorsement by HGS.

ABSTRACT

Flexible rockfall catch fences are an effective and widely used solution to rockfall problems encountered around the world. Flexible fences undergo considerable geometric deflection in the process of slowing and stopping falling rocks. In some cases, this deflection envelope may be incompatible with the layout of the site and a low deflection solution will be required and in these situations, an embankment may be specified in preference to a flexible catch fence. Embankments have a range of advantageous properties: they have no upper impact energy limit, they can be considered permanent structures and can they have relatively low maintenance costs in the event of an impact.

Where embankments are used, it can be difficult to achieve an engineering balance between the footprint width/volume of embankment required to absorb the impact, against the space available within the site. For this reason, carefully designed, reinforced embankments may be used. These can enable a greater volume of the embankment fill to be engaged in the task of slowing and stopping an impacting body and allow for considerable reductions in footprint width/volume of embankment compared to unreinforced embankments. Additionally, through careful design of the embankment and its internal reinforcements, deflection can be reduced to levels significantly below those required for equivalent impacts on flexible rockfall fences.

This paper presents details of a reinforced embankment designed for use on a transport infrastructure site in Scotland, UK. The project required a low deflection embankment with a narrow cross-section due to very tight space constraints. Additionally, the solution had to have a low aesthetic profile and a design life of at least 60 years. The embankment was designed using parallel laid, high modulus Kevlar® fibre Parafil® ropes to reinforce a free-standing gabion rockfall embankment and create a narrow, low-deflection structure with a long design life and simple post-impact maintenance regime.

The embankment was struck by a large rockfall approximately 6 months after installation. It underwent almost no deflection and only required minimal maintenance works to be returned to the 'as-installed' condition.

INTRODUCTION

Site Location and Situation Overview

In January 2011 URS (now AECOM) were approached to undertake a geo-hazard risk assessment of a proposed cycle way along a section of the Great Glen (Scottish Highlands, UK), by Sustrans (http://www.sustrans.org.uk). The proposed cycle way was to be constructed along an historic railway line on the eastern edge of Loch Oich.

The risk assessment identified an area that was prone to an increased risk of rock fall. The area was deemed to be an old debris flow scar (Figure 1). The site comprises a steep, sparsely vegetated talus slope, located immediately below an extended and heavily eroded scarp face of steep rock cliffs, leading up to steep vegetated moorland slopes above.

The geology of the site (exposed in local outcrops and the rock cliffs) comprises lower amphibolite grade psammitic and semi-pelitic rocks of the late Neoproterozoic "Dalradian Supergroup". The site lies immediately adjacent to, and within the shear zone of, the Great Glen Fault; one of the most laterally continuous strike-slip faults in Europe with a sinistral offset estimated to measure over 100 km. The intensity of fault related deformation of the rocks varies locally, with in-situ rocks classified between fault gouge and mega breccia, however the latter is more common.

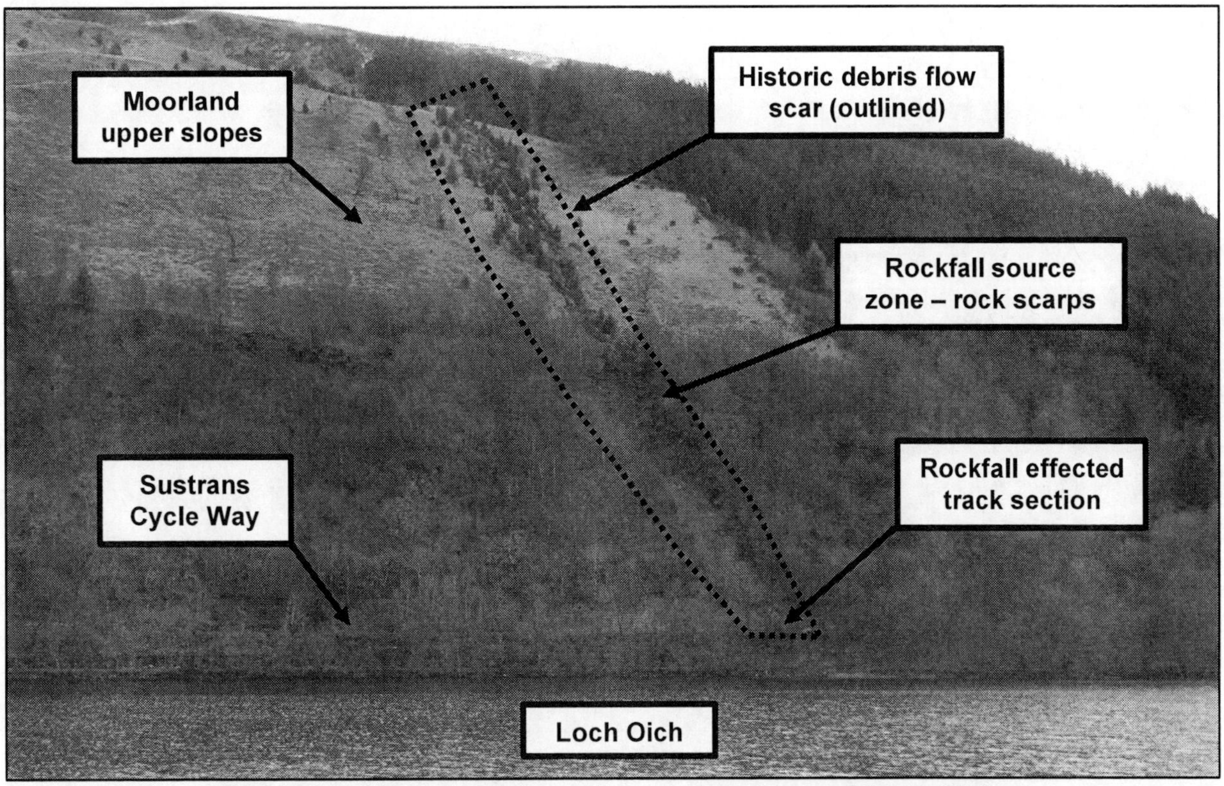

Figure 1 – Site viewed from the western shore of Loch Oich.

During the various assessments required as part of the engineering feasibility studies to open the old railway track as a cycle way numerous repeated fresh rocks were observed at the base of the scar. These suggested a steady flux of material arising from an active rockfall source zone above, specifically the loose and intensively fractured rocks of the rock scarps. The assessment suggested that the active nature of the source zone would result in repeated and regular rockfalls throughout the design life of any engineered protective solution installed on the site. Accordingly, the geo-hazard risk assessment of the cycle way recommended the design and installation of a suitable rockfall protective barrier, continuing along the stretch of track beneath the rock scarps.

GEO-HAZARD RISK ASSESSMENT

A specific risk assessment methodology was developed for the site to assess the risks both during construction of the cycle way and after completion with regular use by cyclists and walkers. The geo-hazard risk assessment identified the relict debris flow scar as the highest risk area, both during construction and once the cycle way was in use. Two sources of rock fall were identified. The first from a rock face at the top of the debris flow scar, the second from a talus slope/cone beneath the rock face.

A stereonet analysis was undertaken which indicated that wedge, plane and toppling failures were likely. Given the site's location, within close-proximity to the broader brittle fracture zone associated with the Great Glen Fault, it would be expected that a large number of failures would occur at the site.

Figure 2 - Photograph looking up talus cone with active rock face at the crest.

Rockfall modelling done for the assessment indicated that there was a high probability that falls would reach the track. This was corroborated by the evidence of fresh rockfall with fallen blocks adjacent to the path varying in size up to a maximum of $0.5m^3$. It was considered that there was a risk of both single-block rock falls occurring as well as a multiple-block rock falls.

Figure 3 – Photograph of newly fallen rocks on the track.

It was also considered that the historic debris flow scar could act as source for a channelized debris flow. The Scottish Road Network Landslide Study (Ref. A) suggests lower and upper slope angle limits for debris flow development of 26 and 50 degrees respectively. The angle of the scar was within this range and, given the lack of vegetation it was considered likely that a failure could occur on these slopes.

During construction of the cycle way pavement, there would be a short period of time (approximately an 8 hour period, 5 days a week for 2 weeks) when workers would be directly at the toe of the slope. During this period, the risk from rock fall was deemed to be high.

During its lifetime, the risk to the cycle way infrastructure e.g. the pavement, signage etc. was deemed to be relatively high, as it is permanent and therefore frequent repair work was likely to be required. The cycle way users, either riding bikes or on foot, were considered to be at lower risk because they would typically have a limited exposure from falling rocks as they would only be passing through the site. Additionally, there is a long straight approach to the hazard area, so it was assumed that fallen blocks on the track could be readily seen in advance so cycleway users would have sufficient time to stop or avoid them.

The nature of the risks identified in the assessment dictated that a rockfall protective measure would need to be recommended, on the grounds of both safety and ongoing infrastructure repair costs.

DEVELOPMENT OF THE REMEMDIAL SOLUTION

Following discussions with the client a design brief was developed for the site and it had to incorporate the following:

- Narrow footprint width – As shown in figure 4 there was limited room to incorporate the barrier at this location without compromising the width of the cycle way.
- Low deflection during and after impact - to enable the cycle way to remain open and operational after a rock fall event and to allow users of the cycle path to pass safely.
- Reliable protection suitable for repeated impacts - the site is not inspected on a frequent basis. It is located 1km from the nearest road and it is possible that the client may not be aware of an impact for some time after it has occurred so the system would have to survive a number of rockfall events without needing maintenance to stay operational.
- Low installation costs and easy to install – the project had a limited budget and a civil engineering contractor had already been appointed. The client's preference was that the appointed contractor would install the barrier, and therefore a system that was easy to install without specialist techniques was required.
- Low maintenance costs with simple works scope – the barrier would be required to absorb numerous impacts during its design life. It was deemed necessary to remove the fallen material and undertake repairs easily without the requirement for specialist materials.
- Long design life –many engineering structures in the local area are subject to increased rates of corrosion due to the harsh environment. The barrier materials needed to resist corrosion, with minimal maintenance operations, over the required 60 year design life.

Figure 4: View North showing the close proximity of the cycle path to the loch.

Following analysis of the impact conditions and taking into account all of the site-specific requirements, it was decided that a rockfall/debris flow protection embankment would provide the best remedial solution for the Loch Oich site. Considering the critical dimensional requirements of the site, it was decided that a steep faced, free-standing gabion basket embankment would be likely to offer the best solution.

Impact Protection Embankments

The use of embankments to protect people and infrastructure from rockfall, debris flow and avalanche impacts, is a well established practice with a very long and successful history. Within the last few decades the commercial sector has driven forward the development of flexible fences designed to offer protection from both rockfalls and debris flows. While these are highly effective protective structures, which are effective across a range of scenarios, they do have limitations. They are costly to buy, can be expensive to install (with installation typically requiring highly skilled specialist contractors) and they often require complex maintenance operations following an impact.

Figure 5: Debris flow protection fences in Argyll, Scotland (courtesy Maccaferri UK).

In contrast to fences, protective embankments can be designed to absorb an impact of almost any energy, at any velocity and of any volume. Embankments can offer multiple protective capabilities such as the example shown in Figure 6, where the structure is designed to provide protection from rockfalls, debris flows and avalanches.

Figure 6: 22m high x 750m long reinforced soil embankment construction, Iceland.

One of the primary reasons for the specification of a protective embankments is their capacity to absorb multiple impacts without the need for expensive and complex maintenance operations. Embankments can be designed to operate for years without any maintenance at all such as those pictured in Figure 7a and 7b below in Valle D'Aosta, Italy.

Figure 7a and 7b: avalanche and rockfall debris accumulated behind a 12m embankment.

FINAL DESIGN

Gabion Embankment Design

The final design of the gabion barrier was undertaken to enable construction. A detailed survey of the area was undertaken to enable rockfall modelling and debris flow assessment for the design of the barrier. The results of the rock fall modelling suggested that the impact velocity and bounce heights of the falling rocks were calculated to be relatively low; mostly due to the amount of loose talus present on the slope immediately above the impact point. Assessment of the potential risks from debris flow indicated that a low volume failure moving at relatively low speeds was the mostly likely scenario.

The design assessment indicated that a 3m high barrier would be sufficient to contain the majority of the predicted rock falls and debris flows and the trajectory analyses showed that the barrier would need to span an area of track approximately 20m in length. Accordingly, a 20m long barrier was developed comprising three rows of 1m high gabions. The base row was 2m in width, the middle row was 1.5m in width and the top row was 1m in width. A 100mm step was incorporated into the impact face, between courses, to increase entrainment capacity at the back of the gabion. Local stone was specified to ensure that the visual continuity of the embankment with the surrounding landscape.

The detailed design of the gabion barrier confirmed that the gabions were not sufficient in themselves to absorb the forces arising from the design impact. Additionally, analysis showed the deflection of the gabion structure during impacts would be unacceptable given the confined space on the site and proximity of the cycle path to the interception structure. The decision was therefore taken that, in order to achieve the design impact capacity while satisfying the geometric operating requirements, the gabion embankment would require the inclusion of additional reinforcement.

Reinforcement Design Development

Linear Composites were approached to develop a reinforcement specification that would meet the project requirements, could be delivered to the site within the time available and could be incorporated into the embankment construction with the minimum of impact on the installation programme.

Following a detailed assessment of the impact analysis developed by Aecom and the geometric limitations of the site it was necessary to develop a calculation to demonstrate the physical parameters of the impact that would need to be addressed by the reinforcement within the embankment, taken as a whole. It was then necessary to develop a mathematical method to establish the optimum mechanical properties of the reinforcement required to counteract the forces arising from the impact and the (semi-) permanent loading after the impact and then to balance this against the number of reinforcements required within the structure and the amount of deflection that could be safely undergone by the embankment in the process of stopping an impact.

The fundamental physics involved in the engineering design of impact systems require a detailed knowledge of both the strength but also the displacement behavior of any reinforcement or strength member. Given the tight geometric limitations (small acceptable deflection envelope) on the site, it was obvious that a high modulus reinforcement would be required. This in-turn would give rise to relatively high forces during impact. In order to avoid the need to use a large quantity of reinforcement to resist these forces (which would be highly impractical to use on site) it would be necessary to use a technical fibre with a very high tenacity. For this reason the 'bullet proof' aramid fibre Kevlar® was chosen.

Given the additional temporal and practical criteria pressing on the project, it was decided to use the parallel laid Parafil® rope product offered by Linear Composites rather than a high strength 2D or sheet-like 'geogrid', such as Paralink, as is frequently used in reinforced soil embankments. Parafil® is a cylindrical "rope" comprising a core of parallel laid technical fibres encased within an extruded protective polymer sheath; in this case of UV stabilized polyethylene. By using Parafil® rope as the reinforcement, ducts could inserted into the gabions during filling and the ropes could then be inserted after construction was complete.

The optimization process for the ropes was made easier by the design of the baseline gabion structure with its three-course construction. The final design included one Type F Kevlar® Parafil® rope, with a nominal breaking load of 22.5 tonnes, per course; giving a total of three equally distributed over the total height of the embankment. The ropes were inserted along the length of the embankment, each within a small diameter polyethylene duct. The duct was inserted along a specifically defined profile to allow the rope to offer the best possible deflection behavior and take maximum advantage of the high modulus of the core fibres. Extending out from the lateral ends of the working portion of each course of gabion baskets (see Figure 8), the ropes were taken down and connected into the anchorages. The ropes were terminated with proprietary "wedge and socket" style terminations designed by Linear Composites. These act to transmit 100% of impact related forces from the rope core fibres into the anchorages.

Anchorage Design

Due to the calculated low deflection of the barrier, the forces on the anchors were expected to be high. If the Parafil ropes were to be retained by anchors then a long bonded length would be required because they would need to be installed into the on-site talus. The talus also suggested a range of other problems would be encountered with the use if drilled anchorages (bit breakage, grout loss etc.). Drilled anchorages were therefore discounted in favour of anchorages based on a mass concrete "deadman" system utilizing self-weight of the concrete and passive resistance from the site soils.

These anchors could be constructed easily on site with the plant available and located within the limited area available adjacent to the track. The final deadman anchor design comprised a 500mm x 500mm by 2.0m deep mass of concrete reinforced with A252 mesh. The terminations one the ends of the Parafil tendons were then connected to 32mm diameter steel bars located in the center of each anchor.

Figure 8: Photograph north showing the completed Parafil® reinforced gabion embankment adjacent to the cycle path. Rope ducts are visible to the right.

ROCKFALL IMPACT

A site visit was undertaken following the reported impact and before the relevant maintenance works. The embankment was found to be in fully serviceable condition despite being marginally overtopped around the impact site (see Figure 9), centered at around 3m to 4m from the northern end.

The impact was assessed as having involved two phases, a preliminary dynamic impact by approximately 30 tonnes of more blocky material followed by a secondary impact comprising lower velocity filling/overtopping by another 20 tonnes of loose material (see Figure 10). The secondary phase of the impact caused some of the material to spill over the top of the gabion embankment (at approximately 2.5m above ground level) and around 5 tonnes of material to spill around the end of the embankment. It is estimated that this over-spilling material had little velocity remaining by the time it passed the gabion structure.

The embankment underwent minimal rotation or deflection during the impact event (as is visible in Figures 9 and 10). A tension increase was noted within the uppermost Parafil® rope with a similar, although slightly lower, increase in the middle rope. No damage was noted to the rope ducts or gabion baskets and the gabion fill (on the non-impact faces of the gabions) showed no obvious displacement.

Figure 9: Northern end of the embankment after the impact, showing overspill of debris.

Figure 10: View of the northern end of the embankment showing the overspill of material arising from the second phase of the debris flow impact

CONCLUSIONS

The specially designed reinforced gabion embankment was successfully installed, on time and on budget in August 2015. The unique design, using high specification technical-fibre reinforcements, allowed a significant increase in the capabilities of an otherwise conventional gabion embankment, both increasing capacity and decreasing post-impact deflection. The combination of gabion embankment and Parafil rope reinforcements will enable the structure to absorb multiple impacts with minimal maintenance and satisfy the long design life required by the client.

In early February 2016, the embankment was struck by a large impact. A subsequent site inspection revealed that the northern end of the gabion structure had been hit by approximately 50 tonnes of material from a two-phase debris flow. The structure remained completely intact and underwent no notable deflection or rotation following the impact event. The only maintenance operation required was for the impact material to be removed from behind the embankment.

REFERENCES:

DESIGN AND CONSTRUCTION OF A TEMPORARY ROCKFALL MITIGATION SYSTEM AT THE BELLWOOD QUARRY RESERVOIR TUNNEL, PHASE 1 WATER SUPPLY PROGRAM, ATLANTA, GEORGIA

David J. Scarpato, P.E.
Scarptec, Inc.
P.O. Box 326
Monument Beach, MA 02553
(603) 361-0397
dave@scarptec.com)

Nick H. Strater, P.G.
Brierley Associates
167 S. River Road, #8
Bedford, NH 03110
Phone: 603-206-5775
(603) 918-0606
nstrater@brierleyassociates.com

Brandon Manahan
Apex Rockfall Mitigation, LLC
2660 I Rd.
Grand Junction, CO 81506
(970) 250-0008
bmanahan@apexrfm.com

Prepared for the 68th Highway Geology Symposium, May 2017

Acknowledgements

The author would like to thank the following individuals/entities for their contributions in the work described:

Mr. Bob Lyne (GeoBrugg North America, LLC)
Mrs. Lizan Gilbert (Guy F. Atkinson Construction)
Mr. Brent Laminack (Guy F. Atkinson Construction)

Disclaimer

Statements and views presented in this paper are strictly those of the author(s), and do not necessarily reflect positions held by their affiliations, the Highway Geology Symposium (HGS), or others acknowledged above. The mention of trade names for commercial products does not imply the approval or endorsement by HGS.

Copyright Notice

ABSTRACT

The City of Atlanta is commissioning a new 1-mile-long, approx. 13-ft. diameter, lined, water conveyance tunnel as part of Phase 1 of the Water Supply Program. The tunnel will be excavated through bedrock with a TBM and will provide the City with potable water from the soon-to-be-filled Bellwood Quarry Reservoir. Construction of the tunnel and ancillary features was initiated in spring of 2016 and is expected to be complete in 2018. The previously mined Bellwood Quarry will serve as a reservoir to impound and distribute the water.

Prior mining activities have resulted in steep pit slopes, some as high as 350 ft., with an abundance of loose rock. In order to help maintain a safe and functional site for site access and tunneling, a temporary rockfall mitigation system was constructed (and is currently being maintained) above the main water supply tunnel and a secondary adit. Critical elements of the temporary system included post-scaling design and construction of draped netting, rock dowels, and two rockfall canopies. The draped netting and canopies were connected as part of a "slot" system, where falling rocks will be contained behind the drape and subsequently guided into (and arrested by) the canopy system.

This paper details the elements of the temporary rockfall mitigation system being utilized during tunnel construction, and will describe the challenges associated with installation of near-horizontal rockfall canopies at elevated, difficult access locations.

KEYWORDS

Rockfall, Barrier, Canopy, Design, Deflection, Temporary, Quarry, Tunnel, Reservoir

INTRODUCTION

The City of Atlanta is commissioning a new 1 mi.-long, approx. 13-ft. diameter, concrete lined, water conveyance tunnel as part of Phase 1 of the Water Supply Program. Subsequent phases of tunneling will result in the construction of another 4 mi. of tunnels to tie the underground water conveyance system together. Tunneling commenced from within the Bellwood Quarry Reservoir (currently drained), which is located approximately 2 miles from downtown Atlanta. The Phase 1 main tunnel consists of an initial short segment of drill and blast starter tunnel, with the remaining drive completed by means of a bedrock tunnel boring machine ("TBM"). Additionally, two pump station shafts were excavated within 200 ft. of the pit slope face, and a series of adits are being excavated that will connect the main tunnel to the pump station shafts. The previously-mined Bellwood Quarry will serve as a proposed 2.4 billion gallon reservoir with which to impound and distribute the water. The new tunnel system will provide the City with potable water from the soon-to-be-filled Bellwood Quarry Reservoir. The area surrounding the reservoir will be landscaped with walking trails, in what will eventually be designated as Westside Reservoir Park. Construction of the tunnel and ancillary features was initiated in spring of 2016 and is expected to be complete in late 2018.

The Bellwood Quarry has been the site of active rock extraction for over 100 years, providing a source of construction stone and aggregate. This principal lithology exposed in the quarry is biotite gneiss of the Clairmont Formation; however, the bedrock is also frequency referred to as granitic gneiss in public domain geologic literature. Prior drill and blast mining activities were utilized to develop the quarry, and have resulted in steep pit slopes, some locally as high as 350 ft. These exposed slopes have been subject to weathering processes, and as such, presented an abundance of loose rock. The exposed silica-rich bedrock is generally highly fractured, and very hard which can present a challenge to drilling operations.

In order to help maintain a safe and functional site during tunneling, a temporary rockfall mitigation system was constructed (and is currently being maintained) above the main water supply tunnel portal and secondary adit. Critical elements of the temporary system included scaling, post-scaling design and construction of draped netting, rock dowels, and two rockfall canopies. The overall site, rock slope and underground features are shown in Figure 1.

TEMPORARY ROCKFALL MITIGATION ELEMENTS

The temporary rockfall mitigation work at the site consisted of initial highwall scaling, followed by installation of a wire mesh rockfall drape, rock dowels, and two individual segments of rockfall canopy. In addition, system monitoring and maintenance efforts are also being conducted over the tunnel to maximize performance over the construction period. The temporary rockfall mitigation system was designed by Scarptec, Inc. ("Scarptec") and Brierley Associates Corp. ("Brierley"), and was constructed by Apex Rockfall Mitigation, LLC ("Apex"). Periodic field engineering visits during installation of the temporary system were also completed by the design team. The underground workings are being constructed by Guy F. Atkinson Construction.

Highwall Scaling

Initial rock slope scaling took place in the spring and early summer of 2016, prior to mobilization of tunneling equipment, with efforts being highly productive. Previous blasting activities and exposure to the forces of weathering resulted in an abundance of loose rock prior to construction activities. In order to minimize the quantity of potentially unstable rock material, Apex completed scaling efforts using manual methods; (e.g., scaling bars, rope access techniques) and mechanical methods; (e.g., pneumatic air bags) which were employed using specialized rope access techniques.

Figure 1 – Northerly view of quarry highwall, tunnel and adit (on bench) (Photo by Scarptec, Inc.)

Draped Rockfall Netting

Draped steel netting was used for both temporary and permanent rockfall mitigation purposes. The temporary application was installed in June 2016 and was intended to mitigate rockfall potential during the 3-yr. period of tunnel construction. The permanent netting application, put forth by the tunnel designer and engineer-of-record (Stantec), considers mitigation of long-term rockfall occurrence to prevent large quantities of rock from clogging the tunnel entrance and impeding the flow of water. Transition from temporary to permanent protection systems will require a series of field-determined retrofits at the end of the tunnel construction period, and are described later in this paper.

Draped netting consists of galvanized G65/3mm Tecco® Mesh manufactured by Geobrugg supported at the crest of the slope by a series of 20-ft. long, ¾-in. dia. IWRC-EIP wire rope cable anchors and a top rope. The draped segment of slope in the vicinity of tunneling operations measures approx. 365-ft. in plan length along the slope crest by approx. 315-ft. in slope height. The top set cable anchors were subject to pull testing at both axial (i.e. vertical) and angled (45 deg.) loading configurations in order to verify minimum load-carrying capacity.

In order to maximize rockfall capture, the temporary draped netting was locally tied into the canopy system. The intent of the connection between the canopy and drape was to create a "slot" with which falling rocks could be contained within the system and could not exit the limits of netting; in other words, the canopy formed the lower limit of the temporary drapery system.

Figure 2 – Constructed rockfall drape (Photo by Scarptec, Inc.)

Rock Dowels

In order to design the temporary canopy-drape netting system, the Scarptec-Brierley design team needed to define the upper limits of rock block size and energy that could potentially compromise the system. Rock blocks greater than this critical size, conservatively assumed to be falling from near the slope crest, would require bolting if such blocks appeared to be loose based on field observations. Based on kinematic calculations and rockfall analyses of rock block free fall from 285 ft. in height, the critical rock block size that could exceed the maximum barrier deflection criterion of 28 ft. was estimated to be a cubic block measuring approx. 2.5-ft. (or the equivalent of 15-c.f.). Rock blocks greater than this size required rock reinforcement to arrest potential movement.

Passive rock dowels were chosen to reinforce potentially unstable rock blocks above both canopy systems due to their relative speed and ease of installation; however, to stiffen up the rock mass and pin down suspect key blocks without the benefit of tensioning requires that additional steel be installed. As such, the initial phase of rock reinforcement called for installation of 74 rock dowels that were marked-out in the field (Figure 3) and submitted to the Owner on plan sheets with calculations.

Figure 3 – Rock dowel layout with paint using rope access (Photo by Scarptec, Inc.)

Rock dowels were comprised of 1-¼-in. dia. grade 75 epoxy coated bars fabricated by Williams Form Engineering. Minimum embedment depths by location were provided to rock remediation technicians from Apex, who then drilled and installed the dowel bars using wagon drills. In two instances, temporary wire rope cable lashing was required as a precaution to stabilize rock blocks prior to drilling. Rock dowel lengths generally ranged from 10 to 20-ft. in total embedment length.

Rockfall Canopies

Initially, a traditional barrier approach was considered whereby a barrier would be constructed along the crest of an intermediate bench slope; however, it quickly became apparent that vertical posts would not work for all locations given the complex geometry of the slope and need for access by tunneling personnel. Therefore, the design team opted for use of two rockfall canopy barrier arrangements, located above the tunnel and adit portals. Both canopies were adapted to the field conditions and would also not restrict construction access by the tunneling crews.

The temporary canopy barriers were constructed with GBE-1000-A rockfall barrier components from Geobrugg that includes segments of G65/4mm Tecco® mesh fabric spanning between the posts. Posts consist of 13.1-ft. long steel sections that are set at 25 ft. centers for a total of four posts with an effective length of 75 linear ft. above the main tunnel portal and adit (Figure 4).

Both canopies were connected to the draped netting as part of a “slot” system, so that falling rocks remain behind the drape and are subsequently guided into (and arrested by) the canopy system. To establish a “closed system”, a cut line was established along the Tecco® drape and an additional segment of Tecco® mesh was connected between the drape cut line and the upper portion of the barrier post top cables (Figure 5).

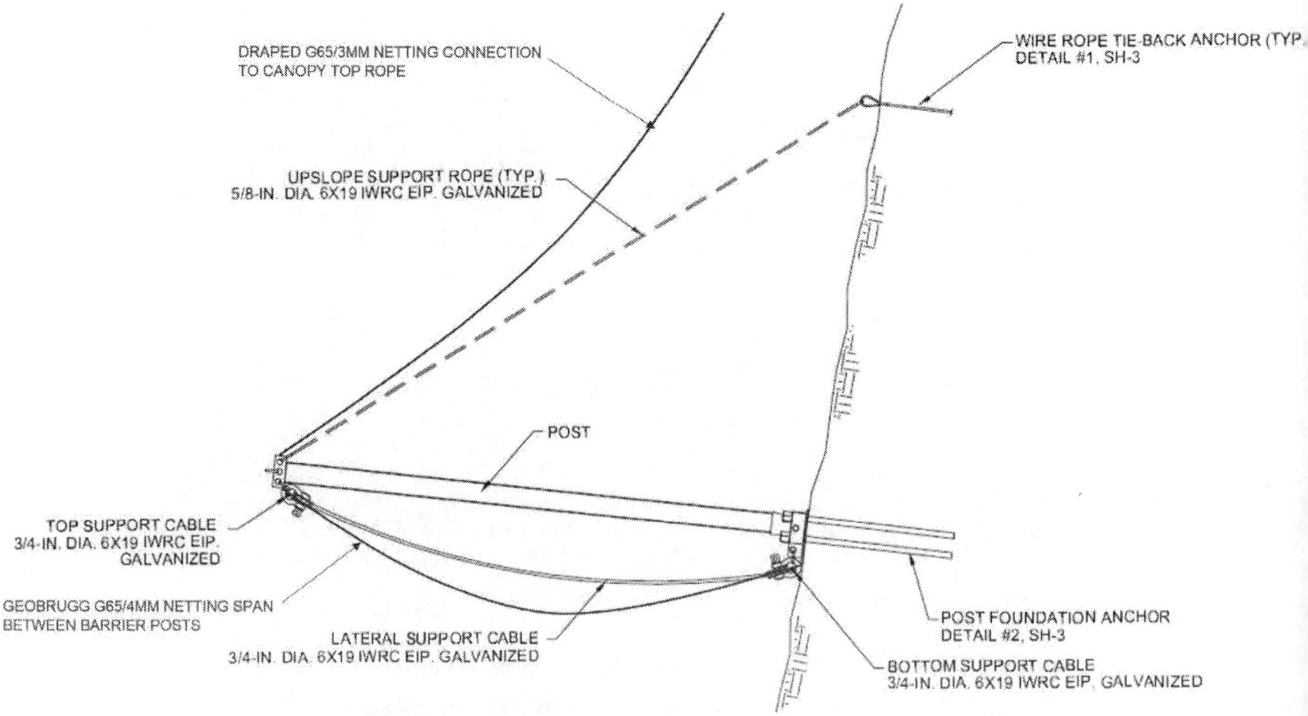

Figure 4 – Canopy post section detail
(Image adapted from Scarptec- Brierley construction drawings)

Figure 5 – Canopy construction and drape tie-in (Photo by Scarptec, Inc.)

Monitoring & Maintenance

In order to maximize the reliability of the temporary rockfall mitigation system, the slope and constructed elements described herein are subject to periodic monitoring and maintenance efforts at the frequency of one visit every 6 months unless specific observations or events dictate more frequent monitoring. Geotechnical monitoring efforts generally consist of assessment of the following:

- the capacity and need for cleaning of rock debris within the canopies and drape;
- need for additional slope scaling;
- condition of canopy anchorage elements;
- need for additional spot rock dowels;
- condition of drape anchors;
- assessment of drape damage/over-stressing; and,
- canopy system tensioning and netting sag adjustments

Maintenance of the system over the tunneling construction period was (and continues to be) completed by Apex, based on monitoring visit observations. Small fragments of rock debris were removed from the tunnel canopy system in January of 2017 (Figure 6), and minor adjustments to the canopy system cabling were also completed.

Figure 6 – Adit canopy rock debris subject to maintenance cleaning (Photo by Scarptec, Inc.)

TRANSITION TO PERMANENT CONDITION

The permanent condition will consider the effects of nearly full-submergence of the rock slope as the old quarry transitions to a long-term water supply reservoir for the City of Atlanta. Upon completion of the approx. 3-yr. construction period (temporary condition), the two rockfall canopy segments will be removed from service. Any interim connections between the drape and the canopies will be disassembled. The temporary rockfall drape will be converted to a permanent system through a series of minor repairs (if needed) and localized geometric reconfigurations which will be field-fit around the tunnel, adit, and any hard slope breaks. Within approx. 6 months of project completion, the temporary rockfall mitigation design team will consult with the tunnel designer regarding the transition from temporary to permanent system.

CONCLUDING REMARKS

The construction of temporary rockfall mitigation features during Phase 1 of the Water Supply Program are critical to site safety and will help provide for minimized down-time while tunneling continues from below. Design-during-construction efforts required the Apex-Brierley-Scarptec team to evaluate and adapt to field conditions "on-the-fly". Development of the canopy system concept initially posed some challenges given the complex slope geometry, height-related difficult access conditions and multitude of other construction priorities directly below the canopies. These initial challenges were overcome with solid field engineering input from the team during construction.

Although most surface and underground blasting is now complete, additional destabilizing forces from construction vibrations (e.g. TBM advancement), surface water and fracture-controlled drainage, and bedrock weathering may result in periodic rockfall at the site, all of which underscores the importance of this temporary rockfall mitigation system. To-date, the system has performed as intended and will be maintained as necessary to mitigate both the frequency and effects arising from potential rockfall events.

REFERENCES:

1. Turner, A.K., Schuster, R.L., *Rockfall – Characterization and Control*, National Academy of Sciences, Transportation Research Board, Washington, D.C., 2012.

2. Geobrugg, *Rockfall Product Range GBE-series 150303e.*

3. Geobrugg, *GS-1104us 140827 GBE-1000A System drawing, Layout1*

4. Geobrugg, *Tecco® Properties*, Doc. No. 1.402.29.EN 1403

Rockfall Mitigation at the Pillar Mountain Slide, Kodiak, Alaska

Eric C. Cannon, CPG
Golder Associates Inc.
2121 Abbott Road, Suite 100
Anchorage, AK, 99507
(907) 865-2525
ecannon@golder.com

Wesley Ashwood, P.E.
Golder Associates Ltd.
37702 3rd Avenue
Squamish, British Columbia, Canada V8B 0B4
(604) 815-0768
Wesley_Ashwood@golder.com

Paul Schlotfeldt, P.Eng., PhD
Golder Associates Ltd.
37702 3rd Avenue
Squamish, British Columbia, Canada V8B 0B4
(604) 815-0768
Paul_Schlotfeldt@golder.com

John D. Thornley, P.E.
Golder Associates Inc.
2121 Abbott Road, Suite 100
Anchorage, AK, 99507
(907)-865-2538
John_Thornley@golder.com

Prepared for the 68th Highway Geology Symposium, July, 2017

Acknowledgements

The author(s) would like to thank the individuals/entities for their contributions in the work described:

Aaron Hughes – Alaska Department of Transportation & Public Facilities
Howard Weston – Previously with Golder Associates Inc., Anchorage, Alaska

Disclaimer

Statements and views presented in this paper are strictly those of the author(s), and do not necessarily reflect positions held by their affiliations, the Highway Geology Symposium (HGS), or others acknowledged above. The mention of trade names for commercial products does not imply the approval or endorsement by HGS.

ABSTRACT

The Pillar Mountain slide in Kodiak, Alaska, has been a site of persistent rockfall hazard for several decades. At the base of the slide, Rezanof Drive traverses the site. Rezanof Drive is a critical lifeline representing the only road linking the community of Kodiak to the northeast with the airport and U.S. Coast Guard Base to the southwest. In addition, piers are located on the downslope side of Rezanof Drive. The slide is located on a southeast-facing slope. The maximum width of the slide is approximately 1,400 feet, and the maximum height of the slide reaches to approximately 760 feet elevation. Steep bedrock cliffs form a complex geometry to the crown and flanks of the slide. The main body of the slide consists of a talus slope. Pillar Mountain is composed of argillite that steeply dips into the slope. The bedrock is part of the accretionary prism complex associated with the Aleutian subduction zone. The bedrock is mantled by loess and vegetative mat.

As part of a Highway Safety Improvement Project, the Alaska Department of Transportation and Public Facilities (ADOT&PF) completed a rockfall hazard mitigation project with the primary objectives of reducing rockfall hazards associated with the slide, reducing maintenance efforts associated rockfall cleanup, and maintaining traffic flow along Rezanof Drive. This paper focuses on providing a historic review of the slide, a detailed look at the field investigation and the use of Light Detection and Ranging (LiDAR) elevation data to characterize the slide, and discussion of the rockfall modeling completed to support the development of rockfall mitigation design alternatives. The results of these efforts supported the selection and ultimate construction of an innovative, mechanically-stabilized earth (MSE) rockfall berm to intercept rockfall and protect travelers and workers along Rezanof Drive.

INTRODUCTION

The Pillar Mountain slide in Kodiak, Alaska, is a prominent geological and geotechnical feature in the community (Figure 1). The slide has been a site of persistent rockfall hazard for several decades. At the base of the slide, Rezanof Drive traverses the site with piers located downslope of the road. Rezanof Drive is a critical lifeline representing the only road linking the community of Kodiak to the northeast with the airport and U.S. Coast Guard Base to the southwest. Rockfall has previously reached the road and motorists have collided with rockfall blocks. Larger volumes of rockfall material reaching the road could result in highway closure, causing disruptions to the local economy, transportation services, and public safety services. As part of a Highway Safety Improvement Project, the Alaska Department of Transportation and Public Facilities (ADOT&PF) completed a rockfall hazard mitigation project with the primary objectives of reducing rockfall hazards associated with the slide, reducing maintenance efforts associated with rockfall cleanup, and maintaining traffic flow along Rezanof Drive. The rockfall mitigation structure constructed was an innovative, mechanically-stabilized earth (MSE) rockfall berm designed to intercept rockfall and protect travelers and workers along Rezanof Drive.

In this paper, we discuss the steps involved with the selection and design of the rockfall berm. This paper discusses an overview of the setting of the Pillar Mountain slide, including discussion of the geography, bedrock and surficial geology, glacial geology, seismicity, and climate of the Kodiak area. A summary of the background on the history of the Pillar Mountain slide is provided. We then introduce the rockfall mitigation project, discuss the approaches to conducting a geological investigation for the project, and present our approach to modeling rockfall from the slide as a design input. Finally, we present several mitigation options and focus on the rockfall berm that was selected and constructed for the project.

PROJECT SETTING

The community of Kodiak is located on Kodiak Island within the Kodiak Archipelago, a series of islands situated to the southeast of the Alaska Peninsula in southern Alaska (Figure 1). Kodiak Island has mountainous terrain dissected by a series of northeast-trending and northwest-trending valleys. Mountain peaks generally reach elevations ranging from 2,000 to over 4,000 feet in elevation, resulting in steep topographic relief down to sea level. The community of Kodiak is located on the southeast (Pacific Ocean) side of Kodiak Island.

Kodiak Island is composed predominantly of Mesozoic and Cenozoic sedimentary rock intruded by granitic rocks and plutons (Wilson, 2013). The bedrock in the project area is part of the Late Cretaceous Kodiak Formation, composed of arkosic wacke, shale, and pebbly conglomerate. These sedimentary rocks are interpreted as turbidite deposits. The Kodiak Formation represents the Cenozoic accretionary prism of the Aleutian subduction zone. This accretionary prism has been faulted and uplifted along the convergent plate margin. Bedrock is typically mantled by soil deposits including glacial till deposits, silt, and volcanic ash originating from volcanoes along the Aleutian volcanic chain.

As mentioned above, Kodiak Island is located along a convergent plate margin. The Pacific oceanic plate is actively subducting to the northwest beneath the Alaska mainland (North

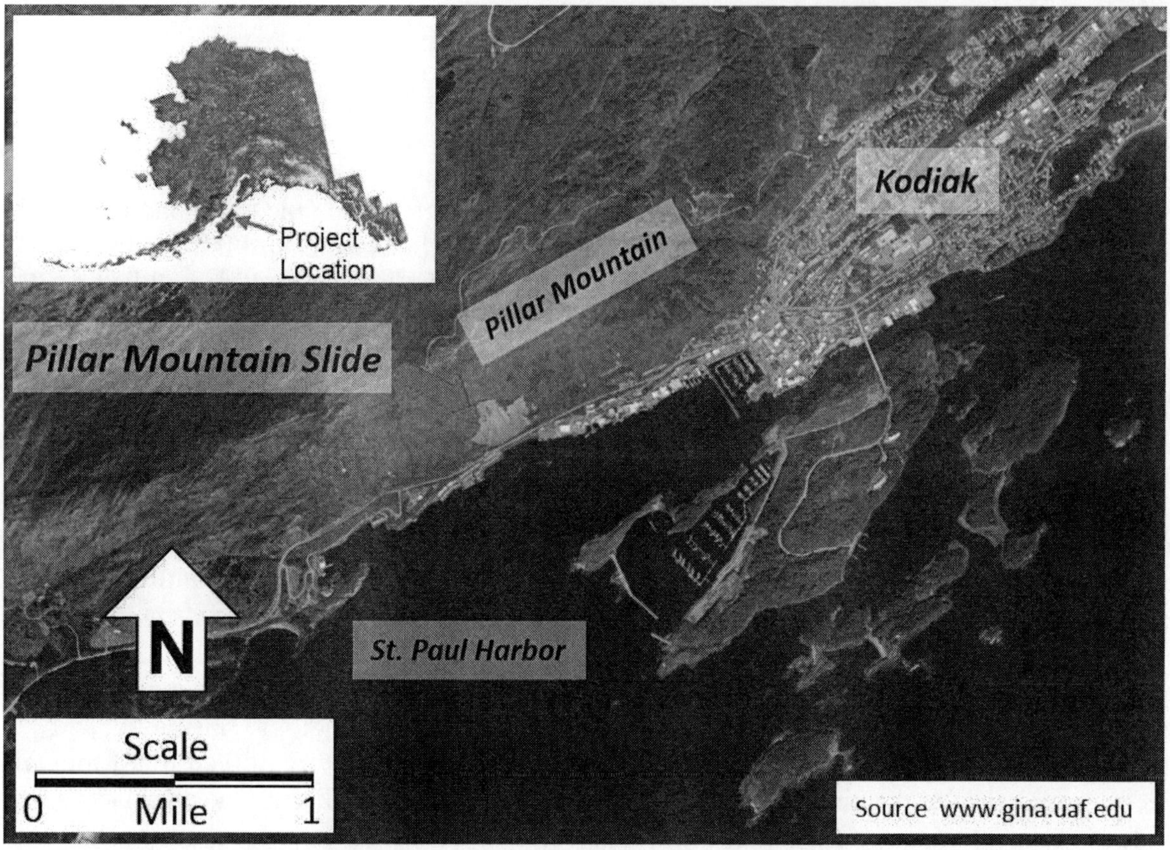

Figure 1 – Location Map of Kodiak, Alaska and the Pillar Mountain Slide

American plate) at a rate between about 2.2 to 2.5 inches/year (Haeussler and Plafker, 2008). The Kodiak Archipelago was located adjacent to the rupture zone of the moment magnitude (M_W) 9.2 Great Alaskan earthquake of 1964 (Plafker and Kachadoorian, 1966). This earthquake resulted in subsidence of Kodiak Island up to a maximum of approximately 6 feet, followed by a series of tsunami waves with runup heights that reached a maximum of approximately 31.5 feet.

In addition to tectonic processes, glacial processes have shaped the topography of Kodiak Island. During the Pleistocene, glacial ice advanced south from the Alaska Range and other mountain ranges in southern Alaska, and flowed south towards the Gulf of Alaska coast. During the Last Glacial Maximum, glacial ice advanced across the Kodiak Archipelago and southeast across the continental shelf (Coulter and others, 1962). The combination of faults and glacial erosion has resulted in a network of valleys and fjords across the archipelago.

Kodiak Island has a maritime climate which supports the temperate rainforest vegetation found on the island. Low pressure weather systems in the Gulf of Alaska generate storm systems that impact the island. Table 1 provides summaries of average annual climate statistics for the Kodiak vicinity (WRCC, 2017). The average annual minimum and maximum temperatures are approximately in the mid-30s °F and mid-40s °F, respectively. Rainfall exceeds over 70 inches a year. Snowfall can be greater than 60-70 inches a year.

Table 1 – Annual Climate Data

Climate Station	Climate Station Number	Average Annual Maximum Temperature (°F)	Average Annual Minimum Temperature (°F)	Average Total Precipitation (inches)	Average Total Snowfall (inches)	Period of Record
Kodiak Airport	504988	46.7	35.5	77.04	76.1	1/1/1931 to 6/9/2016
Kodiak WWTP	504991	44.9	35.5	72.45	65.9	8/1/2005 to 5/3/2016

PILLAR MOUNTAIN SLIDE

The Pillar Mountain slide is located on the southeastern flank of Pillar Mountain (Figure 2). Pillar Mountain is a northeast-trending mountain ridge that rises from sea level to approximately 1,200 feet in elevation, with a broad, elongated summit located along the ridgeline. The main scarp of the slide forms a complex trace rather than the more typical single, curved scarp. The crown reaches a maximum elevation of about 760 feet. The main body of the slide is a talus slope that extends down to about 20 feet elevation at the toe (Figure 3). The main talus has a slope of approximately 37°. Due to the stepped nature of the main scarp, secondary talus cones are present on the flanks of the slide. Overall the maximum width of the slide is about 1,400 feet.

The sources of the talus slopes are bedrock cliffs located in the upper reaches of the slide. Bedrock consists of argillite with bedding that dips at moderate angles (e.g., 45°) in the westerly direction into the slope. A subvertical joint set dips to the north to northeast. A discontinuity set of spaced cleavage dips at shallow angles (e.g., 20°) to the southwest out of the slope. The surface of the talus slopes are generally composed of gravel- to cobble-sized, tabular fragments of argillite.

Rockfall at the Pillar Mountain slide has been a persistent problem for decades. Prior to about 1957 to 1958, Rezanof Drive traversed the flank of Pillar Mountain across the lower portion of the slope at an elevation ranging from 120 to 160 feet elevation (Kachadoorian and Slater, 1978). In part due to rockfall problems, the road was relocated to the toe of the slide. This old road can be seen on aerial imagery by the contrast in vegetation (Figure 2).

On December 5, 1971, the landslide was reactivated during excavation of material to be used for fill at a Kodiak city dock (Kachadoorian and Slater, 1978). Fill was generated by quarrying rock at the level of the abandoned highway, pushing the material downslope, then excavating the material at the toe of the slide. The total volume of material excavated prior the slide reactivating was approximately 300,000 cubic yards. The sequence of events that occurred during the reactivation of the slide suggest that initial rockfall occurred upslope of the abandoned highway at about 300 feet elevation and progressed upslope to an elevation of about 700 feet

Figure 2 – Pillar Mountain Slide

over four days. Ground fractures upslope of the crown were then observed above the slide later in December 1971.

To mitigate the rockfall hazard along the highway, gabions in up to four tiers were installed along the north side of the highway after 1973 as a protective berm to intercept rockfall. In the intervening decades, the catchment between the toe of the slide and the gabions filled with rockfall material, lessening the effectiveness of the gabions to stop blocks from falling and sliding onto the road.

ROCKFALL MITIGATION PROJECT

In 2014, ADOT&PF pursued a Highway Safety Improvement Project (HSIP) to reduce rockfall hazard along Rezanof Drive. Rockfall blocks had landed on the road resulting in vehicular accidents. The primary objectives of the HSIP included reducing rockfall hazards associated with the slide, reducing maintenance efforts associated rockfall cleanup, and maintaining traffic flow along Rezanof Drive.

Understanding the slope characteristics and failure mechanisms are important first steps in mitigating rockfall hazards. In the 1970s and 1980s, investigations suggested a deep-seated, circular failure mechanism for the slide (Kachadoorian and Slater, 1978; R&M, 1982). Upon further investigation in the 2000s including high-resolution Light Detection and Ranging (LiDAR) data acquired by the city of Kodiak in 2013 (Figure 4), flexural toppling was identified as an alternate failure

Figure 3 – View Northeast to Left (East) Flank of Pillar Mountain Slide. Rezanof Drive at Base of Slide (Right Side of Image).

mechanism (Schlotfeldt et al. 2014). In this process, the argillite beds that dip moderately into the slope overturn and topple down the slope. As the buttressing effect is removed downslope by slope instability, the toppling mechanism migrates further upslope. The flexural toppling model allows argillite bedding to topple through tensile failure of the beds. The resulting blocks are then transported down the talus slope via tumbling or sliding mechanisms.

GEOLOGICAL INVESTIGATION

To support rockfall modelling efforts, fieldwork was conducted in November 2014. The objectives of this field effort included interpretation of LiDAR data, conducting a ground reconnaissance of the slide, and investigating the positions and dimensions of rockfall blocks that had been transported down the slide.

The LiDAR dataset acquired by Kodiak Mapping Inc. in 2013 for the city of Kodiak provided a valuable dataset to further define the morphology of slide (Figure 4). In comparison to aerial imagery (Figure 2), the LiDAR image has greater contrast to define the boundaries between the crown, scarps, and talus slopes. The eastern main scarp appears to be primarily

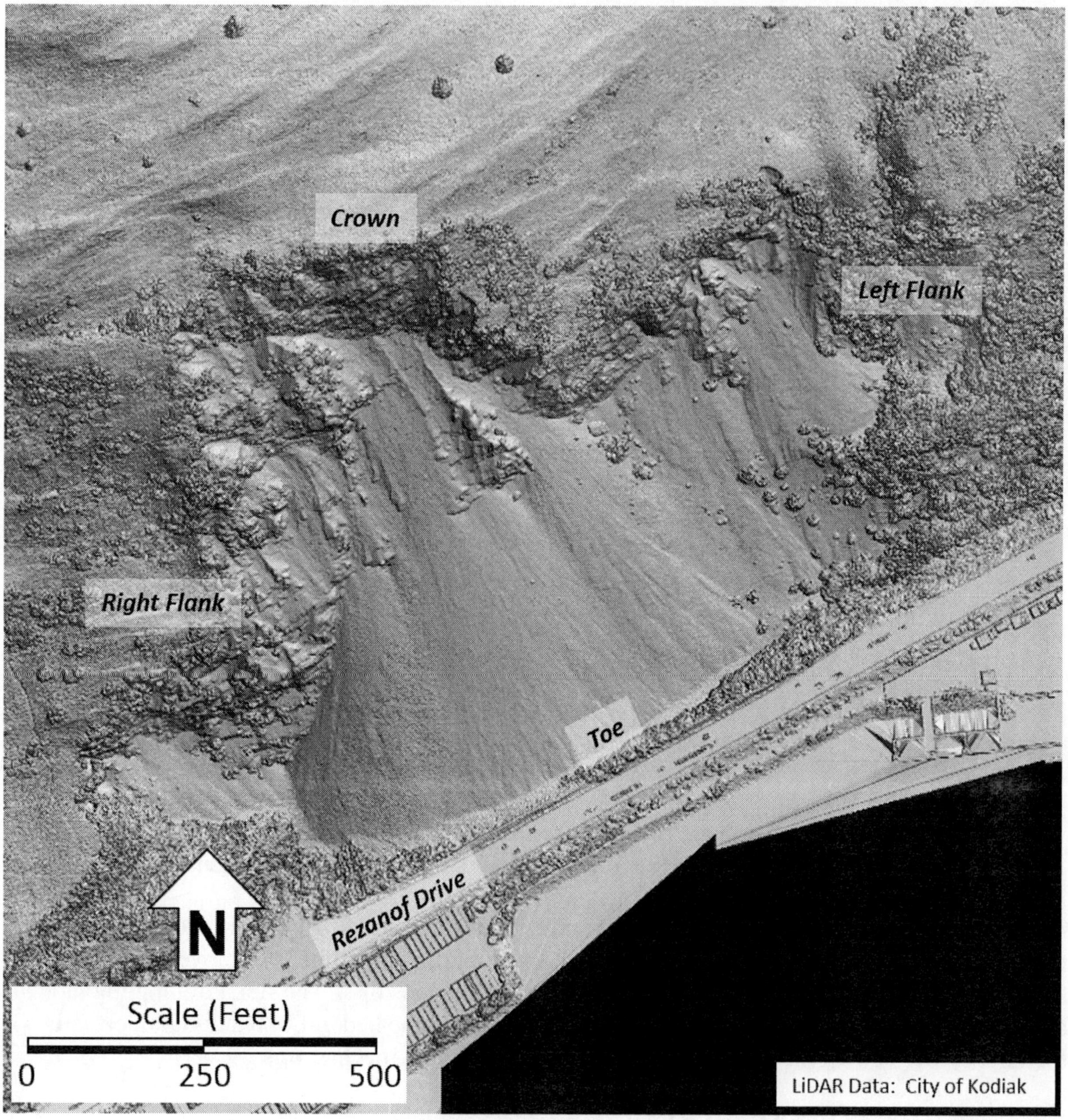

Figure 4 – LiDAR image of Pillar Mountain Slide

cliffs, whereas the western main scarp appears to be composed a sequence of cliff bands and talus slopes. The LiDAR also shows textural evidence that material on the main talus slope is derived from three regions: eastern, central, and western sections of the main scarp.

The slide was circumnavigated as part of a ground reconnaissance to further evaluate the failure mechanisms and source material. The field team investigated bedrock outcrops to acquire rock mass and discontinuity data. Tension cracks were also observed. On the west flank, a tension crack approximately 6 feet wide was encountered. At the crown of the slide, a tension crack at one of the main scarps was approximately 10 to 15 feet wide, and 5 to 10 feet deep.

At the base of the slide, the locations, shape, and dimensions of rockfall blocks were noted. Some blocks were located on or downslope of the gabions (Figure 5). Some gabions had damaged wire presumably from rockfall impacts. On the downslope (southeast) side of the road, there was likely evidence of rockfall damage to the guard rail, and blocks downslope of the guardrail were noted. It is possible that damage to the guardrail may have occurred by mechanisms other than rockfall, and that blocks found downslope of the guardrail may have originally landed in the road but were then moved off the road.

Figure 5 – Example of Rockfall Block Downslope of Gabions at Base of Pillar Mountain Slide (Rock Hammer Approximately 1 Foot Long for Scale)

The data collected from the geological investigation regarding the slide morphology, and block dimensions and locations, were used as a basis for preparing rockfall runout analysis.

ROCKFALL ANALYSIS

Statistical rockfall runout analyses were used to assist in the design of hazard mitigation schemes. The software program RocFall v4 by Rocscience provided a basic understanding of falling block trajectories down the slope, potential bounce heights, impact energies, and likely retention rates provided by design alternatives. A 2D analysis was considered sufficient for this problem considering the relatively uniform nature of the talus slope along the base of the slide.

Initially, a back analysis of observed rockfall conditions was used to calibrate appropriate slope parameters controlling runout. These parameters were then applied to predictive models comparing rockfall mitigation features including width, height, location relative to the highway, and upslope catchment shape.

Only a single section through the center of rock outcrop and talus slope was considered necessary to characterize rockfall (Figure 6). The section used was considered to be the most conservative in terms of producing rockfall with the highest potential for impacting the highway. The selected section intersected the steepest area of exposed rock in the main scarp, had the longest runout distance along the talus, transected an area of the gabion berm where the most amount of material had accumulated behind the berm, and was located where the most blocks were observed to have escaped over the berm onto the highway.

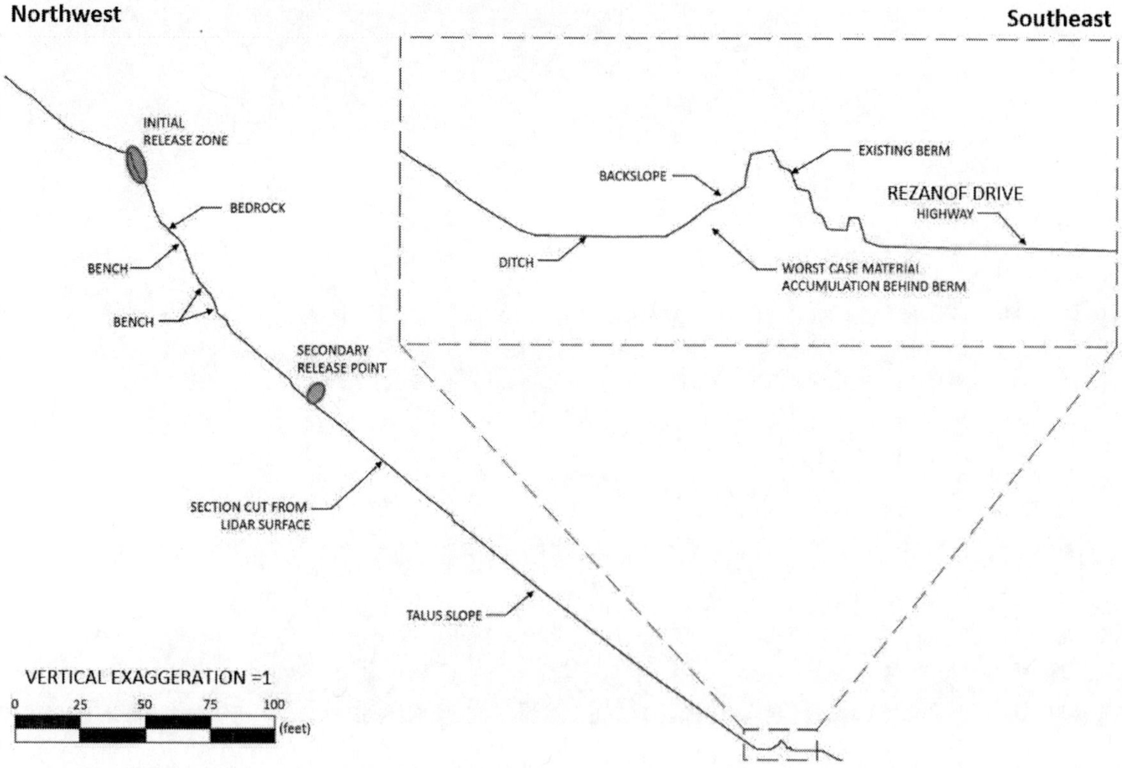

Figure 6 – Rockfall Modeling Section and Analysis Setup for Calibration

In an effort to characterize typical rockfall affecting the highway, only blocks with considerable size that had come to rest in a position that could have posed a risk to the public were considered (e.g., Figure 5). The number of blocks and dimensions were recorded and averaged. The result was three categories of blocks—typical, large, and worst case—with mean volumes of 2.3 ft^3, 18 ft^3, and 54 ft^3, respectively, for the block categories. All three categories were used in the back analysis to characterize modeling parameters. In addition, by visually comparing photographs of the talus slope, it was estimated that 85% of the slope was comparable or smaller than the defined typical block size, 13% could be considered large, and only 2%

would be characterized as worst case. These observations were incorporated into the predictive models.

Field observations suggested that the blocks most likely to topple from the crown of the slope were comparable in size to the worst case blocks, but subsequently fracture before deposition at the base of the slope. Therefore, the rockfall modelling was performed in two steps. Initially, large blocks were released from the main scarp, and observations of translational and rotational velocities were made at the point labeled as Secondary Release Point (Figure 6). The second step used these velocities as initial conditions for smaller blocks; and energy, velocity and bounce height observations were made at the toe of the slope. The decision to maintain the translational and rotational velocities built conservatism into the analysis, as it is likely that energy would be dissipated through the fracturing process.

The final resting points of the blocks observed in the field were used to calibrate the slope parameters. Normal (R_N) and tangential (R_T) restitution values, as well as the normal restitution scaling factor (K), friction angle (f), and slope roughness (r_o) were estimated for the upper bedrock and talus below. Recent studies by Wyllie (2013) were used as guidelines for selecting values for initial normal restitution. Values for R_N were based on average incident angle observed in preliminary model runs (Wyllie, 2013), while the other parameters were varied slightly until appropriate deposition behavior observed in the field was simulated for the different sized blocks.

The predictive models were focused on helping dimension the berm structure. In addition to the rockfall trajectories, the top width of the berm was sufficient for a medium-sized excavator to traverse and clean out debris from the upslope side. The vertical height of the upslope face was set by the maximum reach of the excavator. Sloped and vertical faces upslope were investigated for a series of scenarios representing various stages of debris accumulation. Each model included a range of block sizes the distribution of which was based on the percentages described above. The results suggested that a berm with the upslope catchment approximately 5 feet lower than the road height and the face 15 feet high could retain nearly all blocks behind or on top of the berm regardless of the face orientation. Under the worst case debris accumulation scenarios (no regular maintenance), a near 90% retention rate could still be expected which met the ADOT&PF design criteria for catchment.

ROCKFALL BERM

With the rockfall analysis completed, the focus moved to developing and reviewing several berm mitigation options. The base case preferred by ADOT&PF was a reinforced berm with either gabion facing or wire facing, due to ease of construction. The base case was similar in height and width to the berm used in the rockfall modeling. Other mitigation options explored included a flexible rockfall protection fence, dampening wall structure (high-density foam faced to absorb rockfall impact), and a rockfall shed. In addition, a non-structural option was considered: flattening the main scarps to remove the rockfall sources.

In considering the benefit and drawbacks to each of these options, constructability, cost, and maintenance and operations factors were considered. The flexible rockfall protection fence would require specialized installation and repair efforts. The dampened wall structure and rockfall shed would require significant construction efforts. Excavation of the main scarps would require considerable effort to remove material from the upper slopes. In comparing and contrasting each of these potential solutions, the reinforced berm option was ultimately selected by ADOT&PF to replace the gabions based on cost and ease of construction. The berm was constructed and completed in 2016.

The final reinforced berm selected incorporated innovative design concepts (Figure 7). The berm is faced on both sides with interlocking concrete blocks. The MSE wall geogrid is sandwiched between each level of concrete blocks. Overall, the berm is approximately 870 feet long, 20 feet wide, and at least 15 feet high on the upslope side of the berm. At each end of the berm, the berm is tapered to form ramps to allow access the top of the berm. The ramps allow a medium-sized excavator to travel along the top of the berm.

Figure 7 – Completed Rockfall Berm (Photo Courtesy of Howard Weston)

During periodic maintenance operations to clean out the rockfall material that accumulates between the toe of the slide and the berm, the equipment operator can excavate the talus from the catchment and then swing 180° to place the talus in a dump truck parked along the highway. By positioning the excavator on top of the berm, this allows the operator and the excavator to avoid being located in the catchment below rockfall paths. The concrete facing protects the catchment berm from being damaged by the excavator bucket.

In addition, the catchment can act as a temporary holding basin for rockfall debris, for example triggered during freeze-thaw or precipitation events. The catchment can then be cleaned out after the triggering event passes, to provide improved safety to the equipment operator, rather than cleaning out the catchment during the triggering event which may subject the equipment operator to greater risks.

SUMMARY & CONCLUSIONs

The Pillar Mountain slide in Kodiak, Alaska has been a site of persistent rockfall hazard for several decades. On December 5, 1971, the landslide was reactivated as a result of excavating material for fill. In 2014, the Alaska Department of Transportation & Public Facilities (ADOT&PF) pursued a Highway Safety Improvement Project (HSIP) to reduce rockfall hazard along Rezanof Drive at the toe of the slide. A geologic investigation and rockfall modeling were conducted to support the selection of a rockfall mitigation solution. Mitigation options explored included a reinforced berm with either gabion facing or wire facing, a flexible rockfall protection fence, dampening wall structure (high-density foam faced to absorb rockfall impact), rockfall shed, and excavation of the main scarps to remove the rockfall sources. Statistical rockfall runout analyses using the software program RocFall v4 by Rocscience were used to understand the falling block trajectories down the slope, potential bounce heights, impact energies, and likely retention rates provided by design alternatives.

After consideration of the benefits and drawbacks of the different mitigation options, the reinforced berm option (the preferred option) with concrete block facing was selected by ADOT&PF. Construction of the reinforced berm was completed in 2016. The reinforced berm has innovative features including the ability for an excavator operator to work from the top of the berm to clean out the catchment between the toe of the slide and the berm, rather than working in the catchment and below rockfall paths. The catchment also provides a temporary holding basin for rockfall debris such that the catchment can be cleaned out after a triggering event passes, to reduce risks to the equipment operator. The combination of reviewing the historic slide activity, conducting the geologic field investigation, and completing rockfall runout analyses, were necessary in the design process for the reinforced berm, to address the persistent rockfall hazards associated with the complex geometry and geologic characteristics of the Pillar Mountain slide.

REFERENCES

Coulter, H.W., Hopkins, D.M., Karlstrom, T.N.V., Péwé, T.L., Wahrhaftig, C., and Williams, J.R. *Map Showing Extent of Glaciations in Alaska*: U.S. Geological Survey Miscellaneous Geologic Investigations Map 415, 1962.

Haeussler, P.J. and G. Plafker. *Earthquakes in Alaska*: U.S. Geological Survey Open-File Report 95-624, 2004.

Kachadoorian, R. and W.H. Slater. *Pillar Mountain Landslide, Kodiak, Alaska*: U.S. Geological Survey Open-File Report 78-217, 1978.

Plafker, G. and R. Kachadoorian. *Geologic Effects of the March 1964 Earthquake and Associated Seismic Sea Waves on Kodiak and Nearby Islands, Alaska*: U.S. Geological Survey Professional Paper 543, 1966.

R&M. *Pillar Mountain Slope Stability Study, Final Project Report, prepared for the City of Kodiak*: R&M Consultants, Anchorage, Alaska, July 1982.

Schlotfeldt, P., Tart Jr., R., Panton, B., and Dugan, R.G. *Pillar Mountain Landslide: Failure Mechanism Update:* American Rock Mechanics Association, 48th US Rock Mechanics/Geomechanics Symposium, ARMA 14-147, 1-4 June 2014.

Wilson, F. H. *Reconnaissance Geologic Map of Kodiak Island and Adjacent Islands, Alaska*: U.S. Geological Survey Scientific Investigations Map 2999, 2013.

Wyllie, D.C. *Calibration of Rock Fall Modeling Parameters*: International Journal of Rock Mechanics and Mining Sciences, 67, p. 170-180, 2013.

Rockfall Forecasting:

A Probabilistic Approach

Michael F. George, Ph.D., P.E.
BGC Engineering, Inc.
710 10th St.
Golden, CO 80401
720-598-5982
mgeorge@bgcengineering.ca

Cole M. Christiansen, M.Sc., E.I.T.
BGC Engineering, Inc.
710 10th St.
Golden, CO 80401
720-598-5982
cchristiansen@bgcengineering.ca

Ryan Kromer, Ph.D.(c).
Dept. of Geological Sciences and Geological Engineering
Queen's University
Kingston, OT
ryan.kromer@queensu.ca

Jean Hutchinson, Ph.D., P.Eng.
Dept. of Geological Sciences and Geological Engineering
Queen's University
Kingston, OT
hutchinj@queensu.ca

Prepared for the 68th Highway Geology Symposium, May, 2017

Acknowledgements

The author(s) would like to thank the individuals for their contributions in developing the broader framework for rockfall forecasting presented in this paper:

Dr. Dave Gauthier – BGC Engineering, Inc.
Mr. Alex Strouth – BGC Engineering, Inc.

Disclaimer

Statements and views presented in this paper are strictly those of the author(s), and do not necessarily reflect positions held by their affiliations, the Highway Geology Symposium (HGS), or others acknowledged above. The mention of trade names for commercial products does not imply the approval or endorsement by HGS.

Copyright Notice

ABSTRACT

Rockfall along transportation corridors presents a significant danger to passing motorists in addition to potentially large capital cost associated with maintenance and road closures. Existing rockfall hazard rating systems, implemented by numerous state transportation agencies across the United States, serve to identify key slopes most prone to rockfall but do not provide any information regarding specific rocks that may fall or the timing of rockfall events.

High resolution remote sensing technologies (such as LiDAR, photogrammetry, and GB-InSAR), however, have permitted detailed identification of problematic blocks and their associated displacements leading up to failure. Such information has facilitated an ability to forecast the time to failure of specific rocks through repeated measurements at localized sites. Key drawbacks, however, are the high degree of uncertainty associated with these forecasts and the limited frequency at which measurements can practically be made. This ultimately undermines the ability to provide meaningful information that could be used, for example, to shut down a corridor precursory to a rockfall event.

This paper presents a rockfall forecasting methodology cast within a probabilistic framework to assess probability bounds on the time to failure as well as update time to failure estimates as new information is gathered. This work is being conducted within a broader framework to incorporate remote sensing technology for effective and efficient characterization and prioritization of numerous sites on a network scale spanning several hundreds of miles.

INTRODUCTION

Rockfall along transportation corridors presents a significant danger to passing motorists in addition to potentially large capital cost associated with maintenance and road closures. Existing rockfall hazard rating systems, implemented by numerous state transportation agencies across the United States, serve to identify key slopes most prone to rockfall but do not provide any information regarding specific rocks that may fall or the timing of rockfall events. Accordingly, significant benefit can be achieved through reliable forecasting of future rockfall events.

For this paper, probabilistic forecasts for block time to failure are estimated using the inverse velocity method (Fukuzono 1985) along with Bayesian linear regression to assess parameter uncertainty arising from the inherent variability of the rockfall process and due to measurement noise. High resolution remote sensing technologies are used to efficiently identify problematic blocks and monitor their displacements over time. As new observations of block displacement are gathered, model parameters and forecast predictions are updated to reflect the current state of knowledge in the system. An example analysis is also presented, but first a discussion of the broader rockfall forecasting framework is provided below.

ROCKFALL FORECASTING FRAMEWORK

A framework for evaluation of rockfall hazard across broad spatial scales has been presented by Gauthier et al. (2017) (Figure 1). The first phase consists of searching for problematic sites to focus monitoring/remediation efforts given that it is impractical (and even undesirable) to monitor every rockfall at every location over a large region. At the highest level, the "network" scale, spanning several hundreds of miles (e.g., statewide), high rockfall risk corridors are identified and prioritized based on historic performance, geology, topography, presence of high consequence infrastructure, etc. At the "corridor" scale, spanning several tens of miles (e.g., along a highway section), historical observations combined with remote sensing technology (at a lower resolution) is utilized to rapidly identify problematic sites with moving and/or missing blocks. Finally, at the "site" scale, typically spanning less than a mile (e.g., across a rock slope), high resolution remote sensing methods are used to repeatedly monitor the slope face to detect moving blocks and track their displacement through time or identify blocks that have fallen in between monitoring intervals.

For specific sites that have been identified, the second phase consists of taking actions to access potential consequences associated with the rockfall hazard, forecasting future rockfall events and implementing response measures to mitigate or reduce rockfall risk (if any are required). The nature of forecasts for future rockfall events is dependent on the ability to capture block movements precursory to failure as for some rock slopes this may not be feasible (see discussion below). Furthermore, application of the failure forecast method requires the identification of the accelerating phase of failure. Accordingly, two distinct tracks within the "action" phase of the framework are outlined. The focus of this paper is on forecasts for blocks with detectable displacements and with an accelerating period leading up to failure (highlighted within the red-dashed section, Figure 1).

Finally, the third phase consists of learning, reviewing and updating methods for prioritization, detection, monitoring, forecasting and response. This occurs continually throughout the framework such that as new information is gathered, these methods are updated to improve future rockfall assessment.

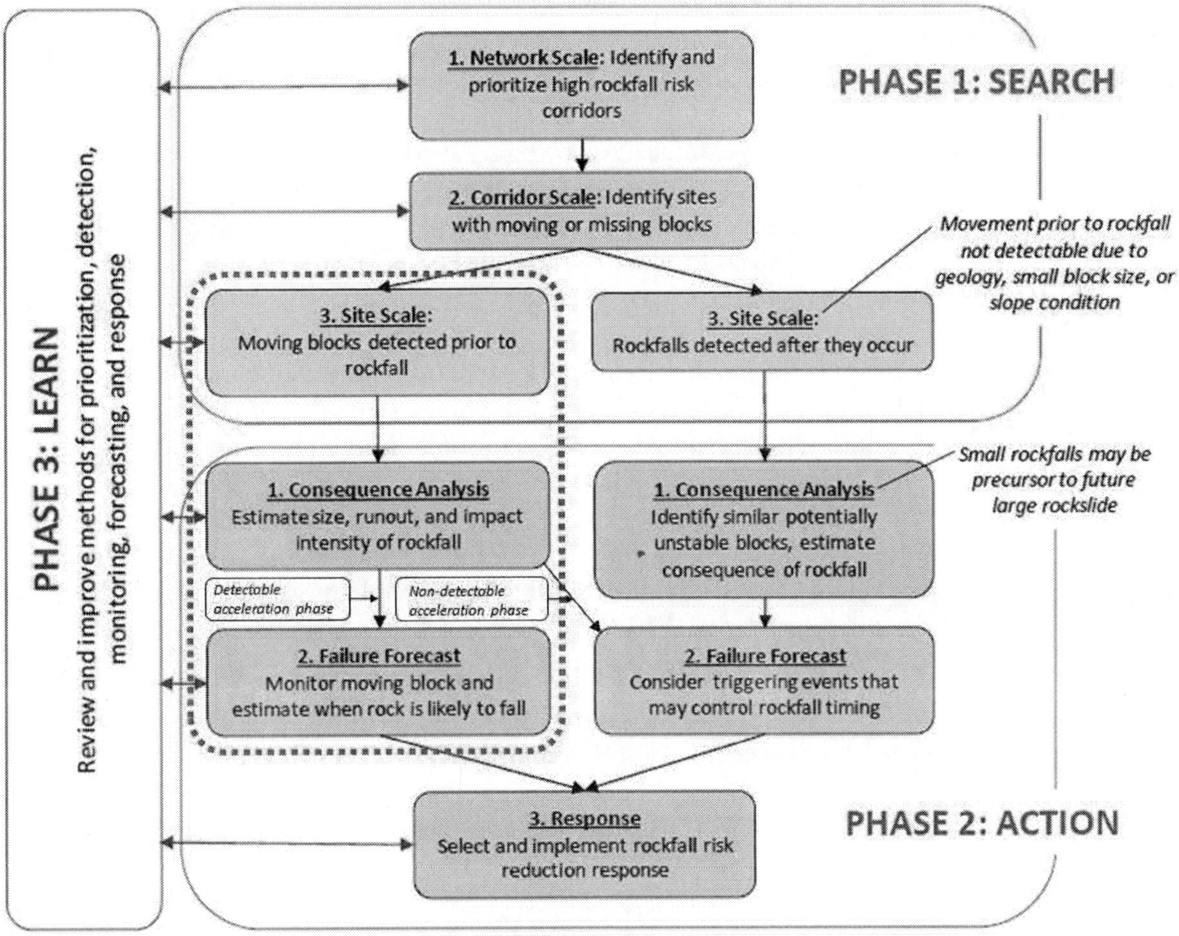

Figure 1 – Framework for rockfall forecasting (adapted from Gauthier et al. 2016)

BLOCK IDENTIFICATION & MONITORING

Block Kinematics

As mentioned above, two rock block classifications exist with regard to identification and monitoring for rockfall forecasting: 1) blocks with detectable change and acceleration precursory to failure, and 2) blocks without detectable change precursory to failure or blocks with detectable change but without detectable acceleration precursory to failure. For the purposes of the methodology presented herein, our principal interest is the former. A number of factors can potentially influence this, one being block kinematics.

It is widely known within the rock mechanics community that the 3D orientations of discontinuities defining block boundaries have a significant influence on block removability and stability. Accordingly, for removable blocks (i.e., blocks that are physically capable of moving from the rock mass into an open space) several kinematic failure modes exist. These consist of lifting (moving away from all discontinuity planes), sliding (on 1 or 2 planes), rotation (about a corner, an edge or an arbitrary point) or some combination of sliding and rotation (e.g., slumping, torsional sliding) (Goodman 1995).

From a rockfall monitoring perspective, recent research by Rowe et al. (2016) related block kinematic failure mode to detectable deformation prior to failure. In general, blocks with translational sliding modes experience the most deformation leading up to failure, followed by blocks with rotational modes. Intuitively, blocks failing in the lifting mode (e.g., overhanging blocks) provide little to no detectable change. Accordingly, certain rock slopes in certain geologic settings yield blocks with failure modes more suitable to forecasting methods presented in this paper. Without any prior knowledge of block kinematic modes for a given rock slope, a block theory framework can be utilized to identify removable block types and assess likely failure modes (Goodman & Shi 1985).

Another consideration is for blocks where the kinematic failure mode may change in response to external loading beyond the self-weight of the block (e.g., due to pore pressure). Figure 2 shows the path of the active resultant force vector on a limit equilibrium stereonet for a block subject to increasing hydraulic pressure on block faces for various scenarios. On the stereonet, the various kinematic modes are shown. Initially (under gravity loading only), the applicable kinematic failure mode is 2-plane sliding on Joints 4a and 4b. As hydraulic pressure is increased, the resultant vector rotates outwards until the applicable mode is 1-plane sliding on Joint 4a. Blocks of this nature may prove difficult to forecast with methods presented herein.

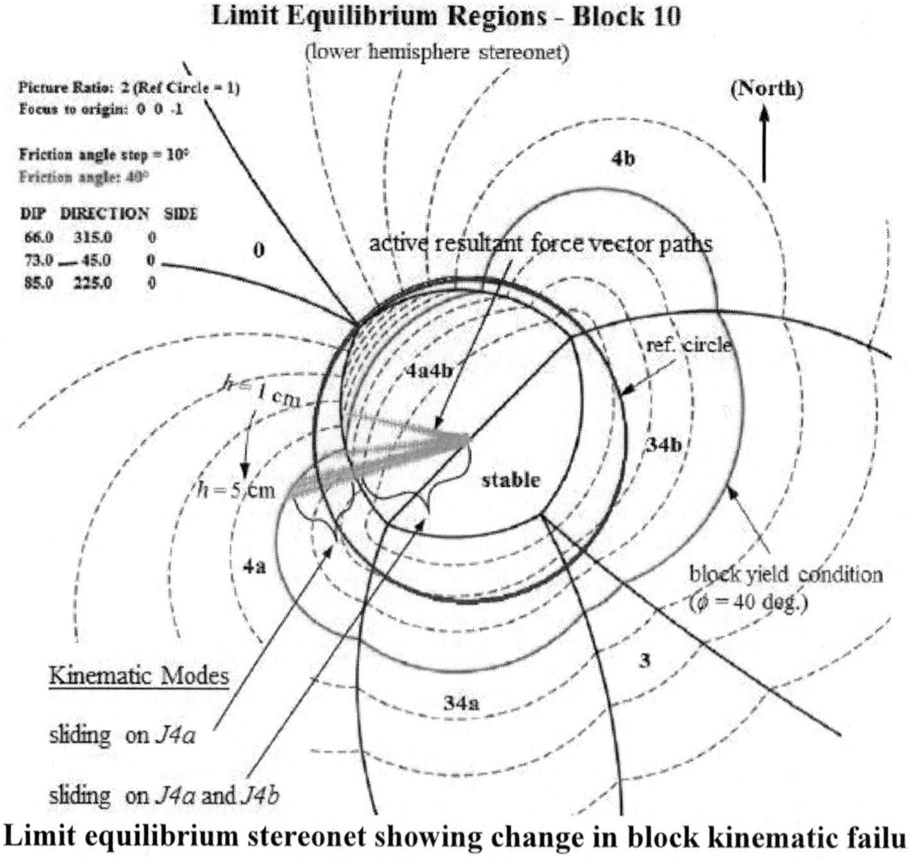

Figure 2 – Limit equilibrium stereonet showing change in block kinematic failure mode in response to external loading (adapted from George 2015)

Remote Sensing and Change Detection

Where traditional slope movement monitoring techniques (such as extensometers or survey prisms) are not practicable at inaccessible or hazardous sites, high resolution remote sensing technologies (such as LiDAR, photogrammetry, and GB-InSAR) have permitted detailed identification of problematic blocks and their associated displacements leading up to failure.

Repeated slope scans provide a collection of unique snapshots in time. For each scan, a digital elevation model is created and compared to the previous. Models differences are interpreted as geomorphological processes such as rockfall, creep-type displacement precursory to rockfall, debris slides, or spurious change (e.g., change in vegetation).

An example of rockfall monitoring using oblique aerial photogrammetry (OAP) is presented in Figure 3 and 4. Oblique aerial photographs were collected from a moving helicopter on five occasions over a three-month period. From each OAP survey, 3D models were generated and 3D quantitative change detection was performed to determine model differences as small as 5 cm (Christiansen et al. 2016). Analysis showed several small rockfalls and debris slides occurred over the course of the study. Additionally, a large block composed of at least three independent and detached blocks, ranging from 10 m^3 to 100 m^3, underwent downslope displacements from 5 to 15 cm (Figure 4).

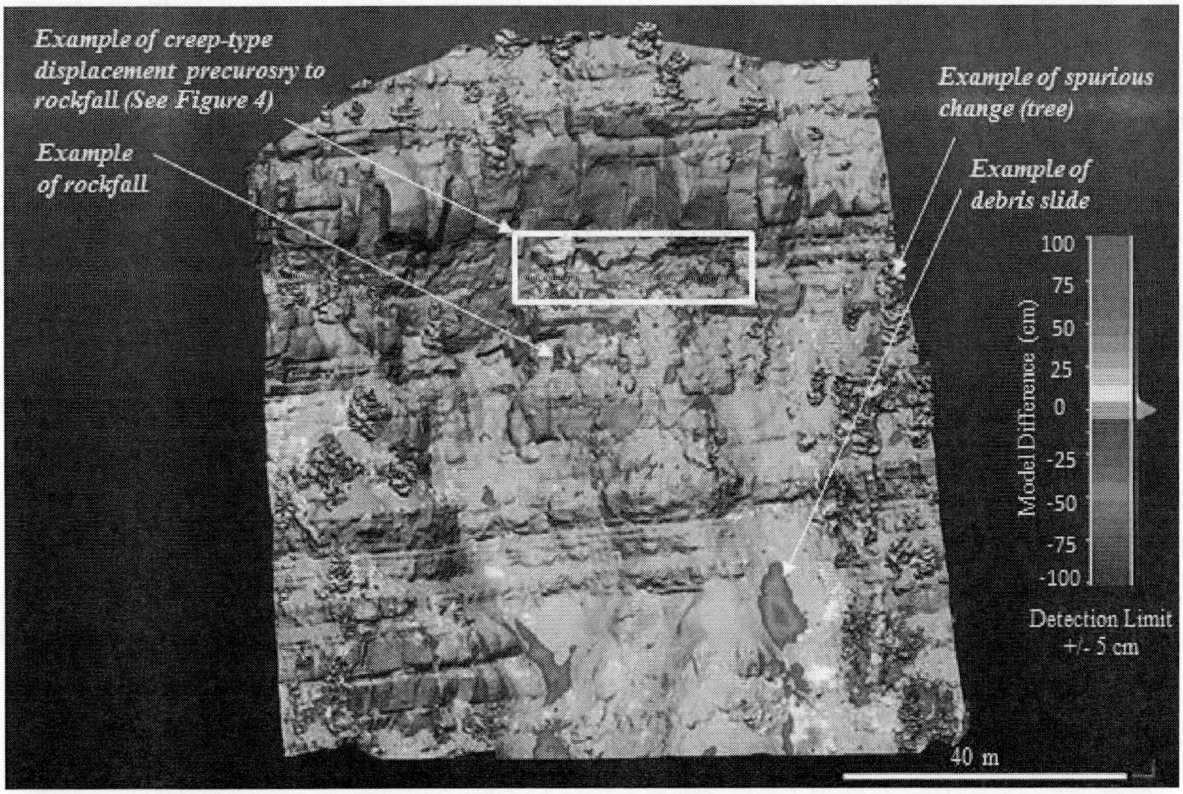

Figure 3 – OAP change detection map and examples of rockfall, precursory rockfall displacement, debris slide, and spurious change (adapted from Christiansen et al. 2016).

Blocks with measurable creep-type displacements, as shown in Figure 4, are ideal for use with the inverse-velocity method of rockfall forecasting (as presented below) should an accelerating period prior to failure also be detectable. The monitoring frequency needs to be adjusted to appropriately capture the onset of the accelerating phase and to provide sufficient number of measurements during the acceleration phase to obtain reliable forecast results.

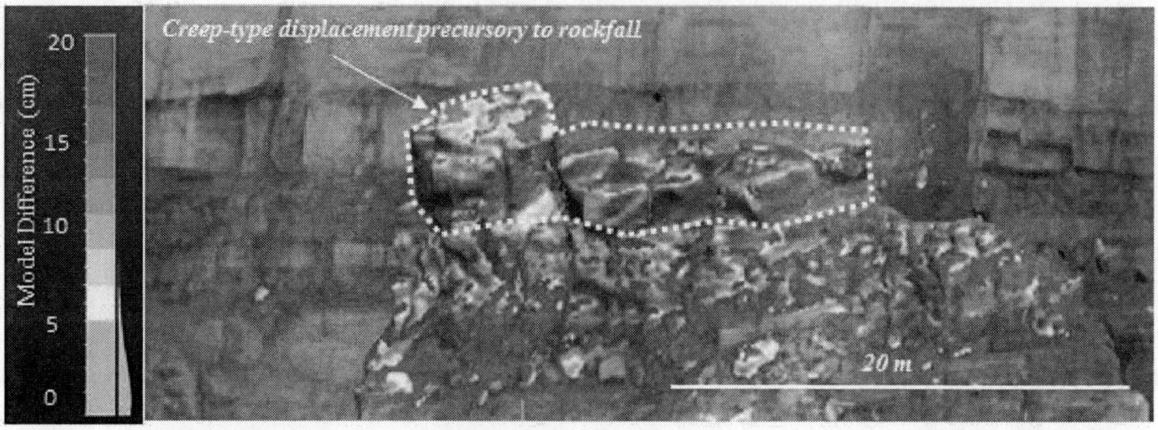

Figure 4 – Inset from Figure 3. OAP change detection map showing model difference between 5 and 15 cm (adapted from Christiansen et al. 2016)

PROBABILISTIC APPROACH FOR FORECASTING TIME TO FAILURE

For blocks identified with detectable displacement and an onset of accelerating creep, a forecast of the time to failure of the block can be obtained using the inverse velocity method (Fukuzono 1985). The basic concept is that measurements of block displacement over time (obtained from any number of means, e.g., remote sensing, GPS, survey prisms, extensometers, etc.) are converted to a rate (velocity), with the assumption that failure is preceded by increasing rates of displacement (acceleration). Accordingly, as the velocity approaching failure becomes large, the inverse of velocity approaches zero. Projecting the inverse velocity to zero with respect to time allows for an estimate of time to failure (Figure 5). In reality, the inverse velocity never reaches zero. Should actual velocity thresholds for block types be known based on experience, those limits can be used instead.

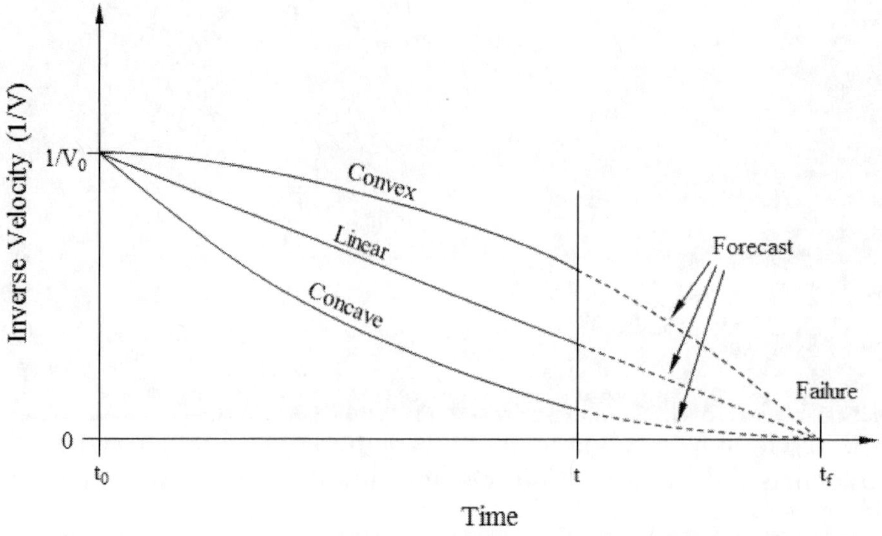

Figure 5 – Schematic for inverse velocity versus time plot leading up to failure (adapted from Fukuzono 1985).

Fukuzono proposed three potential trends for inverse velocity time series data (convex, linear and concave), but concluded a linear trend provided a reasonable time to failure estimate, particularly in the period shortly before failure. A linear trend has also been shown to be theoretically applicable for accelerating creep scenarios for a variety of materials including rock under constant stress conditions (Voight 1988, 1989, Cornelius & Scott 1993 and Kilburn & Petley 2003) as well as for actual field predictions (e.g., Rose & Hungr 2007, Carla et al. 2016). Accordingly, a linear model is adopted herein, although extension of the presented methodology is feasible to the other scenarios. For completeness, the linear model is presented below:

$$\frac{1}{V}=\frac{1}{V_0}+A\cdot(t-t_0) \tag{1}$$

where V = block velocity at time t, V_0 is the initial measured block velocity at the time t_0 , and A is the slope of the linear trend. Rearranging to solve for t given $t_0 = 0$ yields:

$$t=\frac{1}{A}\cdot\left(\frac{1}{V}-\frac{1}{V_0}\right) \tag{2}$$

Forecasts for the time to failure can be difficult given uncertainty of displacement/velocity measurements due to the inherent variability associated with the rockfall process, measurement noise, and sampling frequency. To overcome this, researchers have applied various filter techniques (e.g., Rose & Hungr 2007, Dick et al. 2015, Carla et al. 2016) to improve deterministic estimates of time to failure.

In this paper we present an alternative probabilistic approach to incorporate this uncertainty into the rockfall forecast as well to update forecast estimates as new information (measurements) are made. To do so, a Bayesian linear regression model is implemented, as in Boué et al. (2015). The Bayesian approach considers any uncertain quantity (variable or parameter) as a random variable with a corresponding probability distribution that reflects the likelihoods of its various possible outcomes (Der Kiureghian 2009). This model takes on the form:

$$y_i=\theta_1+\theta_2\cdot x_i+e_i \tag{3}$$

where y_i and x_i are the ith observations of random variables, θ_1 and θ_2 are unknown parameters, and e_i is the associated error assumed to be independent and normally distributed, $N(0, \theta_3^2)$ where θ_3 is the unknown standard deviation. The Bayesian updating rule is used to improve knowledge of the uncertain quantities given our prior knowledge and any new observations that occur. This is stated as:

$$f'(\boldsymbol{\theta}|\mathbf{x},\mathbf{y})\propto L(\mathbf{x},\mathbf{y}|\boldsymbol{\theta})\cdot f(\boldsymbol{\theta}) \tag{4}$$

where f is the prior probability distribution containing all existing knowledge of parameters $\boldsymbol{\theta}$ (i.e., θ_1, θ_2,...), L is the likelihood function relating the probability of variables $\mathbf{x}$ (x_1, x_2,...,

$x_i,\ldots x_n$) and **y** ($y_1, y_2,\ldots, y_i,\ldots y_n$) given parameters $\boldsymbol{\theta}$, and f' is the posterior probability density distribution representing our updated knowledge of parameters $\boldsymbol{\theta}$ given **x**, **y**. Ultimately, given updated knowledge of $\boldsymbol{\theta}$, it is desirable to make a future prediction for a value of y given some value of x. The corresponding posterior predictive probability distribution (p') for y is expressed as:

$$p'(\mathbf{y}|\mathbf{x}) = \int p(\mathbf{y}|\mathbf{x},\boldsymbol{\theta}) \cdot f'(\boldsymbol{\theta}|\mathbf{x},\mathbf{y})\, d\boldsymbol{\theta} \tag{5}$$

where p is the prior predictive distribution conditional on $\boldsymbol{\theta}$.

Example Analysis

An example analysis is presented to demonstrate the probabilistic rockfall forecasting model. For the example, published inverse velocity time series data was used from Barrick Gold's Betze-Post open pit mine in Nevada as presented in Rose & Hungr (2007). Figure 6 shows measurements taken from one of the survey prisms (S-221) on the large rock wedge (~ 18 M m^3) beginning 45 days before the actual failure of the rock.

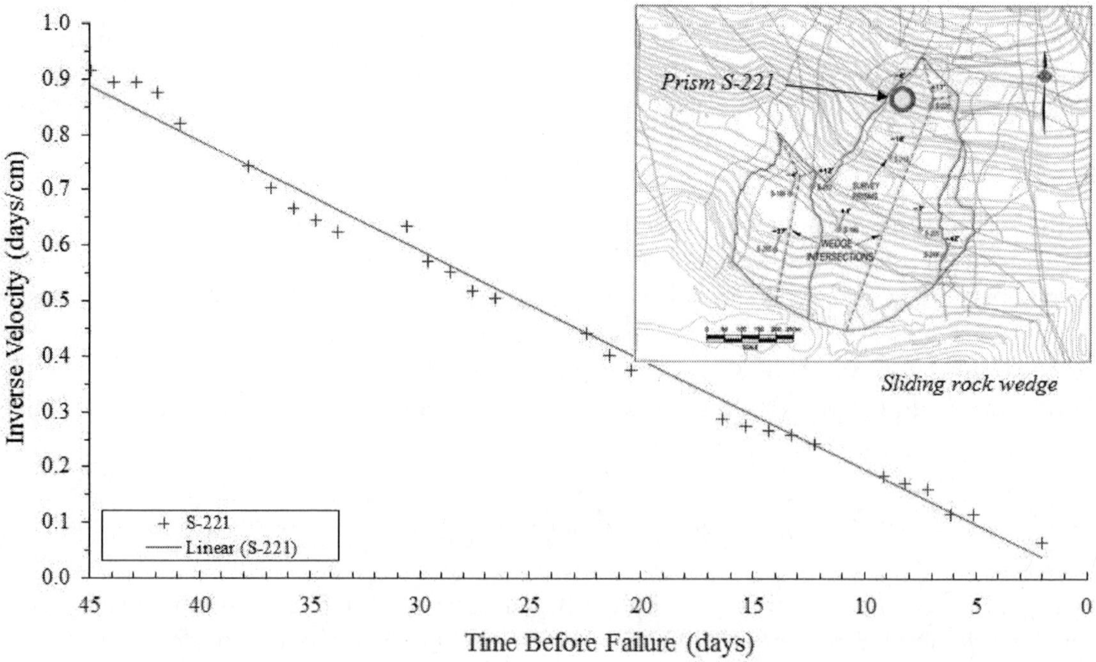

Figure 6 – Inverse velocity time series for rock wedge leading up to failure for Barrick Gold's Betze-Post open pit mine (adapted from Rose & Hungr 2007).

It should be noted that each data point presented in Figure 6 represents an average of six daily measurements. Although this provides some smoothing of the data, this is of secondary interest with regard to demonstration of the forecasting model. Results of the Bayesian linear regression analysis are provided in Figure 7 showing the mean linear forecast for the time to failure. The predictive posterior probability density function (PDF) and corresponding cumulative distribution function (CDF) are also shown for the forecasted time to failure given the inverse velocity, $1/V = 0$.

The plot on the left represents conditions where very few observations are available (i.e., $n = 5$). The associated PDF at the time of failure is wide which relates to the limited knowledge of parameter values in the model. This is also indicative of the relatively large distance on the axis between the forecasted point and where the observations have been made. From a decision making standpoint, this result provides no practical benefit for forecasting other than to show that additional time-series data are necessary to provide more certain estimates of failure time.

The plot on the right represents conditions where considerably more observations have been made (i.e., $n = 17$). Accordingly the updated PDF resides within a much narrower credibility region indicating the certainty with regard to the time to failure forecast has improved considerably. This is further exemplified in Figure 8 which shows the predictive probability distributions (normalized to a mean value of zero) for multiple numbers of observations ($n = 4$, 5, 6, 7, 17, and 30) for the forecasted time to failure. As the number of observations increases so does the credibility of the forecast. This may not always be the case should measurements become more variable or diverge from a linear trend.

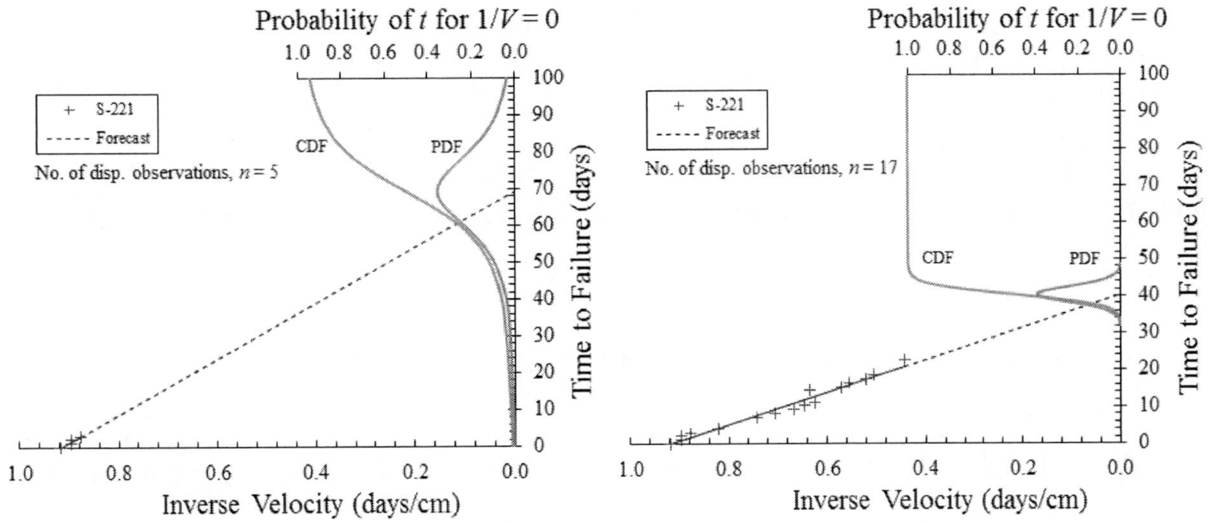

Figure 7 – PDF and CDF for time to failure forecast given $n = 5$ observations (left) and $n = 17$ observations (right) assuming $1/V = 0$.

It should be noted that the posterior predictive PDF carries the form of the Student's t probability distribution. This is a result of selection of a normally distributed likelihood function, L, along with an initial assumption to use a non-informative prior distribution for the $\boldsymbol{\theta}$ parameters. For simplicity in calculations for the example, the latter assumption was carried through the analysis. The use of an informative prior beyond the initial observation is possible although analytical evaluation of the Bayesian updating and predictive probability equations can be difficult and/or impractical. In such instances the use of numerical sampling techniques, such as Markov Chain Monte Carlo (MCMC) simulation, can be implemented.

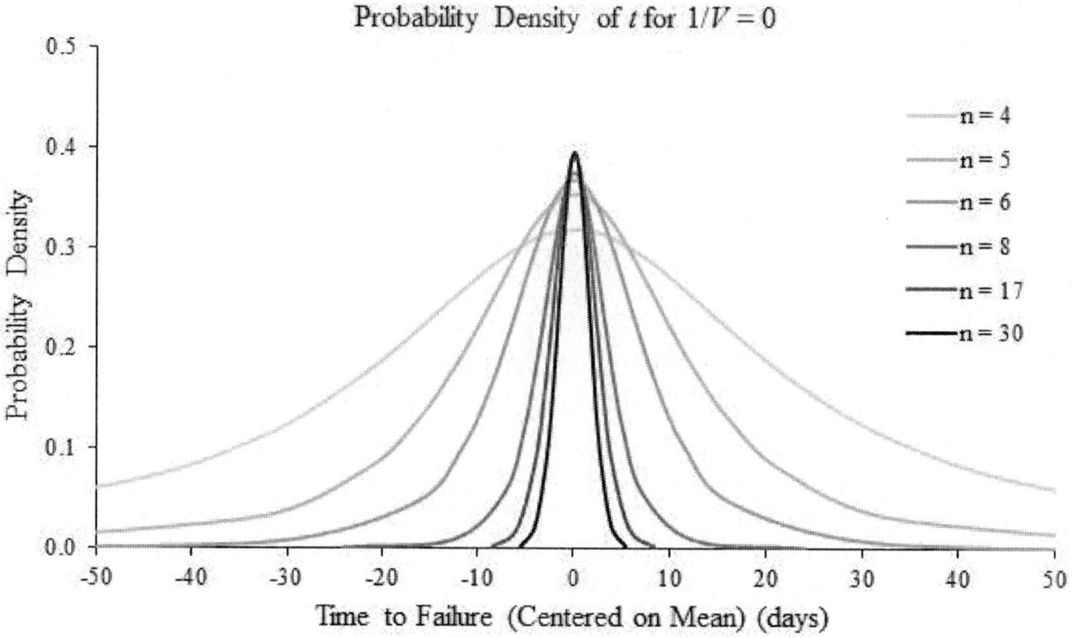

Figure 8 – Progression of posterior predictive PDF for time to failure forecast with increasing number of observations.

CONCLUSIONS

In this paper we present a probabilistic rockfall forecasting model cast within a broader framework to efficiently and effectively characterize and prioritize numerous sites subject to rockfall hazards potentially spanning several hundreds of miles. The model is predicated on detectable displacement and onset of accelerating creep of problematic blocks leading up to failure. Probabilistic forecasts for block time to failure are estimated using the inverse velocity method along with Bayesian linear regression to access parameter uncertainty arising from the inherent variability of the rockfall process and due to measurement uncertainty. As new observations of block displacement are gathered, model parameters and forecast predictions are updated to reflect the current state of knowledge in the system using the Bayesian updating rule.

In the example provided, credibility of time to failure forecasts greatly improved with increased numbers of observations. Initial forecasts, based on limited data, yielded probability densities that were too wide to allow for any practical decisions regarding rockfall response. Subsequent forecasts, however, have much narrower credibility regions which can be beneficial from a decision maker standpoint. With this information, for example, justification for remediation expenditures could be made given that a rockfall event is likely to occur in some number of days with an associated probability. Alternatively, this information could be used to provide some degree of credibility for requests to temporary shut down a section of highway should an event be imminent.

The limitations of the inverse velocity method are acknowledged in that the influence of external triggers, such as rain and freeze/thaw cycles, can result in deviations from the linear model causing unreliable forecasts. The relative ease afforded by high resolution remote sensing technologies to collect large amounts of data economically, however, can make it feasible to develop relationships between inverse velocity model parameters and trigger events. Even if this is only done in a subjective manner, model forecasts could be updated to include the anticipation of future trigger events. This is the advantage of the Bayesian approach in that any information, quantitative or subject can be included in the estimation of model parameters.

REFERENCES:

Boué, A., Lesage, P., Cortés, G., Valette, B., and G. Reyes-Dávila. Real-Time Eruption Forecasting Using the Material Failure Forecast Method with a Bayesian approach. *Journal of Geophysical Research: Solid Earth* vol. 120 (4), 2015, pp. 2143-2161.

Carla, T., Intrieri, E., Di Tragila, F., Nolesini, T., Gigli, G., and N. Casagli. Guidelines on the Use of Inverse Velocity Method as a Tool for Setting Alarm Thresholds and Forecasting Landslides and Structure Collapse. *Landslides*, 2016.

Christiansen, C., Gauthier, D., and N. Oester. 3D Monitoring of Rockfall Sources in Colorado. *Proc., 67th Highway Geology Symposium*, Colorado Springs, CO, 2016, pp. 431-445.

Cornelius, R.R. and P.A. Scott. A Materials Failure Relation of Accelerating Creep as Emperical Description of Damage Accumulation. *Rock Mechanics and Rock Engineering* vol 26 (3), 1993, pp. 233-252.

Der Kiureghian, A. *CE193 Probabilistic Methods for Engineering Risk Analysis*. Lecture Notes, Department of Civil & Environmental Engineering, University of California, Berkeley, 2009.

Dick, G., Eberhardt, E., Cabrejo-Lievano, A.G., Stead, D. and N.D. Rose. Development of an Early-Warning Time-Of-Failure Analysis Methodology for Open-Pit Mine Slopes Utilizing Ground-Based Slope Stability Radar Monitoring Data. *Canadian Geotechnical Journal* vol. 52, 2015, pp. 515-529.

Fukuzono, T. A New Method for Predicting the Failure Time of a Slope. *Proc., 4th International Conference and Field Workshop on Landslides*, Tokyo, Japan, 1985, pp. 145-150.

Gauthier, D., Strouth, A., Hutchison, D.J., and T. Ortiz. Network-Scale Rockfall Forecasting: Pilot Studies and Framework Development in Colorado and Canada. *Proc., 3rd North American Symposium on Landslides*, Roanoke, VA, 2017.

George, M.F. *3D Block Erodibility: Dynamics of Rock-Water Interaction in Rock Scour*. Ph.D. Dissertation, Department of Civil & Environmental Engineering, University of California, Berkeley, 2015.

Goodman, R.E. Block Theory and Its Application. *Geotechnique* vol. 45 (3), 1995, pp. 383-423.

Goodman, R.E. and G. Shi. *Block Theory and Its Application to Rock Engineering*, Englewood Cliffs, N.J. Prentice Hall, 1985.

Kilburn, C.R.J and D.N. Petley. Forecasting Giant, Catastrophic Slope Collapse: Lessons from Vajont, Northern Italy. *Geomorphology* vol. 54, 2003, pp. 21-32.

Rose, N.D. and O. Hungr. Forecasting Potential Rock Slope Failure in Open Pit Mines Using the Inverse-Velocity Method. *International Journal of Rock Mechanics and Mining Sciences* vol. 44, 2007, pp. 308-320.

Rowe, E., Hutchinson, D.J., Kromer, R., Van Veen, M., and T. Edwards. Analysis of Structural Controls on Rockfalls in White Canyon using LiDAR Change Detection. *Proc., GeoVancouver, Canadian Geotechnical Society,* Vancouver, BC, 2016.

Voight, B. A Method for Prediction of Volcanic Eruption. *Nature* vol. 332, 1988, pp. 125-130.

Voight, B. A Relation to Describe Rate Dependent Material Failure. *Science* vol. 243, 1989, pp. 200-203.

Coal Mine Subsidence Below Streets, Highways and Public Buildings: A Special Concern for Southwest Indiana

Terry R. West
Purdue University
Department of Earth, Atmospheric, and Planetary Sciences
550 Stadium Mall Drive
West Lafayette, IN 47907-2051
(765)-494-3296
trwest@purdue.edu

Prepared for the 68th Highway Geology Symposium, July, 2017

Acknowledgements

Marvin Ellis, Indiana Department of Natural Resources

Disclaimer

Statements and views presented in this paper are strictly those of the author and do not necessarily reflect positions held by their affiliations, the Highway Geology Symposium (HGS), or others acknowledged above. The mention of trade names for commercial products does not imply the approval or endorsement by HGS.

ABSTRACT

Homeowners in Indiana can purchase insurance to protect their property from subsidence that occurs when older coal mines collapse. However, such insurance is not available to federal, state, county, and city governments when collapse under highways and structures occurs. Fortunately, the Abandoned Mine Lands (AML) section, Division of Reclamation, Indiana Department of Natural Resources, has a mitigation program to resolve some of these occurrences. Under the AML program, coal mining conducted before the 1977 Surface Mining Act took effect, is eligible for funding support. The AML program traditionally supports the reclamation of old strip mining areas, consisting of rows of cast over mine spoil piles, tailings ponds, steep final cut high walls, and gob piles of coarse material separated from the coal during the final operation. Within the last several years, the AML program has been extended to take care of subsidence below roads and structures of government entities. To date 14 projects have been completed under highways, bridges, an airport and a school building. Only mining operations conducted prior to 1977 are eligible for treatment, but later mines appear to be less susceptible to subsidence. Different techniques have been used in the past to stabilize room and pillar mines such as the use of grout columns. A different approach has been found to be more effective in Indiana, consisting of filling the entire opening with cement grout. The selection of projects, exploration, completion of the grouting operations and verification of success for several of the 14 projects is provided in the full paper.

INTRODUCTION

Coal mining in Indiana has been conducted for the last 100 years, beginning in the eastern portion of the outcrop where the depth to coal seams were fairly small. To date coal has been mined in 17 of the 92 counties, located in southwest Indiana with 186,000 acres of underground mines. Figure 1 is a bedrock map of the state, showing the Pennsylvanian aged rocks where the coal measures are located. Figure 2 shows the locations in southwest Indiana where underground coal mines exist. Surface strip mining also prevailed and currently 284,000 acres of these mines are found in the state. With time, mine roofs and pillars have weakened followed by mine collapse and surface subsidence. Home owners are able to obtain mine subsidence insurance from the Indiana Department of Insurance for a reasonable cost (Website: Indiana Department of Insurance, 2017). Up to $500,000 worth of coverage is available at an annual cost of $325 for $500,000, and $24 for a $25,000 coverage. In FY 14/15, 28 claims were filed for a total of $1,780,077.

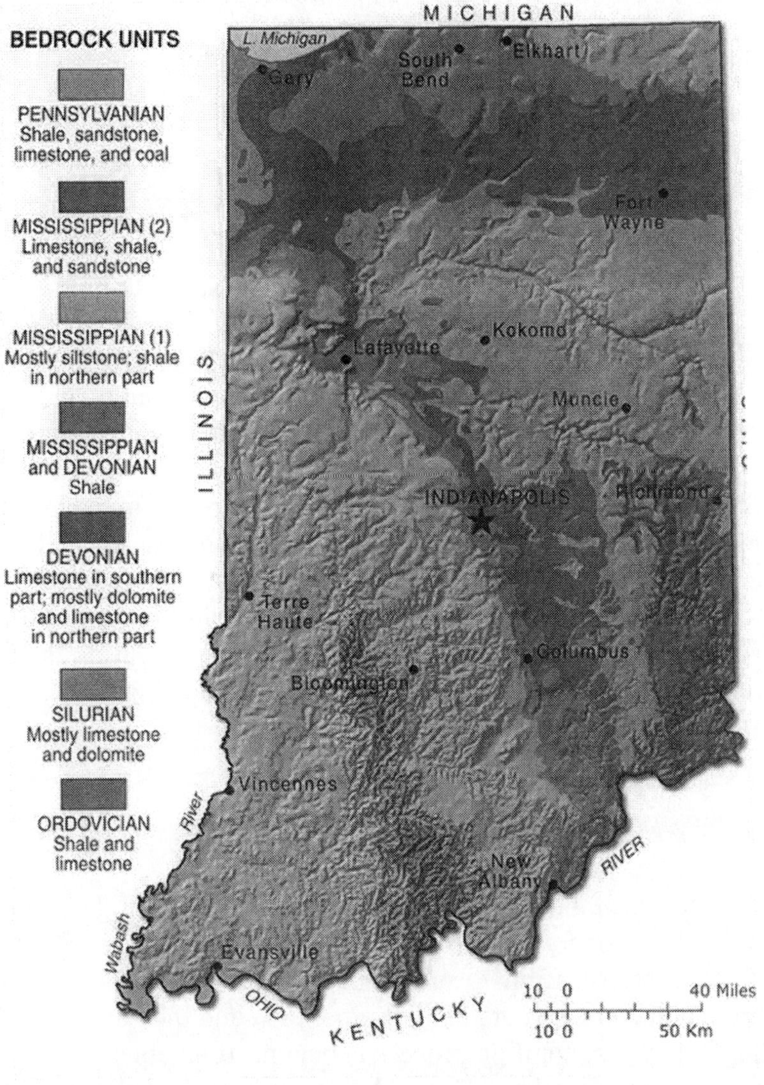

Figure 1. Bedrock Geologic Map of Indiana

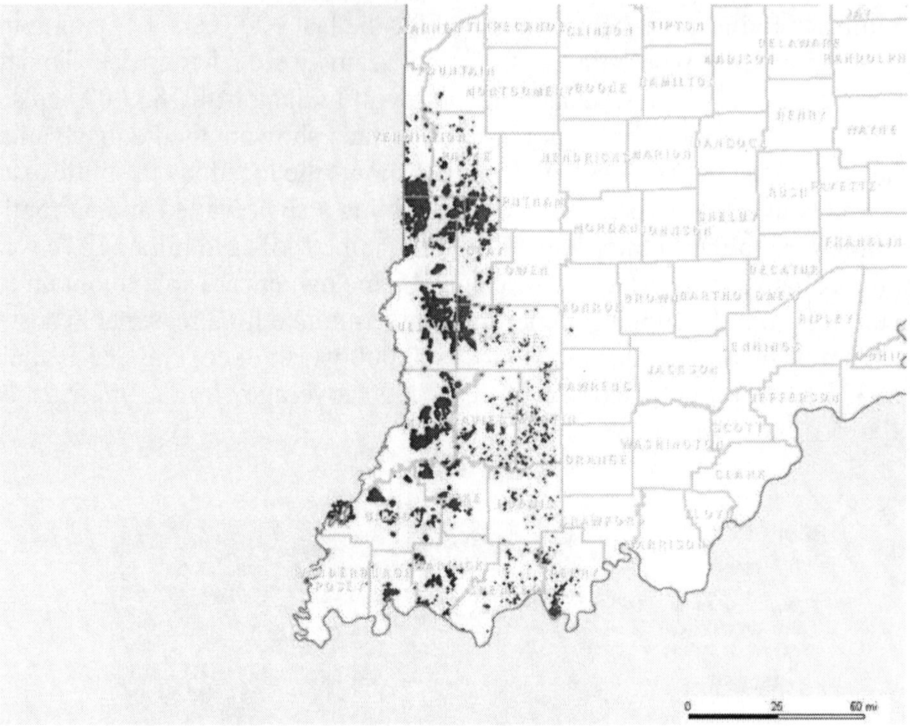

Figure 2. Underground Coal Mines in Indiana.

However, such insurance is not available to federal, state, county and city governments, when collapse under highways and structures occurs. Fortunately, the Abandoned Mine Lands (AML) program through the Indiana Division of Reclamation of the Department of Natural Resources, has a mitigation program to resolve some of these problems. Under the AML program, coal mining conducted before the 1977 Surface Mining Act (SMCRA) took effect, is eligible for reclamation funding. The AML program traditionally supports the reclamation of old strip mining areas, consisting of rows of cast over mine spoil piles, tailings ponds, steep final high walls and gob piles of coarse material separated from the coal during the final operation. Within the past several years, the AML program has been extended to take care of subsidence below roads and structures of governmental entities.

Background

To date, 14 projects have been completed under highways, bridges, an airport and a school building. Only mining operations conducted prior to 1977 are eligible for treatment, but later mines appear to be less susceptible to subsidence. The sites are selected on the basis of concern to the general public using a detailed GIS approach and they are carefully investigated by subsurface drilling before a restoration process is begun. If an emergency case is involved they can mobilize to the site within three days. The areas of voids, collapse and solid coal are determined before the grouting program is initiated. In general, a mining pattern that removes

60% of the coal with 40% left in the pillars is assumed and the thickness of the coal seam is estimated in advance. The Division of Reclamation has provided detailed information to the author that provides the basis for this discussion (Ellis, Marvin, 2017).

To begin the reclamation, exploration drilling is spaced 200 to 800 feet apart, voids and fall down areas are noted and used to focus in on closer spacing of drill holes. Both vertical and angle drilling is performed to outline the potential collapse zone. Quite commonly, a linear pattern is found as related to highways and airport runways, whereas the repair of a school foundation yielded a more complicated pattern.

The 14 projects are listed in Table 1 and shown on Figure 3, a map of southwest Indiana. The sites are clustered into three general locations, located from north to south. On the north, the first cluster is near Terre Haute, in Vigo County, and includes the grouting operation for the Terre Haute airport. The next cluster is near Linton in Greene County and the third is in Warrick County east of Evansville. The Terre Haute airport project is Site 2078, the bridge foundation stabilization is State Road 62, Site 1615 in Warrick County, Loge School is Site 2147 in Warrick County and the remaining sites are roads and highways in the three areas. Limited space prevents a detailed presentation of all 14 projects, so a select number has been included to represent the group. Terre Haute airport, the bridge on SR 62, the Loge school and the street repair for the town of Linton are considered below.

Table 1 – Sites Stabilized by Grouting Procedures

	AML Site
SR 62 Bridge	1615
TH Airport	2078
Loge School	2147
Parkwood	3009
Coalmont	2140
Linton	2146
CR 1600W N	2212
CR 100S	2213
Boonville E. Main	2144
US 40	2237
Boonville McElroy A	2148A
CR 750N	2219
Island City/CR 1100 West	2251
Boonville McElroy B	2148

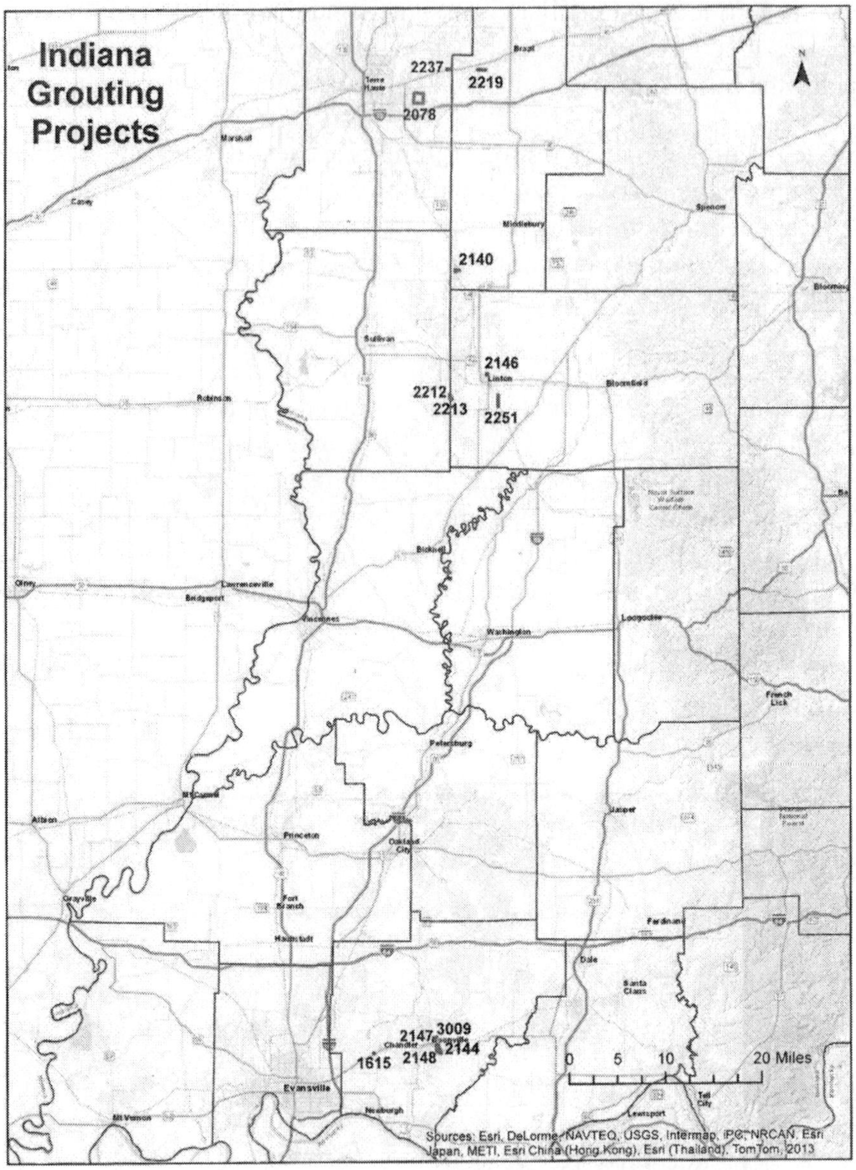

Figure 3. Fourteen Indiana Grouting Projects

Discussion

Figure 4 is a map showing the completion of the grout and concrete placement for the Terre Haute Airport. The location and grout or concrete take is shown along with the numerous exploration borings and the results obtained from them. As observed, four separate areas were involved, Building 2, the terminal, apron and the very extensive treatment of the runway. Table 2 provides the quantities of exploration and injection drilling, grout and concrete placement, and the acreage involved.

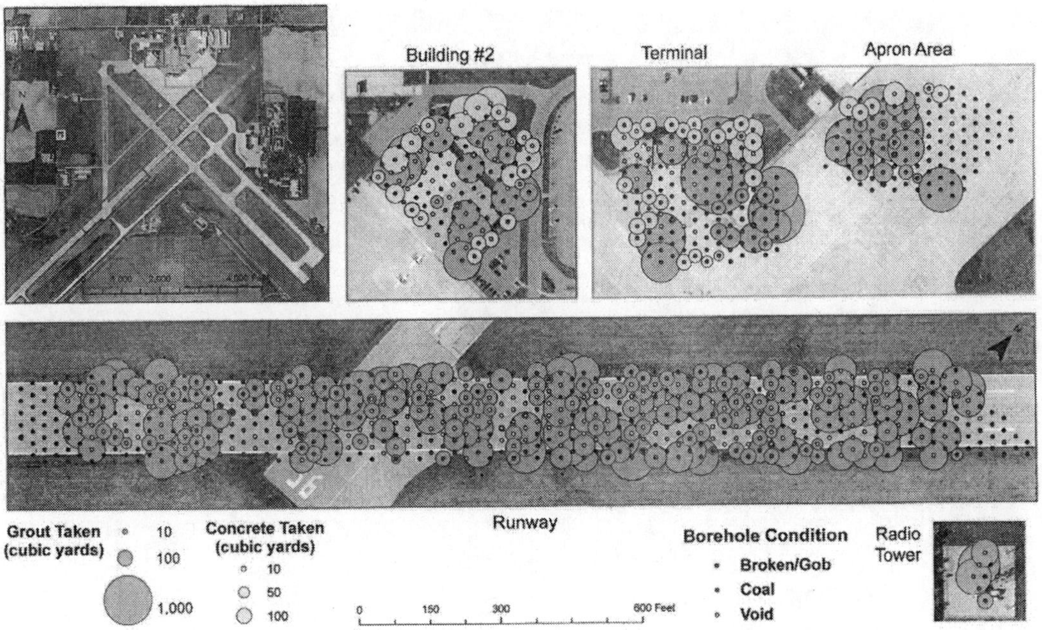

Figure 4. Borehole 3 Condition, Grout and Concrete Distribution at Terre Haute Airport

Table 2 – Quantities for Site 2078 Terre Haute Airport Project		
Feature	Unit	Quantity
Angled Injection Holes,	Number of	521
Vertical Injection Holes,	Number of	612
Angled Drilling Feet,	Linear Feet	73900
Vertical Drilling Feet,	Linear Feet	42387
Grout Utilized,	Cubic Yards	101048.4
Concrete Utilized,	Cubic Yards	8181.7
Acreage Grouted,	Acres	20.1

Figure 5 is a map showing the completion of grout placement for the State Road 62 bridge in Warrick County. The location and grout take are shown, along with the exploration borings and the results obtained. Voids and broken collapse area are shown. A zone measuring 240 by 330 feet was treated for the bridge and adjacent abutments for this project. Table 3 provides the quantities of exploration and injection drilling, grout placement and acreage involved.

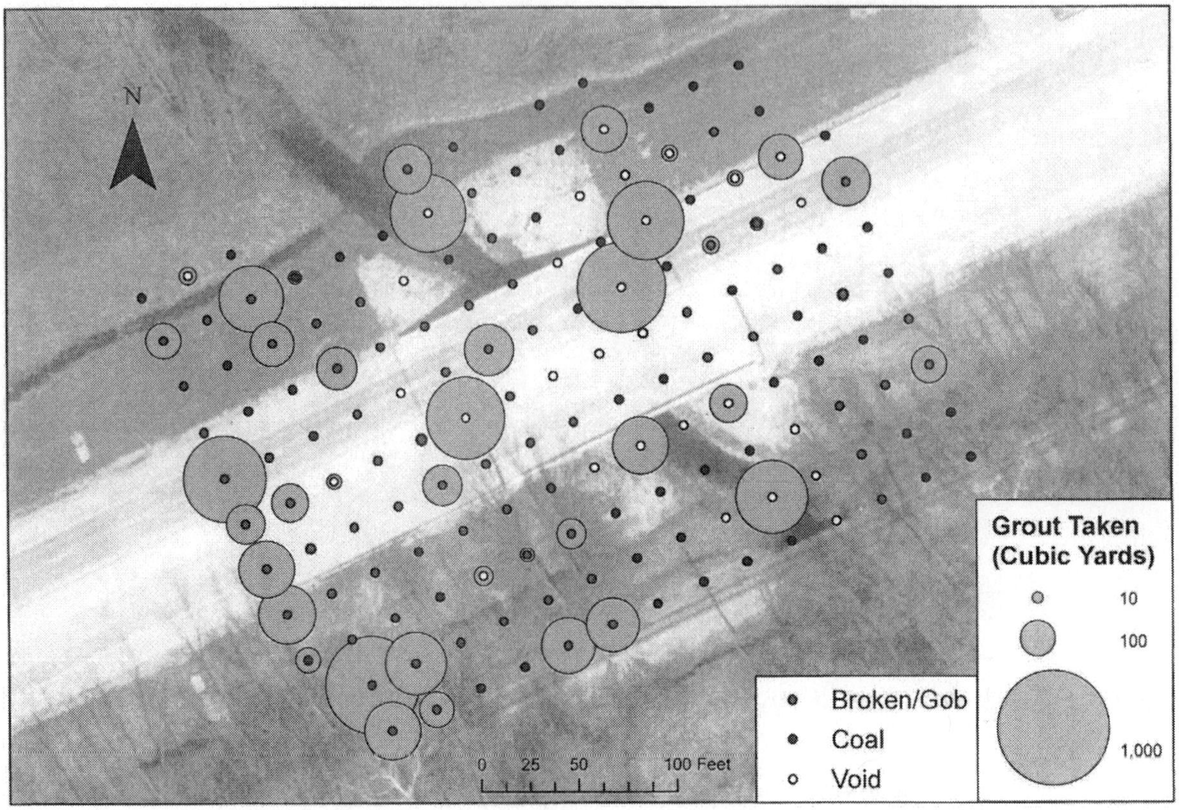

Figure 5. Borehole Condition and Grout Distribution at the State Road 62 Bridge in Warrick County, IN

Table 3 – Quantities for Site 1615 SR 62 Bridge Support		
Feature Unit		Quantity
Angled Injection Holes,	Number of	130
Vertical Injection Holes,	Number of	37
Angled Drilling Feet,	Linear Feet	19598
Vertical Drilling Feet,	Linear Feet	3601
Grout Utilized,	Cubic Yards	8299
Concrete Utilized,	Cubic Yards	156
Acreage Grouted,	Acres	1.3
Average Thickness of Coal	Linear Feet	5
Average depth of Hole,	Linear Feet	135

Figure 6 is a map of the Loge School in Warrick County in the town of Boonville. Many grout holes were utilized along the perimeter of the building, however, several extensive grouting sites were needed in the interior of the structure. Many drill holes intercepted solid coal and a total of more than 44,000 feet of vertical and inclined exploratory drilling was performed.

Table 4 provides the quantities of exploration and injection drilling, grout and concrete placement, and the acreage involved.

Figure 6. Borehole Condition and Grout Distribution at Loge School

Table 4 – Quantities for Site 2147 Loge Elementary School		
Feature Unit		Quantity
Angled Injection Holes,	Number of	400
Vertical Injection Holes,	Number of	450
Angled Drilling Feet,	Linear Feet	22332
Vertical Drilling Feet,	Linear Feet	21796
Grout Utilized,	Cubic Yards	9422.51
Concrete Utilized,	Cubic Yards	1699.41
Acreage Grouted,	Acres	4.1

Figure 7 is a map of the fourth and final example to be considered, the roadways in the town of Linton. The exploration drilling, and grout and concrete placement are shown along four orthogonal roads in the town. Prior to this project, subsidence occurred under several

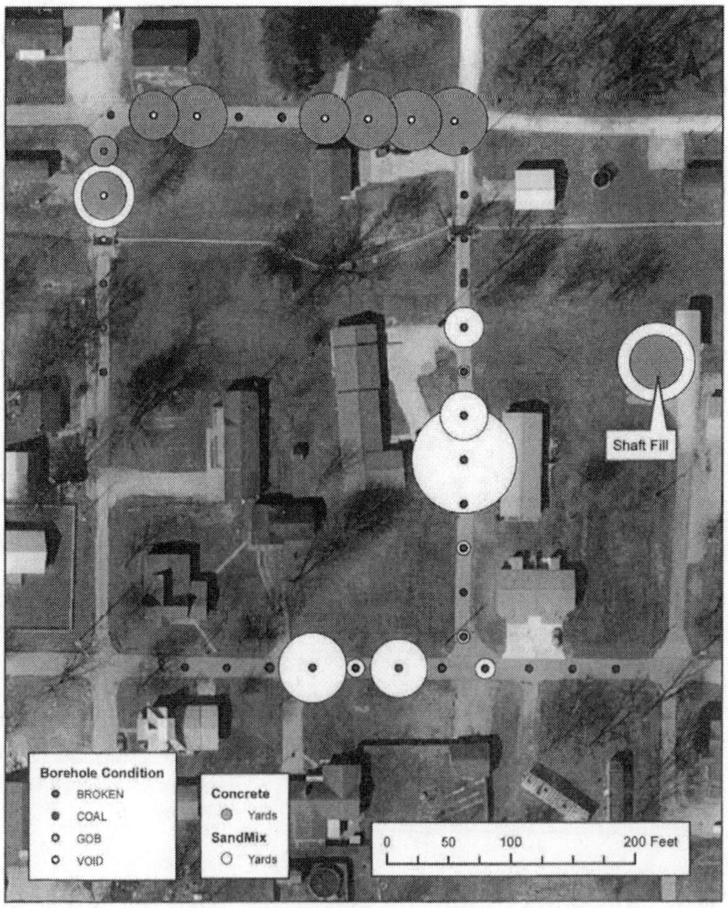

Figure 7. Borehole Condition, Concrete and Sand Mix Footprints at Site 2146 Linton Grouting.

homes with mitigation accomplished by the Indiana insurance coverage. Sand mix grout was used for the southern part of the project and concrete along the northern portion. A mine shaft was also grouted with sand mix grout. Table 5 provides the quantities of exploration and injection drilling, and grout and concrete placement. This project involved smaller quantities of filling materials than the other sites considered, but it was selected because the author had the opportunity to observe this site during construction and more detailed information was made available for this site. A boring log for the site is presented as Figure 8. The depth to bedrock is 17.0 feet, with shale extending to 41 feet. The coal seam is found from 41.0 to 46.5 feet, and shale extending to 50.2 feet at the bottom of the boring. 100 % recovery was obtained for the coal seam showing that the coal is intact at this point. Another boring in an area that had been mined showed a void from 25.3 to 30.2 feet, with broken material from 30.2 to 31.0 feet and shale from there to 35.3 and the bottom of the boring.

The author obtained a series of photos of the Linton grouting operation during the site visit. These are presented in Figure 9. These show the road where grouting was conducted and the material used to make concrete, cement grout, and sand mix grout that were used to fill the voids under the road.

Table 5 – Quantities for Site 2146 Streets in Town of Linton

AML Site Information

Site#	Site Name	County	Quadrangle	Pol Township	Open?	Site Notes
2146	Linton Subsidence - Grouting	Greene	Linton	Stockton	Y	This is one of the sites that was identified using GIS. It is made up of high priority improvements in a subsidence-prone area. It may be considered for grouting in the future.

Program Codes	AMLIS Pad Numbers	Location
AML Abandoned Mine Land	IN2146	T 7 N R 7 W S 15

Features Per Site After Construction

Contract: 008-190

2012	Acreage Groute	Acres	1.6
2012	Angled Drilling F	Linear Fee	69
2012	Angled Injection	Number O	2
2012	Concrete Utilize	Cubic Yar	250
2012	Concrete-Sand	Cubic Yar	1700.29
2012	Grout Utilized	Cubic Yar	2446.19
2012	Vertical Drilling	Linear Fee	1270
2012	Vertical Injection	Number O	36

TEST BORING RECORD

CLIENT : Indiana Department of Natural Resources - Division of Reclamation — BORING NO. : **BH- 1**

PROJECT : Mine Investigation - Linton Site 2146 — SHEET : 2 OF 3

Stratum Elevation	Sample Depth	SOIL/MATERIAL DESCRIPTION	Stratum Depth	Sample Number	SPT per 6"	SPT per 12" (N)	Recovery (%)	Moisture Content (%)	Total Unit Weight (pcf)	Unconfined Compression (ksf)	Atterberg Limits LL	PL	PI
	22.5	Gray, Highly Weathered, **SHALE** (Visual)											
505.0			24.3	SS-7	48 50/1"		100						
	25.0			RC-1 RQD= 100%			100						
	27.5			RC-2 RQD= 22%			96						
	30.0												
	32.5	Gray, Weathered, **SHALE** (Visual)		RC-3 RQD= 26%			100						
	35.0												
	37.5			RC-4 RQD= 31%			95						
	40.0												
488.2			41.0										
	42.5	**COAL** (Visual)		RC-5 RQD= 0%			100						
	45.0												

Continued on next page

CTL Engineering, Inc.
4343 Saguaro Trail
Indianapolis, Indiana 46268
Phone: 317-295-8650

BORING METHOD	SAMPLING METHOD	ABBREVIATIONS
HSA - Hollow Stem Auger	SS - Split Spoon Sample	* - Hand Penetrometer
SFA - Solid Flight Auger	ST - Shelby Tube Sample	LL - Liquid Limit
RC - Rock Coring	CR - Rock Core Sample	PL - Plastic Limit
MD - Mud Drilling	BS - Bag Sample	PI - Plasticity Index
WD - Wash Drilling	AC - Auger Cuttings	SPT - Standard Penetration Test
HA - Hand Auger		

INDOT_TEST BORING RECORD 11050042IND.GPJ INDOT_4.GDT

TEST BORING RECORD

CLIENT : Indiana Department of Natural Resources - Division of Reclamation   BORING NO. : **BH- 1**

PROJECT : Mine Investigation - Linton Site 2146   SHEET : 3 OF 3

Stratum Elevation	Sample Depth	SOIL/MATERIAL DESCRIPTION	Stratum Depth	Sample Number	SPT per 6"	SPT per 12" (N)	Recovery (%)	Moisture Content (%)	Total Unit Weight (pcf)	Unconfined Compression (ksf)	Atterberg Limits LL	PL	PI
		COAL (Visual)											
482.7			46.5										
	47.5	Gray, **SHALE** (Visual)		RC-6 RQD= 55%			88						
479.0	50.0		50.2										
		Bottom of Boring at 50.2 feet Auger refusal at 24.3 feet Boring backfilled with cement bentonite grout.											
	52.5												
	55.0												
	57.5												
	60.0												
	62.5												
	65.0												
	67.5												
	70.0												

CTL Engineering, Inc.
4343 Saguaro Trail
Indianapolis, Indiana 46268
Phone: 317-295-8650

BORING METHOD	SAMPLING METHOD	ABBREVIATIONS
HSA - Hollow Stem Auger	SS - Split Spoon Sample	* - Hand Penetrometer
SFA - Solid Flight Auger	ST - Shelby Tube Sample	LL - Liquid Limit
RC - Rock Coring	CR - Rock Core Sample	PL - Plastic Limit
MD - Mud Drilling	BS - Bag Sample	PI - Plasticity Index
WD - Wash Drilling	AC - Auger Cuttings	SPT - Standard Penetration Test
HA - Hand Auger		

INDOT_TEST BORING RECORD 11050042IND.GPJ INDOT_4.GDT

Figure 8. Boring Log, Linton Project

CAT
257B

HOWARD
CONCRETE
PUMPING, INC.

Figure 9. Grouting of Voids Under Roads in Linton, Indiana

Final Considerations

The mine grouting operations discussed in this paper were conducted under the direction of the Indiana Department of Reclamation, AML Program of the Department of Natural

Resources. Grouts of different constituents were used to fill the voids completely thereby supporting the mine roof to prevent further subsidence from occurring. In other areas of the United States, grout columns spaced at intervals are used for such support. The Reclamation staff concluded that filling voids with grout was a better solution to this problem of mine subsidence. Various constituents were considered, cement, concrete and sand mixtures to fill the voids as needed. This yielded the right combination of viscosity and strength to accomplish stabilization. The 14 projects considered in this paper have yielded successful results and others are planned for the future to alleviate similar problems. This aids the government agencies responsible for maintaining engineering structures used by the general public.

References

Personal Communication

1. Ellis, Marvin, 2017, Personal communication with supply of reference data on Mine Void Reclamation

Website

1. Indiana Department of Insurance Website, 2017

A Legacy of Value Added – Long-Term Contributions of the Highway Geology Symposium

Peter Ingraham, PE
Golder Associates Inc.
670 North Commercial Street
Manchester, NH 03101
(603)-668-0880
pingraham@golder.com

Harry Moore
Golder Associates Inc., (Tennessee DOT, retired)
7409 Somerset Rd
Knoxville, TN 37909
(865) 603-7652
hlaamoore@aol.com

Jeff Dean

Terracon Inc., (Oklahoma DOT, retired)
4701 North Stiles Avenue
Oklahoma City, OK 73015
Phone: (405) 445-3280
Fax: (405) 557-0549
Email: jeff.dean@terracon.com

Prepared for the 68th Highway Geology Symposium, July, 2017

Acknowledgements

The author would like to thank the following individuals for their contributions to this paper:

John D. Duffy – Yeh & Associates (Formerly CalTrans)

John Szturo – HNTB

Michael Vierling – NY State Thruway (retired)

Disclaimer

Statements and views presented in this paper are strictly those of the author(s), and do not necessarily reflect positions held by their affiliations, the Highway Geology Symposium (HGS), or others acknowledged above. The mention of trade names for commercial products does not imply the approval or endorsement by HGS.

Copyright Notice

INTRODUCTION

"What makes a good engineer?" the student asked. "Good judgement" the engineer replied. "What makes good judgment?" the student queried. "Experience" came the reply. "And what makes experience?" persisted the student. The engineer paused and replied with a smile, "bad judgement". The Highway Geology Symposium (HGS) was started with an eye toward sharing experiences, learning from each other's failures and successes, and working toward applied solutions to geological problems in transportation. The authors hope that if you are reading this, you are planning to attend – or you are reviewing the merits of having one of your subordinates attend the HGS. With that in mind, the HGS has met annually for almost seven decades. The technologies and roadways have changed, but the concept is as valid today as it was in 1950 when transportation geologists got together from eight states: Georgia, South Carolina, North Carolina, Virginia, Kentucky, West Virginia, Maryland and Pennsylvania to meet and discuss issues they faced. Working to make better highways through applied geology – it was a new era then: widespread expansion of the nation's highway networks, a new Interstate System coming - and the group of transportation geologists, engineers, geological survey, university and Corps of Engineers participants that gathered at that first HGS recognized that it was better to learn from each other, *and each other's mistakes*, than to face the entire learning curve on one's own.

With this paper and the 68th set of proceedings we would invite you to look at the full body of work – juried papers – in sixty-eight sets of proceedings that are available as a free resource at www.highwaygeologysymposium.org – and recognize that the run of symposia could only have happened *because the HGS adds value and saves agencies and projects money*. The technical challenges today are greater in number, and now include aging transportation corridors, more complex constructed assets within the systems, more recently recognized issues where geology affects aggregate longevity, and new and more efficient methods of investigation, analysis and design. The cost of maintaining and constructing new transportation assets is a key challenge, and staying abreast of the technical challenges with diminishing transportation budgets is why the HGS meets each year.

This paper's theme was suggested by Dick Cross, the NY State Thruway Geologist for many years, who estimated that his attendance at the HGS had likely saved the Thruway over $2 million dollars over his career. The 68th HGS proceedings are dedicated to Dick Cross, and we felt it fitting to put down on paper some examples of where the HGS has saved transportation agencies and projects money and time. Irrespective of the perception of the cost of sending an employee to attend a conference out of state, the cost of learning a hard lesson can be far greater. The rest of this paper includes examples from Dick Cross, and other transportation geologists who saved their agencies money and time through new approaches, connections made and others' "lessons learned" over the years by attending the HGS.

WHERE'S THE MONEY?

Firstly, the money spent for attending the HGS is travel cost, hotel cost at state rates, and around $300 for registration that covers most meals, a field trip and the cost of the meeting. The meeting costs are usually a break-even prospect, sometimes in the red and sometimes in the black depending on sponsorship and exhibitor enthusiasm. The HGS does not operate the event for profit and there are no dues to join the group. The benefits are many:

- Networking and engaging with peers, academia, industry and leading practitioners who participate;
- Learning about new approaches and technology that can save money;
- Learning from other peoples' mistakes to avoid having to learn by making the same mistake yourself (bypassing the learning curve…);
- Professional development and engagement for employees – this helps with retention and career growth, and
- The big enchilada – when calamity comes your way, you know who to call, how to tackle the problem - and the cost of the calamity is reduced through efficiency and minimized down-time.

On to the money -

Case History 1

The first example comes from Dick's idea for this paper. In 2003 the Thruway was re-cutting a 110-foot slope in Nyack, New York, in diabase. The ADT was on the order of 90,000 and there was little room for catchment, so an innovative combined concrete barrier and temporary rockfall catch fence (developed by Dick and GeoBrugg through a collaboration developed at HGS) was used to contain rocks being brought down during blasting (Figure 1). The savings this barrier design has produced over the years may be as much as $500k. The slope excavation contract was let and work was proceeding with removal of vegetation and soil removal along the crest in preparation for the cut. And the phone call came – "hey Dick, the soil here is thicker than we thought". Dick asked what the borings had indicated along the crest of the cut – there were no borings, someone had saved money by skipping the borings along the crest due to poor access and not much room there.

Figure 1 – Temporary Traffic and Rockfall Barrier

In Nyack there isn't a lot of open space, and the thick section of soil (15 feet thick) included a neighbor's swimming pool that came within 5 feet of the Right-Of-Way fence. There was no room to grade the soil to the neat line of the new cut and maintain soil slope stability, so what could be done? The conversation turned to constructing a soil nail wall in the thick soil section that was roughly 100+ feet long, but it would be interesting finding a contractor to install ground anchors and apply reinforcing and shotcrete with a hundred-foot drop at their backs, and keep the rock removal moving. Through his connections developed at HGS meetings, Dick called on Golder Associates to rapidly develop the soil nail wall design, and Golder and Dick identified Janod Contractors – a high-angle rock slope contractor to do the construction of the soil nail wall working on rappel.

Figure 2 – Soil Nail Wall Being Constructed

And it didn't end there – the work would require use of dry-mix shotcrete to facilitate the access difficulties and provide a uniform, quality product - and the proposed prepackaged shotcrete was not a New York State DOT-approved product. After a quick series of phone calls, Dick discovered that the Ministry of Transportation Ontario (MTO) had done extensive testing on the proposed shotcrete and it was an approved product in Ontario. He called the NYDOT materials laboratory and asked the state lab if they would accept the MTO certification - and they said "sure". After another call to Steve Senior of the MTO (and an HGS steering committee member) he sent the full shotcrete testing lab package to Albany by courier and the soil nail wall could move forward.

Figure 3 – Completed Wall, note presplit holes at the toe of the wall.

The soil nail wall was constructed using rope access methods and the project stayed on schedule (Figures 2 and 3), but the slope had one more problem to be solved. A lower portion of the slope needed to be constructed steeper than the soil would allow. Reaching out again, GeoBrugg (another HGS sponsor) offered a solution using a mesh-retained stabilization solution called "Pentafix". The lower slope was stabilized with the lashed cable-net system and the project completed in the planned construction season.

Savings – So what did the connections and knowledge of possible fixes provided by attendance at the HGS bring to this project?

- Temporary barrier – this barrier has been used system-wide on the Thruway and conservatively Dick estimated it has saved the Thruway over $500k by facilitating projects that would require unacceptable closures or construction of site specific barriers – the system is re-usable and compatible with their standard concrete barrier system.
- Soil Nail Wall – without the soil nail wall approach, made possible by the acquaintance of the specialty contractor and MTO folks, alternatives ranging from additional Right-Of Way acquisition (and the attendant glacial pace of ROW acquisition), to conventional retaining wall design and postponement of the project were all on the table. Dick estimated that simply not having the project delayed into the next construction season saved well over $100k, and it could have been far more if property had to be acquired.

Case History 2

The exchanges of information between professionals working within HGS's community of transportation practitioners are also of benefit, albeit perhaps not as readily quantifiable in terms of dollars. Kirk Hood of Wyoming DOT and Pete Ingraham of Golder Associates have had many discussions at HGS over the years and when Kirk had some questions regarding rock bolting and specifications for bolting, he gave Pete a call (not knowing Pete did his thesis on rock bolts and he had just uncorked the Genie…). The discussion was long but lively, and Pete emailed Kirk three or four specifications in DOT formats, all of which were in the public record having been published in bid documents by New York, Vermont and other states. It was a pure professional exchange and no money changed hands, the specifications were in WORD, and saved the recipient time and effort in assembling a specification for Wyoming. A couple years later, the tables were turned and Pete was asking if WYDOT had a recent specification for horizontal drains – and Kirk sent the published spec for a recent WYDOT project to Pete.

Savings – Specifications from scratch take time to develop and even modifying an existing specification can take time. From a dollar perspective, it may only be a week or two of time and labor saved, but that is enough to send someone to the HGS a few times. The drain specification probably saved Golder's DOT client $4-$5k in labor and review costs. So the combined exchange was likely worth $10k or so – roughly 5- to 10 times the cost of Kirk or Pete attending the HGS depending on location.

Case History 3 - from Harry Moore, Tennessee DOT (retired)

The Tennessee Department of Transportation's (TDOT) involvement with the HGS began in the early 1970s when then Division Director, David Royster, began attending the symposium. He later joined the National Steering Committee for the HGS in the mid-1970s and served until his death in 1985. Harry Moore began attending the HGS in 1976 (held in Orlando, Florida.) at Mr. Royster's suggestion and took his position on the HGS Steering Committee in 1985, following Mr. Royster's passing.

Early-on, David Royster saw the HGS as a great tool for the TDOT engineering geologists to acquire exposure to new and different ideas, concepts and practices in the still evolving field of highway engineering geology. Harry Moore, following his lead, continued this support for the HGS as a representative of the TDOT organization. In more recent times, Vanessa Bateman became involved in the HGS upon Harry Moore's retirement from TDOT in 2009.

Over the years, both David and Harry wrote and presented many technical papers on issues of highway engineering geology regarding landslides, rockfalls, karst, structure foundations, rock slope design, acid producing rock, and remedial measures associated with those issues. This participation was in an effort to share TDOT's experiences and procedures with DOT personnel from other states – and other state DOTs involved with the HGS did the same over the years.

The value in TDOT's participation in the HGS meetings was the valuable exchange of ideas and experiences in dealing with the geotechnical issues of highway engineering. We would gain ideas from other state DOT's regarding technical problems with say landslides or karst that would enable us to remediate similar problems in our state road system. Being able to say "Hey - that is what we can do on our subsidence problem on such and such road…" is definitely an "ah-ha" moment - but it is difficult to place a finite dollar amount on that benefit because it is the difference between that solution and a more expensive solution (or trial and error) that would have occurred and did not take place.

Providing actual cost savings in dollars is difficult to produce, however the sharing of the ideas and work experiences is invaluable. Just being able to call a fellow HGS member in another state and ask how they handled a geotechnical problem is worth the cost of attending and being involved with HGS.

Figure 4 – I-40 Horizontal Drain Installation

A good example took place in the 1970s when TDOT was dealing with very large landslide issues on I-40 and I-75 in East Tennessee. We were able, thru HGS contacts, to design and find a qualified horizontal drilling contractor to provide the necessary experience and assistance to implement large scale horizontal drilling procedures (Figure 4) to de-water several large, million-cubic-yard landslides for stabilization purposes. This saved several million dollars <u>per</u>

landslide as compared to the previous practice of total removal of the landslide mass requiring waste disposal costs.

Another example where TDOT benefited from information gathered at HGS meetings is regarding the use of railroad rail walls in repairing small scale embankment failures (Figure 5). The Kentucky Transportation Cabinet geotechnical office presented a technical talk on the repair of small embankment failures on rural and remote roads in eastern Kentucky. Seeing that Tennessee has very similar conditions as Kentucky, the TDOT geotechnical personnel quickly evaluated the Kentucky DOT information as a possible application to TDOT highways.

Figure 5 – Railroad Rail Wall Embankment Stabilization

As a result, TDOT was able to save several thousand dollars on each of several embankment failures in East Tennessee. The typical repair of using rock buttresses to stabilize embankment failures was much more expensive and time consuming and also requiring some road closures to complete. The savings varied from 10 to 30 percent by using the railroad rail walls. On one particular project that the author was involved in the savings were over $200,000.00.

There are many more examples where information gleaned from HGS meetings has both provided solutions to complex problems and has saved money in the process. TDOT's continued support of the HGS meetings over the years shows that the intrinsic value to our profession is unequaled.

Savings – Looking at just one of several large landslides stabilized using horizontal drain installation and adding the one rail wall installation the total *savings to TDOT amounts to over $1.2M* – or the cost of attending the HGS over 600 to 1000 times depending on location.

Case History 3 - from Jeff Dean, Oklahoma DOT (retired)

Southern Oklahoma experienced an extreme weather event in June of 2015. Upwards of 10 inches of rain fell over a 24-hour period in addition to several previous days of steady rainfall. Consequential to this heavy rain were several washed out bridges, flooded highways, and a rockslide that closed a section of Interstate 35 that passed through the Arbuckle Mountains. Coincidentally the slide occurred at the same location as one of the HGS field trip stops visited when Oklahoma hosted the 61st HGS in 2010. Oklahoma DOT personnel were mobilized early the next morning to a meeting with personnel from Division 7 where the rockfall occurred, to discuss a plan of action. The northbound lane of Interstate 35 was closed because of large boulders that had fallen out onto the travel lanes and shoulder during the night. This was relatively easy to clean up using DOT maintenance crews to reopen the highway. However, the remaining issues included several isolated areas of loose rock still on the slope along with a large section of rock that had become dislodged and was in danger of toppling out onto the highway (Figure 6).

Rockfall events of this magnitude are uncommon in Oklahoma and contractors

Figure 6 – I-35 Rockslide and destabilized slope, Arbuckle Mountains

experienced in this type of specialized remediation are not readily available. A contractor specializing in large earth moving operations was called to visit the site with the DOT personnel and after sizing up the project offered their best recommendation of laying the rock slope back to 3 to 1. This would also involve the added expense of purchasing considerable Right of Way from the land owners beyond the crest of the slope in order to construct this suggested solution.

The Arbuckle Mountains are the oldest known formation in the United States between the Appalachian and Rocky Mountains. They are noted for their unique geologic features which can be seen in their folded and faulted limestones, dolomites, sandstones, and shales that are very steeply dipping to near vertical in orientation within the Arbuckle Uplift region. The project section where I-35 cuts through the Arbuckles has long been a valued resource for geologists throughout the United States who come to study the exposed geology along the highway. The issues with the solution proposed by the contractor were not only the overall expense but the idea of potentially ruining a valued location used for geologic study.

After attending the Highway Geology Symposium for the past 20+ years, I knew there were better and more economical solutions available to stabilize the slope and retain as many of the geological features of the site as possible (In my opinion, this proposal was like splitting a grape with an axe!). I encouraged members of the ODOT Senior Staff to look for a contractor with experience in rock scaling and rock bolting as well as trim blasting. There are several of these specialty contractors who regularly attend the HGS and present case histories of similar if not more complicated rock stabilization projects each year. Members of the Senior Staff, in turn, called on Ty Ortiz with the Colorado DOT to visit the site and offer his thoughts on how to address the situation. Ty was very experienced with rockfall remediation projects in Colorado, having managed their rock slopes program, and had attended several HGS events in the past. After reviewing the slope conditions, Ty reinforced what had already been suggested about using a specialty contractor to stabilize the slope.

With the decision made to adopt the repair approach, the project was placed out for emergency bidding and GeoStabilization International (GSI), a regular attendee and exhibitor at the HGS, was awarded the contract. This was reassuring knowing that a competent contractor was on the job. GSI worked through the hot weeks of July to clear the slope of loose rock and potential falling rock sources. Interstate traffic was occasionally stopped to allow for trim blasting to remove rock that was too large for scaling and GSI completed the slope stabilization project using rock bolts and anchored rock netting as needed.

Savings – The specialty contractor's expertise in addressing the needs of the project saved ODOT *several million dollars over the more traditional approach* proposed by the local contractor, and accelerated the opening of the interstate to the regular flow of traffic.

Case History 4 - from John D. Duffy, CalTrans (retired)

My first HGS meeting was in Park City, Utah back in 1988. My supervisor at the time, Marvin McCauley, had been a long time HGS participant and encouraged me to attend. In my relatively new job I had been assigned to develop an understanding on the subject of rockfall. I was there on a shoestring budget but in the long haul, huge dividends were realized from that initial HGS attendance and the connections and networking it facilitated. The theme of that years HGS was "rockfall" and largely centered on the development of the new Colorado Rockfall Simulation Program (CRSP). Bob Barrett (Colorado DOT) was showcasing the program and its attributes. Of course it was a very exciting new development and CRSP was well received. It was there that I first met Robert Thommen - the USA Geobrugg General Manager. Geobrugg had a booth, tucked away in a far corner, and they were showcasing flexible rockfall barrier fences. Think about this - in this one simple HGS meeting the course of rockfall evaluation and analysis, and design of mitigation measures made a great leap into the future. It was not necessarily obvious at the time, although CRSP was well underway to becoming a mainstay in the soon to be rockfall profession and the inspiration of subsequent rockfall programs. And with Robert Thommen was the seed that started another landmark leap in rockfall mitigation – the development of flexible rockfall barrier fences. There at the HGS, Robert and I discussed and planned a strategy to test fences under actual field conditions. Within a year's time we managed to obtain funding and field test the fences (Figure 7) proving their effectiveness - and, as they say "the rest is history."

Figure 7 – Early CalTrans Rockfall BarrierTesting

What were the savings to the state? Well, that is very difficult to quantify. Within a year's time California DOT was investigating and analyzing rockfall at a new level - now using computers and mathematical modeling to study rockfall trajectories. A new and cost-effective rockfall mitigation measure was now tested and evaluated. Under federal funding a "Construction Evaluation" project using these flexible rockfall fences was underway at a difficult rockfall site in Santa Barbara County called the Gaviota Pass – and to top it off, right around this time California's new environmental regulations were developing and these new lightweight, less intrusive fences were an alternative to more robust mitigation measures and were favored in the regulatory world. The Gaviota project was awarded to a fence company whose superintendent was Howard Ingram. His crew consisted of Howard's son Chris Ingram, and Jim Roth. Soon after, these men started Hi-tech Rockfall, now a leader in rockfall mitigation construction. To quote Bob Barrett "The New Science of Rockfall" was born, and the nucleus of its formation was the HGS meeting.

It has been my experience that situations like those described above happen regularly at HGS. The HGS format encourages this type of collaborative innovation. New developments in geohazard mitigation have been planted or nurtured at HGS many, many times over the years. Anyone attending would always take home a new idea in rockfall mitigation or investigation *or for that matter what not to do or what did not work.* Debris flow mitigation, rockfall draperies, attenuators and many more new solutions to old problems are presented at HGS each year.

Savings- How many new innovative ideas did I bring back to the state? - too many to count. And what were the savings? – in terms of safety, immeasurable. In terms of dollars, certainly several million dollars given that rockfall mitigation using flexible barrier fences is far less expensive than mitigation by excavation or with rigid structures. By any measure the cost of attending HGS pales in comparison to the savings realized over the years.

A FEW PARTING THOUGHTS

A wise man once said "It is best to learn from other people's mistakes because you don't have time to make them all yourself." Dr. Raymund Spang, a globally known authority on rockfall mitigation from Witten, Germany, was greatly impressed by the HGS when he attended in San Luis Obispo, California, in 2002. Dr. Spang told John Duffy (the HGS host committee chair that year) that it was perhaps the best conference he had attended in decades because the speakers _talked about what didn't work as well as what worked_ – and he said what doesn't work is "often more important".

Beyond the potential for dollars saved, the HGS provides a strong positive venue where professionals can get engaged in their discipline, interact with their peers (leaving the dreaded hand-held device behind) and have conversations with real flesh and blood people. Too often we think of everything being available on the internet – but when was the last time an internet site you went to rang you up to say "…you know that problem you have that we talked about at HGS last September? - I think I found the answer for you".

The symposium lasts two and a half days, is generally preceded by a half day Transportation Research Board gathering with 4 to 6 presentations on a specific topic (yes, the TRB folks are staunch supporters of the HGS), and there is a full-day field trip looking at projects and geologic hazards in the host state each year – Applied Geology (Figure 8). The event is inclusive and attended by practitioners in industry, academia, state DOTs, FHWA, TRB, and consulting.

Figure 8 – Rockfall Barrier Live Demonstration Cow Cliffs (US-1, Big Sur) – 53rd HGS Field Trip 2002

Irrespective of perceptions ("optics") and a reluctance to send people out of state to attend a conference, the HGS comes through and adds value every year. New technology demonstrations, case histories, and new theories and trends – the HGS has been putting them forward _every year for almost 70 years_. Dick Cross, who provided the theme for this paper attended every HGS from 1988 through 2015, and had perfect attendance on the Steering

Committee for 22 years. Over his career with the NY Thruway, he estimated that the HGS had saved the Thruway several million dollars by providing answers, methods, connections and support for emergencies and standard projects at critical times.

The symposium has been held in 34 states, many states several times. If you still cannot get permission to travel out of state for the symposium, please consider hosting the HGS. The authors and Steering Committee will help with the heavy lifting, just give us a call. And remember, this symposium is for you - and the goal of "Better Highways through Applied Geology". Check us out at www.highwaygeologysymposium.org

Emergency Response: Fossil Cut Rockslide Investigation and Repair BNSF Spokane Subdivision, WA

Dale Moore
&
Thomas Pallua
McMillen Jacobs Associates
1109 1st Ave #501.
Seattle, WA 98101
(206)-588-8169
dmoore@mcmjac.com
Pallua@mcmjac.com

Prepared for the 68th Highway Geology Symposium, May 2017

Acknowledgments

Thank you Bill for showing us the ropes!
Dr. William Gates – McMillen Jacobs Associates
We also extend thanks to the following personnel with BNSF who made this project possible:
Glen Gaz, Director of Engineering - BNSF Northwest Division
Bruce Sparling – Engineering
Wesley Pruitt – Engineering
Logan Serra – Road Master
Eric Nixon – Maintenance

Disclaimer

Statements and views presented in this paper are strictly those of the author, and do not necessarily reflect positions held by their affiliations, the Highway Geology Symposium (HGS), or others acknowledged above. The mention of trade names for commercial products does not imply the approval or endorsement by HGS.

Copyright Notice

ABSTRACT

In May 2016, a complex rockfall and rotational slide event occurred along the BNSF railway at Fossil Cut, a through-cut near Spokane, Washington. The event fouled the track and interrupted rail traffic. At BNSF's request, McMillen Jacobs Associates provided emergency engineering and rock slope evaluation services. Within Fossil Cut, jointed and spheroidal weathered flows of the Columbia River Basalts are conformably overlain by jointed, blocky claystone of the Latah Formation. While under live-track conditions, 700 cubic yards of loose unstable rock were removed from the slope during a 44-hour period of slope scaling. During the same time, the authors mapped selected vertical scanlines using limited access rope techniques to characterize the rock mass. Kinematic and global stability analysis of field data supported the following conclusions. First, the head scarp and gaping tension fracture on the brow of the slope, the drunken (tilted) trees, and the stressed roots suggest the slope failed as a rotational rockslide. The rupture surface appears to have developed along the weak interface between weathered basalt blocks. Second, rockfall was derived from toppling and wedge failures of weathered basalts and Latah claystone, exacerbated by freeze thaw weathering and mechanical wedging by pervasive root systems. Third, surface runoff percolating through discontinuities and root systems within the rock mass, suggested by surface staining and weathering of discontinuity surfaces, decreased slope stability. To manage rockfall and stabilize Fossil Cut, the authors recommended installation of subhorizontal drains and a wire mesh slope stabilization system, combined with routine clearing of drainage ditches.

INTRODUCTION

The Fossil Cut is located 3 miles southwest of Spokane, Washington, along the west side of State Highway 195, near the junction with Interstate-90 (Figure 1). This line is owned and operated by Burlington Northern Santa Fe (BNSF) and shared by AMTRACK. A topographic high in the project area required BNSF to excavate the Fossil Cut to provide a consistent grade for rail traffic heading south out of Spokane, Washington.

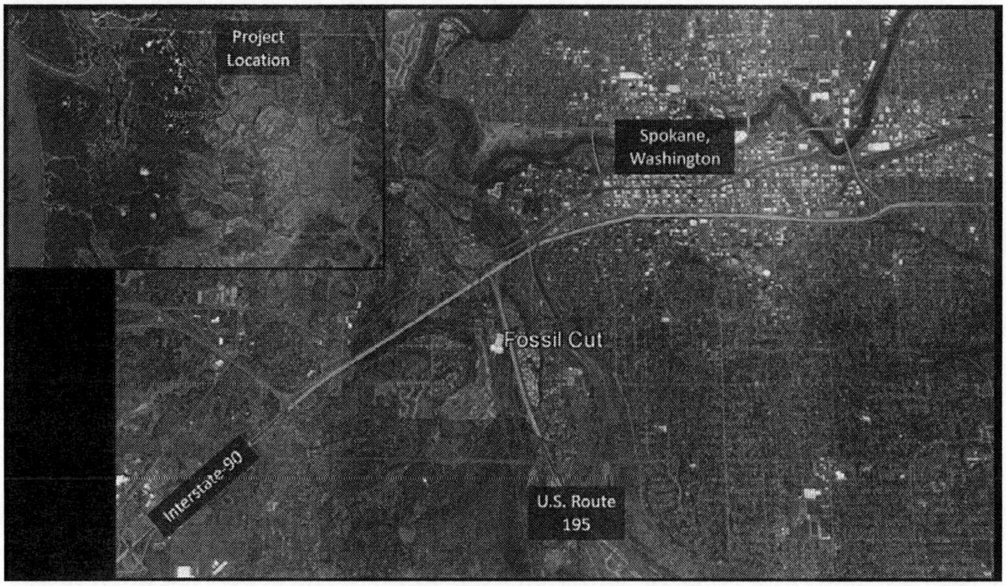

Figure 1 – Project Location.

The through-cut is approximately one-half-mile in length, and ranges in height from 20 to 60 feet (Figure 2). Numerous single family homes exist on the west side of the cut, along the entire length. The brow of the slope is moderately vegetated with grasses, shrubs, and ponderosa pines less than 100 feet tall. The through-cut crosses volcanic rocks of the Columbia River Basalt Group, which is locally overlain by the Latah Formation.

The Fossil Cut has a history of instability. Approximately four years ago, a rockslide occurred at the project location. A geotechnical consultant investigated the slide and produced a report. The consultant recommended scaling of the loose rock and debris on the slope, but no other mitigation was recommended. In May 2016, the BNSF road master noted rock slides and raveling forming a talus deposit in the ditch, blocking the drainage. The slide material consisted of basalt blocks and small clasts of the Latah Formation. BNSF crews cleaned the ditch, but additional rock fall occurred the week of May 15, blocking the drainage ditch again, and potentially jeopardizing and fouling the track with rock debris. BNSF immediately contacted Dr. William Gates from McMillen Jacobs Associates (Gates, pers. comm., 2016) to conduct an initial assessment of the unstable slope. Based on his assessment, the slope failure appeared to be an active complex rotational slide with a large tension fracture at the brow of the slope. The slide material consisted of basalt blocks and small clasts of Latah claystone. Because of the high traffic volume and continued risk of fouling the track with rock debris, BNSF immediately implemented a cautionary slowdown of rail traffic passing through the Fossil Cut. At the same time, BNSF had contracted with McMillen Jacobs Associates to conduct a detailed geotechnical assessment of the unstable slope; oversee rock scaling operations to mitigate the immediate rockfall hazards, and complete a rock slope investigation in support of the conceptual design of long-term mitigation measures.

Figure 2 – Fossil Cut looking north toward Spokane, WA.

PROJECT GEOLOGIC CONDITIONS

Regional Geology

Geologic mapping of the area was completed at a 1:24,000 scale in 2004 by Derkey et al. The mapping focused on the Spokane southwest 7.5-minute quadrangle, in Spokane County, Washington. The regional geology is composed of Quaternary sedimentary deposits, volcanic rocks from the mid-Miocene Columbia River Basalt Group, and sedimentary rocks from the mid-Miocene Latah Formation.

Columbia River Basalt flows are ubiquitous throughout eastern Washington, Oregon, and western Idaho. The Columbia River Basalt Group forms a high plateau of sheet flows between the Cascade Range to the west and the Rocky Mountain Range to the east. Eruptions began approximately 16.6 million years ago in east-central Oregon, near the present-day Steens Mountains. The eruptive focus moved progressively north along the eastern borders of Oregon and Washington, so that by 6 million years ago, the youngest flows were erupting from vents in southeast and central Washington (Hooper et al., 2007).

The Priest Rapids Member of the Columbia River Basalt Group crops out in the project area, as resistant cliff-forming beds (Figure 3). Bedrock failures are most commonly in the form of very large slumps, slump flows, and translational landslides, controlled by weak interbeds or palagonite zones between flows (WSDOT, 2013). Basalt flows throughout the northwest are typically massive and columnar jointed. Columnar jointed basalt flows are generally susceptible to toppling rock falls due to block geometry. Basalt in the project area is spheroidal weathered, and absent distinct columnar jointing. However, basalt blocks are elongate spheroids (prolate), making them susceptible to toppling failure. Priest Rapids basalts (Figure 3) are dark gray to black, fine-grained, and consist of plagioclase (20–30%), pyroxene (10–20%), and olivine (1–2%) in a mostly glass matrix (40–60%) (Derkey et al., 2004).

Figure 3 – Priest Rapids basalt (left); Laminated Latah deposits (right).

The Latah Formation (Figure 3) occurs discontinuously throughout eastern Washington and Idaho. Lacustrine and fluvial deposits of finely laminated siltstone, claystone, and minor sandstone are characteristic of the Latah Formation. Regionally, the Latah Formation overlies pre-Miocene rocks, or is interbedded with the basalt flows of the Columbia River Basalt Group (Derkey et al., 2004). The Latah Formation is notorious for producing unstable slopes and adversely affecting structural foundations (WSDOT, 2013). The unit has presented constant design challenges for construction projects for the Washington State Department of Transportation (WSDOT), Idaho Transportation Department (ITD), and other government and private endeavors.

The Latah Fault is the only significant mapped structure in the Spokane southwest 7.5-minute quadrangle, located east of the project area. The Latah Fault dips steeply to the southwest, and is interpreted to be a normal fault with approximately 200 feet of vertical offset (Derkey et al., 2004).

Project Engineering Geology

Two distinct rock types crop out in the project area: (1) bedded claystones and siltstones of the Latah Formation; and (2) massive, spheroidal weathered basalts of the Priest Rapids Formation (Figure 4). In the Spokane area, near the project location, individual flows of the Columbia River Basalt group, including the Priest Rapids Formation, are separated by lacustrine deposits from the Latah Formation (WSDOT, 2013).

The Latah Formation in the project area varies in thickness and conformably overlies the weathered basalts of the Priest Rapids Formation. At the Latah-basalt interface, a baked contact is evidenced by irregular coloration and relatively brittle consistency of the Latah claystones and siltstones. The main engineering concern for the Latah Formation is its potential for rapid deterioration by softening and eroding when exposed to water and cyclic wetting and drying (Hosterman, 1969).

The Priest Rapids Formation in the project area is highly weathered, forming discrete prolate spheroids, surrounded by a weak and brittle weathered rind. Spheroidal weathering of the Priest Rapids Formation leads to increased aperture width of discontinuities in the rock mass, facilitating water flow and penetration from vegetative root systems above. Mechanical weathering is known to result from these two processes, making spheroidal weathered basalts susceptible to failure.

Figure 4 – Spheroidal weathering of Priest Rapids basalt; note the size of basalt boulders (left). Mapping scanline #3 with rope access techniques (right).

Field Mapping

To evaluate the stability of rock slopes of the Fossil Cut, representative slope sections were selected and mapped with vertical scanline rope access techniques (Figure 4). This evaluation established the orientation of structural discontinuities in the rock mass, and allowed for evaluation of the rock mass quality. Discontinuity and rock mass data were collected using a geologic hammer, a Brunton® geo transit compass or a CLAR® geo-stratigraphic compass, and a GeoID® (smartphone inclinometer technology).

The authors' overall approach to evaluate and assess the rock slopes consisted of four primary methods to classify the quality and the stability of the rock mass:

- Geomechanical rock mass classifications
- Uniaxial compressive strength (UCS) using a geologic hammer
- Rock Quality Designation (RQD) using Palmström's method
- Friction analysis using the tilt method

Geomechanical Rock Mass Classification

As part of the field assessment, the authors developed the geomechanical rock mass classification for the rock slope outcroppings in the Fossil Cut. Two widely accepted classification systems were utilized in the assessment, the Rock Mass Rating System (RMR) and the Geological Strength Index (GSI).

RMR, also referred to as the geomechanics classification system by Bieniawski (1989), is based on the algebraic sum of six rock mass property ratings, namely:

- Strength of intact rock material
- Rock Quality Designation (RQD) – Palmström's volumetric method
- Spacing of discontinuities
- Condition of discontinuities
- Groundwater conditions
- Orientation of discontinuities relative to the rock slope

Field data were compared to tables published by Bieniawski (1989) to estimate RMR_{89} (base RMR). RMR_{89} values can range from 0 to 100. Bieniawski's RMR_{89} classification can be related to Hoek's (1997) Geological Strength Index (GSI). GSI = RMR_{89}-5, where RMR_{89} has the groundwater rating set to 15 and adjustment for joint orientation set to zero. In addition, the GSI rating can also be estimated directly from information collected from field mapping.

Uniaxial Compressive Strength

The authors used a geological hammer to indent or break rock specimens, estimating rock strength in the field. The results were compared to published tables by ISRM (1981) and Hoek and Bray (1981) on field estimates of rock strength. These values were converted to approximate uniaxial compressive strength of the rock.

Rock Quality Designation

To estimate the rock quality designation (RQD), the authors employed the joint volume relationship by Palmström (2005), where RQD % = 115-3.3*J_v, and J_v is the total number of discontinuities per cubic meter. To evaluate J_v in the field, the number of discontinuities impacting a rock mass in x, y, and z directions were summed, then used to compute RQD.

Friction Estimate

To estimate peak friction, the authors employed the tilt test, cross-checking the results against the angle of repose of the talus slope (~ 35°). The authors selected representative slabs of Latah claystone and Priest Rapids basalt for use in the tilt testing. The Joint Roughness Condition (JRC) of discontinuities in the rock mass was observed to be between 4 and 12. The authors chose slabs for tilt testing with these same engineering properties.

Field Observations

Tension Fracture

The authors observed a gaping tension fracture at the brow of the slope, on the west side of the cut. The tension fracture was 3 to 4 feet wide, 10 feet deep, and 50 feet long. Stressed roots were observed spanning the tension fracture along most of the slope length (Figure 5). In addition, drunken (tilted) trees were observed adjacent to the tension fracture, indicating movement of the slope.

Head Scarp

The head scarp on the west side of the Fossil Cut, in the project area, varies in height to a maximum of 14 feet. The head scarp is composed of well-developed soil, bedded claystones, and siltstones of the Latah Formation. When the head scarp was viewed from the east, a clear offset could be seen in bedding of the Latah Formation (Figure 5). An extensive root system was also observed penetrating several feet into the brow of the slope (Figure 6).

Figure 5 – Tension fracture, west side of Fossil Cut on the brow of the slope (left). Head scarp on the west side of Fossil Cut and offset of Latah bedding (right).

Figure 6 – Exposed root system penetrating the brow of the slope (left). Interpreted offset of Latah Formation along arching failure plane (right).

Failure Trace in the Latah Formation

After the slope was scaled, the authors observed an arching failure surface oriented parallel to the face of the slope. The failure surface appeared to offset bedding of the Latah Formation by approximately 8 feet (Figure 6). The marker bed was identified by its unique joint style and spacing.

Rock Mass Characteristics

Latah Formation

The Latah Formation in the project area varies in color from white on weathered faces, to shades of green and gray (Figure 7). Geologic hammer tests indicate the intact rock strength varies from extremely weak rock (R0) to very weak rock (R1), equating to 36 to 725 pounds per square inch (psi). The rock mass is bedded, forming a blocky to tabular fabric. The RQD calculated with Palmström's volumetric method varies between 26% and 66%. Five distinct discontinuity sets with spacing between 0.8 and 8.0 inches were mapped in the Latah Formation. The discontinuities in the Latah Formation are generally polished with undulating surfaces and aperture widths between 0.02 and 0.1 inch. The rock mass is highly weathered on exposed faces, but the extent of weathering was observed to be less than 1 foot into the rock mass. The peak friction angle, measured with tilt tests in the field, averaged 36°. In summary, the base RMR (discontinuities set to 0 and groundwater set to 15) of the Latah Formation is estimated between 48 and 55, suggesting Class III, fair quality rock. The base GSI of the rock mass is between 43 and 50. Variability in RMR and GSI is largely a result of varying RQD throughout the formation.

RMR and GSI estimates from field data collected while mapping the site in July 2016 represent Latah rock strength in the driest conditions of the year. The slope failures prompting the investigation occurred in May 2016, one of the wettest times of year for Spokane, Washington. As noted by Hosterman in 1969 and the Washington Department of Transpiration in 2013, the Latah Formation is very weak in wet conditions. The authors consider the calculated GSI and RMR of the Latah Formation to be high, and the strongest the formation will be throughout the year.

Figure 7 – Jointed Latah Formation (left). Priest Rapids basalt (right); note the spheroidal weathering of the large blocks.

Priest Rapids Formation – Columbia River Basalt Group

Priest Rapids basalts mapped in the project area are reddish brown (Figure 7). Geologic hammer tests indicate the intact rock strength is strong rock (R4), equating to 7,250 to 14,500 psi. The rock mass is massive, forming a blocky to columnar fabric. The RQD calculated with Palmström's volumetric method is 82%. Five distinct discontinuity sets with spacings between 0.8 inch and more than 20 feet were mapped in the Priest Rapids Formation. Discontinuities in the basalt are generally smooth and undulating with aperture widths between 0.004 and 0.4 inch. A peak friction angle of 40° was measured in the field, and is the average of seven tilt tests. The rock mass is highly weathered on exposed faces, with obvious spheroidal weathering. Weathered rinds are relatively weak (725–3,625 psi) in comparison to the strong cores (7,250–14,500 psi). Where discontinuity apertures are widest (0.4 inch), weathering rinds surrounding basalt blocks are in contact with one another. Also, where weathering rinds link over considerable distances, global stability may be reduced by the propagation of tension fractures into the rock mass. In summary, the base RMR of the basaltic rock mass is estimated to be 72, suggesting Class II, good quality rock. The base GSI of the rock mass is 67.

Groundwater

Groundwater seepage and surface runoff were not observed within the slope in the vicinity of Fossil Cut during the site field mapping; however, stained surfaces (Figure 8) of discontinuities suggest there is a history of water seepage through the slope. Infiltrating surface water runoff from the adjacent properties is likely contributing to the quantity of water flowing through the slope. The homes situated on top of the slope have downspouts from gutter drainage systems discharging into the soil and rock in their backyards (Figure 8). The additional water from these downspouts and other surface water derived from rainfall or irrigation may contribute to the water percolating through the slope.

Figure 8 – Oxide staining evidence from surface water infiltration into the rock mass (left). Downspouts from the residents on the brow of the slope (right).

SLOPE STABILITY ANALYSIS

Kinematic Analysis – Rock Fall

The authors collected 385 dip and dip directions of the discontinuities in the rock mass during the field mapping. These data were plotted on stereonets using the computer program Dips® Version 7.0 by Rocscience (2016a). The mapped discontinuities and the major discontinuity sets include one bedding plane and four joint sets (Figures 9 and 10). Bedding and joint sets mapped in the project area are summarized below in Table 1.

Table 1 – Discontinuities Mapped in the Rock Mass

Set ID	Discontinuity Type	Dip / Dip Direction
1	Bedding	02/282
2	Joint	46/259
3	Joint	90/282
4	Joint	85/063
5	Joint	88/025

Figure 9 – Joint Sets 3, 4, and 5.

Figure 10 – Joint Set 1 (bedding) and Joint Set 2.

The rock slope orientation in the project area varies slightly along its length, zigzagging along the strike. The primary orientation of the slope is parallel to Joint Set 4, with the dip/dip direction of 85°/063°. This slope face parallels the rail road tracks. Joint Set 5 has a dip/dip direction of 88°/025°, with the orientation of the joint face forming an oblique angle (about 40°) to Joint Set 4, completing the zigzag. The kinematic analysis focused on these two slope orientations when considering the potential failure modes. The authors used a friction angle of 32° for the Priest Rapids Formation and the Latah Formation. This varies from the peak friction angle measured in the field (36°) because the authors chose to build conservatism into the kinematic analysis. The results of the kinematic analysis suggest that the basalt blocks failed by toppling (flexural and direct), typical for columnar basalt flows exacerbated by rotational failure in the Latah Formation.

Global Stability Analysis of Fossil Cut

The slope appears to have failed by circular rotation within the Latah Formation and, along weak zones in the basalt flows. The authors observed a large tension fracture and scarp near the brow of the slope. Based on these features, the authors believe this to be the head scarp of the failure surface, and that the slope failed along joints in the rock mass in a stepped circular fashion.

Assumptions

The Slide® (Rocscience, 2016b) model was initially designed with two geologic units, the Latah Formation and the Priest Rapids Formation. Information collected in the field suggests the weathered basalt should be included in the slide model because it represents a significant weakness in the Priest Rapids Formation. Because of modeling limitations, the authors were forced to include the weathered basalt as its own unit, placing it between the Priest Rapids basalt and the Latah Formation (Figure 11).

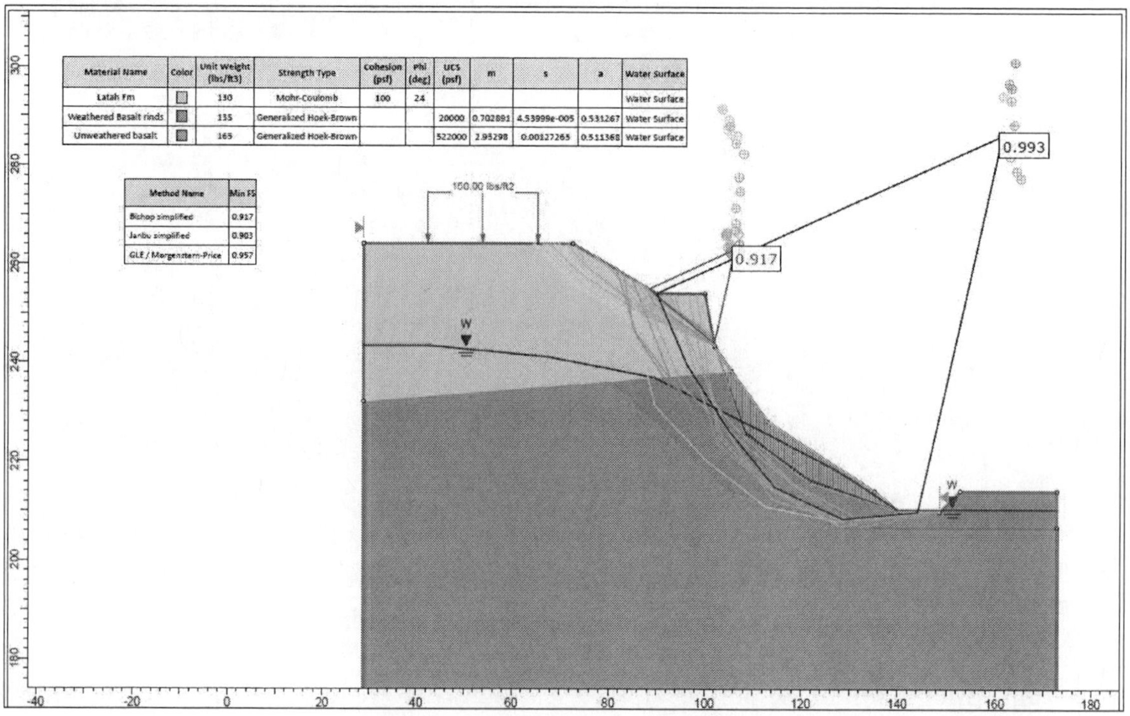

Figure 11 – Base case Slide Model showing Latah (yellow), weathered (orange), and unweathered (green) basalt.

The authors acknowledge the slope is currently in a state of marginal equilibrium and on the verge of failure. These conditions are assumed to represent a factor of safety (FOS) equal to 1.0. To estimate the friction angle of the Latah Formation at slope failure (FOS = 1), the authors decreased the friction angle from 36° (peak friction) to 24°. The same method was used to estimate the Latah Formation cohesive strength at failure. Assuming a friction angle of 24°, the modeled cohesive strength of the Latah Formation at slope failure is 100 pounds per square foot (psf). It is necessary to complete this back analysis of friction angles and cohesive strength because the friction angles measured in the field were in very dry conditions. The authors assume the Fossil Cut's latest failure was under saturated to semisaturated conditions. Under these conditions, the authors expect Latah friction angles to be closer to 24° than 36°.

The authors used publicized information on Geologic Strength Index (GSI) of weathered rocks (Dearman et al., 1978) to estimate the GSI of weathered and unweathered basalts for use in the Slide® model. The authors selected a GSI of 40 for unweathered basalt, and 25 for highly weathered basalt. The GSI of 67, estimated form field data, was not included in the slide model because failures in the formation are expected to travel along weathered contacts (GSI = 25), or adjacent to weathered contacts in unweathered basalt (GSI = 40), not through the core of unweathered basalt (GSI = 67). For this reason, the authors selected a GSI closer to that of weathered rocks from the publicized data instead of very strong rocks (basalt core) as measured in the field.

The highest slope cross section was selected for analysis. Material properties used in this analysis are those shown below in Table 2. The generalized Hoek-Brown and Mohr-Coulomb criterion variables were selected for this slope stability analysis because they are readily available or estimated in the field, and these variables are easily contextualized in engineering geology.

Table 2 – Engineering Properties of Rock Units

Rock unit	Moist Unit Weight (γ)	Cohesion	Phi	UCS	GSI	m_b	s	a
Units	**(pcf)**	**(psf)**	**(deg)**	**(psf)**	**-**	**-**	**-**	**-**
Latah Formation	130	100	24	-	-	-	-	-
Weathered Basalt	135	-	-	20,000	25	0.403	4.54E-05	0.5313
Unweathered Basalt	165	-	-	522,000	40	2.933	0.001273	0.5114

UCS = Uniaxial Compressive Strength.
GSI = Geological Strength Index.
m_b is a reduced value (for the rock mass) of the material constant m_i (for the intact rock).
s and a are constants that depend upon the characteristics of the rock mass.

Table 3 summarizes the predicted seismic base acceleration coefficients for the Fossil Cut site based on AREMA limit conditions (2014): serviceability (GM1), ultimate (GM2), and survivability (GM3) (USGS, 2014). The base acceleration coefficients are calibrated for Site Class B, and are applicable for the average return periods shown in Table 3. Analysis was performed for static and seismic loading conditions. Seismic loading was evaluated utilizing pseudo-static methodology (Kramer, 1996). The horizontal seismic coefficient used in this analysis was equal to one half of the maximum site-adjusted peak ground acceleration during the design seismic event. GM2 ground motions were considered in the Slide® analysis.

Table 3 – Seismic Base Acceleration Coefficients

Ground Motion Level	Return Period (yr)	Base PGA (g)	Base S_s (g)	Base S_1 (g)
GM1	100	0.02	0.046	0.014
GM2	475	0.056	0.131	0.042
GM3	2,475	0.143	0.336	0.096

PGA = peak ground acceleration; Ss = short period (0.2 s) spectral acceleration; S1 = long period (1.0 s) spectral acceleration.

Results

Stability analytical results for the maximum slope height are presented in Table 4. The factor of safety (FOS) for the unreinforced slope equals about 1.0, for the static case. For the seismic case the FOS of the slope equals 0.94. If the slope is reinforced with a high tensile Tecco mesh and 20-foot-long rock dowels, arranged in a 6 x 6 foot pattern, the FOS exceeds 1.3 for both static and seismic cases.

Table 4. Summary of Slope Stability Factors of Safety

Slope Condition	Height (ft)	Factor of Safety	
		Static	Seismic[1]
Unreinforced slope section	55	~ 1.00	0.94 (GM2)
Reinforced slope section	55	1.36	1.31 (GM2)

[1] Seismic analyses based on Level II AREMA (2014) ground motion (see Table 3).

The analysis shows the slope without reinforcement proves to be marginally stable, both under static and seismic conditions.

CONCLUSIONS AND RECOMMENDATIONS

Conclusions

This is an area of frequent rock slides and raveling, with the last major failure occurring in 2012. The rockslide area is approximately 120 feet wide and 60 feet high and is located at MP 74.1 at Fossil Cut. Recently, the slope failed as a subrotational slump in the Latah Formation and along joints within the basalt units, exacerbated by apparent water seepage from uncontrolled surface runoff. Global stability analysis shows that the upper portion of the slide area (claystones of the Latah Formation) and the entire slope are in a state of marginal equilibrium with a static FOS ≤ 1.

The columnar basalt rock units fail primarily by toppling (flexural and direct). Periodic rock falls and slope raveling will lead to continual filling of the drainage ditch, potentially allowing rocks to foul the rail line. Surface water runoff and seepage through the rock mass is facilitated by pervasive discontinuities and an extensive root system, exacerbating the slope instability. After scaling of the unstable rockslope, the slope appears to be marginally stable. Continued rock fall and periodic rockslides are expected if no mitigation measures are implemented.

It appears that slope drainage and surface water runoff play a significant role in the stability of Fossil Cut in the project area. Evidence of water seepage and water staining was observed on discontinuity faces, and along the trace of root systems crisscrossing the soil, claystones, and basalts. Unsealed discontinuities behind the scarp face may carry surface water into the failure zone and reactivate a rockslide.

Recommendations

To control and mitigate the flow of water through the slope, the authors recommended the following to BNSF.

To protect the rail line and manage continual rock fall, the authors recommended the following options or combination of options to BNSF.

Option 1 – Regular Track Inspection and Rock Scaling:

- Conduct regular track inspections by BNSF personnel to monitor the drainage ditch and remove rocks and other materials as they accumulate before fouling the track.
- Conduct periodic rock scaling as needed.

Option 2 – Improve Drainage:

- Install up to nine 40-foot-long radially arrayed, subhorizontal perforated PVC gravity drain pipes along the slope.
- Control or divert storm water runoff at brow of slope from neighboring residences.

Option 3 – Reinforcement and Slope Protection:

- Install approximately twenty 20-foot-long rock dowels in a spot pattern as needed to support identified blocks in the massive basalt.
- Install 3,600 square feet of 4 mm GEOBRUGG® TECCO® system for containing rock falls in combination with a SPIDER® rhomboid mesh wire (S3-130 high-tensile spiral rope net) around unstable boulders. Rock dowels used in the installation of the TECCO® system will be 20 feet in length.

REFERENCES

American Railway Engineering and Maintenance-of-Way Association. *AREMA Manual of Railway Engineering*, 2014.

Bieniawski, Z.T. *Engineering Rock Mass Classifications*. New York: Wiley, 1989.

Dearman, W.R., F.J. Baynes, and T.Y. lrfan. 1978. Engineering Grading of Weathered Granite. Engineering Geology, Vol. 12, pp. 345-374.

Derkey, R.E., M.M. Hamilton, and D.F. Stradling. Geologic Map of the Spokane Southwest 7.5-minute Quadrangle, Spokane County, Washington, 2004.

Geobrugg. *Ruvolum, Online Dimensioning tool for the TECCO® / SPIDER® Slope and Rock Stabilization System*. Geobrugg Ag, Geohazard Solutions, 2016.

Hoek, E., and J.W. Bray. *Rockslope Engineering*. Institution of Mining and Metallurgy, 1981, 357p.

Hoek, E. Rock Mass Classification. *Practical Rock Engineering*. Rocscience. https://www.rocscience.com/education/hoeks_corner, 2007.

Hooper, Peter R., Victor E. Camp, Stephen P. Reidel and Martin E. Ross. The Origin of the Columbia River Flood Basalt Province: Plume versus Nonplume Models, Special Paper of the Geological Society of America, 2007.

Hosterman, John W. *Clay Deposits of Spokane Co. WA*. U.S. *Geological Survey Bulletin* 1270, 1969.

International Society for Rock Mechanics (ISRM) Commission on Standardization of Laboratory and Field Tests. *ISRM Suggested Methods: Rock Characterization. Testing and Monitoring*. Editor E.T. Brown. London: Pergamon Press, 1981, 211 pp.

Kramer, S. *Geotechnical Earthquake Engineering*. Prentice Hall, 1996.

Palmström, A. Measurements of and Correlations between Block Size and Rock Quality Designation (RQD). *Tunnels and Underground Space Technology* 20, 2005.

Rocscience Inc. *DIPS, Plotting and Analysis Software for Geomechanical Structural Data*. Toronto, *Version 7.0*. Canada, 2016a.

Rocscience Inc. *SLIDE Slope Stability Analysis Software, Version 7.0*. Toronto, Canada, 2016b.

United States Geological Survey (USGS). *USGS 2014 Earthquake Hazards Program/Hazard Tool*. Available at: http://earthquake.usgs.gov/hazards/interactive/, 2014. Accessed June, 2016.

Washington State Department of Transportation. *WSDOT Geotechnical Design Manual*, 2013.

Investigation and Mitigation Design for a Highwall Rock Slope in Southwest Virginia

Jay R. Smerekanicz, P.G.
Golder Associates Inc.
670 North Commercial St., Ste. 103
Manchester, NH 03101
(603) 668-0880
jay_smerekanicz@golder.com

Peter C. Ingraham, P.E.
Golder Associates Inc.
670 North Commercial St., Ste. 103
Manchester, NH 03101
(603) 668-0880
pingraham@golder.com

Stu Gitchell
15515 Bull Run Road
St. Paul, VA 24283
(804) 771-3000
stu.gitchell@dom.com

Andrew Salmaso
Ameritech Slope Constructors, Inc.
21 Overland Industrial Boulevard
Asheville, NC 28806
(828) 633-6352
asalmaso@ameritech.pro

Robert Lyne
GeoBrugg North America, LLC
804 Windspray Drive
Summerfield, NC 27358
(336) 337-0795
bob.lyne@geobrugg.com

Prepared for the 68th Highway Geology Symposium, May 1-4, 2017

Acknowledgements

The author(s) would like to thank the individuals/entities for their contributions in the work described:

Rusty Hicks, Golder Associates Inc.

Disclaimer

Statements and views presented in this paper are strictly those of the author(s), and do not necessarily reflect positions held by their affiliations, the Highway Geology Symposium (HGS), or others acknowledged above. The mention of trade names for commercial products does not imply the approval or endorsement by HGS.

ABSTRACT

In 2015 - 2016, a solid waste facility in southwest Virginia mitigated rockfall hazards associated with an old mine highwall adjacent to a lined leachate pond by removing loose rocks from the highwall, stabilizing large metastable blocks in place, and designing and installing a wire mesh rockfall drape. The work was completed in response to frequent rockfalls and raveling of portions of the highwall that could potentially injure personnel working next to the leachate pond, damage a perimeter drainage ditch at the toe of the highwall or damage the pond liner. A portion of the highwall was constructed during past open-pit coal mining, and past blasting exhibited backbreak up to 30 feet (ft) into the highwall. The maximum highwall height is on the order of 320 ft, and the highwall slope is 1,000 ft long, with eight benches and an overall slope angle of 50 degrees. Rockfall mitigation design was initiated in late 2014 consisting of geologic/geotechnical data collection, ground-based light detection and ranging (LiDAR) topographic surveying, and rockfall trajectory modeling to establish a basis for design.

Slope mitigation was started in mid-2015 and consisted of hand and mechanical scaling to remove loose rocks from the slope, along with stabilization of large rock blocks and dental/structural shotcrete placement to fix blocks too large to be safely scaled down to the leachate pond in place. Scaling was performed with a fifteen-foot-high, temporary, movable rockfall barrier placed along the edge of the lined leachate pond to protect the pond liner.

Analysis of controlled rockfall trajectories during scaling indicated two rockfall mitigation approaches could be considered: a rockfall drape placed over the entire slope, and a hybrid rockfall barrier placed at the toe. In September 2015, 50% design packages were developed for each option to allow pricing of each alternative. While the hybrid system was about half the cost and could ostensibly be constructed faster, it did not provide as much worker protection from a large rockfall event as the drape alternative. Based on the reduced performance of the hybrid, a need for barrier post foundation support in blast-damaged ground, and the operator's desire for more comprehensive worker protection, the drape system was selected. GeoBrugg's 4mm high corrosion resistant Supercoating® high-strength steel Tecco™ wire mesh material was chosen for the drape. This heavier-gauge mesh provides the strength to control larger rock blocks, has mesh spacing sized for smaller rock pieces, obviated the need for a smaller secondary mesh, and could be deployed in one pass. From late November 2015 when materials were ordered through June 2016, the drape system anchors and mesh were installed. The completed system includes 144 wire rope anchors and about 372,000 ft^2 of wire mesh drape over the slope. The wire mesh panels were installed using a heavy lift helicopter over ten days in April 2016 and were finish-seamed together after placement. The length of the 11.5-ft wide panels hung with the helicopter ranged from 25- to 225-ft long, averaging about 200-ft. The panels were hung in 187 flights (i.e., "picks"), and weighed between 430 and 1,617-lbs. The helicopter averaged panel placement cycle times on the order of 13 minutes, a testament to the teamwork of the ground crew and the military training of the pilot. Approximately 500,000 clips, placed by hand on rope rappel, were used to finish seam the panels to one another.

INTRODUCTION

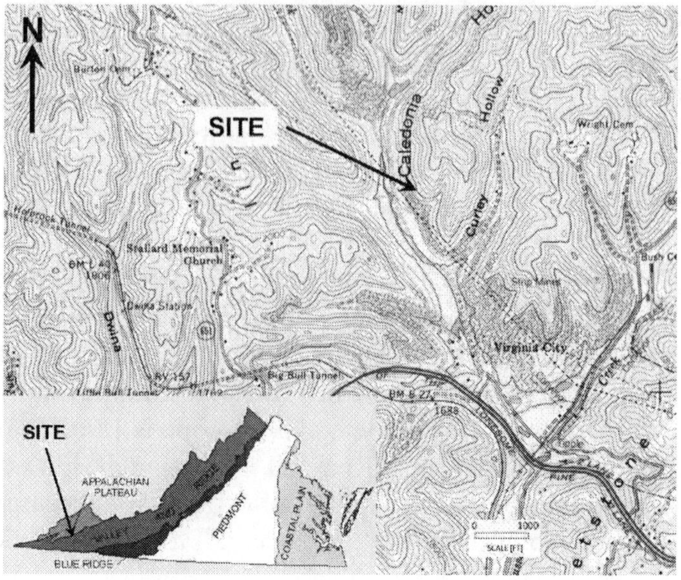

Figure 1 – Site location map.

A portion of a solid waste facility in southwest Virginia included construction of a leachate containment facility at the toe of a highwall slope that was part of an historic coal mine (Figure 1). Past production blasting and mining activities left ragged slopes that were prone to rockfall generation. The constructed leachate pond and drainage system were subject to repeated rockfall strikes requiring rockfall mitigation to protect the facility and site workers responsible for facility maintenance. Rockfalls striking the facility could not only cause damage to the facility and impact site workers, but could also damage the pond liner and release of leachate into the environment. To prevent rockfalls from impacting the leachate facilities, the facility commissioned geotechnical studies and construction of rockfall mitigation measures between January 2015 and June 2016.

The solid waste facility constructed the rock slope above the Final Leachate Pond (FLP) in 2013 and 2014 using production blasting for rock removal and excavation. The rock slope was designed using ¾ horizontal to 1 vertical (3/4H:1V) slopes between benches placed at roughly 50 ft elevation intervals (Figure 2). The benches had a design width of about 10 ft, producing an overall rock slope cut angle of about 50 degrees from horizontal. The overall rock slope vertical height is about 320 ft. Perimeter control blasting techniques were not used to construct the final rock slope grade. A concrete lined diversion channel was constructed in 2014 at the toe of the rock slope on the first bench (Bench #1) to collect and divert runoff around the FLP. A 20-ft wide gravel roadway was constructed around the perimeter of the FLP for access and

Figure 2 – Pre-construction FLP rock slope condition panorama from west side of FLP. View from northeast (left) to southeast (right).

maintenance purposes. The FLP liner system consists of a 4-inch (in) thick grouted fabriform armor layer, underlain by a 16-ounce non-woven geotextile and 60-mil high density polyethylene (HDPE) membrane liner. The HDPE liner lies on a double-sided geocomposite drainage layer and 6 inches of crushed drainage stone. The FLP is designed to hold up to 75 million gallons of leachate generated from storage of coal-ash waste.

Shortly after construction of the concrete-lined perimeter channel on Bench #1 in late 2014, rockfall debris started to accumulate in the channel, requiring removal (Figure 3). Rockfalls continued to occur, filling the benches and accumulating within the channel, on the gravel roadway, and occasionally reaching the FLP. The rock block sizes ranged up to 3 ft in longest dimension. Based on site observations and review of blasting logs and videos, it appeared that the combination of uncontrolled blasting and other mining disturbance, and the lack of perimeter control blasting techniques led to extensive backbreak within the final rock slope, loss of benches, and to generation of rockfalls. Additionally, the previous bench design elevations were not based on lithology, which led to formation of overhangs of more resistant sandstone overlying less resistant siltstone, shale and coal due to differential erosion.

Figure 3 – Rockfall debris accumulating in drainage ditch, perimeter roadway, and concrete liner. Note ponded water in ditch dammed by debris on far left side of photo.

The facility operator was concerned that rockfall could damage the perimeter ditch and FLP liner, present a hazard to personnel, and inhibit required operation and maintenance of the FLP facilities. Hence, in late 2014, institutional controls were established to prohibit personnel working at the toe of the slope. The facility started filling the FLP with leachate in April 2015. If rockfall strikes occurred on the FLP liner while filled, and a leak developed, repairs and mitigation would be difficult to conduct, and site operations and environmental permit requirements could be hampered. The facility then fast-tracked rock slope mitigation design and construction to reduce the potential for rockfall impacts on the FLP.

EXISTING CONDITIONS

Geology

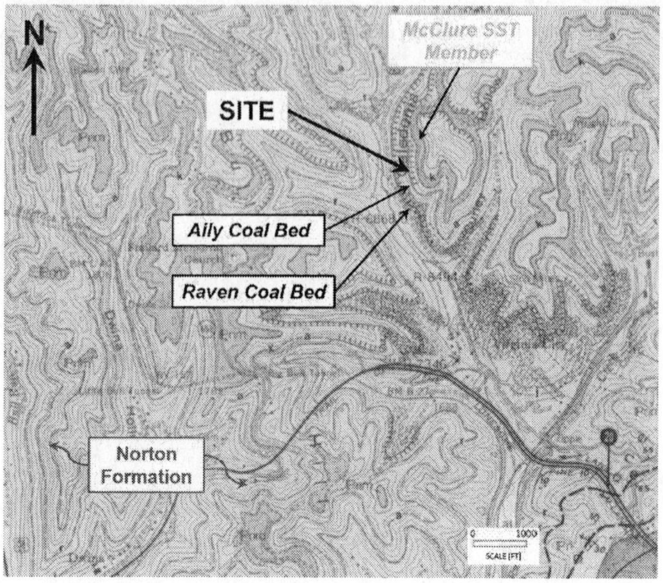

Figure 4 – Site geologic map (Evans and Troensgarrd, 1991).

The project is located at the margin between the relatively flat-lying Appalachian Plateau and the folded/faulted Valley and Ridge physiographic provinces (see inset map in Figure 1). Regional geologic mapping indicates the project highwall slope consists of the Middle Pennsylvanian-aged Norton Formation, including strata from below the Raven No. 1 coal bed stratigraphically upward to above the Aily coal bed (Figure 4; Evans and Troensgaard, 1991; Nolde, J.E., 1996). The lithology of the Norton Formation consists of cyclothemic coarsening-upward sequences of coal, shale, siltstone and sandstone of various thicknesses and extents. The Norton Formation consists of cyclothemic sequences of sandstone, siltstone, shale and coal. The sandstone is light-to medium-gray, fine-grained, thin- to medium-bedded, contains feldspar and mica, with large fragments of plant fossils, and has local conglomeratic lenses. The siltstone is medium- to dark-gray, and laminated, containing siderite nodules and lenses. The shale is dark-gray and laminated. The coal is black, slightly iridescent, and brittle, and forms a local hydrogeologic barrier perching groundwater above. The total thickness of the Norton Formation ranges from 270 to 420 ft. The project rock slope contains at least two coal riders known as the Raven Nos. 2 and 3. Locally, the thick sandstone beds above the Raven No. 3 coal are known as the Dismal and McClure Sandstones (Englund, 1981; Nolde, 1989; Whitlock, 1989). The coal beds are typically 1 to 5 ft thick, and the sandstone beds range from about 3 ft to over 50 ft thick.

Locally, the Norton Formation claystone, mudstone and shale weathers rapidly to clayey soils highly susceptible to landsliding. Additionally, the formation is mapped as the source of numerous active and inactive landslides, debris flows, debris avalanches, and areas susceptible to rockfall. The latter areas contain steep and locally vertical slopes and cliffs, formed dominantly of sandstone, limestone, sandy shale, mudstone and claystone. The interbedded finer grained shale, mudstone and claystone weather rapidly leaving the more competent sandstone and limestone rock faces unsupported (Outerbridge, 1982).

Open-pit and auger mining methods were active at the site into the 1970's. The mining methods included quarry blasting without perimeter control, and construction of adits, shafts and drilling of horizontal auger holes for coal extraction. These subsurface disturbances likely contributed to the slope conditions causing rockfalls.

Overall bedding strikes roughly north-northeast to south-southwest, and dips gently to the east. Several joint sets are present within the sandstone and siltstones (at least four sets exist).

Additionally, highly weathered dikes of very fine grained diabase (likely of Mesozoic age) and/or clastic debris, up to about 0.5 ft thick are present within a north-south trending vertical joint set. The diabase and/or clastic dike debris has weathered almost completely to a clay in some exposures.

The sandstone has an estimated field strength of 15,000 to 36,000 psi (R5 rating, very strong rock), while the shale and coal have much lower estimated field strengths of 35 to 725 psi (R0/R1 ratings, extremely weak to very weak rock). This difference in strength leads to differential weathering, causing the less resistant rocks to weather quickly compared to the more resistant sandstones, leading to rockfalls primarily due to undercutting of blocks and toppling. The extensive backbreak caused by production blasting and past mining activity also contributes to rockfall generation.

STABILIZATION DESIGN DEVELOPMENT

Conditions Requiring Stabilization

The slope above the FLP was cut into the old mine highwall in 2013-2014 as part of the site development plan, with an overall design slope angle of 50 degrees and a series of six benches as shown in Figures 2 and 3. The layout of the slope and benches was based on overall slope geometry and was not designed to account for bedrock lithology by building benches in weak units underlying hard units to limit subsequent undercutting and possible rockfall generation. During production blasting, several prominent joint sets were encountered and the slopes between the benches broke along those joints; however, the dip of the joints was steeper than the bench face angle and daylighting joint faces were not created. Some variation in bench width and in some cases complete loss of benches occurred as a consequence of the backbreak along these joint sets (Figure 5).

Figure 5 – Backbreak along crest of bench above north end of perimeter runoff channel. Grade stake for scale.

During initial site inspections in early 2015, an upper series of hard sandstone beds with undercut shale beds were observed in the middle and upper portions of the slope. In many locations the bench width varies widely, and several benches have collapsed due to the lack of perimeter control and extensive backbreak. A large amount of loose rock and debris was present on the slope that apparently was not removed during excavation of blasted rock, or not scaled after completion of each bench. In consequence, many rock blocks were falling from the full length of the slope and accumulating in the FLP perimeter channel, on the perimeter gravel roadway below, and some rocks rolled down into the FLP. Most of the benches on the slope had

Figure 6 – Heavy bench accumulation of rockfall debris. Bench can no longer retain additional rock blocks and soil, and they are being passed downslope.

accumulated falling rocks from the slopes above, and were filling with debris (Figure 6). Surface water emanating from an adjacent wetland northeast of the highwall drains through strata along the bottom half of the slope at the north end, and is locally perched on the more impermeable cleats at the base of the coal seams. Icefalls and rockfalls are common in these areas throughout the winter season.

To aid rockfall mitigation design, in March 2015 the facility commissioned geologic and geotechnical mapping of the base of the FLP rock slope. This effort included data collected from an adjacent rock slope excavation area to support rockfall bounce analyses and evaluation of potential rockfall remedial design approaches. Geologic mapping was supplemented by initial 3D terrestrial LiDAR survey scans of the slope.

Observations and analysis indicated rockfalls on the FLP highwall fell into three categories:

- Type 1 - Blocky rockfalls that comprised harder competent medium to thick bedded sandstone and siltstone strata formed by differential erosion and undercutting of weaker shales, mudstones and coal strata until individual blocks could fall out of the face.

- Type 2 - Intermediate-sized blocks derived from broken and slaking thinly bedded siltstones, shales and coal that exist in blast-damaged zones; broken benches; blocks that had already fallen; and from joint/weathering controlled blocks that could break up as they fell into smaller pieces (6-in to 10-in size).

- Type 3 - Small shards of slaking shale, commonly from the lower half of the slope, that flaked off the face in small pieces in response to wetting and drying of the shale causing expansion and contraction of the wetting surface. The small shale shards fall nearly continuously during periods of precipitation and during winter months as the face freezes and thaws daily.

The liner of the FLP was considered susceptible to damage by falling rocks of the Type 1 category of rockfalls discussed above. Long term operations could be susceptible to ongoing rockfalls occasioned by weathering of shale strata - either undercutting hard sandstone blocks (Type 1) or generating the second and third types of rockfalls noted above.

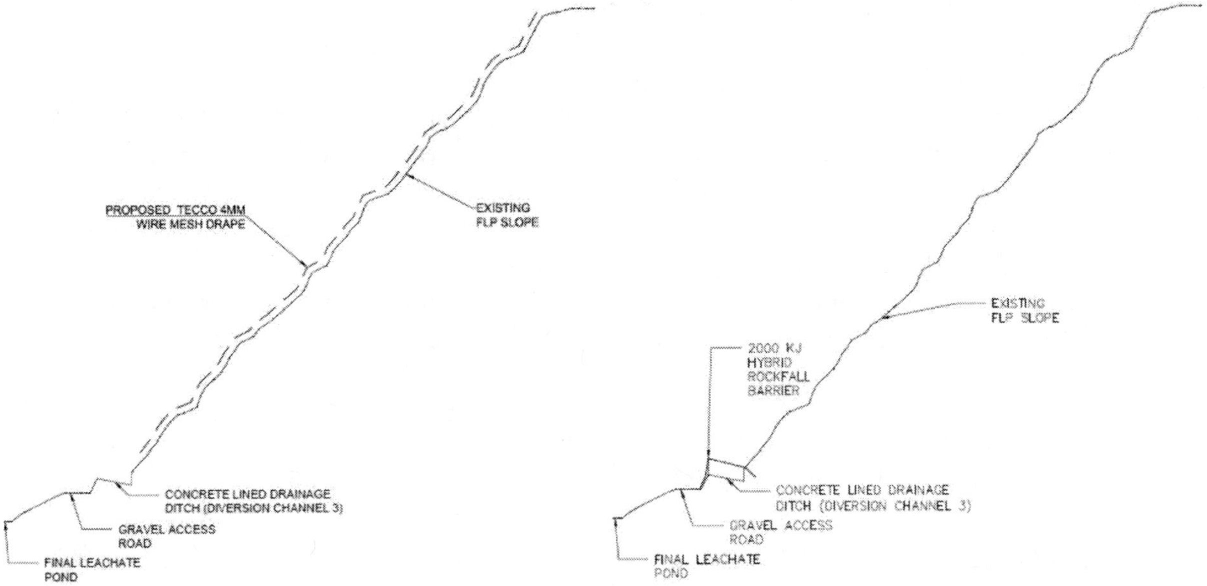

Figures 7A (left) – Drape design section, and 7B (right) – Hybrid rockfall barrier design.

Consideration of Alternatives

Golder developed two alternative mitigation approaches to the 50% design phase to provide sufficient detail to conduct a cost comparison between the options. The first option consisted of a 4 millimeter (mm) Tecco™™ rockfall drape system mantling the FLP rock slope, using approximately 120 wire rope anchors around the perimeter of the drape (Figure 7A). The drape was designed to intercept falling rocks and control their trajectory to direct them to the concrete lined drainage ditch. The second option consisted of a hybrid rockfall barrier/drape positioned on the outboard edge of the first bench (originally the second bench but the second bench was not stable and was missing in places), with a short "tail" constructed of ring nets extending from the top of the barrier to the inboard FLP access road shoulder (Figure 7B). The hybrid drape system was designed to extend approximately 950 linear ft along the edge of the concrete lined drainage ditch on Bench #1. The hybrid barrier/drape would be constructed with 15 ft high posts spaced horizontally every 30 ft with a hybrid rockfall barrier/drape composed of ring nets and 3 mm Tecco™ mesh, requiring approximately 55 wire rope anchors for support. The hybrid barrier/drape was designed to intercept falling rocks at higher velocity and deposit them on the gravel road that surrounds the FLP for removal.

To make conditions safe to work under for both options, the condition of the slope required both mechanical and hand scaling, and installation of spot rock dowels and dental shotcrete to stabilize large sandstone rock blocks in the middle and upper portions of the slope that could not safely be brought down. These areas contained large rock blocks (exceeding 4 ft) that could compromise the integrity of the FLP liner if they fell from the slope. Rock blocks of this size would have an estimated energy of 4,300 kilojoules (kJ), which would also compromise the drape and hybrid systems, and were therefore stabilized in place.

Selected Slope Treatment and Stabilization System

In April 2015 the facility decided to begin work with rock scaling and the installation of the dental shotcrete and dowels. The selection process for the permanent rockfall mitigation method progressed to the 50% design level and occurred simultaneously with the scaling, dental shotcrete and rock dowel activities, with approximate construction costs for both alternatives estimated in September 2015. While the hybrid system was about half the cost and would be faster to build, it did not provide sufficient protection if a large rockfall event were to occur. Based on the reduced performance of the hybrid, its need for foundation support for barrier support posts in questionable ground, and a need for more comprehensive protection of workers, the facility chose to move forward with the drape system option.

Rockfall Drape Design

The rock drape system consists of Tecco™ mesh manufactured by GeoBrugg® Protection Systems (GeoBrugg). The Tecco™ system is constructed of high-tensile steel (256 ksi) coated in a proprietary aluminum-zinc anti-corrosion coating and arranged in a diamond shaped pattern. The drape is secured to the rock slope by primary cable anchors and intermediate anchors used to secure the top, side and bottom cables.

The major components of the rock drape are:

- GeoBrugg Tecco™ mesh, consisting of 0.157-in (4 mm) diameter steel wire, single twisted into diamond-shaped meshes, with a nominal unstrained opening of 3.3 by 5.4-in. The wire is coated with a proprietary zinc-aluminum coating for corrosion protection.

- 3/4-in diameter, galvanized 6x19 extra improved plow steel (EIPS), independent wire rope core (IWRC), double-leg wire rope anchors with a cementitious grout bond zone of at least 15-ft for top anchors and at least 5.5 ft for bottom anchors. The wire rope anchors are constructed using galvanized thimbles and wire rope clips.

- 7/8-in diameter galvanized 6x19 EIPS IWRC wire rope was used for upper and side support ropes, and galvanized 6x19 EIPS IWRC 3/4-inch diameter wire rope was used for the bottom support rope.

- 5/16-in diameter galvanized 7x19 wire rope for seaming the perimeter of the drape to the top, side and bottom support ropes, wrapping the seaming rope through each Tecco™ mesh diamond and around the support ropes.

- Two GeoBrugg® 4 mm diameter T3 connection clips connecting Tecco™ drape mesh at every other diamond overlap. Generally, the mesh was overlapped by at least 6-in between vertical panels.

The mesh design consisted of an evaluation of potential rockfall particle size, rockfall modeling, slope condition, interbench and overall slope angles, interface friction angle, potential debris load, and snow/ice loads. Rockfall modeling and observation of the scaling operations indicated that the average rockfall block diameter is about 1.5 ft, and at most, 10 cubic yards (CY) of rockfall materials are anticipated to fall in any one event. The analysis indicated that a ring net drape system with an overlain finer mesh would adequately retain rockfalls of this size. However due to the very large area requiring drape materials (estimated at 403,000 ft^2, including a 25% contingency), and the difficulty in installing the relatively heavy ring nets on a restricted-access rock slope, the use of lighter drape materials was evaluated. The larger gauge Tecco™ system mesh was chosen as it could hold the assumed rockfall, is lighter than a ring net/mesh system, and could be rapidly deployed by heavy-lift helicopter in relatively long lengths (up to 225 ft) in limited access areas. Additionally, the drape could be installed in one pass, as opposed to a ring net/light drape design which would require installation of the ring nets first, followed by a lighter mesh (e.g., double-twist wire mesh), requiring two passes.

CONSTRUCTION

Construction activities started in April 2015, and included:

- Temporary rockfall barrier installation
- Mechanical and hand scaling
- Rock dowel and shotcrete installation
- Cable anchor drilling and installation
- Cable anchor testing
- Rockfall drape installation

Figure 8 – Temporary rockfall barrier.

The paragraphs below provide brief summaries of these activities.

Temporary Rockfall Barrier Installation

Prior to implementing mechanical and hand scaling operations, a 120-ft long, 500 kJ capacity temporary rockfall barrier was installed at the toe of the north end of the slope, just inboard of the FLP (Figure 8). The barrier consisted of 4 mm Tecco™ wire mesh supported by steel I-beams, 10- to 15-ft tall mounted on steel plates, and anchored by concrete blocks. The barrier was designed to be moveable with minimum disassembly, such that the barrier could be dragged along the FLP slope toe by a front-end loader and placed into position below the area to be

Figure 9 – Scaled debris in concrete ditch resting on blasting mats and contained by temporary barrier.

scaled. The temporary rockfall barrier was used in tandem with rubber tire blasting mats that were laid along the top of the concrete lined drainage ditch on Bench #1 to dampen the kinetic energy of falling rocks (Figure 9). The barrier and mats were moved from a north-to-south direction as the scaling operation moved across the rock slope.

Scaling

The rock slope was scaled to remove loose rock and soils from the rock face and benches (Figure 10). The majority of the rock slope was hand scaled using standard 4-ft long steel mine scaling bars; however, some larger unstable rock blocks were broken apart on the slope using a "boulder buster" and scaled using compressed air bags. The boulder buster uses a small propellant charge placed within a water-filled drill hole and then initiated. The pressure pulse initiated by the propellant charge is directed via a water-tight barrel at the collar of the hole into the incompressible water, resulting in an expansion and hence breakage of the rock in tension. The resulting charge is strong enough to break the rock, but not enough to produce fly-rock. The air bags were rated to produce jacking forces of up to 70 tons, and when coupled with scaling bars are able to push rocks off the slope in a controlled manner.

The northern 100 ft of slope was scaled using a specialized mining slusher. The slusher consists of a compressed-air powered 3-drum hoist, with cables running up to pulleys attached to wire rope anchors at the crest of the slope, and connected to a mini-dragline excavator bucket. The drum hoists are controlled by clutches, which direct the bucket up and down and across the slope, which is in turn used to pull down loose debris to the toe of the slope. Four temporary wire rope cable anchors were installed at the top of slope to support the slusher pulleys: one into the soil and rock on the east side of the soil slope, and three in the rock face above Bench #5 to the west of the soil slope. .

Figure 10 – High scalers scaling rock face.

Dental Shotcrete and Rock Dowel Installation

Following completion of the scaling, inspection of the slope face was conducted to identify rock blocks that needed to be stabilized in place. Ten (10) areas containing medium- to thick-bedded sandstone layers that contained large dilated rock masses (blocks 4 ft or larger) that could not be scaled from the slope without risking damage to the FLP were identified during the rope rappel inspections. The rockfall energy of these masses was calculated to be too great for the temporary rockfall barrier as they may bounce over the barrier or pass through the barrier and thus impact the integrity of the FLP liner. As these rock masses had the potential to fall in the near future, area-specific stabilization was designed, incorporating untensioned rock dowels and fiber-reinforced dental shotcrete (with drainage) to support these rock masses. There were several variations of the specified repairs depending on the degree of reinforcement required. The materials specified for use in the repairs included rock dowels, geotextile drainage board, steel welded wire fabric, and steel fiber reinforced shotcrete.

Rock dowels, shotcrete support dowels, and shotcrete were installed from July 9 to August 27, 2015. The rock dowels were drilled using either a hand held "plugger" drill or a wagon mounted down-hole hammer drill (Figure 11). The dowel spacing and depths were selected to address specific areas of the FLP rock slope where rock blocks were to too large to be safely scaled or retained by the Tecco™ rock drape. The rock dowels were separated into two categories: rock dowels used to pin large rock blocks into place, and rock dowels used to provide support to the areas repaired using buttressing shotcrete. The rock dowels used to pin the large rock blocks were designed to be 8 to 15 ft long and fully grouted along their entire length. Rock dowels used to provide support for the shotcrete buttress areas were designed to be 4-ft long and fully grouted.

Figure 11 – Drilling shotcrete support dowels.

Figure 12 – Dry-mix shotcrete application.

Many of the rock masses had overhangs for which the shotcrete formed a filling buttress for structural support (Figure 12). Dry-mix fiber-reinforced shotcrete was specified due to the large distances between the potential shotcrete plant staging and the application areas, and the difficulty in getting concrete wet-mix trucks close to the slope. Shotcrete was also applied to further stabilize and armor loose portions of the rock slope and limit further crumbling and erosion of the slope. In areas exceeding 5 CY of shotcrete, a welded wire fabric was used to provide additional support for thick layers of shotcrete (generally exceeding 6 in). Additionally, based on field conditions, drainage elements such as geotextile-backed drainage board (Mirradrain®) and PVC pipe drains were used ensure water would not build up behind the repairs.

Cable Anchor Drilling and Installation

Figure 13 – Drilling cable anchor with wagon drill.

The cable anchors for the top and lateral support ropes for the drape system were installed during winter conditions. The primary cable anchors were installed at roughly 16-ft spacings at the crest, intermediate anchors (three each) at roughly 80-ft spacings at the top, and cable anchors at roughly 50-ft spacings on the sides. The anchors were installed using wagon drills (see Figure 13), and ranged from 15.5 to 25.5 ft in length. Most of the wire rope cable anchors consisted of premanufactured anchors, but for instances were weak rock and/or soil were encountered, longer anchors were fabricated in the field, using an identical design as the premanufactured anchors. Cable anchors for the bottom support rope were placed at roughly 50-ft spacings, using a John Henry-style top hammer drill mounted on a mini-excavator. The excavator accessed the bottom anchor locations from the concrete lined ditch on Bench #1.

Cable Anchor Testing

To verify the design pullout strength value for the cable anchors, 13 tension tests on representative anchors were conducted using a calibrated hydraulic jack and pressure gauge. Most of the anchors chosen for testing were based on drill hole logs that had suspected zones of weaker, disturbed rock. Two anchors had to be abandoned, redrilled, reinstalled and retested.

Figure 14 – Drape panel assembly.

Rockfall Drape Installation

Between March to mid-April, 2016 the as-delivered Tecco™ rolls (11.5 x 100 ft) were unrolled, and reassembled to panel lengths ranging from about 25 to 225 ft long to stage the rolls for placement on the slope (see Figure 14). The rockfall drape mesh was installed on the FLP rock slope in mid- to late-April 2016 via airlift using a modified Sikorsky S55(T) helicopter (Figures 15 through 17). The helicopter used a proprietary lifting/ spreader bar attached to the cargo hook, which was attached to the bottom of the drape panel. The panel weights ranged from 430 to 1,670 pounds (lbs). The helicopter lifted the unrolled panels from the staging area and flew them to the drape area, where workers on slope would then secure the panels to the top support rope or other deployed panels using temporary pin screw shackles. Additional workers then guided the panel edges as the helicopter flew down the slope so that the mesh could be laid on the slope and properly overlapped. Once most of the mesh had been laid on the slope, the pilot released the remotely controlled cargo hook, which opened the lifting/spreader bar from the end of the panel, and the remaining panel length fell onto the slope in a controlled manner.

Figure 15 – Drape assembly/staging area and pick.

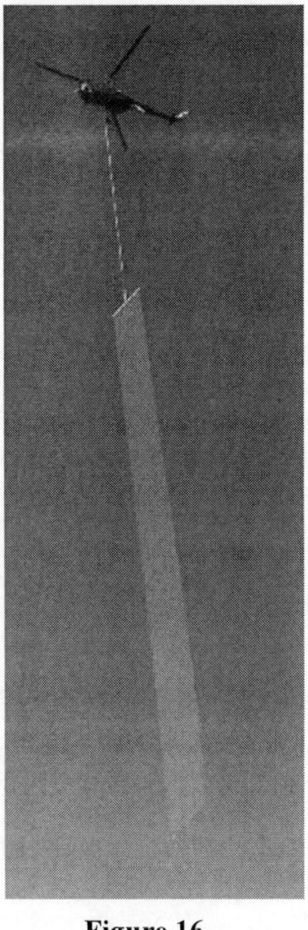

Figure 16 – Helicopter pick

Flights per day ("picks") depended on wind and weather, and ranged from 4 to 27, with daily deployments of 9,200 to 62,940 ft^2. A total of 187 flights were completed to install 371,846 ft^2 of drape. After hanging the mesh, workers clipped the panels together with Tecco™ clips with a lateral overlap of 6-in or greater. A seaming rope was then used to attach the drape to the top and side border ropes. The design included a fold of mesh over the bottom support rope to form a hem. Drape construction was completed on in June 2016 (Figures 18 and 19).

Figure 17 – Deploying drape panel.

Figure 18 – Completed drape.

CONCLUSIONS

The success of this project relied on the collective specialty design and construction experience from the owner, geotechnical design team, specialty rock slope mitigation contractor, general contractor, helicopter operator, and material supplier. Through careful evaluation of the rockfall problem, and evaluation of several alternative approaches, an effective solution was designed and constructed to address a complex rockfall problem. Field design and construction activities were completed on site with no injuries or lost time accidents, contributing to a critically important operations metric to the facility operator. Additionally, the project involved one of the largest helicopter-deployed Tecco™ drape installation projects in the United States.

Figure 19 – Completed drape panorama. Compare with Figure 2.

REFERENCES:

Englund, K.J., 1981. Geologic Map of the Jewell Ridge Quadrangle, Buchanan and Tazewell Counties, Virginia. U.S. Geological Survey Geologic Quadrangle Map GQ-1550, scale 1:24,000.

Evans, N.H. and Troensgaard, K.W., 1991. Geology of the St. Paul and Carbo Quadrangles. Commonwealth of Virginia, Department of Mines, Minerals and Energy, Division of Mineral Resources Publication 106, 2 plates, scale 1:24,000.

Nolde, J.E., 1989. Geology of the Keen Mountain Quadrangle, Virginia. Commonwealth of Virginia, Department of Mines, Minerals and Energy, Division of Mineral Resources Publication 96, scale 1:24,000.

Nolde, J.E., 1996. Geology, Natural Gas, Oil and Other Mineral Resources of Wise County, Virginia. Commonwealth of Virginia, Department of Mines, Minerals and Energy, Division of Mineral Resources Publication 144, 38 p.

Outerbridge, W.F., 1982. Landslides and Related Features of the St. Paul, Virginia Quadrangle. U.S. Geological Survey Open-File Map 82-0422 (sheet H-14), scale 1:24,000.

Whitlock, W.W., 1989. Geology of the Virginia Portion of the Patterson, Bradshaw, and War Quadrangles. Commonwealth of Virginia, Department of Mines, Minerals and Energy, Division of Mineral Resources Publication 99, scale 1:24,000.

Development of design method for rockfall Attenuators

Duncan Wyllie, P.Eng.
Wyllie & Norrish Rock Engineers
850 – 789 West Pender Street
Vancouver, BC V6H 1C2
604-418-4617
dwyllie@wnrockeng.com

Tim Shevlin, PG
Geobrugg North America, LLC
4676 Commercial Street SE
Salem, OR 97302
503-423-7258
tim.shevlin@geobrugg.com

James Glover, Ph.D.
Geomechanics, Mountain Geohazards,
Global Risk Forum GRF
Davos, Promenade 35,
CH-7270 Davos Platz, Switzerland

Corinna Wendeler, Ph.D.
Geobrugg AG
Aachstrasse 11
CH-8590 Romanshorn, Switzerland

Prepared for the 68th Highway Geology Symposium, May, 2017

Acknowledgements

The author(s) would like to thank the individuals/entities for their contributions in the work described:

Disclaimer

Statements and views presented in this paper are strictly those of the author(s), and do not necessarily reflect positions held by their affiliations, the Highway Geology Symposium (HGS), or others acknowledged above. The mention of trade names for commercial products does not imply the approval or endorsement by HGS.

ABSTRACT

A five year research program is nearing completion to develop improved rockfall mitigation structures that only absorb a portion of the impact energy, such that the net deflects the rock into the ground where the balance of the energy is absorbed. The structures are termed Attenuators. The research has involved theoretical studies of impact mechanics, laboratory experiments, and full-scale testing where blocks of rock and concrete cubes weighing up to 950 kg (2200 lb) were dropped down a steep, 60 m (200 ft.) tall rock face. The tests were documented in detail with high speed cameras, load cells on the support ropes recording at 2000 Hz, as well as rock motion sensors with 3D accelerometers and gyroscopes embedded in the blocks. The results have provided a unique insight into the interaction between translating and spinning blocks impacting flexible nets, including the distribution of energy losses in the system and the deflection of the net. It is found that the nets are self-cleaning, which minimizes maintenance costs. This test data with state-of-the-art data acquisition methods are being applied to develop a design tool for estimating the performance of Attenuator Systems. Herein we present the design method concept.

INTRODUCTION

Over the past five years, a research program, comprising theoretical studies, model experiments and full scale testing has been carried out to develop improved rockfall protection structures, termed Attenuators. The principle of Attenuators is that the rock is deflected by the net into the ground such that the net structure only absorbs a portion of the impact energy, with a major portion of the energy being absorbed by the ground. This is in contrast to conventional nets where all the impact energy is absorbed by the net. Significant advantages of Attenuators are that they can be constructed with lighter structures compared with conventional fences, in addition they are self-cleaning which minimizes maintenance costs.

ATTENUATOR PRINCIPLE

Figure 1 shows the typical features of Attenuators comprising a freely hanging, flexible, but impact resistant steel net suspended from a pair, or series, of steel posts with hinged bases bolted to the rock face. Each post is supported with four support cables anchored to the rock face with cable loop anchors and cement grout (Wyllie, 2014).

As can be seen in Figure 1, cleaning of the accumulated rock in the ditch can be readily carried out with equipment located beside the ditch without having to dismantle the structure. Another feature of Attenuators is that the deflection of the net during impact is often limited to about 1 to 2 m, but can be greater depending on impact location and other factors. The limited deflection values means that the structure can be located close to the highway or railway without deflection interfering with traffic. An advantage of this condition is that construction and maintenance costs of Attenuators are significantly less than structures that have larger deflections, and need to be located at a greater distance from the facility being protected.

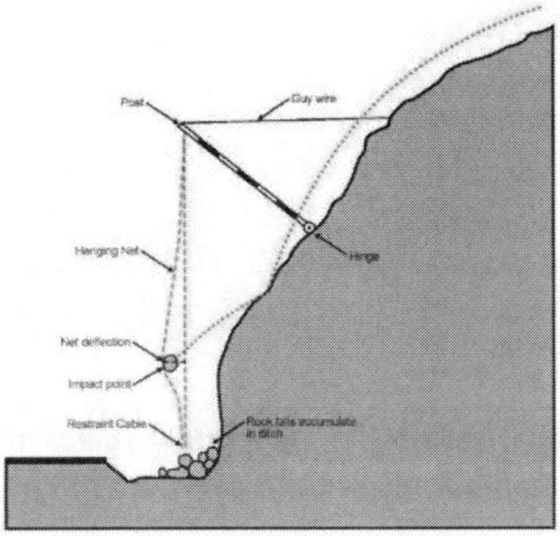

Figure 1.Typical Attenuator configuration

EXISTING ATTENUATOR INSTALLATIONS

The primary author has experience with the design and installation of approximately 24 freely hanging style Attenuators in North America over the last 20 years approximately, and many of these have been impacted hundreds of times over their operational life. This experience has shown that virtually no maintenance is required, and that removal of the accumulated rock can be readily carried out. Significant maintenance was only required when structures were impacted by snow avalanches that encompassed the entire net. However, even in these events, the avalanche was mostly contained in the ditch.

Figure 2 shows an Attenuator installed to protect a railway close to a tunnel portal where the source of rockfalls is about 450 m above railway. The design procedure for the Attenuator was to closely study the rockfall trajectories in the lower 50 m of the slope to identify both the path of the rockfalls, and their trajectories, i.e., height above the ground, in order to position the structure correctly on the slope, and that the top of the net would be high enough to contain rocks that could impact the track. The required momentum capacity of the net was calculated by studying the site geology and existing rockfalls to determine the design mass (m), and the trajectories to determine the design velocity (v).

Construction of the ditch required to contain rockfalls that impact the net required trim blasting to excavate rock at the base of the rock face, and placement of concrete blocks to form a vertical face along the outside of the ditch.

Figure 2. Typical Attenuator installation showing a series of hinged posts with support cables, and a freely hanging net.

FULL SCALE TESTING OF ATTENUATORS

Full scale testing for Attenuators described in this paper has been carried out to verify the detailed mechanics of their behavior during impact with respect to such factors as net impact, load transfer into the support cables, and net deflection. This information provides design parameters for future installations, consistent with the performance of the previously constructed installations described in Section 3 above. The testing was performed by dropping rocks down a natural slope, impacting the rock face as they fell so that they were translating and rotating when they impacted the net.

The test facility was constructed in a quarry where the test blocks could be dropped from heights up to 60 m down an irregular rock face at an overall slope angle of 60 degrees (Figures 3 and 4). The Attenuator was constructed with two, 8 m long steel beams attached to hinged bases bolted to the rock face. The posts were 12 m apart, and the 12 m wide by 11 m long (vertically) net was suspended from a 19 mm diameter steel cable strung between the tops of the posts. Each post was supported with two up-slope support cables and two lateral support cables attached to anchors drilled into the rock face. Figure 4 shows a 40 ton crane lifting the posts into place; a man-lift was used to install the support cables and hanging net.

A laser scan was carried out to produce detailed topography of the site, and to accurately locate the anchors for the support cables. The scanner was located at the base of the slope, and because some areas of the face were in occlusion zones, a drone was used to take images of the slope from above the crest. The point clouds from the scans and the images were combined to produce a topographic plan of the site.

Figure 3. Image of test site

Two types of blocks were used for the testing. First, blocks of rock, with dimensions of between 0.5 and 0.8 m and weighing up to 200 kg were available in the quarry. Although the rock was a very strong, massive crystalline rock, some fragmentation of the blocks usually occurred as they impacted the rock face. Second, heavily steel reinforced concrete cubic blocks, with dimensions of 0.75 m and a mass of 950 kg, were specially fabricated for the tests. These blocks withstood many impacts, with only chipping of the corners. An electronic crane scale was used to weigh each block after the test.

Figure 4. Construction of test Attenuator

The following is a summary of the instrumentation used to document testing, with emphasis on how the impact momentum is transferred into the net and the cables supporting the posts and net.

Cameras

The rockfall motion was recorded with three video cameras. The first camera, running at 30 fps, was located near the drop point to record the trajectories prior to impact with the net. The second camera, running at 60 fps, recorded a face-on view of the test blocks impacting the net. The third camera, running at either 250 fps or 1000 fps, recorded the impacts from a side-on view aligned parallel to the net.

Load cells

The load in each support cable was measured with a Z-type tension load cell, while compression load cells were placed on the bolts holding the hinged base of one of the posts. All the load cells were connected to a data acquisition system running at 2400 Hz; it was necessary to collect data at this rate in order to capture the very short duration impact loads. A trigger was used to start the side view camera and the load cell data acquisition system simultaneously so that observations of the interaction of the test block with the net could be correlated with measured load in the support cables.

Accelerometers and gyroscopes

A sensor, manufactured by Diversified Technical Systems (DTS), incorporating 3-D accelerometers, 3-D gyroscopes (angular rate sensors), a data acquisition system running at 20,000 Hz and a programmable gravity trigger was used to record the translational and rotational motions of the concrete blocks. The sensor, measuring 60 mm by 25 mm, had calibrated sensor range of up to 500 g, but was shock proof up to 1500 g which was necessary to ensure it would survive impact with the rock face. The sensor was used in selected concrete blocks that incorporated a steel pipe in which the sensor could be positioned at the center of the block to minimize centrifugal accelerations. The sensor was mounted in a custom housing that ensured direct transfer of rock's motion to the sensor; minimizing any sensor noise through shaking.

RESULTS OF ATTENUATOR TESTING

During testing carried out in January 2015 and 2016, a total of 46 tests provided dynamic measurements of rockfalls attenuator interaction. These measurements captured rock impacts into attenuator nest filmed with video whereby the system loads were recorded in the support cables and select experiments also included rock motions captured with the embedded sensor device. The following is a brief description of the results.

Translational and rotational velocity

ProAnalyst software was used to analyze the side view video and calculate the translational velocity and where necessary the rotational speed of the blocks from the time just before impact with the net, to time of impact with the ground. The procedure was to track the motion of the block frame by frame (every 0.004 seconds for 250 fps). The video images were scaled using a dimension scale painted on the posts supporting the net. The scaling was also corrected for depth of field with respect to the rock's lateral position passing through the posts. With the correct scaling applied over image frames, it was possible to calculate the velocity (Glover et al., 2012).

On net impact, the rock decreased its velocity rapidly. The rapid deceleration showed correlation with force peaks recorded in the load cells (see discussion below on load cells). During an approximate 0.2 second time interval, the translational speed decreased by about 50 per cent, and from this time the speed remained approximately constant until impact with the ground occurred. Importantly the velocity vector was deflected to the ground during this time.

With respect to the rotational speed, the videos clearly showed that the frictional contact between the rotating, irregular blocks and the openings in the wire mesh, caused the rotational speed to be reduced to zero in a period of 0.15 to 0.2 seconds. Following this rotation of the block is induced in the opposite direction before impacting the ground.

The significance of these velocity observations is that about 50 per cent of the translational momentum and 100 per cent of the rotational momentum of the blocks are lost during the period of 0.2 seconds after impact. During this time period the block is in contact with the net, which means that the loss of momentum of the test blocks is equal to the gain in momentum in the net and load cells.

Loads in support cables

Figure 5 shows typical loads induced in the support cables during impact on the net. It was found that the peak load occurred at a time of about 0.2 seconds after impact, and that most of load is in the up-slope cables. The duration of peak loading is coincident with the most rapid reduction and deflection of translation velocity, and the attenuation of rotational speed. This demonstrates that the greatest momentum transfer from the test block to the Attenuator system occurs within the initial impact period.

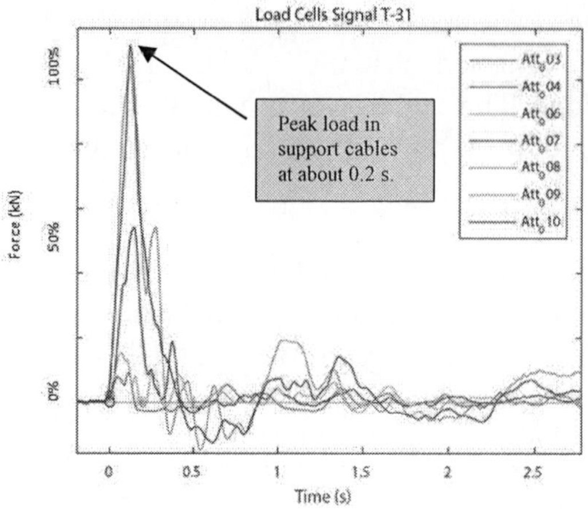

Figure 5. Typical results of forces in load cells during impact.

Accelerometers

Figure 6 shows the acceleration of the test block from the moment of release to impact with the ground in the ditch; a duration of about 6.5 second. The plot on the left shows the impacts with the rock face where accelerations of between 100% and 15% g are generated, and that the acceleration on impact with the ditch is 30%.

The plot on the right shows the acceleration components of the block during a 0.5 second period after impact with the net. Prior to impact during free fall of the block, the acceleration components are greater than zero, representing centrifugal forces of rotation, resulting from the sensor not being precisely in the center of the block. After impact with the net, for a time of 0.2 seconds, accelerations change due to the frictional contact between the irregular, rotating block and the openings in the net.

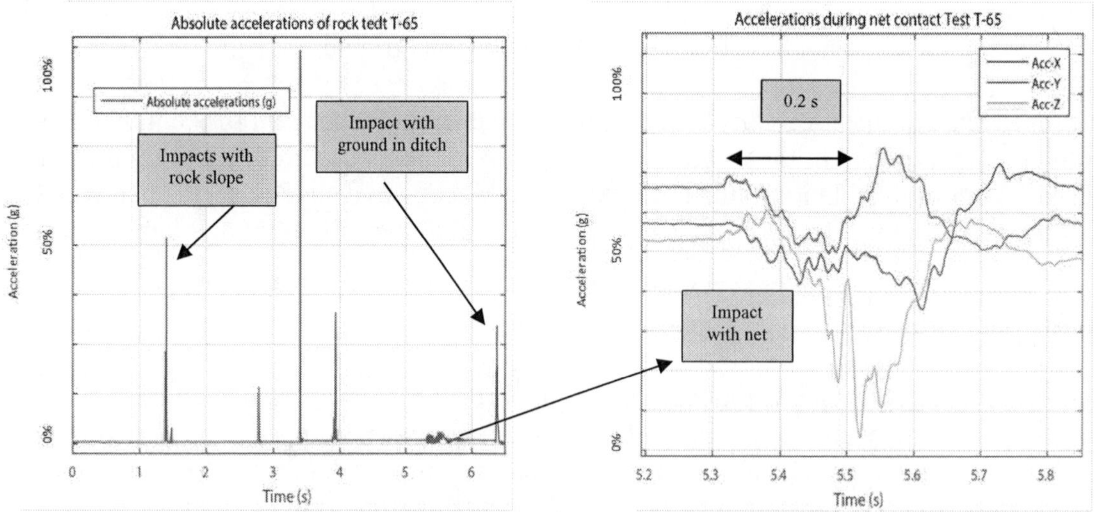

Figure 6. Typical results of accelerometers

After the impact duration of 0.2 seconds, the accelerations of the block decrease as it rolls down the net into the ditch as the gravitational force of the falling block is opposed by the frictional force between the block and the net.

Gyroscopes

The 3-D gyroscopes embedded in the test blocks recorded the rotation of the blocks throughout the 60 m fall. This is respectively, the transfer of translational kinetic energy into rotational momentum on impact with the rock face. The consecutive impacts of the rock with the rock face caused a stepped increase in rotational speed up to a maximum of about 22 rad s^{-1}. The rotational speed during free fall periods remained constant. Once the block impacted the net, the shear force between the irregular block and the net reduced the rotational velocity to zero at 0.2 seconds. The rotational velocity then reversed and increased to more than the impact rotational velocity until it impacted the ground. The significance of the change in rotational direction is first, that all the rotational momentum is absorbed by the net during the initial 0.2 second contact period. Furthermore, the reverse rotation of the block when it impacts the ground causes it to roll back towards the slope and not roll out of the ditch.

DESIGN METHOD FOR ATTENUATORS

The impact of rockfalls with an Attenuator net system as shown in Figure 1 can be analyzed using the conservation of momentum principle as follows.

Momentum lost by the rock body

For an impacting rock body (mass m_r, moment of inertia, I), with translational and rotational velocities ($v_{r(t=0)}$, $\omega_{r(t=0)}$), the total momentum (P_r)at the moment of impact ($t = 0$) is:

$$P_{r(t=0)} = (m_r \quad v_{r(t=0)}) + (I \quad \omega_{r(t=0)}) \quad (1)$$

At time $t = 0.2$ seconds, when the loads in the support cables are at the maximum values, the translational velocity of the body has been reduced to $v_{r(t=0.2)}$ and the body is not rotating ($\omega = 0$) the momentum of the body is:

$$P_{r(t=0.2)} = (m_r \quad v_{r(t=0.2)}) \quad (2)$$

Therefore, the momentum lost by the body to time $t = 0.2$ seconds is equal to the change in velocity and rotational speed during this time period:

$$\Delta P_r = m_r(v_{r(t=0)} \quad v_{r(t=0.2)}) + (I \quad \omega_{r(t=0.2)}) \quad (3)$$

Momentum gain by net system

According to the principle of conservation of momentum, the momentum lost by the impacting body to time $t = 0.2$ seconds is equal to the momentum gained by the net and support cables during this time period. The momentum gain in the net and the load cells is calculated as follows.

First, at time $t = 0.2$, the body and the net are in contact so the velocity of the net as it moves horizontally is equal to the horizontal component of the velocity of the body ($v_{rH(t=0.2)}$) and the momentum of the net is:

$$P_{n(t=0.2)} = (m_n \quad v_{rH(t=0.2)}) \quad (4)$$

where m_n is the mass of the net that is engaged by the impacting rock at $t = 0.2$ seconds.

Second, at time $t = 0.2$ seconds, the total force in the eight support cables, as measured by the load cells, is $\sum_1^8 F$. The force in the load cells acts over time Δt, so the increase in momentum in the net support system over this time interval is:

$$P_{LC} = (\sum_1^8 F) \quad t \; (5)$$

This momentum is made up of two components comprising the sudden movement of the net due to impact, and the oblique, impact between the rotating body and the net that generates a frictional force in the net. Both these actions generate reaction forces in the net, which are recorded by the load cells in the supporting cables (Figure 5).

The resultant acceleration of the block due to the frictional contact between the body and the net is measured by the accelerometers (Figure 6) and is the Euclidean sum of each acceleration axis. The force generated by this frictional contact is the product of the resultant acceleration and total of the mass of the block and the mass of the net engaged by the impact. Therefore, momentum of the contact force is given by:

$$P_{contact} = \left[\left(a_x^2 + a_y^2 + a_z^2 \right)^{0.5} \quad (m_r + m_n) \quad \Delta t \right] \; (6)$$

Based on these equations, the conservation of momentum is given by:

$$\Delta P_r = \left[P_{n(t=0)} + P_{LC} \quad P_{contact} \right] \; (7)$$

Examination of the equations shows that the correct function of Attenuators depends on the ratio of the mass of the rockfall to the mass of the net engaged during impact to deflect the rock into the ditch.

CONCLUSIONS

The integrated test information on the mass and shape of the test blocks and the mass of the net, and data from the video cameras, load cells, accelerometers and gyroscopes has provided a unique insight into the performance of Attenuator net systems under full scale rockfall impact conditions. This information showed that the following features of Attenuator performance:

- Nets fabricated with high strength steel wire and weighing 3 kg m-2 can withstand impact forces generated by translating and rotating blocks of rock and concrete with masses up to 1000 kg.
- The foundation of Attenuator design is to apply an appropriate ratio between the mass of the rockfall and the mass of the net.
- The rotation of the blocks was reversed during impact with the net which helps to contain the rocks in the ditch.
- Because the velocity of the blocks is only reduced by their impact with the net, only a portion of the impact momentum (and energy) must be absorbed by the net and support system.
- The net is self-cleaning because the rocks fall out of the lower edge of the net into the ditch.
- A nearly maintenance free flexible rockfall system, was developed and tested by these tests.

REFERENCES

Glover, J.; Denk, M.; Bourrier, F.; Volkwein, A.; Gerber, W., 2012: Measuring the kinetic energy dissipation effects of rockfall attenuating systems with video analysis. In: Koboltschnig, G.; Hübl, J.; Braun, J. (eds) 12th Congress INTERPRAEVENT, 23-26 April 2012 Grenoble - France. Proceedings. Vol. 1. Klagenfurt, International Research Society INTERPRAEVENT.

Wyllie. D. C. (2014). Rockfall Engineering, CRC Press, Boca Raton, FL., 243 p.

Design-Built Semi-Rigid Rockfall Barrier on U.S. Routes 11/15 in Perry County, Pennsylvania

William K. Petersen, PE
Schnabel Engineering
1380 Wilmington Pike, Suite 100
West Chester, PA 19382
Ph: 610-696-6066
wpetersen@schnabel-eng.com

Giorgio Giacchetti
OFFICINE MACCAFERRI S.p.A.
Via Kennedy 10
40069 Zola Predosa
Ph: 01139051646000
giorgio.giacchetti@maccaferri.com

Ghislain Brunet
Maccaferri, Inc
10303 Governor Lane Blvd.,
Williamsport, MD 21795-3116
Ph: 301-223-6910
gbrunet@maccaferri-usa.com

Lucas Maben
J.D Eckman Inc.
4781 Lower Valley Road
Atglen, PA 19310
Main Office: 610-593-5143
lmaben@jdeckmaninc.com

Prepared for the 68th Highway Geology Symposium, July, 2017

Acknowledgements

For example: The authors would like to thank the individuals/entities for their contributions in the work described:

Craig Welfer, PE – American Engineers Group, Inc.
Pennsylvania Department of Transportation – District 8-0

Disclaimer

Statements and views presented in this paper are strictly those of the author(s), and do not necessarily reflect positions held by their affiliations, the Highway Geology Symposium (HGS), or others acknowledged above. The mention of trade names for commercial products does not imply the approval or endorsement by HGS.

Copyright Notice

ABSTRACT

A series of rock cut slopes along U.S. Routes 11 and 15 (combined) near Marysville, Pennsylvania has presented a chronic rockfall hazard, and the stabilization of these slopes became a high priority for the Pennsylvania Department of Transportation (PennDOT) Engineering District 8-0. The cuts expose steeply inclined clastic sedimentary rocks of Ordovician, Silurian and Devonian age, having variable resistance to weathering. The primary challenge to developing rockfall mitigation measures was the very limited lateral space between the high rock slope to the west of the road and railroad tracks to the east, which are situated along the western shore of the Susquehanna River at a lower elevation than the roadway. With slope heights of up to about 240 feet, very large impact loads (2,000 kJ) had to be resisted without deflecting beyond the roadway edge.

Standard tested flexible rockfall fence as per the European Norm ETAGE 027 couldn't be used because the elongation would have far exceeded the allowable value for the specified impact energies, which were developed by CRSP simulations performed by the District's preliminary design team. As such, the fences for this project had to be custom designed to perform as semi-rigid structures at impact. The design approach for modelling the behavior of the semi-rigid barrier was developed by *Cantarelli and Al* in "Modeling Rockfall Protection Fences". The design theory refers to the elastic deformation of the net under impact in relation with the area of contact surface for a tested barrier.

INTRODUCTION

Two sections of roadway along the U.S. Routes 11 and 15 corridor just north of Harrisburg, Pennsylvania (in Perry County) have posed a chronic rockfall hazard for the Pennsylvania Department of Transportation (PennDOT), District 8-0. Within these sections of roadway, there are five separate rock cuts that were created as part of the roadway construction in the late 1930's. Four of these cuts, referred to herein as the Marysville site and the subject of this paper, are located just south of the Borough of Marysville, as shown by Figure 1. The fifth cut area, located several miles to the north near the Borough of Duncannon, was treated as a separate contract prior to slope remediation at the Marysville site.

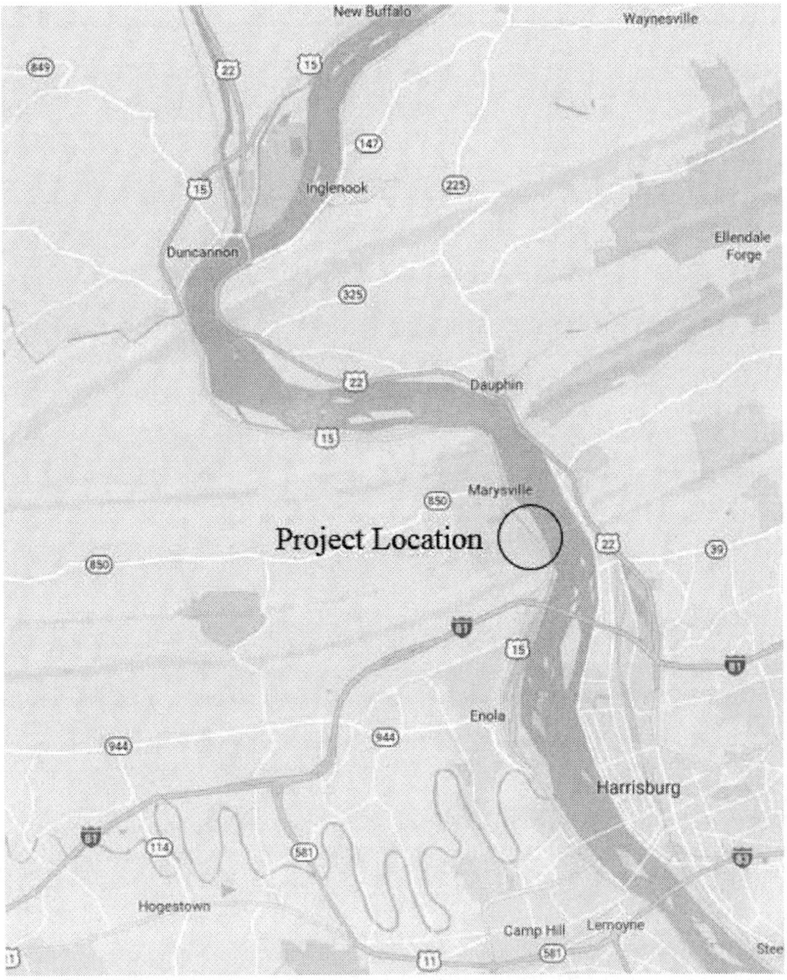

Figure 1 – Project Location of Marysville Site

The distinct cut areas at the Marysville site are designated as Cuts 1 through 4, with Cut 1 subdivided as Cuts 1A and 1B. Table 1 presents a summary of the critical parameters at each separate cut section and the relative locations of each section are shown on Figure 2. Note that the mapped geologic formations are illustrated on Figure 2, and discussed in more detail under the "Geologic Conditions" section below. The cut slopes along the east side of the roadway

(Cuts 2 and 4) form discontinuous "fins" of rock that separate the roadway from a Norfolk Southern rail line situated along the west bank of the Susquehanna River, oriented roughly parallel to and depressed below the level of the Route 11/15 roadway.

Table 1 – Summary of Cut Sections

Cut (Side of Roadway)	Length (ft)	Maximum Height (ft)	Range of Slope Angles (deg)
1A (west)	750	240	38 to 68
1B (west)	372		
2 (east)	275	40	55 to 73
3 (west)	610	132	54 to 78
4 (east)	790	30	54 to 75

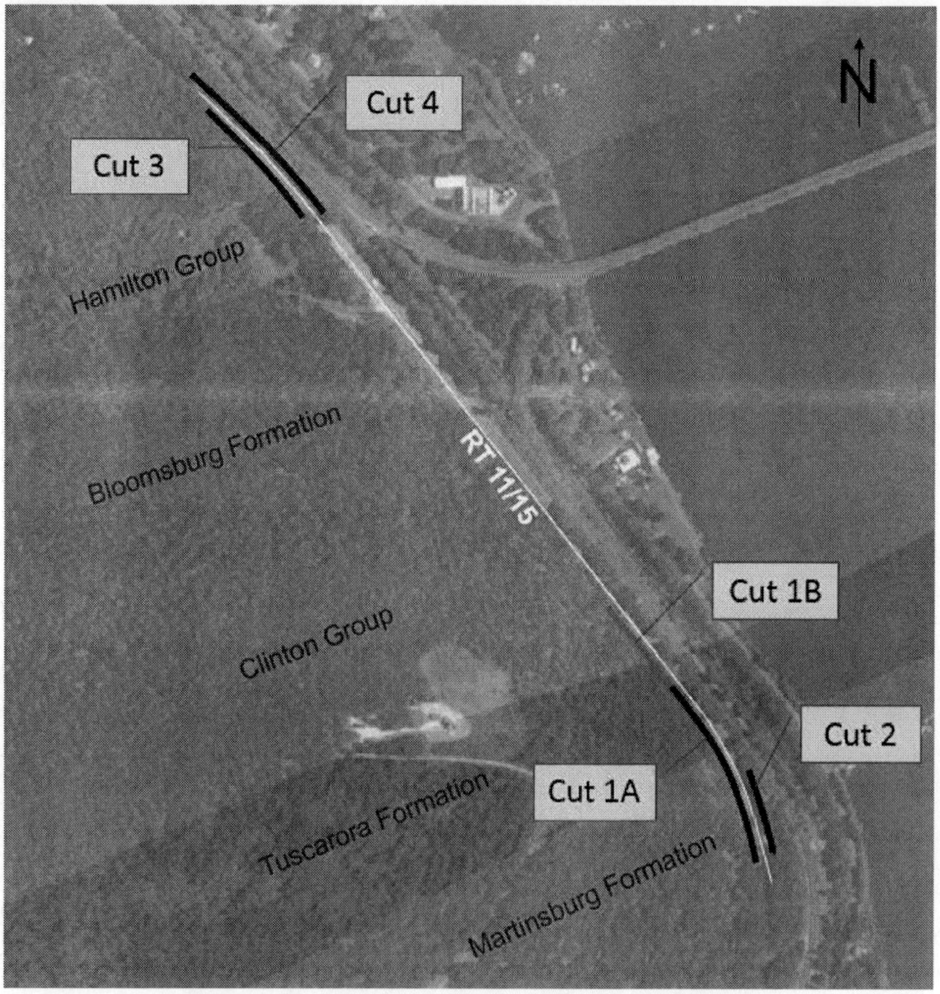

Figure 2 – Cut Locations and Site Geology

As shown by Table 2, the slopes throughout the Marysville site are very steep, ranging up to approximately 78 degrees. This factor, combined with the narrow roadway shoulders, creates a very tight roadway configuration with almost no catchment area for falling rocks (see Figure 3). The heavily jointed, blocky nature of the sedimentary rock units along with the steep slopes and lack of catchment, have combined to make this site one of the highest priority rockfall mitigation projects in District 8-0.

Figure 3 – Photo Looking South (Cut 2 on left and Cut 1A on right)

GEOLOGIC CONDITIONS

The Marysville project site is located within the Ridge and Valley Physiographic Province of Pennsylvania. This region is characterized by nearly parallel anticlines and synclines having axial orientations trending generally west-southwest to east-northeast. This underlying structure has resulted in the formation of long, narrow ridges separated by valleys of variable width. Ground surface elevations in the site vicinity range from about 1,100 feet at the northern project limit (crest of Blue Mountain) down to about 300 feet along the Susquehanna River. The bedrock consists of Devonian-aged clastic rocks (shale, siltstone, and fine-grained sandstone with minor carbonates), Silurian-aged ridge-forming rocks (sandstone and conglomerate), and Ordovician fine-grained clastic rocks.

The project area coincides with the southern limb of the Cove Syncline, with generally steeply dipping beds having a strike orientation roughly perpendicular to the roadway alignment, varying from N20W and N45W. The bedding is typically overturned, with dip angles of between about 62 and 89 degrees to the south (see Figure 4). Three unique joint sets had been identified during the preliminary design reconnaissance, and stereonet analyses concluded that wedge failures and toppling were the primary rockfall mechanisms at the slopes. Sandstone beds were observed to exhibit very prevalent well-developed, near-vertical jointing. These relatively hard beds are often undermined by differential weathering of adjacent shale beds, resulting in the accumulation of detached sandstone blocks along the roadway shoulder.

Figure 4 – Rock Bolt Installation on Steeply Dipping, Thinly Bedded Sandstone

As shown by Figure 2, the rock cuts at the project site are situated within the mapped limits of the following geologic formations:

- Martinsburg Formation, Ordovician (Cuts 1A and 2) – Well bedded, dark gray shale with thin interbeds of siltstone, metabentonite, and fine-grained sandstone
- Tuscarora Formation, Silurian (Cuts 1A and 1B) – Cross-bedded, light-colored sandstone and quartzite.
- Clinton Group, Silurian (Cut 1B) – Well bedded, light to dark gray, fossiliferous sandstone with minor shale and limestone.
- Hamilton Group, Devonian (Cuts 3 and 4) – Well bedded, olive-gray fossiliferous shale and siltstone with interbedded sandstone.

Laboratory testing performed on rock core samples during the plan development phase yielded unconfined compressive strengths varying between about 3,500 and 30,500 pounds per square inch (psi).

ROCK SLOPE MITIGATION PROJECT

Background

In late 2015, PennDOT District 8-0 advertised the construction bid documents for the Marysville project, shortly after completing similar rock slope mitigation work at the Duncannon site under a separate contract. The major components of both projects included scaling, installation of pinned and draped mesh, rock bolting, shotcrete buttresses (overhang support), subhorizontal drains and roadway improvements consisting of repaving, shoulder widening, and placement of new guiderail. However, unlike the Duncannon project, the Marysville site also required the construction of roughly 1,400 lineal feet of rock fence barriers because the slope geometry and tight shoulder areas were not always conducive to providing adequate protection with only bolting and mesh.

GEO-Technical Services, Inc. (GTS) of Harrisburg, Pennsylvania performed the preliminary design work and bid documents for both projects, and also served as the Department's reviewer for contractor design submittals (operating as American Engineers Group, LLC (AEG) at the time of construction). Although traditional design-bid-build project delivery was utilized for these projects, they also contained significant contractor-design components. In the case of the Marysville project, the contractor was responsible for final design of the pinned/draped mesh systems as well as the rockfall barrier fences, the latter being the subject of this paper.

General contractor J.D. Eckman, Inc. of Atglen, Pennsylvania was awarded the contract with a bid price of about $18.9MM. Hi-Tech Rockfall Construction, Inc., of Forest Grove, Oregon, was brought in as a major subcontractor to perform the slope work, including scaling, rock bolting, and mesh installation. Design of the mesh systems was performed by Kane Geotech, Inc. of Stockton, California. Eckman retained Schnabel Engineering, LLC of West Chester, Pennsylvania to prepare the rock fence designs. As described below, Schnabel joined forces with the fence material provider, Maccaferri, Inc. of Williamsport, Maryland to interactively design fence systems meeting the stringent energy and deflection criteria that were established for this project.

Unlike the Duncannon project, which allowed for slope mitigation work to be conducted under single lane closures, the Marysville project required a full detour due to the challenging physical characteristics of the site. Since Route 11/15 serves as the only major north-south corridor along the west side of the Susquehanna River, closing this road for an extended period of time was extremely unpopular with the local residents, businesses and politicians. As such, a compressed window of 90 days was allotted for the detour, with a $160,000/day penalty assessed to the Contractor for exceeding this schedule.

Rock Fence Design Criteria

As part of preliminary design, rockfall simulations had been performed using the Colorado Rockfall Simulation Program (CRSP) in order to develop fence design criteria, namely fence height and energy rating. The energy rating of the rockfall fence was determined based on an assumed largest size block that could be expected to detach from the slope at a given location. The allowable deflection was selected to limit deflection to the available shoulder width at each respective cut area in order to prevent possible encroachment into the travel lane.

Table 2 summarizes the design specifications for fence height, minimum energy rating, and maximum deflection for each segment of the rock fences on the Marysville project.

Table 2 – Rock Fence Design Criteria

Cut	Design Case	Segment Length (feet)	Minimum Height (feet)	Minimum Energy Rating (kJ)	Maximum Deflection (feet)
1B	1	190	11	500	9
	2	110	25	1000	9
2	1	65	12	100	10
	2	90	20	200	10
	3	20	12	100	10
3	1	40	18	300	6
	2	80	18	800	6
	3	50	25	800	6
	4	150	25	1500	6
4	1	85	11	150	10
	2	230	16	2000	10
	3	180	16	100	10
	4	60	11	100	10
	5	40	8	100	10

During the bidding phase of the project, contractors became concerned about the fence specifications because standard pre-designed flexible fence systems cannot meet these deflection limits given the required impact energies. Furthermore, it was recognized that designing rigid posts to meet these energy and deflection criteria would likely require very large steel sections at relatively tight spacings.

Following project award, J.D. Eckman retained Schnabel to design the rock fence systems. Upon inspection of the graphical CRSP output generated from the analysis files provided by AEG, the design team observed that the rock trajectories tended to be very steep for the majority of the design sections - not surprising given the slope heights steep slope angles. Schnabel then proposed to the Department's review team that the fence energy rating requirements be reduced to the horizontal component of the trajectory vector, arguing that this would be the controlling factor for horizontal deflection. This proposal was approved on the condition that the fence supplier could demonstrate that the rating for the proposed fence material satisfies the actual specified energy.

Table 3 shows both the specified and revised energy ratings for each fence segment. The revised energies were calculated as the horizontal component of the specified energy based on the assumed minimum impact angles (measured from horizontal) determined graphically from the CRSP output for each respective fence segment. For some short fence segments, it was considered more efficient to utilize a single design energy rating for multiple adjacent segments even though a smaller rating could have been substantiated; these cases are signified by an asterisk on the revised energy rating values in Table 3.

Table 3 – Revised Energy Ratings

Cut	Design Case	Assumed Minimum Impact Angle (deg)	Specified Energy Rating (kJ)	Revised Energy Rating (kJ)	Percent Energy Reduction (%)
1B	1	0	500	500	0
	2	40	1000	766	23
2	1	62	100	47	N/A
	2	47	200	136*	32
	3	No CRSP run	100	100	N/A
3	1	51	300	189	N/A
	2	64	800	351	N/A
	3	62	800	376*	53
	4	52	1500	924	38
4	1	No CRSP run	150	150	0
	2	59	2000	1030	48
	3	No CRSP run	100	100	0
	4	65	100	42	N/A
	5	No CRSP run	100	100*	0

*Revised energy was also applied to adjoining segments shown as "N/A" under Percent Energy Reduction

Fence Design: A Collaborative Effort

Early in the fence design process, Schnabel realized that it would be a difficult undertaking to gain Department approval for a rigid fence system designed completely "from scratch", especially under the time constraints and disincentive penalties for completing the work on schedule. Because the energy ratings and deflections for various fence systems are largely empirical (developed from full-scale field testing as opposed to just numerical calculations), the designers were concerned that it might not be possible to substantiate the design loadings and deflections without having any field testing or other industry data to back up their calculations. Thus, it was decided work interactively with a fence material provider, using their expertise and specific product data to evaluate deflections and required post spacings for the rigid fence system.

Schnabel then contacted Maccaferri, and was encouraged to learn that they had previously designed a similar rigid fence system for a New Jersey Department of Transportation (NJDOT) project several years earlier. Although the energy and deflection criteria for the NJDOT project were less demanding, Maccaferri believed that a design could be developed to satisfy the revised energy ratings.

Deflection of the barrier

Using the fence height and deflection criteria, along with the reduced energy requirements developed by Schnabel, Maccaferri designed the barrier using an analytical formulation developed by Cantarelli and al (2008).

The design theory is based on the fence behavior during the impact in relation with the net elongation and the breaking time. It basically compares the barrier to a "huge spring" so that the deflection required to absorb the energy of the boulder via the deflection of the barrier throughout the impact increases in accordance with the second principle of Newton. In the case of a rockfall barrier impacted by boulder, it becomes:

$$m\frac{d^2s}{dt^2} - m\,g \cdot sen\alpha = -k\,s$$

or:

$$\frac{d^2s}{dt^2} - g \cdot sen\alpha = -\omega^2 s$$

Where:
s = deflection of the barrier
g = acceleration gravity
α = angle between the mesh plane and the trajectory
k = elastic coefficient of the barrier compared to a spring. It is related to the footprint and the mass of the block and the stiffness of the mesh.
m = mass of the block, and

$$\frac{k}{m} = \omega^2$$

The solution of the previous differential equation becomes:

$$s(t) = \frac{g \cdot sen\alpha}{\omega^2}(1 - \cos\omega t) + \frac{v_0}{\omega}\cos\omega$$

Such equation describes the motion of the block impacting the barrier, and the deflection of the barrier accordingly. The first instant (after the impact) in which the velocity of the boulder becomes zero is:

$$t_c = \frac{1}{\omega} arctg\left(\frac{\omega \cdot v_0}{g \cdot \sin\alpha}\right)$$

With this formula, it possible to obtain the value of ω

$$\omega^2 = \frac{2 \cdot s_m \cdot g \cdot \sin\alpha + v^2{}_0}{s^2{}_m}$$

The maximum elongation of the mesh barrier becomes:

$$s_m = s(t_c) = \frac{g \cdot \sin\alpha + \sqrt{(g \cdot \sin\alpha)^2 + \omega^2 v_0^2}}{\omega^2}$$

The crash test provides the values of s_m , v_0 and t_c , whereas ω is obtained by reiteration of the equations. The approach describes a harmonic motion, which is valid in case of impact between homogeneous frames in the elastic field. But in reality, the deflection of the barriers is mainly plastic, so that the equations are valid only for the period of impact. In any case, it reflects very well the behavior of the barrier as per full scale testing validation from boulder impacts. The relation between the deflection and the dissipated energy is nonlinear. We can assume that in the initial deflection (just after the impact) the dissipated energy is negligible whereas in the intermediate it becomes more appreciable, and finally in the last third the dissipated energy is a lot larger. This behavior of the barriers under impact will change depending on the configuration of the structure and more important the arrangement of the netting. As an example, the net made from rhombohedra cable panels will start dissipating the energy almost immediately after the initial impact, where netting made of ring nets will dissipate progressively the energy from the intermediate comportment of the deflection. Because of these different behaviors, the formula written above needs to be calibrated for each barrier in relation with the barrier behavior during the maximum energy level (MEL) and/or the service energy level (SEL) tests carried out in accordance with ETAG 027 "Guideline for European Technical approval of falling rock protection kits" (EOTA, 2008).

As per the Cantarelli and al. model, not only the energy level is important, but the area of the contact surface between the rockfall block and the netting is very important. As an example, a tested rockfall barrier may be suitable to stop a rockfall block with a given mass and velocity but not another block impacting with the same energy but having a different mass and velocity. It means that the value of "ω" depends also upon the stiffness of the barrier and the ratio between the footprint of the block and its mass. That is why the impact area "A" and the mass "m" are related to "k" respectively by the positive constants "μ" ($k = \mu A$), and "η" ($k = \eta\, m^{2/3}$). Because the deflection in the time "s(t)" is known from the crash tests, it is possible by reiteration to obtain "ω^2" value. With some simple operation of the formula, it is also possible to know the elastic coefficient of the mesh when the post spacing is changed, or the deflection of the barrier under smaller energy levels, as well as to appreciate the effect of the shape of the block on the deflection.

As per Cantarelli and al., during the full scale test the elasto-plastic deformation of the structure is interrelated with boundary conditions, which determine the way the barrier can absorb the energy by transferring the forces to the supporting and connecting components like wire rope cables and wire ropes anchored to the ground. The absorption of the energy is amplified by the insertion of energy dissipating devices.

The design procedure based only on full scale tested energy level barriers may not be sufficient for certain types of impact where the block parameter differs from the test parameter like area of contact, mass of the block and the velocity at impact. Especially the area of the contact surface between the rockfall block and the geohazard netting could be crucial for the barrier deformation. According to the mathematic model used to design the fence for a given energy level, the greater the area of contact with the netting the smaller the maximum net elongation.

The design of the rockfall barriers used on Route 11/15 was based on tested 500 kJ and 2000 kJ impact barrier that were also used to develop the mathematic model from Cantarelli. The standard barriers tested as per ETAG had a maximum deflection for the 500 kJ of 3.4 m (11.5 feet) and for the 2000 kJ, it was 4.3 m (14.1 ft). For both barriers, cable net panel was found to have better deflection performance than other types of netting, such as ring net. For comparison, a barrier tested with ring net as interceptor net will have a maximum elongation of 5.25 m (17.2 ft) at 2000 kJ.

On the area where the 500 kJ barrier model was used the maximum elongation required was 6.5 ft (2.0 m) and 10 ft (3.05 m). As an example, for the section with 10 ft elongation, the maximum block velocity was 37.9 ft/s (11.55 m/s) with a mass of 4,528 lb (2,054 kg) for an energy of 137 kJ (50 ft-ton). For comparison, the standard 500 kJ is tested at 82 ft/s (25 m/s) with a mass of 3,633 lb (1,648 kg). The tested fence section was 9.28 ft x 33 ft (3 m x 10 m) where for the project the netting section was 17 ft x 29 ft (5.2 m x 8.9 m). As per the mathematic model, the maximum calculated elongation for the barrier was 7.2 ft (2.2 m), compared to 11.5 ft (3.4 m) for the tested barrier.

For the highest design energy impact of 1,030 kJ (376 ft-ton), the maximum defection required was 10 ft. The maximum block velocity was 23.1 ft/s (7.05 m/s) that is only 28% of the tested barrier velocity with a mass of 41,446 kg (91,374 lb) with a volume of 15.35 m3 (18.3 yd3). As a comparison, the standard 2000 kJ is tested at 87.3 ft/s (26.6 m/s) with a mass of 15,112 lb (6,855 kg). The tested fence section was 16.1ft x 33ft (5 m x 10 m) where for the project the netting section was 14 ft x 30 ft (4.27 m x 9.1 m). As per the mathematic model, the maximum calculated elongation for the barrier was 5.52 ft (1.68 m) compared to 14.1 ft (4.3 m) for the tested barrier.

Table 4 – Barrier Defection Calculation						
Cut	Revised Energy Rating (kJ)	Maximum Velocity (ft/s)	Ratio Tested Block vs Design Block	Ratio Tested Area vs Design Area	Ratio Tested forces vs Design Forces	Calculated Maximum Defection (feet)
1B	499	51	0.71	0.67	1.30	6.4
	766	56	0.84	0.46	1.46	6.0
2	47	18	3.01	1.08	0.32	3.5
	137	38	2.34	1.54	0.37	7.2
	100	34	2.10	1.55	0.30	6.5
3	189	31	3.80	0.68	1.19	4.3
	351	35	4.83	0.68	1.81	5.2
	376	38	4.62	0.68	1.82	5.5
	923	47	1.19	0.41	1.63	5.3
4	150	28	3.77	0.84	0.97	4.5
	1030	23	3.32	0.78	1.22	5.5
	100	11	10.10	0.78	1.68	3.1
	42	11	5.66	0.84	0.78	2.5
	100	11	10.10	0.78	1.68	3.1

As there is geometrical proportion between tested barrier and the modified barriers, it becomes possible to know the forces on the foundations at any energy level smaller than MEL. The forces were calculated for each barrier in relation to the respective tested barrier and modified to comply with the high forces if required. As an example, the forces for the 1,030 kJ impact were 1.22 times higher than the tested barrier even if the barrier was tested for 2,000 kJ, mainly due to the fact that impacting block was 3.32 times larger than for the tested barrier. Therefore, the interceptor netting couldn't dissipate the energy by deformation as with the full scale test, therefore the energy will have to be absorbed by the foundation systems.

Post and Anchor Design

Maccaferri provided Schnabel with the required post spacing and mesh layout, including cables and braking systems, for each fence segment. In addition, the perpendicular force reactions for the posts and cables were provided based on the mesh analysis, which allowed for the determination of the required steel post sections, rock socket embedment depths, and lateral anchor design. The fence posts, caissons, and anchors were designed in accordance with AASHTO LRFD load factors and resistance factors for an extreme event impact load case. Bearing capacity and bond stress values for bedrock were provided by the project specification for rockfall fence design, and the specified ultimate bearing capacity value was used in the computation of required caisson embedment depth. Calculated deflections were checked using the LPILE computer program to confirm that they were within the specified range of allowable movement.

Since the original test borings performed as part of preliminary design indicated that the depth to competent rock varied somewhat across the length of the proposed fences, the Contractor opted to perform a boring at each post location (under temporary lane closures) in order to develop unique bedrock surface elevations to be used for the design of each post. This measure allowed for the determination of total post lengths in advance of shaft drilling, while providing sufficient depth for the required rock socket embedment depths without the risk of post length changes in the field.

Table 5 outlines a summary of the final fence designs for each segment.

Table 5 – Rock Fence Design Summary

Fence	Design Case	Post Size	Caisson Diameter (in)	Rock Socket Length (ft)	Lateral Anchor Rope Diameter (in)	Lateral Anchor Bond Length (ft)
1B	1	W12x79	24	9	0.875	20
	2	W18x143	30	12		
2	1, 2 & 3	W10x49	18	5	0.75	10
3	1, 2 & 3	W14x99	24	10	0.75	20
	4	W18x175	30	15	0.875	20
4	1	W10x49	18	6	0.75	20
	2	W18x175	30	15		
	3	W12x79	24	7		
	4&5	W10x49	18	6		

Construction

Construction of the rockfall mitigation items was carried out during a full road closure between May 1 and June 24, 2016, with crews working around the clock for seven days a week during this 55-day period. All of the fence posts, cabling and anchors, as well as roughly 80% of the cable mesh installation, was completed with the detour in place. The remainder of the fence construction was finished over the following two weeks using single lane closures.

Concurrently with the scaling and bolting operations being carried out by Hi-Tech up on the rock face, Eckman forces began drilling (Figure 5) and installing wire rope ground anchors (Figure 6) for the lateral fence supports. The ground anchors consisted of double-leg wire rope grouted into drilled holes oriented between 30 and 45 degrees from horizontal, with rope diameters and minimum bond lengths (in bedrock) as summarized in Table 4. As shown by Figure 6, the anchors were tremie-grouted through sacrificial PVC grout tubes, and the anchor heads were fitted with galvanized thimbles for shackle connection to the longitudinal fence cables and bracing cables.

Each fence included a minimum of two pairs of ground anchors (i.e., one at each end of the fence) for connection to the fence cables. For redundancy, the fence system design specified that longitudinal cable connections are alternated between the two ground anchors in a pair such that a rock impact anywhere on the height of the fence will engage both ground anchors in that pair. For runs greater than about 200 feet, and/or when the fence changed in height, intermediate anchor pairs were installed to allow the cables to terminate about a given interior post.

Figure 5 – Drilling Ground Anchors with Articulated "Spider" Drill

Figure 6 – Anchor Pair After Tremie Grouting

As a result of the steeply-dipping sedimentary beds striking nearly perpendicular to the roadway, bedrock conditions were noted to change dramatically along the wall alignments. To accommodate the wide range of rock hardness, several different caisson drilling tools were used for the post installations. As shown in Figure 7, the shaft drilling operation utilized core barrels with roller bits (left) and cutting teeth (center), as well as rock augers (right). In the very hard sandstone, a downhole hammer bit was necessary to maintain adequate production.

Figure 7 – Caisson Drilling Tools

Some caissons required the use of steel casing to prevent soil caving from above the rock socket interval. However, in most cases competent rock was shallow enough so that the holes could remain uncased prior to post and concrete placement (Figure 8). Following completion of drilling a group of shafts, the posts were set in the holes and temporarily braced during concrete placement and curing. The photo in Figure 9 is oriented looking north at completed posts in Cuts 3 (left) and 4 (right). Note temporary bracing still in place on the Cut 3 fence posts.

Figure 8 – Typical Caisson With Shallow Bedrock

Figure 9 – Completed Posts Prior to Mesh Installation

Following the completion of Hi-Tech's work on the slope and the installation of all fence posts and ground anchors with the full detour, the placement of mesh, cables, braking systems and related hardware could be performed during single lane closures. Figure 10 is a photograph taken at Cut 4 during the final inspection, showing a height transition between different design cases. Note longitudinal cables sloping to ground anchors in both directions about post in center of the photograph.

Figure 10 – Completed Fence Section at Height Transition Post

Conclusion

This project was successfully completed ahead of schedule to the satisfaction of all parties involved. With the aggressive construction approach employed by Eckman and Hi-Tech, the duration of the Route 11/15 detour was kept as short as possible to minimize impacts on the local communities. For their patience, regular users of this busy route were rewarded with greatly improved safety conditions throughout these rock cut zones. Furthermore, this project is an excellent example of what can be accomplished when public agencies, contractors, designers, reviewers and product suppliers fully cooperate to quickly solve challenging problems.

REFERENCES:

1. Cantarelli G., Giani G. , Gottardi G. and Govoni, L., 2008*, Modelling Rockfall Protection Fences,* World Landslide Forum, Tokyo
2. GEO-Technical Services, Inc., November 26, 2014, *U.S. Routes 11 and 15 Rock Slope Protection Project, Cumberland and Perry Counties, Pennsylvania.*
3. Geyer, A R. and J. P. Wilshusen, 1982. *Engineering Characteristics of Rocks of Pennsylvania,* Pennsylvania Geologic Survey, Environmental Geology Report No. 1.

Augercast Landslide Stabilization in Potomac Clays

Thomas Monaco, P.E.
3340 W. Market St. First Floor, Akron OH 44333
(330) 436-7121 tmonaco@gfnet.com

Erik Schuller, E.I
2500 Corporate Exchange Drive, Suite 230, Columbus, OH 43081
(614) 794-9424 eschuller@gfnet.com

Prepared for the 68th Highway Geology Symposium, May, 2017

Acknowledgements

Gannett Fleming Engineers and Architects, PC
United Foundations Inc.
William "Kyle" Nelson, for outstanding field work and construction observation
Terry Downs, P.G., for help with construction management
Behzad Amir Faryar PHD, PE, Alexandria County engineer
Mitch Weber, PG, for the crash course in Potomac clays

Disclaimer

Statements and views presented in this paper are strictly those of the author(s), and do not necessarily reflect positions held by their affiliations, the Highway Geology Symposium (HGS), or others acknowledged above. The mention of trade names for commercial products does not imply the approval or endorsement by HGS.

ABSTRACT

The opportunity to collect performance data of a structural landslide remediation allows for reflection on the analyses and the potential for improvements in future design projects. This project involved the design, implementation, and post-construction inclinometer data collection of augercast piles used in the remediation of a landslide along the Potomac River in Virginia. Two rows of augercast piles, one tangent and one spaced at 1.5D, were installed perpendicular to the slope to arrest the continuing movement. Inclinometers were installed in two of the augercast piles, one in each row, to collect data of actual deflections within the piles. This paper summarizes the project from exploration to post-construction data collection; discusses the data interpretation; and identifies potential implications on future designs of a similar nature.

During the exploration of the slide, inclinometers were installed to identify the failure plane. Within two weeks of installation, the inclinometers had sheared to the point that no further readings were possible due to movement of the landslide mass. The use of augercast piles allowed for rapid construction using lightweight equipment that would not exacerbate the movement during construction. 138 augercast piles in total were installed with construction lasting 5 weeks. Inclinometer data collected in the days following the installation indicate the deflections observed are significantly less than expected.

INTRODUCTION

During the construction of a private residence located along the Potomac River in Fairfax County, Virginia a landslide occurred. The rapid progression of the slope movement, sensitive nature of the location along the Potomac River, and the location of the residence within the Potomac Formation of the Atlantic Coastal Plains physiographic region, which has been studied for its susceptibility for slope instability, presented geotechnical challenges for remediation. In this paper, we discuss the interim remedial measures put in place prior to the final design; geotechnical exploration program and results; the auger cast soil improvement (ASI) system; and discusses data collected from inclinometers installed within the ASI for use in future design applications.

GEOLOGY AND SITE HISTORY

Site Geology

The project area is a residential property along the Potomac River located in southeast Fairfax County, Virginia. The project site falls within the Potomac Formation, a Cretaceous Age geologic unit found within the Atlantic Coastal Plain physiographic region along the eastern coastline of the United States. Due to its susceptibility to landslides, both naturally occurring and construction induced, this formation was studied by the United States Geologic Society to identify and discuss relevant geologic factors of marine clays that affect slope stability and their use in engineering design (USGS, 1984). (USGS, 1984) generally describes the formation as clay-rich sediments that are highly overconsolidated with stiff to hard consistency, the top 20 ft. of which may be weakened by physical or chemical weathering including fractures, joints, and parting planes.

The search of *Natural Resources Conservation Service* Soil Mapping records by the United States Department of Agriculture (NRCS, web) and the Fairfax County Soil Map 102-2 record (Fairfax, 2015) revealed that the majority of the slopes at the site falls under the "Sassafras-Marumsco complex" soil unit. The soil group is defined by the sandy and gravelly sandy fluviomarine deposits in the Sassafras units, which is stratified with thick layers of the highly plastic clays of the Marumsco unit. Perched water tables are common where the sand and gravel meet the underlying clay layers. The Marumsco Soils are classified as Class III (most problematic) soil by the Fairfax County geotechnical guidelines.

Site History

Ongoing construction at the site, as of the submission of this paper, consists of a single-family residence with an approximate footprint of 120 feet by 60 feet. The home is located atop a slope approximately 250 feet west of the western bank of the Potomac River with the long dimension oriented roughly parallel to the river. The basement elevation of the structure is approximately 68 feet above mean sea level. The river level is approximately at mean sea level (El. 2).

After completion of the preliminary geotechnical exploration in the early spring of 2016 a Drilled Shaft Wall (DSW) was installed. The DSW consists of 15 caissons installed roughly parallel to the eastern side of the house, offset approximately 55 to 60 feet downslope of the house. The caissons are 36-inch diameter caissons spaced at 9 foot centers. At that time the slope appeared to be stable, and the intent of the DSW was to prevent possible movement during construction. However, excavation material from the foundation of the house and the material removed during the installation of the caissons was used to fill and regrade the "upper slope" and "mid-slope". This regrading resulted in loading the slope beyond the intent of the DSW design. Following the completion of the caissons and regrading, a period of extensive rainfall occurred during March and April 2016. As a consequence of the regrading and the intense spring rains, it is hypothesized that tension cracks formed down slope of the DSW and activated the slide. Slope movements were observed starting in May 2016. As movements progressed, Gannett Fleming Architects and Engineers, P.C. (GF) was contacted to provide further investigation of the slope movements and provide remedial measures for slope stabilization.

Site Description

For the purposes of this report, the site has been divided into four parts:

1. The "upper slope" between the house and the DSW.
    a. This area was not vegetated due to the spring regrading.
    b. This area generally appeared stable;
2. The "mid-slope" between the DSW and the silt fence installed upslope of the RPA;
    a. This area was only gently sloped a the initiation of GFs involvement in the project;
3. The "lower slope" between the upper and lower super silt fence protecting the RPA; and
4. The "Potomac Bank", is the area between the lower RPA silt fence and the river.

The RPA, or "Resource Protection Area", is defined as a 100-foot buffer from the shoreline of the Potomac River; this area was to be preserved with construction activities minimized to the extent practical.

Site Reconnaissance

Site reconnaissance was performed between June and July 2016 prior to initiation of the subsurface exploration. During the initial reconnaissance on June 23, 2016 to assess the current condition of the slope, two main slope observations were identified:

1. Soil ablation and sloughing of the fill material placed over and upslope of the DSW; approximately 100 feet long (paralleling the DSW) and 4 feet high.
2. Head scarp immediately downslope of the DSW line; approximately 150 feet long and 2 to 3 feet exposure.

Based on observations made, it was concluded that the head scarp below the DSW line caused the soil ablation and sloughing observed upslope. Standing water was observed at the mid-slope at head scarp below the DSW in several locations. Seepage was also noted in the head scarp below the top of the caissons. Tension cracks were observed throughout the mid-slope area and into the lower slope area. See Photo 1.

The lower slope area encompassed most of the RPA. Super silt fence was constructed to encapsulate the extents of the surface movement on the slope. The "Potomac Bank" area showed little or no signs of slope movement but showed typical shoreline erosion at the normal high water mark.

Photo 1: June 23, 2016 - Looking uphill with the house construction in the background and the DSW in for ground.

A supplemental reconnaissance was performed on July 11, 2016, immediately prior to the initiation of the subsurface exploration. In general, observations made were similar to those made during the initial reconnaissance. Tension cracking and signs of surface moisture at the DSW line, in the mid-slope, and in the RPA were still present. No further advancement of the sloughing above the DSW line was observed. The head scarp immediately below the DSW line had dropped another approximately 2 to 3 feet vertically. See Photo 2 on the next page.

For reference, the caissons of the DSW line were numbered 1 to 15 south to north. The location of caissons 1 through 12 were visible due to the ablation of soil between the caissons creating an arch-like pattern. This indicates the failure was tied to the spacing of the caissons exceeding the arching effect of the soil between caissons. The concrete and top of shaft were visible in caissons 10 and 11.

Photo 2: July 11, 2016 - Looking uphill with the house construction in the background and the DSW in for ground. Scale rod with 1-foot increments placed at Caisson 1.

EXPLORATION

Historical Exploration

The historical exploration was performed in September 2014 by another engineering firm. The exploration consisted of five borings; four within the footprint of the proposed residence and one in the slope to the east of the household; located in the now defined "mid-slope", downslope of the DSW. The borings B-1 to B-4 were outside the slide area. Boring B-5 was completed in the mid-slope and encountered layers of clayey and silty sands stratified with layers of lean and fat clays. These conditions are similar to those described in the NRCS Sassafras-Marumsco soil unit discussed above. Groundwater was not encountered in any of the five borings completed during the historical exploration. Twenty four hour readings were also dry in all five borings.

Project Exploration

Between July 13 and July 21, 2016, under full-time oversight by a GF engineer, the exploration and instrumentation program was completed at the site. Seven borings, labelled GF-1 through GF-7, were planned. Two borings (GF-1 and 2) were located at the top of the slope near the house. Three borings (GF-3, 4, and 5) were located in the mid-slope area. Two borings (GF-6 and GF-7) were proposed in the RPA at the toe of the slope. Due severe instability of the ground surface conditions within the RPA, it was determined that borings GF-6 and GF-7 were relocated to the mid-slope immediately above the RPA.

Subsequent testing within the RPA was completed using a Wildcat Dynamic Cone Penetrometer (DCP). The DCP consists of a sacrificial 1-inch diameter cone-point attached to steel rods that are driven into the ground by repeatedly dropping a 35-pound hammer from a height of 18-inches. The number of blows it takes to drive the cone ten centimeters is recorded, and can be converted to SPT N-values. No samples are taken during DCP tests as the test is used to provide subsurface profile information based only on blow counts. The two DCP test holes (labelled P-8 and P-9) were completed in the RPA along the same alignment as GF-2, GF-5, and GF-7.

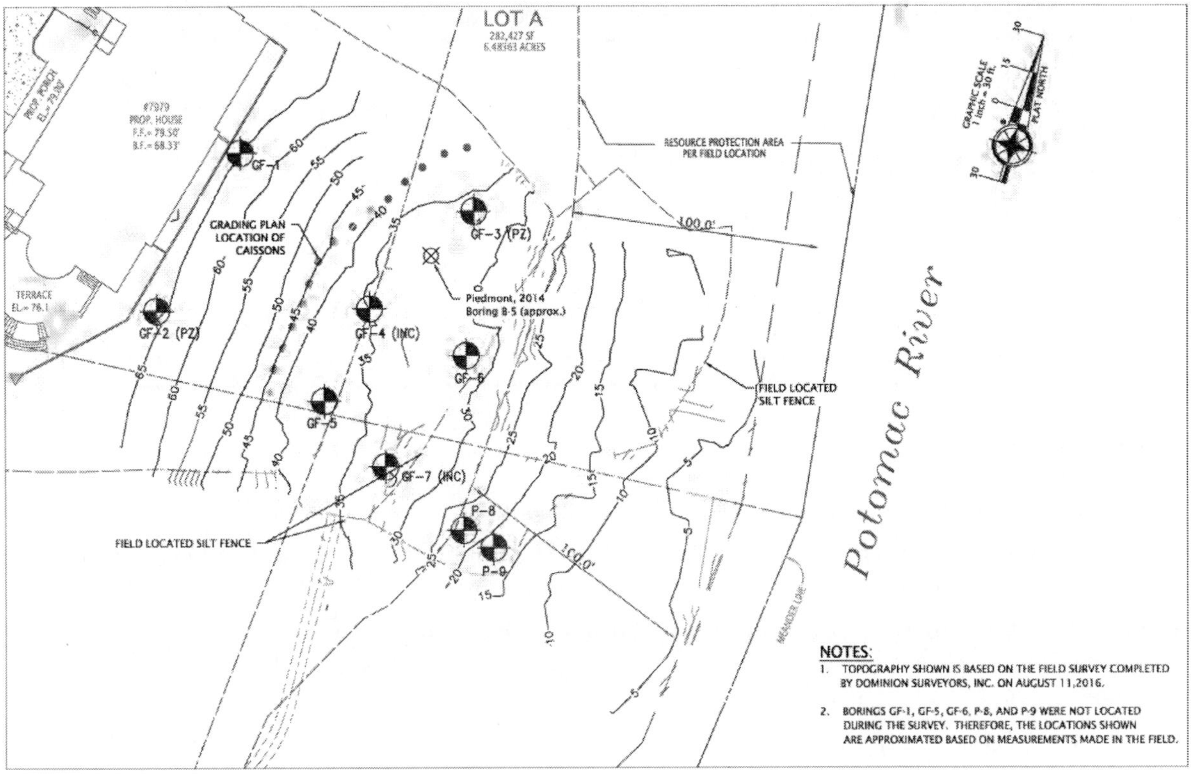

Figure 1: Boring Location Plan

Instrumentation

Following the completion of Borings GF-2, GF-3, GF-4, and GF-7 instrumentation was installed at these boring locations. Standpipe piezometers were installed in offset holes of GF-2 and GF-3 in the upper slope and mid-slope, respectively. Inclinometers were installed in GF-4 and GF-7 in the mid-slope. Forty feet of 2.75-inch diameter inclinometer casing was installed in both borings. This allowed for the identification of any slide planes along the depth of installed casing. During the installation of the inclinometer at GF-4, a cement-bentonite grout mix was used. However, grout was lost at approximate depth of 33 feet below ground surface. This correlated with the location of a sand layer encountered in the boring. A bentonite chip seal was placed at that depth and cement-bentonite grout was used the remaining length of the inclinometer. Due to the loss of grout in GF-4, the inclinometer in GF-7 was backfilled with dry sand the full length of the inclinometer. Generally sand is not the preferred backfill media for long term inclinometer installation; however, given the movements observed at the site, it was determined that sand would be an acceptable backfill under short term conditions. A summary of the borings completed and instrumentation installed on site is provided in the Table 1.

Table 1: Summary of Borings

Boring	Location	Total Depth (Feet)	Notes
GF-1	Upper Slope	40.0	Open for 72 hrs.
GF-2	Upper Slope	30.0	Standpipe Piezometer
GF-3	Top of Mid-Slope	40.0	Standpipe Piezometer
GF-4	Top of Mid-Slope	40.0	Inclinometer
GF-4A	Top of Mid-Slope	14.0	Shelby Tube
GF-4B	Top of Mid-Slope	14.0	Shelby Tube
GF-5	Top of Mid-Slope	36.0	Open for 72 hrs.
GF-6	Bottom of Mid-Slope	40.0	Open for 72 hrs.
GF-7	Bottom of Mid-Slope	40.0	Inclinometer

LABORATORY TESTING

A total of 14 jar samples were selected for classification testing including grain-size analysis, Atterberg Limits, moisture content, and hydrometer analyses. Direct Shear Testing, including peak and residual shear strength testing was completed on each of the two undisturbed samples collected. Full classification testing similar to the jar samples was also completed on the undisturbed samples. A summary of the laboratory testing results is provided in Table 2 on the next page.

Table 2: Summary of Laboratory Testing

Boring	Sample	Depth (ft)	Lab USCS	Natural Moisture Content	Liquid Limit (%)	Plastic Limit (%)	Consolidated Drained Direct Shear Strength (Peak / Residual)	
							Φ' (deg)	c' (psi)
GF-1	S-15	28 to 30	CH	21.5	52	19	-	-
GF-2	S-4	6 to 8	GW-GM	6.8	-	-	-	-
GF-3	S-4	6 to 8	SC	35.0	35	17	-	-
GF-3	S-15	28 to 30	CL	17.1	48	19	-	-
GF-4	S-3	4 to 6	CL	22.3	30	16	-	-
GF-4	S-10	18 to 20	CL	19.6	49	19	-	-
GF-4	S-17B	33 to 34	SM*	21.5	-	-	-	-
GF-4A	ST-1	10 to 12	CH	37.1	59	22	23.1 / 6.6	2.0 / 1.1
GF-4B	S-1	8 to 10	SM*	10.9	-	-	-	-
GF-4B	ST-1	10 to 12	CH	29.6	67	24	17.1 / 7.1	2.1 / 0.2
GF-5	S-5	8 to 10	GM*	12.4	-	-	-	-
GF-5	S-11	20 to 22	CL	19.3	45	20	-	-
GF-6	S-15T	28 to 28.5	SM	28.2	NA	NA	-	-
GF-6	S-16	30 to 32	CL	22.8	48	25	-	-
GF-7	S-4	6 to 8	Cl*	18.0	-	-	-	-
GF-7	S-12	22 to 24	SC	12.9	35	13	-	-

FINDINGS

Subsurface Stratigraphy

Based on the seven borings completed, a generalized profile of the slope consisting of 5 stratum was developed. Near surface Stratum 1 and Stratum 2 is composed of highly weathered marine clay as described in (USGS, 1984). The Strata were differentiated due to engineering properties not deposition. Likewise Strata 3 and 5 are unweathered marine clay. While these two stratums do have similar engineering properties, they are separated by a thick water bearing stratum of granular soil (Stratum 4). The generalized subsurface profile within the mid-slope is summarized below.

- Stratum 1: Located from the ground surface to a depth of approximately 10 feet. This stratum is highly variable with materials ranging from granular sands and gravels (SW/SM) to sandy clays (SC/CL/CH). The relative densities for granular materials ranged from very loose to loose. The consistencies of the cohesive materials ranged from very soft to soft in the moving mass and soft to medium stiff in the "upper slope" above the DSW.
- Stratum 2: Located below Stratum 1 is a one to two foot thick layer. It is described as being wet, indicating the presence of a perched water table in this layer. The materials encountered ranged from silty sand (SM) to sandy clay (CL/CH). The relative densities

for granular materials ranged from very loose to loose and consistencies for cohesive materials ranged from very soft to soft.

- Stratum 3: is an approximately 20-foot thick layer underlying Stratum 2. This stratum generally consists of very stiff to hard clay and sandy clay (CL/CH) characterized as dry to damp.
- Stratum 4: A 2 to 4-foot thick layer starting from depths of approximately 28.0 to 30.0 feet. The stratum generally consists of medium dense silty sand with some intermixed layers of clayey sand; moisture conditions within this stratum were described as moist to wet with free water being encountered in GF-3, 4, 6, and 7 at depths around 30 feet.
- Stratum 5: Below the sand lenses encountered in Stratum 4, a layer of very stiff to hard clay was described as being damp indicating the water encountered in Stratum 4 represents a static water table lying above Stratum 5.

Groundwater

In GF-3 (piezometer), GF-5, and GF-6 the water levels were generally between 25 feet to 32 feet below ground surface, similar to the water table encountered in Stratum 4. In GF-4A, terminated at 14 feet, water levels were recorded between 1.5 and 11 feet, similar to the water table encountered in Stratum 2. These observations reiterate the presence of perched water tables in Stratum 2 and natural ground water elevation in Stratum 4. It is expected that groundwater conditions from the perched water table in Strata 1 and 2 vary across the site and are likely associated with the permeability of the soil at various locations.

Mineralogy

Professor Casagrande developed a procedure for determining mineralogy of clays without the use of microscopy based on the Liquid Limit and Plastic Index of clays. Figure 2 is a Plot of Liquid Limit vs. Plastic Index on the Casagrande's Plasticity Chart developed in 1948 for the project classification testing (Casagrande, 1948). The Liquidity Index and Activity are based on the Skempton formula and were calculated based on the classification data. Based on the results of the calculations, no samples would be classified as Active (Activity greater than 1.25) (Skempton, 1953). The unweathered zone has an average liquidity of -0.016 meaning it will behave as a brittle solid. The weathered zone of marine clay has an average liquidity Index of 0.497 but a maximum value of 1.000, this means that the zone generally behaves plastically with some 'liquid' zones. This wide range of liquidity between the weathered and unweathered marine clay strata supports the observation of the two local water tables, a perched water table and a lower static table. These values were also used to develop and justify the soil parameters used in the slope stability analyses and LPILE analyses discussed in the Analysis Section.

Many soils within Sassafras-Marumsco soil unit are chemically classified as Montmorillonites. Montmorillonites are highly expansive, which allows fracturing of the clay that subsequently becomes saturated and resulting "fully softening" of what would normally appear to be a very hard soil. Holtz and Kovacs Chart, see Figure 2 on the next page, indicates that the soil is chemically a Illites or Kaolinites, not a Montmorillonites. Furthermore, the exploration found that soil fracturing is at a vertical or near vertical inclinations. This is not consistent with Montmorillonites that tend to fracture more horizontally. However, it was assumed that the clays exhibit Montmorillonites behavior per Fairfax Geotechnical Guidelines, Section 4-0303.8E. This assumption may, in part, be the reason for the discrepancy in mathematically predicted deflection of ASI walls and the actual field determined deflection.

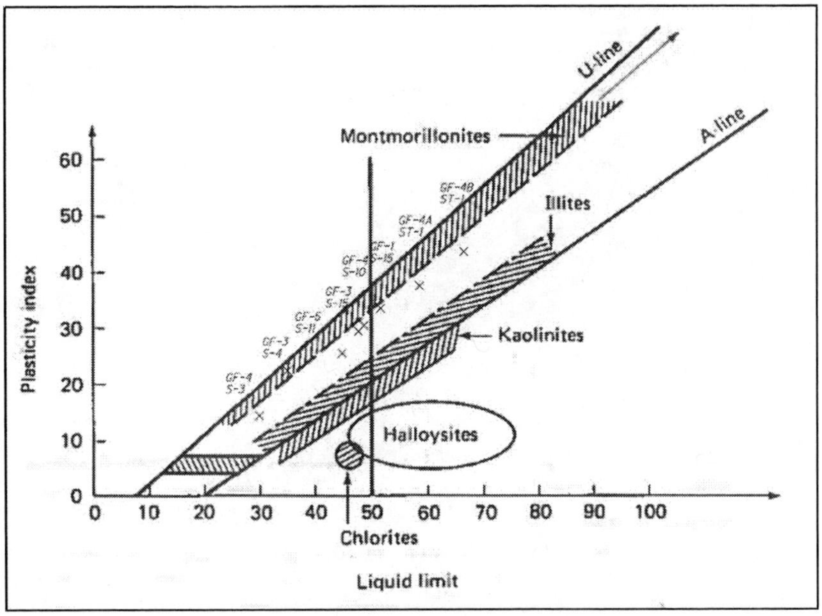

Figure 2: Mineralogy of Clays Based on Atterberg Results (adapted from the Holtz and Kovacs Chart, 1981)

After consulting with Alexandria County engineer, Behzad Amir Faryar, PHD, Gannett Fleming used tables from a publication by Timothy D. Stark, F.ASCE, D.GE, Ph.D., P.E., for the design of the ASIs (Stark, 2013). Values from classification data are plotted with respect to the (Stark, 2013) empirical correlation on Figure 3 on the next page. These values, ranging from approximately 14 to 20 deg, exceed Fairfax County maximum allowable effective angle of friction of 12 degrees for soils not tested by direct shear; therefore, an effective angle of friction of 12 degrees was assumed. This assumption may, similar to the mineralogy assumption, may, in part, be the reason for the discrepancy in mathematically predicted deflection of ASI walls and the actual field determined deflection.

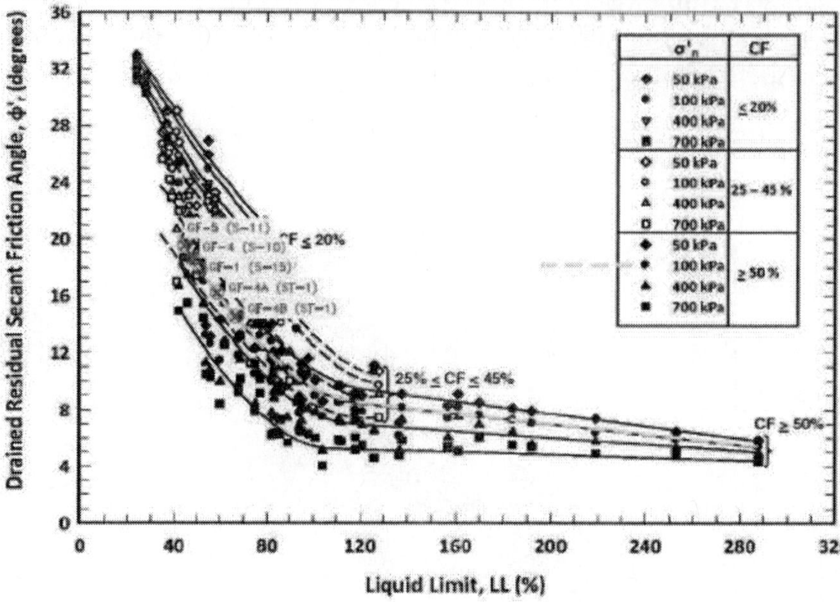

Fig. 2. Updated empirical correlation for drained residual secant friction angle based on liquid limit (LL), clay-size fraction (CF), and effective normal stress (σ'_n) for seventy-three natural soils

Figure 3: Empirical Correlations based (Stark, 2013)

INCLINOMETER READINGS

Preconstruction inclinometer readings were initiated on July 27, 2016. The initial reading in GF-7 on July 27th indicated lateral movements of approximately 4-inches in the top eight feet of the profile with movement initiating at a depth of 10 feet. Between July 27th and August 3rd, the incremental movements indicate the slide plane is located along the interface of Strata 2 and 3 at a depth of around 7.5 to 10.0 feet. The August 3rd reading is the last data point collected because the inclinometer casing deflected to a point where the probe could not be advanced to take readings. A block movement was identified and used as the basis for the slope stability modeling. The moving soil block occurred between the surface and approximately 7.5 feet below ground surface. The inclinometer casing manufacturer does not provide specific lateral deflections at which the probe is no longer usable. However, the four inches of movement in the initial reading and the subsequent 0.75 inches of movement between July 27th and August 3rd suggests the movement along the failure plane was ongoing and significant. See Figure 4 on the next page for the inclinometer readings.

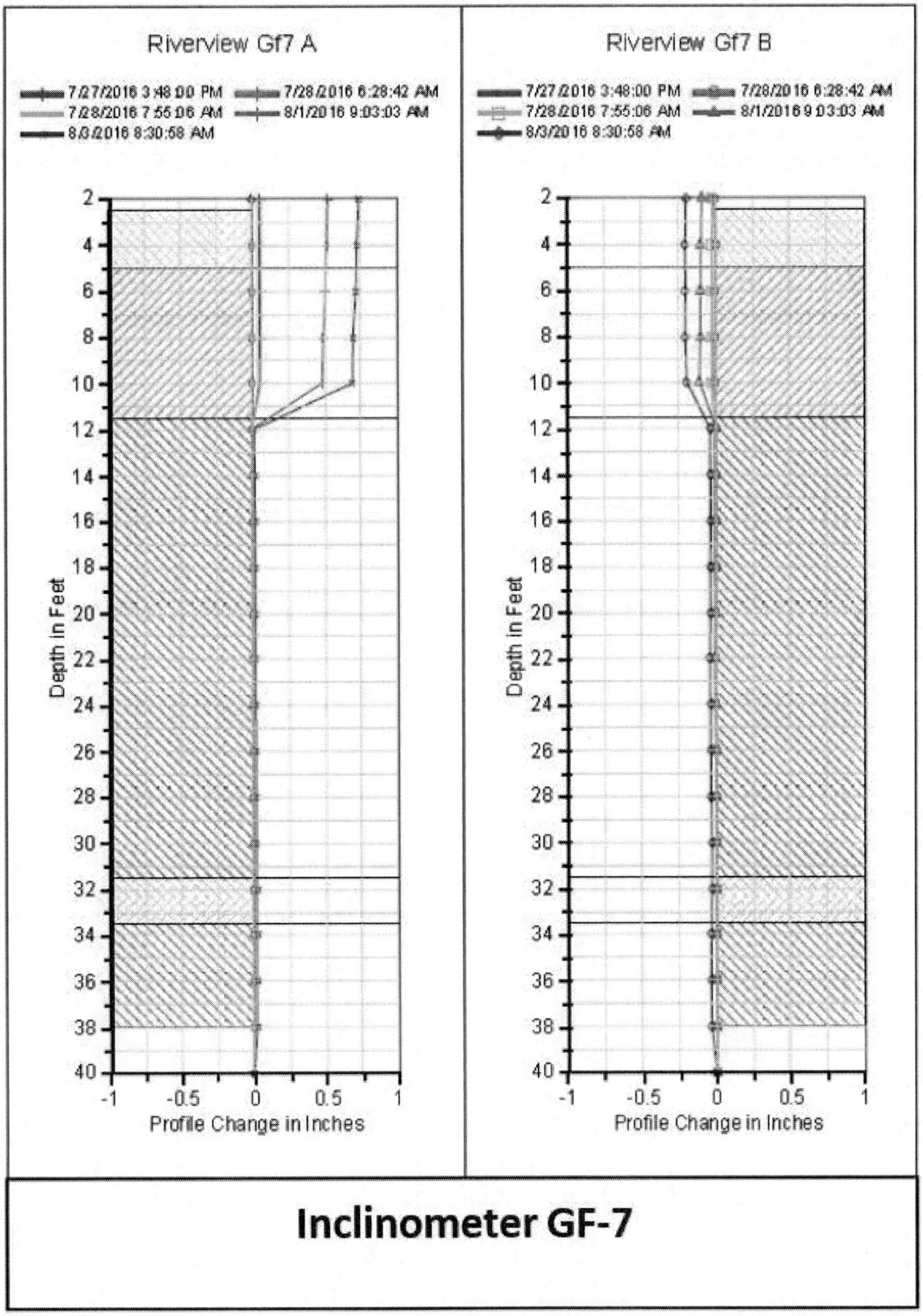

Figure 4: Inclinometer GF-7

In GF-4, from the time of installation on July 15th to the initial reading on July 27th, the inclinometer casing deformed to the point where the probe could not be advanced beyond a depth of 8.5 feet below ground surface. Based on the initial reading in GF-7 in which 4 inches of movement was observed, this suggests that movements greater than 4 inches occurred in the top 8.5 feet and reiterates the presence and general depth of the failure plane observed in GF-7 and is consistent with being a block failure since the depth is close to GF-7 despite GF-4 being closer to the head scarp.

Inclinometers were installed in piles 145 and 233 during construction on November 15th and 16th, 2016. The inclinometers were read again March 11, 2017. The inclinometer readings are shown in figure 5, based on the deflection readings, they appear to have stabilized. Continued reading is planned.

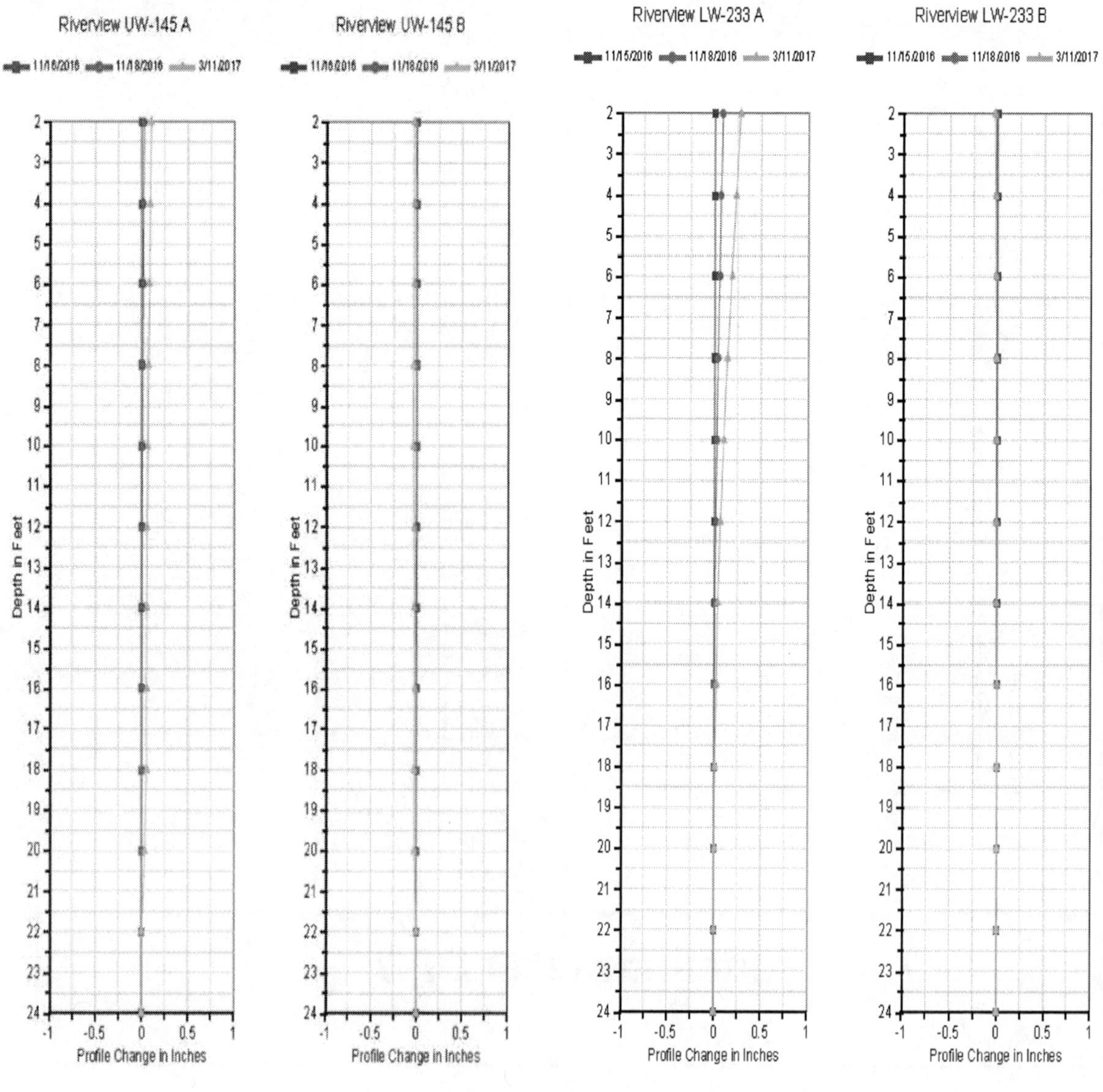

Figure 5: Post Construction Inclinometers

ANALYSES AND RECOMMENDATIONS

Working on a slope in active failure always faces some difficulties; however, this project had significant constraints including:

1. Poor access to the work area – there is only one, 12-foot wide access road;
2. The access point is through the mid-slope from the south - restricted the entry of heavy equipment to a location where the machines will add to the driving force of the slope;
3. Soft soils and undulating ground in the RPA - made machine access difficult;
4. Challenges of obtaining permits to work within the RPA - limited the potential methods of slope stabilization; and
5. The site was an active slide that was spreading – the longer construction was delayed the larger and further downslope the slide would have moved.

Design alternatives analyzed included Excavate and Replace, Geofoam or Light Weight Fill, Micropile Wall, Soldier Pile Wall, Soil Nails, Geopier, and Auger Cast Soil Improvement. Based on the alternatives analysis a tangent or nearly tangent ASI wall to remediate the slide was recommended. Auger Cast Piles are essentially small diameter drilled shafts that, when placed tangent (or nearly tangent) they create a continuous soil improvement increasing the shear strength of the sliding mass.

This method of construction had the advantage that the existing grade outside the ASI walls' access roads alignments could be left undisturbed, or with limited disturbance to allow planting. This method was preferred as it allowed for smaller equipment, reduce the amount of soil hauling materials more invasive methods would require, and be a cost effective and have a relatively fast installation. Still, the installation process faced significant difficulties due to the daily movement of the slope. The slope access was improved before installing the ASI. This was accomplished by removing 2 feet of the soil and replacing it with crushed #1 and #2 aggregate (approximately 0.5-inch to 1.5-inch diameter stone) placed on a woven geotextile. Following the ASI installation, a drainage system immediately upslope from the ASI walls was installed to reduce or cut off the flow from the perched water table.

Slope Stability Modeling

It is hypothesized that the landslide was activated by the opening and subsequent water filling of tension cracks below the DSW. Formation of tension cracks is a common occurrence in the weathered zone of the Sassafras-Marumsco soil unit (Potomac Marine clays) and is further exacerbated by the granular lenses separating the clay layer that allowed water to enter the clay in multiple paths resulting in a saturated weak zone in the expansive soil. The expansive nature of the clay allowed more tension cracks to develop into scarps and weathered marine clays to the fully soften condition. Essentially, the soil shear strength below the DSW, and in Stratum 2 went from nearly peak to residual quickly. The movement defined by the inclinometer readings indicate a translational failure (or a very flat circular failure) occurring at the interface of the highly weathered clay and the unweathered clay (strata 2 and 3). There is no indication that the deep unweathered clay of stratums 3 and 5 have been soften or penetrated by the perched water table or static water table. The lowest short term shear strength value was used in the analysis of the short term stability of the slide in the disturbed mass. The limit equilibrium stability program GSTABL7 Version 2 by Gregory Geotechnical Software was used for the analyses. This produced a safety factor that was less than one at the slide plane, and supports the observation of the depth of the slide as shown in the inclinometer, the locations of the toe bulge and head scarp observed in the field. See Figure 6.

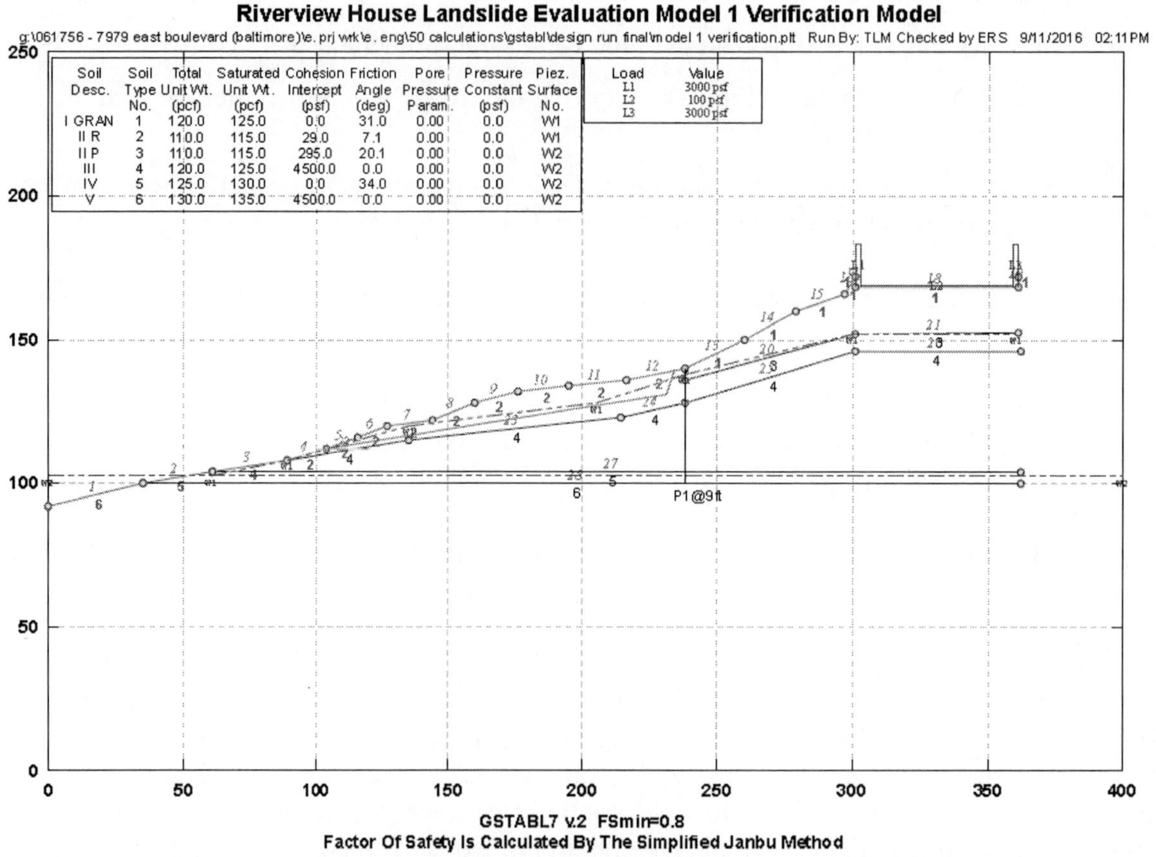

Figure 6: GSTABL Analysis of Slide

The long term value was conservatively taken as the average shear strength with the effective cohesion ignored (φ' = 6.9 degrees). The long term strength of the undisturbed soil above the DSW was determined based on the Stark correlations. The correlations determined that the clay would have a shear strength greater that than the maximum shear strength value allowed by Fairfax County of 12 degrees. This assumption may be impart the reason for the discrepancy in mathematically predicted deflection of ASI walls and the Field determined deflection. See Figure 7 on the next page.

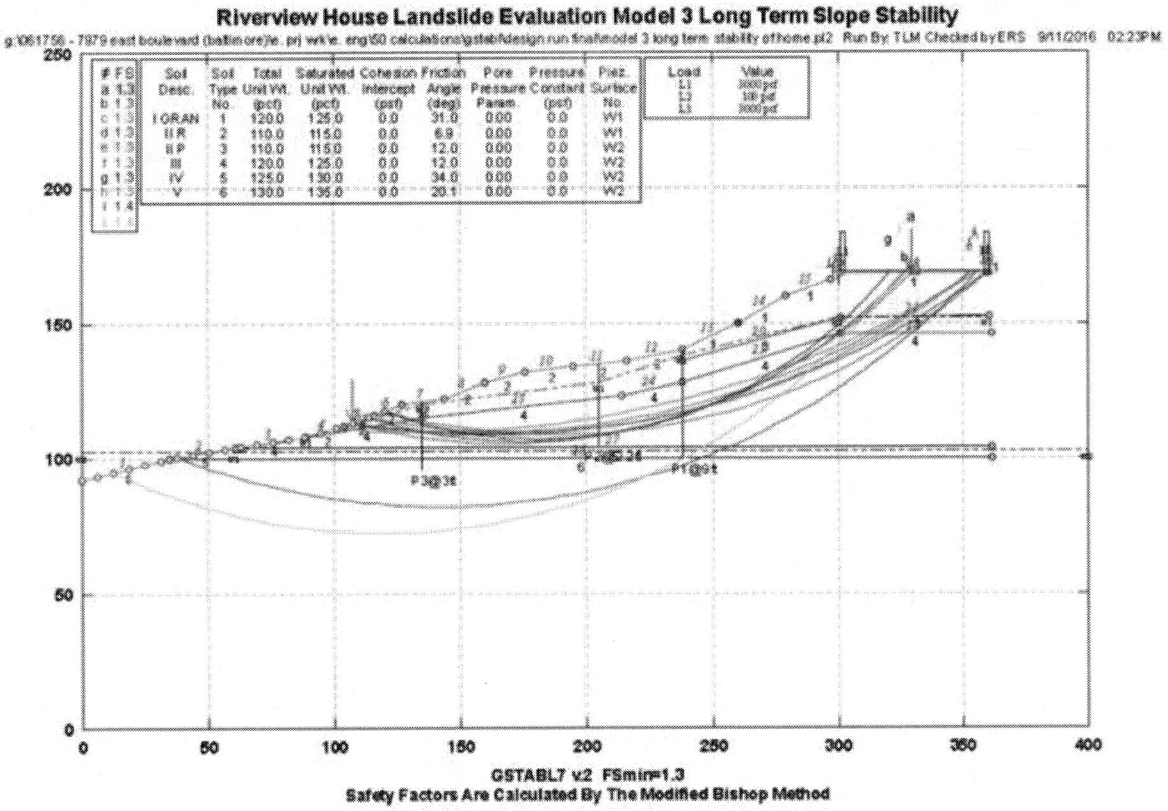

Figure 7: GSTABL Analysis Landslide Stabilized with ASI Wall 1 and 2

LPILE Analysis

The design of the auger cast piles was completed using LPILE v2015 by Ensoft, Inc. The LPILE models were created using the design loads and slide plane determined by the slope stability model and the inclinometer data. The design load is modeled as a distributed load over the depth of the slide plane shown in the inclinometer.

Based on the LPILE analysis both ASI 1 and ASI 2 were sized to be two foot diameter auger cast piles reinforced with four, vertical, epoxy coated #9 bars. The bars were tied at one foot centers with epoxy coated #3 bars. ASI 1 (upper wall) was 30.5 feet in depth and spaced 26-inches center to center. ASI 2 (lower wall) was 25.5 feet in depth spaced 36-inches center to center. The maximum deflection of was determined to be 3.05 inches and 2.65 inches for ASI 1 and ASI 2, respectively. See Figure 8 on the next page for the calculated deflection in ASI 1 and ASI 2.

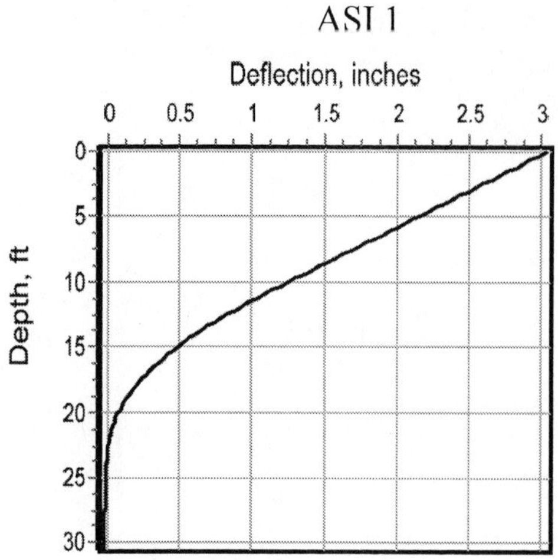

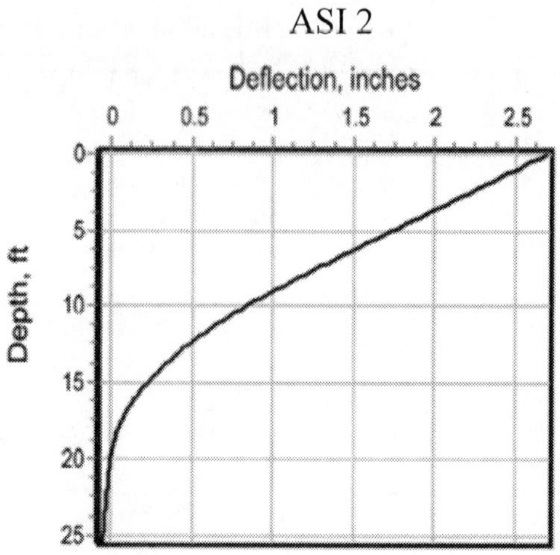

Figure 8: LPILE calculated Deflection of ASI 1 and 2

CONCLUSIONS

Sufficient time has passed for the active slide to have imposed loading on the piles. The resulting deflection is approximately 1/10th the predicted deflection. One of the challenges of the engineer is to perform a meaningful analysis of a slope failure that takes into consideration conservative assumption imposed by review boards and design codes, while also considering the cost to the client of these conservative assumptions. The difference between the calculated deflection and actual deflection of the piles might be the result of the following:

- calculation of loads that was too conservative;
- the use of p-y (load induced displacement) inputs and shear strength that do not accurately actual soil properties; or
- the assumption that the lower clays will develop their residual strength

Clearly the calculations of the load on the pile are tied to shear strength of the soil and thus the first two possible explanations are related. The latter explanation can only be dismissed after long term monitoring of the inclinometers; however, the former explanations are hard to ignore. The conservative, perhaps overly conservative, assumptions made by the author are the following:

- Ignoring Cohesion from all strata;
- Setting the maximum effective angle of friction of stratum 3 and 5 is 12 degrees; and
- The sliding mass clay mineralogy is Montmorillonites.

All these assumptions, while justified individually, may have made the design overly conservative when used in combination. However, the use of ASI walls were still cost effective when compared to other alternatives examined for this project. As with all landslide stabilizations only time will tell if all of the slide causes have been remediated and long term stability has been obtained.

REFERENCES

- Federal Government Reports
    1. U.S. Geological Survey. *Engineering Geology and Design of Slopes for Cretaceous Potomac Deposits in Fairfax County, Virginia, and Vicinity*, Geological Survey Bulletin 1556. U.S. Department of Interior, 1984.

- Local Government Reports
    1. County of Fairfax, Virginia. *Soils Map 102-2*. Department of Information Technology, 2015.

- Website
    1. Natural Resources Conservation Service. *Web Soil Survey*. U.S. Department of Agriculture, websoilsurvey.nrcs.usda.gov, Accessed August 25, 2016.

- Journal
    1. Casagrande, A. Classification and Identification of Soils. *Transactions of the American Society of Civil Engineers* vol. 113 (1), 1948, pp. 901-930.
    2. Skempton, A.W. The Colloidal Activity of Clays. *Selected Papers on Soil Mechanics*, 1953, pp. 106-118.
    3. Stark, T.D. and M. Hussain. Empirical Correlations – Drained Shear Strength for Slope Stability Analyses. *Journal of Geotechnical and Geoenvironmental Engineering* vol. 139 (6), 2013, pp. 853-862.

- Book
    1. Holtz, R.D and W. Kovacs. *An Introduction to Geotechnical Engineering*, No. Monograph, 1981.

Improved Geophysical Imaging for Engineering and Infrastructure Projects Using the Multi-Electrode Resistivity Implant Technique (MERIT) Case Studies in Florida

David Harro PG
G3 Group
2509 Success Drive, Suite
Odessa, FL 33556
(727)-376-7833
David. Harro@geo3group.com

Prepared for the 68th Highway Geology Symposium, July, 2017

Acknowledgements

The author(s) would like to thank the individuals/entities for their contributions in the work described:

Dr Henok Kiflu G3 Group
Dr Sarah Kruse University of South Florida
Dr M H Loke Getomo Software
Dr Paul Wilkinson British Geological Survey

Disclaimer

Statements and views presented in this paper are strictly those of the author(s), and do not necessarily reflect positions held by their affiliations, the Highway Geology Symposium (HGS), or others acknowledged above. The mention of trade names for commercial products does not imply the approval or endorsement by HGS.

ABSTRACT

ABSTRACT: The Multi-Electrode Resistivity Implant Technique (MERIT) (Harro and Kruse 2013) involves rapid installation of parallel surface and buried arrays of electrical resistivity electrodes. Implanting deep electrodes increases the depth of investigation of a resistivity survey by a factor of two or more effectively decreasing the required array length by one-half, and enhancing resolution capabilities of electrical resistivity tomography at depth. This technique utilizes electrical resistivity using a novel approach that can provide greatly enhanced subsurface images of rapid recharge basins, leaky aquifers in karst areas, and saltwater intrusion. This paper will focus on case studies performed using this technique

INTRODUCTION

The Multi-Electrode Resistivity Implant Technique MERIT

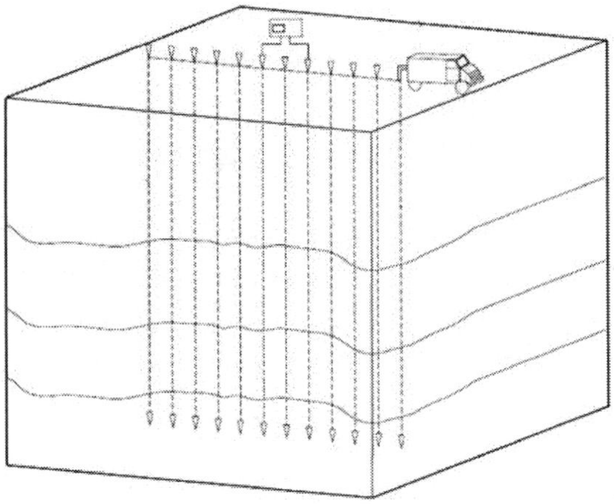

Plate 1: MERIT method schematic.

MERIT is a unique technology that increases the depth of penetration of resistivity surveys to approximately twice the depth as surface geophysical application. MERIT system also has higher resolution than surface methods due to its tomographic arrangements of surface and buried electrodes (Plate 1) and its proprietary mathematically optimized data collection process. Electrodes are placed at the surface and at depth with direct push technology effectively increasing the depth of electrical resistivity.

<u>Case Study Landfill Tampa Florida</u>

Project Information

The site of the geophysical survey was in a self-storage facility that was constructed over a former landfill. Large amounts of differential settlement were observed in the western most storage unit building and at the adjacent building to the east (See Figure 1)

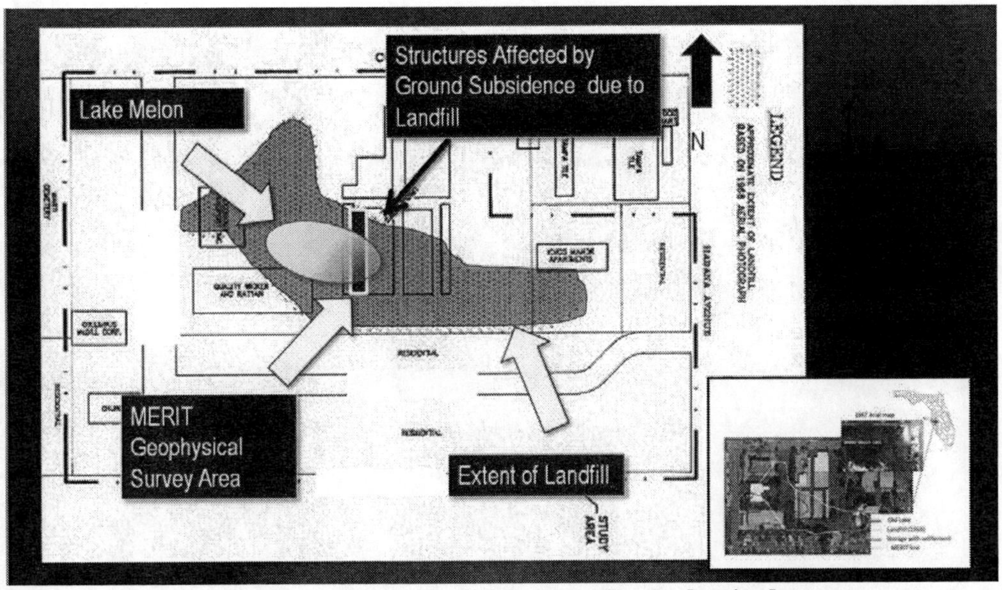

Figure 1: Location of landfill, Lake Melon, and geophysical survey area

The subject property currently has an on-going landfill gas removal system. Concerns over subsidence of the structure due to the landfill resulted in test borings being performed, however due to difficulties in drilling into the landfill adequate information was not obtained to make a determination as to the cause of the ground subsidence. The selection of this geophysical method was based on these concerns and difficulties encountered during the initial shallow exploration drilling.

Previously conducted ground penetrating radar survey (GPR) was not able to identify geophysical anomalies associated with sinkhole activity due to existing landfill conditions and attenuation of the GPR signal.

Based on the inability to perform Standard Penetration Testing and poor resolution of ground penetrating radar survey, the use of the patent-pending geophysical technology Multi-Electrode Resistivity Implant Technique (MERIT) was proposed. This technique has been tested for the application of imaging for sinkholes.

Purpose of the Geophysical Investigation

The objective of the MERIT survey will be to document any deep geophysical anomalies associated with the existing landfill and possible karst features using MERIT including the offset method.

Conclusions

The application of the MERIT geophysical survey was superior to the surface methods as can been seen in the comparison of geophysical images. The MERIT geophysical survey revealed the existence of a deep depression along the profile of the east building.

Figure 2 Picture of MERIT installation locations

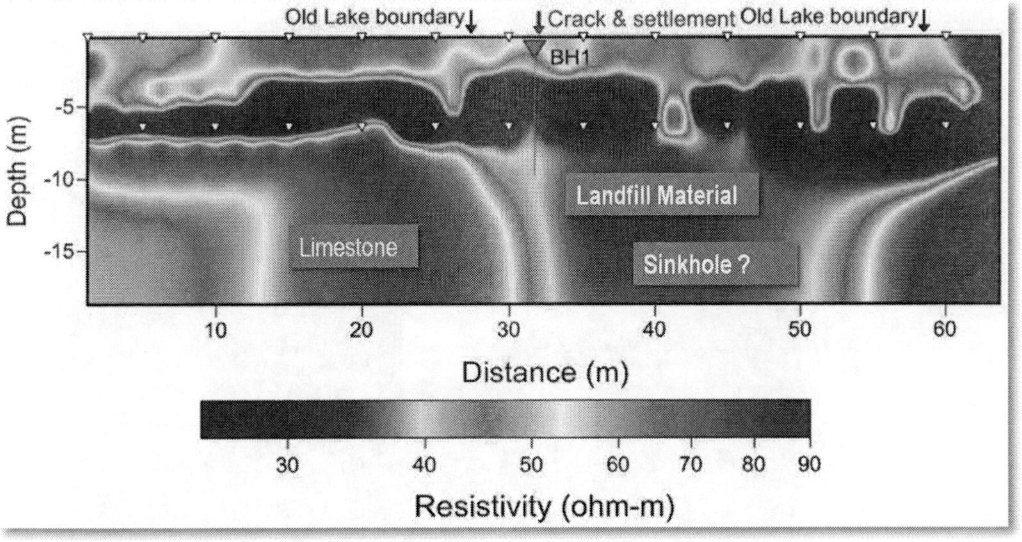

Figure 3 MERIT image showing large depression feature of the former Lake Mellon

Optimized configuration of the data collection for MERIT was utilized for this survey (Loke M.H, Kiflu H.G., Wilkinson P.B., Harro D., and Kruse S. 2015)

This feature is considerable in size, 22.5 meters (67.5 feet) in length and over 18.8 meters (56.4 feet) deep. This feature appears to be filled with materials that correspond to the landfill materials as identified in the geotechnical borings.

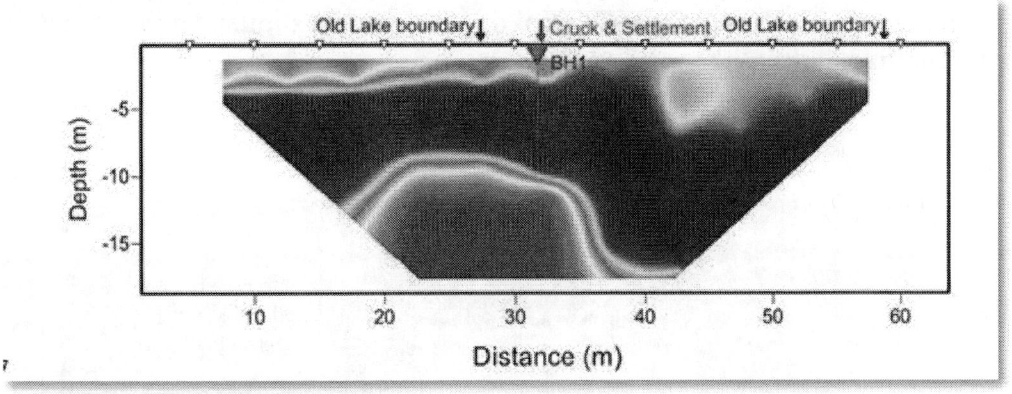

Figure 4: Surface geophysical image taken over the primary MERIT

Comparison of surface the geophysical result clearly shows the edges of the lake/sinkhole feature not as defined as in the MERIT image (Kiflu, H., S. Kruse, M.H. Loke, P. Wilkinson, and D. Harro,2016)

A new application technique for MERIT is the offset method. This method allow for lateral imaging to be conducted by movement of the upper electrodes to known distance and angle from the implanted electrodes. This technique is unique to the MERIT method.

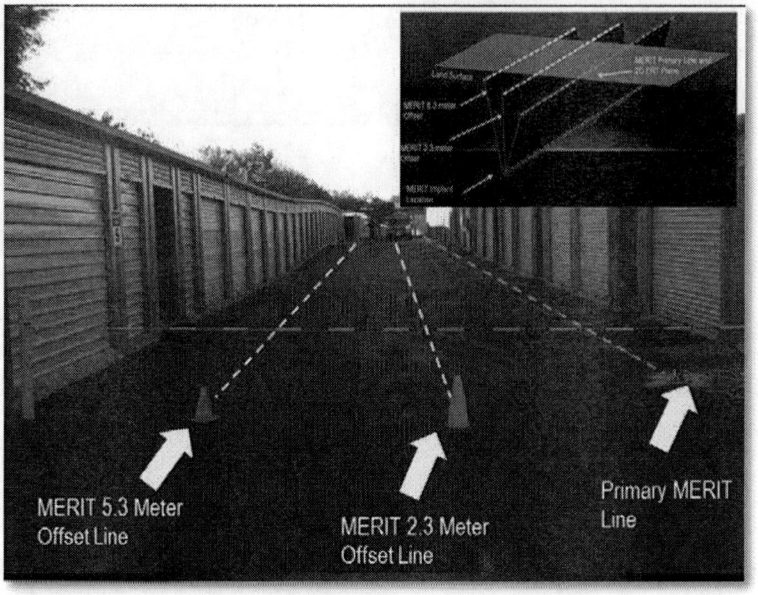

Figure 5: Location of MERIT offset positions

In the offset measurements the feature has lateral dimensions and similar depth profiles extending from the east building to near the west building.

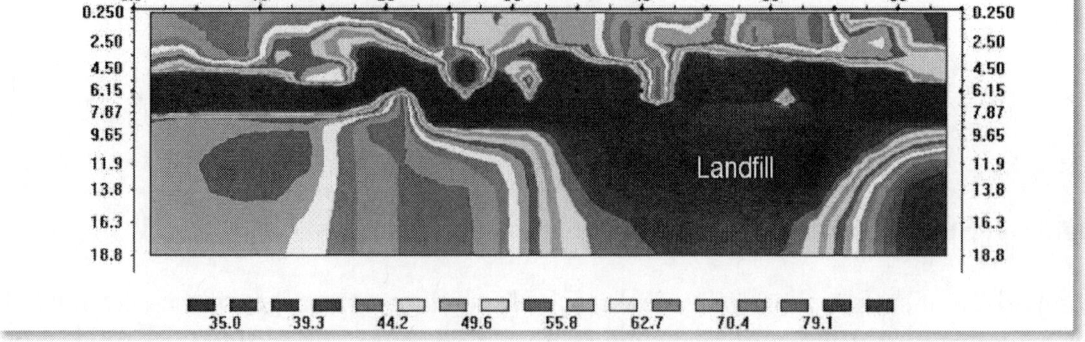

Figure 6: Image of the 2.3 meter offset

Based on the geophysical results the landfill material occupies a significantly sized area into the deeper stratums and is most likely a karst related sinkhole or paleo sinkhole feature.

Case Study Sinkhole Florida Turnpike Right –of-Way

Project Information

A large sinkhole developed on a residential property located on Salmon Drive in Orlando Florida. The sinkhole feature was located on the eastern side of the residential property adjacent to the existing sound barrier of the Florida Turnpike. After the sinkhole development occurred, the sound barrier adjacent to the residential property and two of the south bound lanes of the Florida Turnpike began to show signs of distress related to ground subsidence. This was expressed as slumping of the two lanes and up to 6 inches of differential movement of the sound barrier. A section of a 54 inch pressurized reclaimed water transmission main is located in the area of ground subsidence, which has approximately a 200 foot radius, between the roadway and the sound barrier [See Figure 1].

Figure 1: Impact to roadway and infrastructure due to sinkhole

The site of the geophysical survey was the R-O-W of the Florida Turnpike in Orlando, Orange County Florida. The R-O-W contains utility corridor which is utilized by the City of Orlando and Orange County for a pressurized 54 inch reclaimed water transmission main operated by Water Conserv II. The area of investigation was located approximately 1300 feet south of the turnkey in the R-O-W of the southbound lane of the Florida Turnpike (91) between toll road 408 and US I-4 [See Figure 1]. In the area of investigation the R-O-W was estimated to be 35 feet in width with the eastern boundary the roadway and the western boundary consisting of concrete sound barriers.

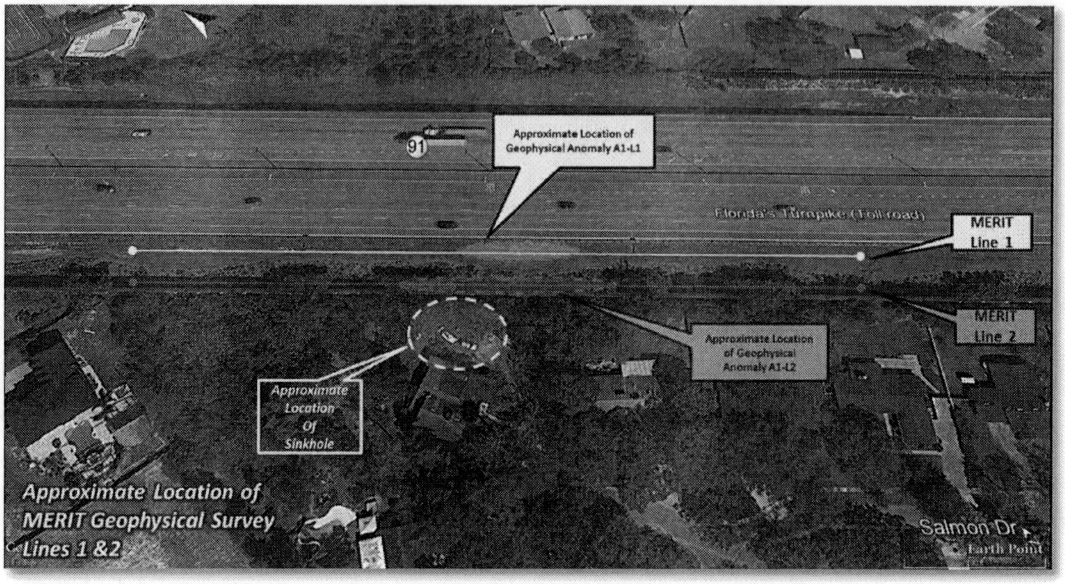

Figure 2: Image of the MERIT Lines, sinkhole and anomalies

Concerns of sinkhole development as a cause of ground subsidence which could pose a potential risk to the 54 inch transmission main prompted the geophysical survey. The selection of the MERIT geophysical method was based on these concerns and on the difficulties of surface geophysical methods reaching the same depths of over 100+ feet as the exploration drilling performed by the geotechnical consultant.

G3 proposed the use of its patent-pending geophysical technology MERIT. This technique has been extensively tested for the application of imaging for sinkholes.

Figure 3: Shows the amount of movement of the sound barrier and DPT installation

Two (2) MERIT geophysical surveys were performed to address the potential sinkhole development in the R-O-W. The first MERIT survey (Line 1) was designed to provide subsurface profile that would intersect the previous three geotechnical standard penetration test borings to depths of 130 feet below land surface (bls). In addition, Line 1 was designed to encompass and extend past the 200 foot long affected area of the R-O-W. After the results of Line 1 were reviewed by Water Conserv II it was decided to perform a second MERIT survey. MERIT Line 2 was conducted paralleling Line 1 but positioned along the sound barrier [See Figure 3].

Purpose

The objective of the MERIT survey will be to document any deep geophysical anomalies associated with the possible karst features using G3's patent-pending geophysical Multi-Electrode Resistivity Implant Technique (MERIT).

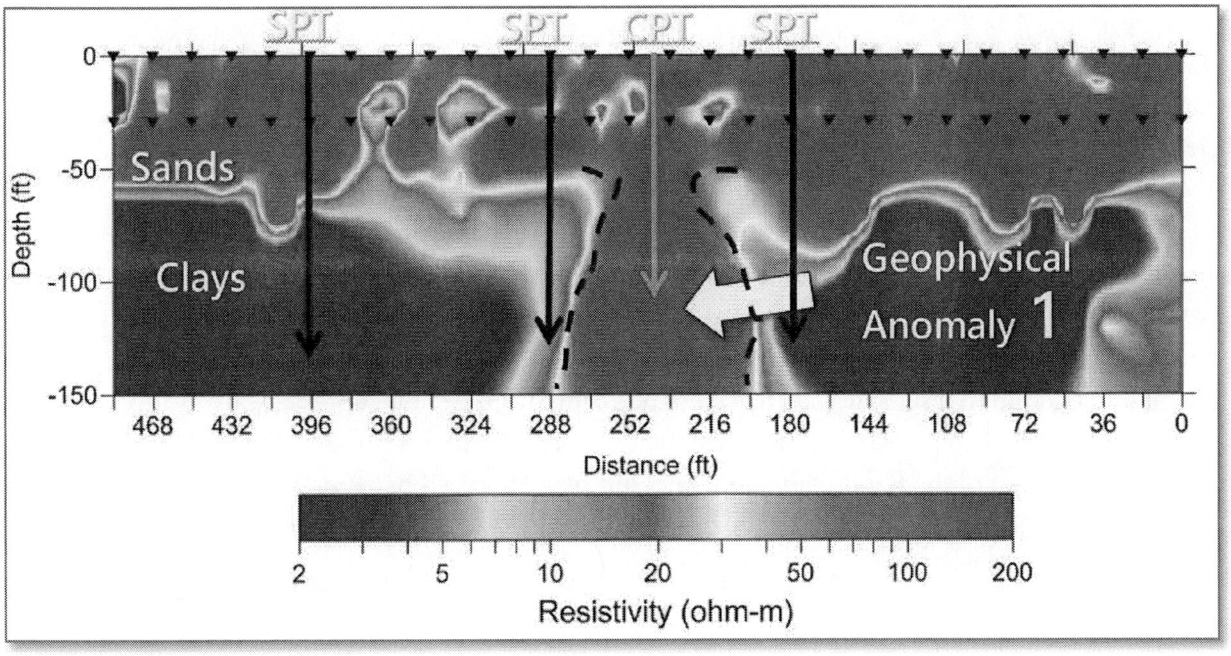

Figure 4: Results of MERIT Line 1 and location of geotechnical testing

Analysis of Geophysical Data

- The (Red) in the MERIT images is consistent with highly resistive undifferentiated sediments consisting of siliciclastics referred to as a sand unit from the surface to depths of 50 to 60 feet bls. This unit is believed to be associated with Undifferentiated Quaternary Sediments (Qu). This unit penetrates the clay unit to depths of greater than the 150 foot depth provided in the MERIT image.

- The (Yellow) areas within the sand unit are lower resistive areas believed to be due to either water or possibly lower compaction or a combination of both.

- The lower (Blue) in the MERIT images is consistent with poorly to moderately consolidated, slightly sandy, silty clay referred to as clayey unit. This unit is believed to be associated with the Hawthorn Group and ranged in depths from 50-60 feet bls to a maximum depth of penetration of approximately 150 feet bls.

Conclusions

The results of the MERIT geophysical survey identified two distinct geophysical anomalies located in the subsurface. These anomalies are located within the areas of the highest concentration of distress/ground subsidence observed on the roadway and the sound barriers. In addition, the geophysical anomalies identified by the MERIT surveys are adjacent to the corresponding sinkhole development on the adjacent property.

The geophysical anomaly identified suggests that the sand unit has in the past or has recently moved downward into the underlying clay unit. This would correspond with the sinkhole development type called cover-collapse. Cover-collapse sinkhole formation occurs when the underlying limestone is covered by a significant layer of clay. Dissolution of the limestone creates a void in the clay which will eventually collapse. If the clay has significant amounts of sand material covering it the sand will infill the voids created.

Site Geologic Conditions from Soil Borings

Standard penetration test (SPT) borings were performed at the subject property by the geotechnical consultant and will be reported in the consultant's report. From our review of the SPT's performed, a good correlation of the Sand Unit and Clay Unit was observed between the SPT borings and the MERIT image for Line 1.

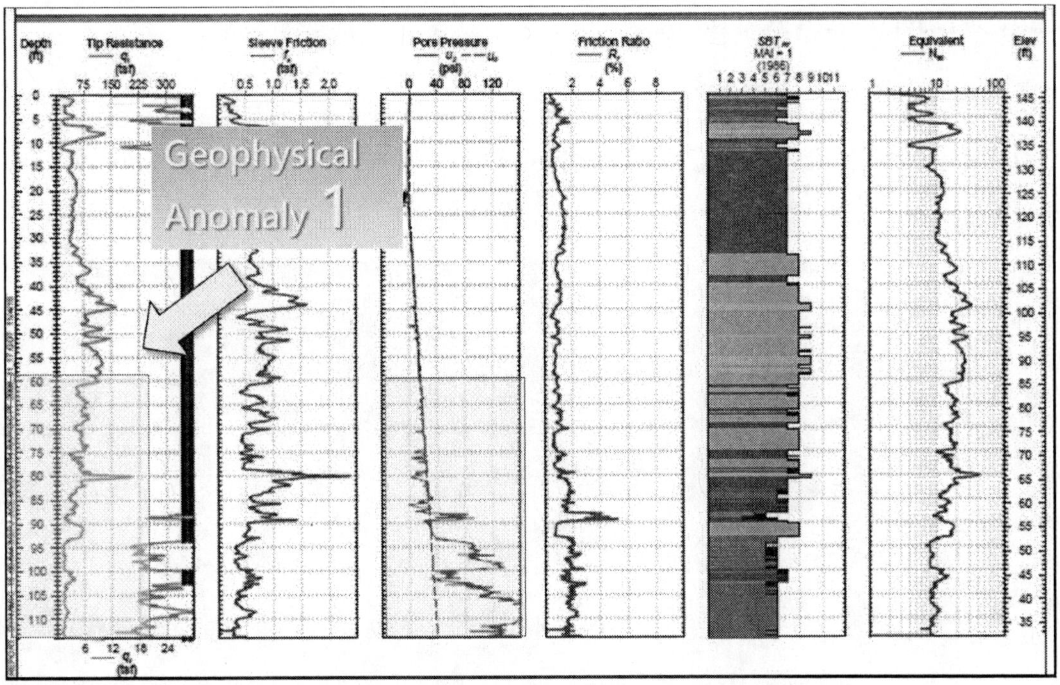

Figure 5: Results of CPT at the location of the geophysical anomaly Line 1

Results

The SPT and CPT did not indicate sinkhole conditions or anomalous geotechnical conditions; this may be due to the lateral movement of the soils and limitations of geotechnical testing to evaluate such conditions. Based on the results of the MERIT survey an engineering proposal was put forward to create an above ground bridge spanning the affected subsurface area identified in the MERIT images.

Case Study Wekiva Parkway CR46A

Project Description

The MERIT geophysical survey is located within an area identified as a relic sinkhole called site B of Section 5 of the Wekiva Parkway CR46A [Figure1].

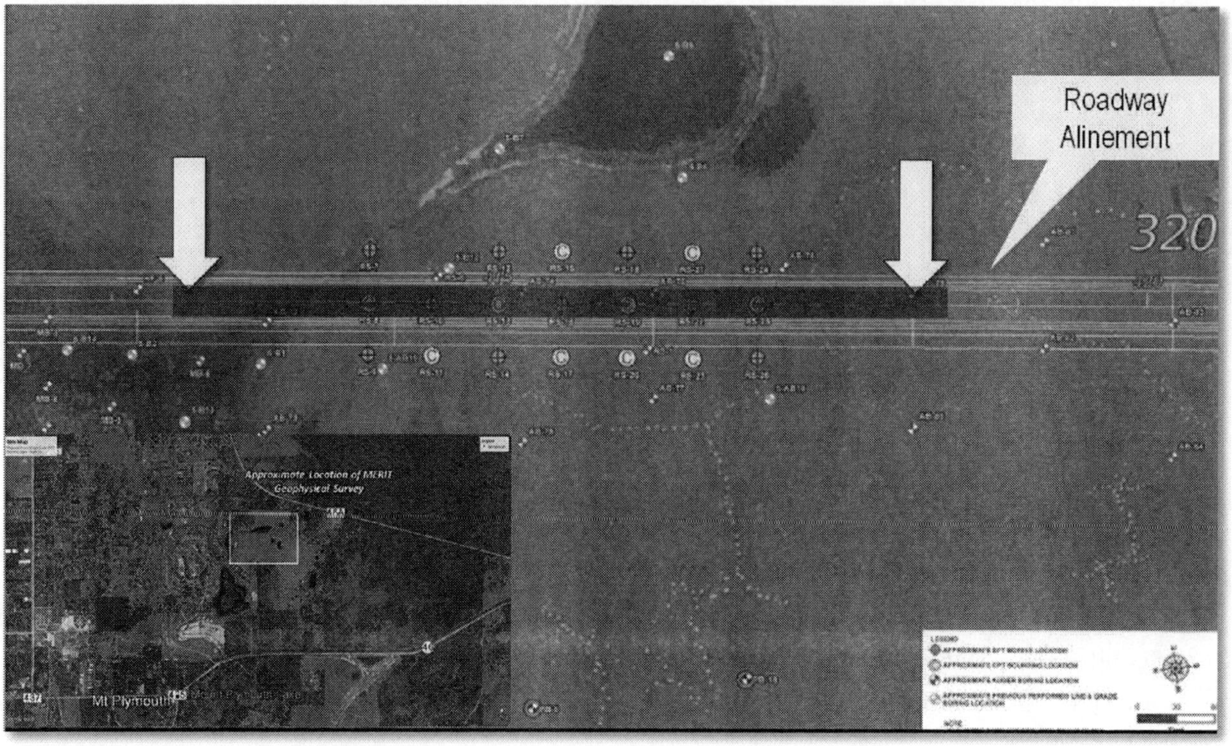

Figure 1: Shows project location

A preliminary geotechnical investigation was conducted by others at Site B which included: Ground Penetrating Radar (GPR) the drilling of 26 Standard Penetration Borings (SPT) and Cone Penetrometer Tests (CPT's) to better define the depth and extent of the relic sinkhole.

Based on the geotechnical investigation performed by a consultant and the FDOT, the limestone formation was encountered at depths between 60 and 150 feet below land surface (bls). The high degree of variability of the limestone formation encountered in the geotechnical investigation identified the potential for karst or sinkhole conditions which are considered to be a concern for this project. The application of the MERIT system abilities to provide deep geophysical images was deemed beneficial to the project.

Purpose

The objective of the MERIT survey was to document any deep geophysical anomalies associated with the possible karst features using G3's patent-pending geophysical Multi-Electrode Resistivity Implant Technique (MERIT).

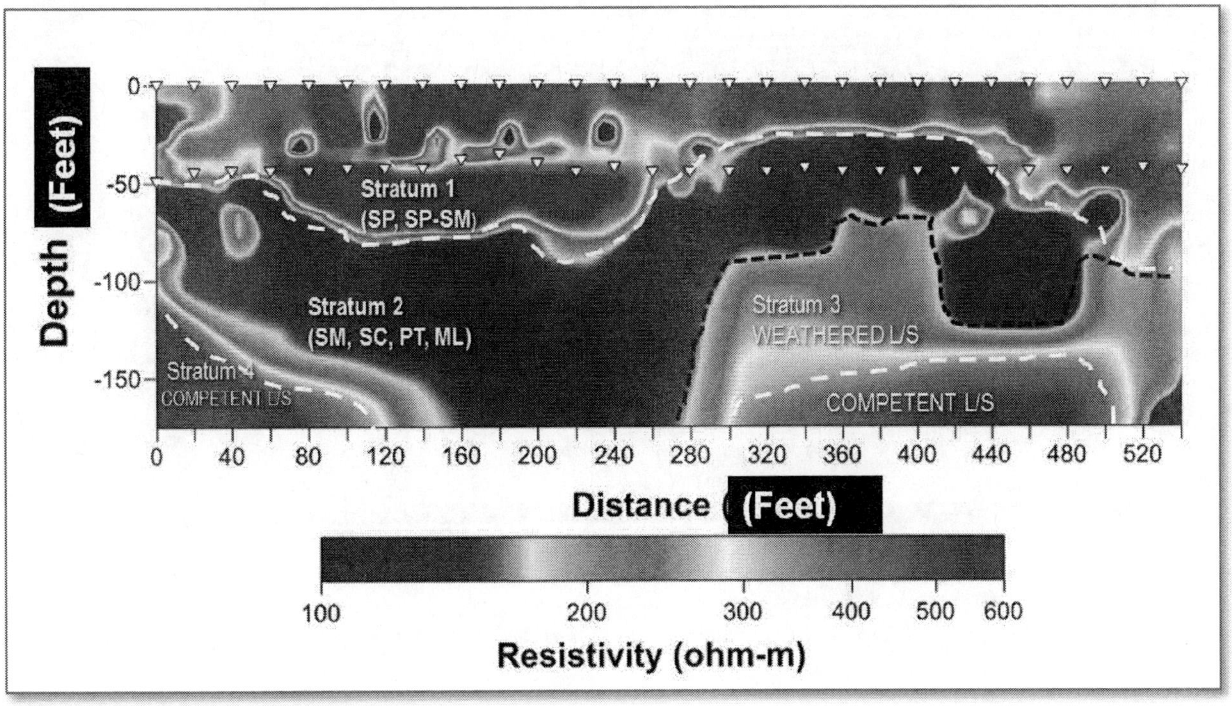

Figure 2: Results of MERIT geophysical survey

Investigation Conclusions

The MERIT geophysical survey achieved several objectives.

- A complete profile of across the relic sinkhole where a geotechnical investigation was performed to 170 feet bls using a 540 foot long array instead of the required 850-1000 foot surface electrical resistivity to reach similar depths
- The MERIT profile achieved significantly higher resolution at depth than what can be achieved with surface electrical resistivity
- The MERIT profile depth extended to and beyond the depths of the geotechnical investigation or the capacity for typical surface geophysical methods
- The results of the primary MERIT profile indicated a very good correlation with SPT and CPT data obtained during the Geotechnical investigation. A comparison of deep SPT borings indicated similar depths of all stratums as the MERIT profile
- MERIT geophysical survey shows a very good potential to correlate CPT and SPT data

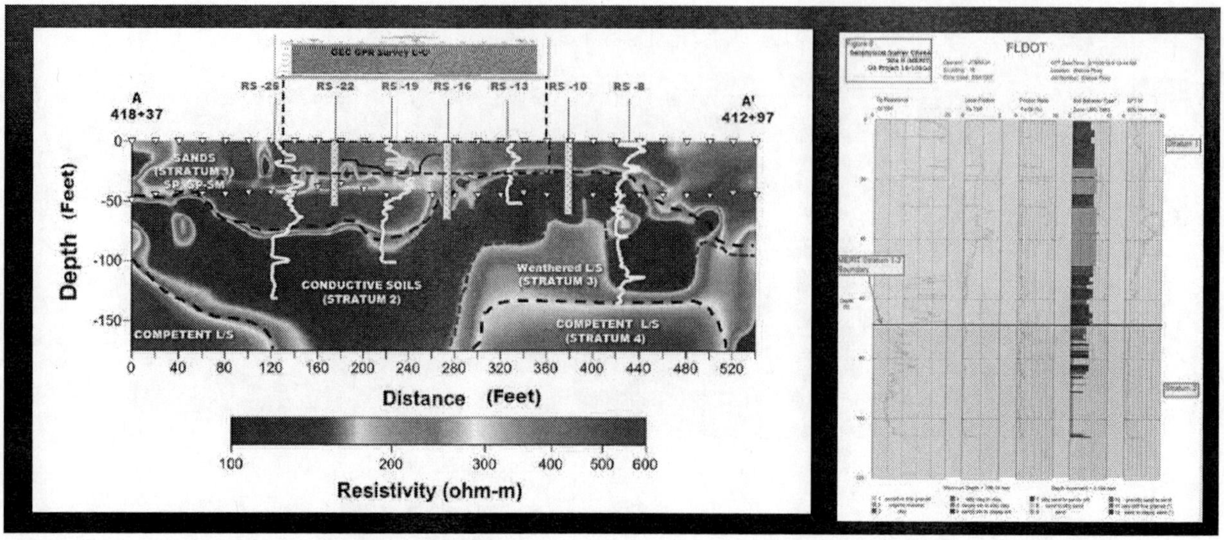

Figure 3: Comparison of the geotechnical SPT and CPT results compared to MERIT

Summary of Geotechnical Evaluation Compared Primary MERIT Line Results

The CPT results taken along the center line were compared with the results of a MERIT along the primary line. There is a good correlation between the MERIT Stratum boundaries and the results of the CPT's soil behavior type and noticeable changes in tip resistance.

Results

Based on the totality of the geotechnical testing and deep geophysical imaging and risk modeling was performed to determine the potential activity of the sinkhole was low. Thus no additional ground modification would be required to support the proposed roadway

REFERENCES:

- Conference Proceedings

1. Harro, D., Kruse, S. (2013). Improved imaging of covered karst with the multi-electrode resistivity implant technique. NCKRI Symposium 2. Proceedings of the 13th Multidisciplinary Conference on Sinkholes and the Engineering and Environmental Impacts of Karst. Carlsbad, New Mexico, US.
2. Loke, M.H., H. Kiflu, P.B. Wilkinson, D. Harro, and S. Kruse, Optimized arrays for 2-D resistivity surveys with combined surface and deep arrays, Near Surface Geophysics,13,505-517, 2015.
3. H. Kiflu, S. Kruse, M.H. Loke, P.B. Wilkinson, D. Harro, Improving resistivity survey resolution at sites with limited spatial extent using Research Article; Pages 338-355;. Journal of Applied Geophysics (2016)

Atlanta's Latest Mega-Tunnel

Adam Bedell

Stantec

2310 Parklake Drive NE Suite 400

Atlanta Georgia 30345

(470) 249-0289

Tao Jiang

Wayne Warburton

Konner Horton

Don Del Nero

Stantec

2310 Parklake Drive NE Suite 400

Atlanta Georgia 30345

Brian Jones

City of Atlanta Department of Watershed Management

72 Marietta Street 5th Floor

Atlanta Georgia 30303

Prepared for the 68th Highway Geology Symposium, July, 2017

Acknowledgements

Ade Abon-City of Atlanta Department of Watershed Management
George Ajy-River 2 Tap
George Reynolds-PRAD Group
Steve Fradkin-Stantec
Julian Prada-Stantec
Yong Wu-Stantec
Rick Ponti-Stantec

Disclaimer

Statements and views presented in this paper are strictly those of the author(s), and do not necessarily reflect positions held by their affiliations, the Highway Geology Symposium (HGS), or others acknowledged above. The mention of trade names for commercial products does not imply the approval or endorsement by HGS.

Copyright Notice

ATLANTA'S LATEST MEGA-TUNNEL

The City of Atlanta is currently constructing a $300M water supply system, known as the Water Supply Program (WSP), which includes the conversion of a century-old rock quarry into a 2.4 billion-gal raw water storage facility, 4.5 miles of 400 ft.+ deep, 12 ft. diameter tunnel bore with ten shafts of various types. The tunnel is being mined from a portal at the base of the quarry and will connect the quarry to two water treatment plants and three pump stations. The project is being delivered using the Construction Manager at Risk (CMAR) model, which is an innovative contracting method that is fairly new to the tunnel industry. Multiple aspects of the project will be highlighted in this paper, including subsurface investigations, design elements, ground conditions, and tunnel lining.

PROJECT OVERVIEW

Resiliency is now in the lexicon of the global community and has become one of the primary goals of many urban centers. To that end, major urban centers like the City of Atlanta are building resiliency into their water, wastewater, and transportation infrastructure.

The City of Atlanta, like most municipalities, is estimated to have just a three-day backup supply of clean water and most of the world is experiencing some type of drought. Now more than ever, forward-thinking communities are seeking to build resiliency into their infrastructure. For Atlanta that meant the purchase of the Bellwood Quarry some years ago from the Vulcan Materials Company with the intent to create a 2.4 billion-gallon raw water storage facility that would extend the City's back-up water supply to 30 days at full use, and around 90 days with emergency conservation measures. This is truly a "mega project" that involves getting water from the Chattahoochee River to the Quarry, pumping it up multiple vertical shafts to two water treatment plants, and then pushing it into the City's water distribution system.

The Atlanta region is well known for its "firsts" in the world, and this project is no different. From the start of construction to project buildout, it will feature the:

- First blind bore shaft, over 400 feet deep in hard rock, in the Southeastern United States;
- Deepest tunnel in Georgia; and
- Largest quarry repurposed as a raw water storage facility in North America.

The current water supply program operated by the City's Department of Watershed Management (DWM) consists of four aged raw water pipelines, one of which dates to 1893. Based on previous assessments completed by the DWM, the entire water system is at, or will soon reach, its recommended useful life. As such, the City acquired the Bellwood Quarry in 2006 with the intention to create a water storage facility with a volume of approximately 2.4 billion gallons to serve approximately 1.2 million people.

Using the Quarry (Figure 1) as a water storage facility greatly enhances the reliability and security of the drinking water supply to the greater Atlanta metropolitan area. For many years it was used for mining granitic gneiss and crushed-stone aggregate production. The Quarry has nearly vertical sides and ground elevations around the rim ranging from approximately 850 feet to 970 feet above mean sea level based on the NAVD 88 vertical datum. Quarry floor elevations range from about 520 feet to 540 feet. The proposed full pool level for raw water storage is at an elevation of 840 feet.

Figure 1. Bellwood Quarry prior to construction.

The project will connect the Quarry to the Hemphill Water Treatment Plant (HWTP), the Chattahoochee Water Treatment Plant (CWTP) and Chattahoochee River. Raw water will be supplied to the Quarry storage facility from the Chattahoochee River. Stored raw water will be withdrawn from the Quarry for treatment at the Hemphill and/or Chattahoochee water treatment plants, with treated water subsequently pumped to the City's treated water distribution system. This offline operating mode includes routine withdrawals and replenishments.

The project location is shown in Figure 2 on the following page, which is generally in the Northwest part of downtown Atlanta, Georgia. The overall project has been divided into two phases. The Phase 1 project connects the Quarry and the HWTP, and the Phase 1 Extension project connects the HWTP to the CWTP and the Chattahoochee River.

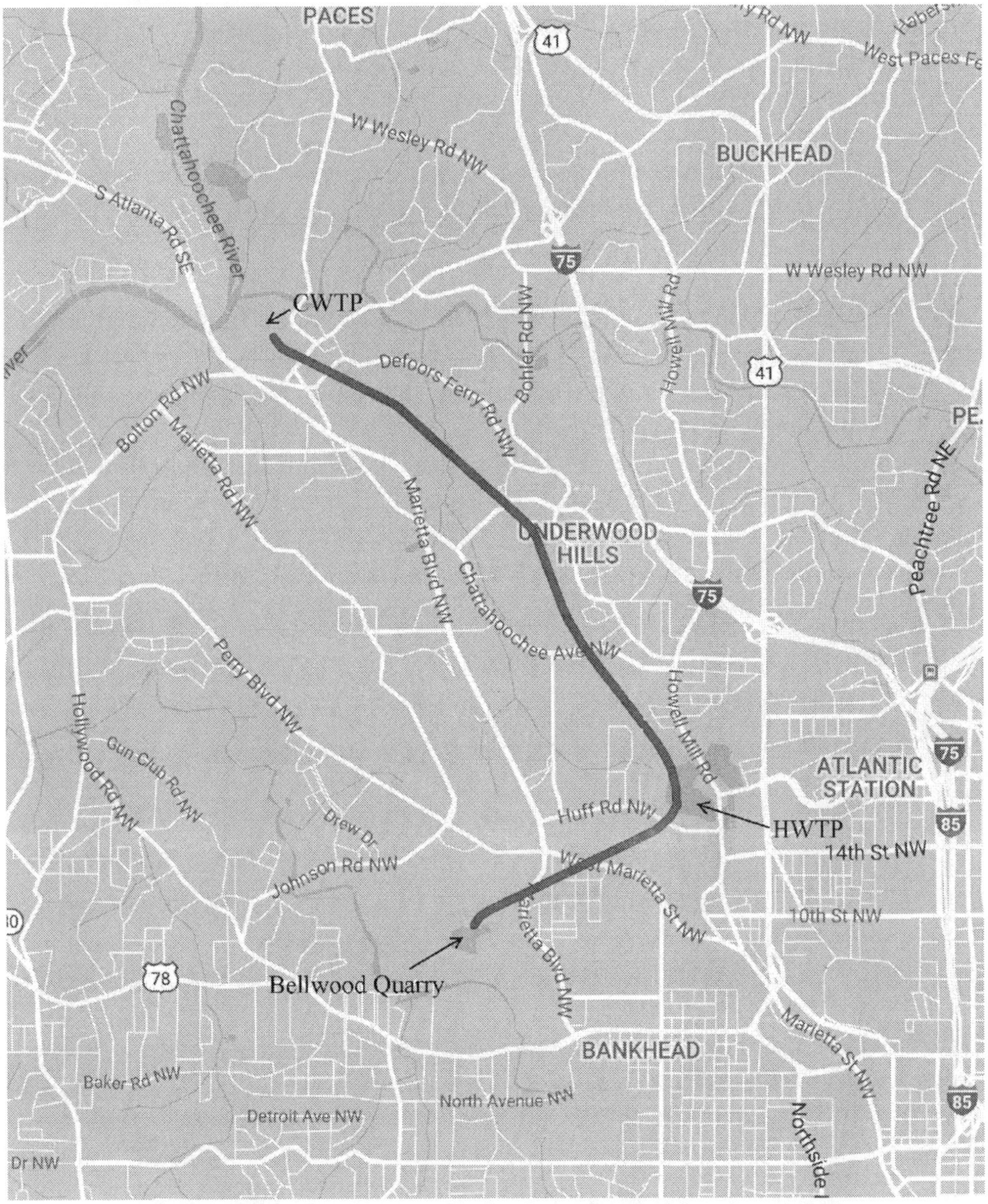

Figure 2. Project location map. Project is northwest of downtown Atlanta.

The main features of the project include a TBM-excavated tunnel, seven pump station shafts, a drop shaft, a riser shaft, one combined drop and construction shaft, and a Quarry highwall rockfall protection system to provide long-term protection of the tunnel inlet. Two of the seven pump station shafts will be constructed using conventional shaft excavation methods (including drill-and-blast in rock) while the remaining five will be excavated using blind bore methods. A 3-D rendering depicting the general arrangement of shafts, tunnel and adits at the Quarry site is presented in Figure 3 on the next page. These components, along with the other project components generally noted above include:

- A TBM tunnel that is approximately 24,000 feet long and partially concrete-lined with a finished diameter of 10 feet.

- A primary pump station shaft at the Quarry that is approximately 250 feet deep with a finished diameter of 35 feet. The low level pump station shaft has a finished diameter of 20 feet and is approximately 340 feet deep. The primary and low level pump station shafts are connected to the tunnel and Quarry via adits.
- A drop shaft at the Quarry that is about 320 feet deep with a finished diameter of 25 feet above El. 805 feet and 4.5 feet below El. 805 feet. The drop shaft is connected to the Quarry low-level pump station shaft, the riser shaft, and the main tunnel through adits. The drop shaft provides a flow capacity of 90 million gallons per day.
- A riser shaft at the Quarry that is about 320 feet deep with a finished diameter of 25 feet above El. 805 feet and 12 feet below El. 805 feet. The Quarry riser shaft is connected to the Quarry drop shaft and the main tunnel through adits. Five pump station shafts at the HWTP that are about 420 feet deep and 9.5 feet in bored diameter.
- Each of the five blind bored pump station shafts will have a 76-inch diameter grouted steel casing to house the pump, and are connected to the main tunnel by five, 8-foot diameter adits with lengths ranging from 20 feet to 30 feet.
- A construction/drop shaft at the CWTP site that is about 250 feet deep with a finished diameter of 30 feet.

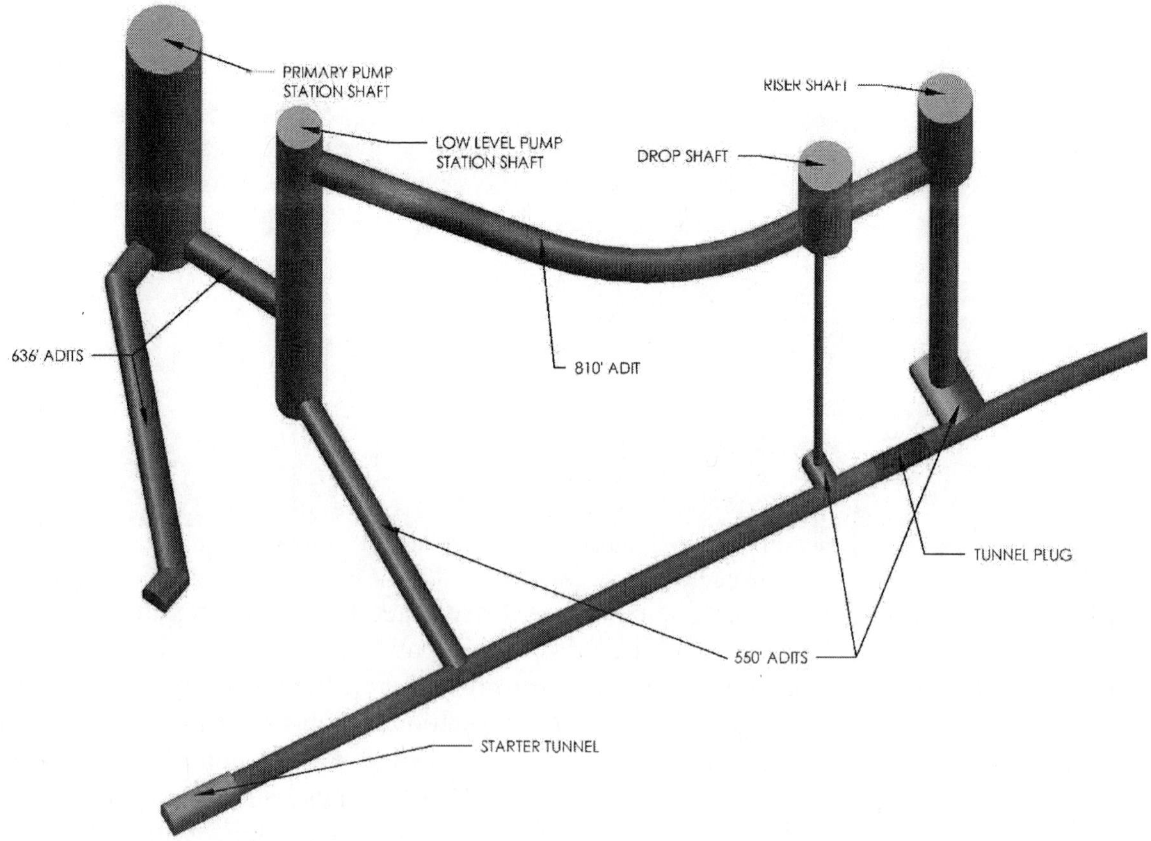

Figure 3. Arrangement of the structures at the quarry site.

The Construction Manager at Risk (CMAR) model was used as the contracting method, with the City selecting the joint venture PC Russell JV as the CMAR. Other important players include the

Atkinson/Technique JV (ATJV) as the tunnel contractor and the joint venture design team of JP2. Stantec Consulting acted as the tunnel designer for JP2. At the time of writing, both pump station shafts at the Quarry have been completed, along with the 636' adit and its breakout structure. The upper portions of the drop and riser shaft have been excavated and are being prepared for the start of raise bore operations and pilot holes for the five blind bore shafts are being drilled. The TBM tunnel is nearly 13% mined and the Peachtree construction/drop shaft is underway.

GEOLOGIC AND GEOMORPHOLOGIC CONDITIONS

The project is located in the Piedmont Physiographic Province. The geology of the Piedmont in the greater Atlanta area generally consists of medium-grade metamorphic rocks with granitic intrusions. These crystalline rocks are some of the oldest rocks in the Southeastern United States, ranging in age from some 275 million to over 1 billion years ago, with the youngest forming during the series of orogenic events that culminated in formation of the Appalachian Mountains. Since their origin, the rocks have undergone a complex history of metamorphism, weathering, and deformation. More specifically, the rocks in the greater Atlanta area have undergone episodes of both progressive and retrogressive metamorphism, with the peak regional metamorphism occurring in the Paleozoic Era, 360 to 380 million years ago.

As a result of this complex geologic history, structural features of the rocks include folds, fractures, and lineaments. The high pressures and temperatures at great depths resulted in a full range of deformational styles, ranging from medium-grade metamorphism, through fully-welded ductile shearing and mylonite formation, to brittle fracturing with rocks that commonly contain hydrothermally deposited minerals. At shallower depths, structures like exfoliation fractures were formed in the rocks due to erosion of overburden and unloading. The exfoliation fractures occur mainly along the foliation "planes" of the rocks. The foliation "planes" tend to act as areas of weakness within the rock mass, and the exfoliation fractures tend to be open and act as conduits for water movements through the rock mass.

Lineaments, which are surface topographic expressions of underlying rock mass or crustal structure, occur throughout the Piedmont. The lineaments are often controlled by weathering associated with discontinuities in the bedrock. In many cases, the lineaments represent fracture zones in the underlying bedrock. At greater depth, the fracture zones are typically cemented with minerals. At shallower depths, erosion of these weathering minerals (primarily micas) often results in zones of broken, water-bearing rocks and topographic features such as valleys and draws.

A key characteristic of the Piedmont region is the mantle of residual soils, derived from weathering of the parent metamorphic rocks and localized granites in the area. These residual soils grade downward into the underlying unweathered bedrock. The humid climate promotes chemical weathering of the parent material. Degradation of the parent crystalline rock begins at the grain boundaries and progresses inward through the rock mass producing residual soil. The residual soil resembles the original rock in appearance, but its physical characteristics such as strength and permeability are more similar to a micaceous sandy silt (ML) or silty sand (SM).

Within the Southeast United States, saprolite is the term used to describe a soft, thoroughly degraded rock that is clay rich, while retaining the original parent rock structure.

For this project, as well as a number of previous tunnel projects in the Atlanta area, the subsurface is divided into three zones:

- **Soil Zone.** Residual soils in the project area are the result of continued chemical breakdown of saprolite. All relict structure is absent and the resulting soil mass is reddish-brown in color and is either a silty clay (CL or CH) or a clayey silt (ML or MH).

- **Transition Zone**. The transition zone consists of partially weathered rock and highly fractured rock, underlying the overburden soils. The top of this zone occurs where rock and partially weathered rock begin to predominate over soils, and the bottom of this zone is defined where slightly weathered or fresh rock takes control of the rock mass.

- **Bedrock Zone**. The bedrock zone lies below the transition zone. This zone is dominated by fresh rock and faintly weathered rock, with local occurrences of more weathered material typically along discontinuity planes.

Groundwater occurs in all three zones of the subsurface described above. The depth of the groundwater table varies significantly along the proposed tunnel alignments, ranging from less than 10 feet to over 200 feet. The soil zone is generally considered to be a good producer of groundwater. The transition zone typically contains abundant open fractures and can become a major storage source for groundwater where its thickness is significant. The bedrock zone in the Piedmont generally has fewer open fractures with depth than the transition zone. However, large fractures with the ability of producing large volumes of water do exist in the bedrock. High-yield wells have been reported to produce sustained yields up to nearly 500 gallons per minute.

Potentiometric gradients may be steep in the Piedmont. Seasonal fluctuations in the water table are common in response to rainfall. Local observations of the water table rising and falling between 8 feet to 14 feet are common. Perennial streams are fed by bank seepage and upwelling groundwater along the course of their lengths.

SUBSURFACE INVESTIGATION PROGRAM FOR THE TUNNELS AND SHAFTS

The geotechnical and hydrogeological field investigations for the Water Supply Program comprised 25 deep borings and 30 shallow borings. The deep borings were advanced along the proposed tunnel alignment with the main purpose of characterizing the bedrock conditions near the tunnel horizon. The shallow borings were drilled at the locations of proposed shafts and surface structures with the primary purpose of characterizing the overburden soil conditions, including information on the transition from soil to rock. Drilling occurred in phases from August 2014 through August 2016 in concert with an evolving design.

Prior to initiation of the geotechnical investigation, readily available, relevant geologic data was summarized and reviewed, and some field work was performed. Ground conditions along previously constructed tunnels proximate to the WSP tunnel were also reviewed. In addition, data provided by geologic field mapping and other available background information were used to complete lineament and structural geologic analyses as shown in Figure 5.

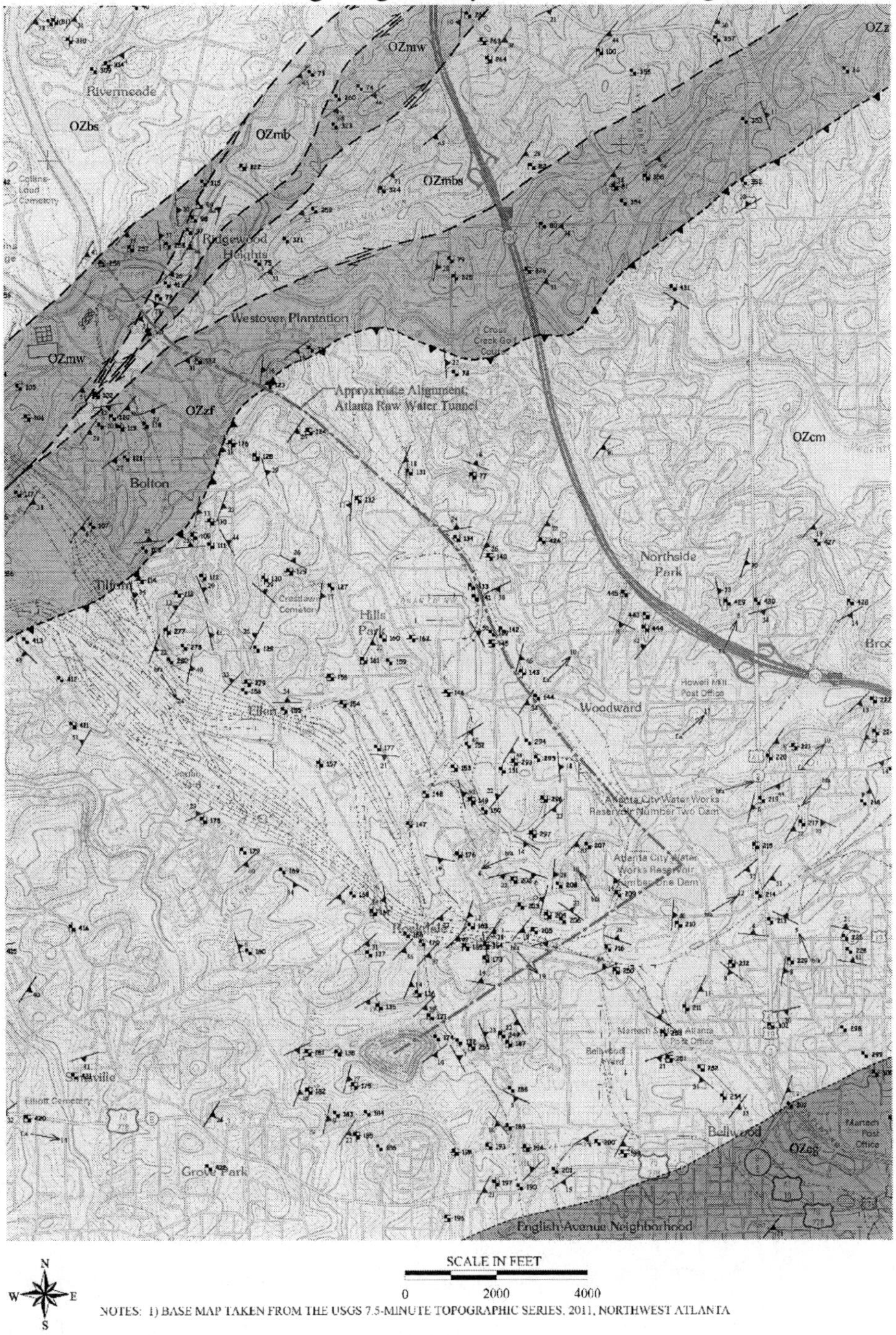

Figure 5. Geologic map prepared during the initial field mapping for the Water Supply Program tunnel project.

The geotechnical investigation was developed based on information contained in the background reports developed from the geologic mapping and associated investigative work. Triple-tube HQ coring was selected to obtain rock samples. In addition to coring, double-packer permeability testing was performed on most of the deep vertical boreholes. Once cores were extracted, they were logged and photographed.

Once drilling was complete, a suite of borehole geophysical tests was run in 21 of the deep borings. This provided the following information: optical and acoustic televiewer logs, full wave sonic logs, fluid temperature and conductivity logs, natural gamma logs, single point resistance logs, three-arm caliper logs, and EM flowmeter logs. These tests helped to further characterize the in-situ geologic conditions at depth while also providing hydrogeologic information and joint orientation data used to create stereoplots.

Following core analysis and geophysical testing, pumping test locations to determine overall hydrogeologic conditions were selected. The locations were selected based on the completed geologic mapping and proximity to identified geologic controls that were expected to influence groundwater movement once tunneling began. Of the three locations chosen, two yielded insufficient groundwater (as determined through air lift testing) to conduct the tests, and the pumping test holes were abandoned. Consequently, only a single pumping test was performed. It was run for 24 hours, and recovery was measured immediately following shutting off the pump.

The depth of the tunnel (greater than 400 ft. in areas) warranted in-situ stress testing. Agapito and Associates conducted the in-situ stress testing in three of the deep borings and attempted 12 tests, of which 7 were successful. They used the over-coring method as developed by Sigra, Pty of Brisbane, Australia. The purpose of this testing is to determine the magnitude and direction of the horizontal principle stresses. The results were factored into tunnel excavation support design.

Subsequent laboratory testing to determine the properties of the observed rock types was performed. These tests include unit weight, unconfined compressive strength, Cerchar abrasivity, Brazilian tensile strength, acoustic velocity, point load index strength, petrographic analyses, x-ray diffraction, and abrasivity/drillability tests.

Two of the three main project sites were scrutinized during the last phase of the geotechnical subsurface exploration program: The Peachtree Drop/Construction Shaft and the Hemphill sites. During the initial site investigation, deep boring RWB-15 at the Hemphill site encountered degraded rock conditions and borehole stability was a constant issue. During the evolving design, 3 additional borings were drilled to help characterize this site. These included permeability testing and borehole geophysics. Additional tests were run as RWB-15 was considered too risky to place any tooling in the borehole. During this time, while shaft configurations evolved, potential impacts to the existing HWTP reservoir were constantly evaluated.

Construction records for the R.M. Clayton Construction Shaft, built for the North Avenue tunnel as part of the West Area CSO Storage Tunnel were reviewed, as the Peachtree Drop/Construction Shaft is approximately 125 ft away. Construction photographs of the R.M. Clayton Construction Shaft depict deep weathering in the shaft. So, shallow borings were drilled around the perimeter of the Peachtree Drop/Construction shaft to determine the thicknesses of the subsurface zones. Typical of the Piedmont, depths to different subsurface zones may vary substantially over short distances.

Lithologies Along the Tunnel Alignment

The majority of the proposed tunnel alignment is located in the Clairmont Melange, with the latter portions in a zoned feldspar gneiss followed by Brevard Zone black and white mylonites. The descriptive text that follows is taken from the Geologic Report (*1*) prepared by PetroLogic Solutions as part of the preliminary geotechnical investigation. The order of the four geologic unit descriptions (Clairmont Melange, Zoned Feldpsar Gneiss, Black Mylonite, and White Mylonite) are from the Quarry to HWTP and then through to the CWTP .

> The majority of the Contorted Unit [of the Clairmont Melange] consists of a sphene-epidote-muscovite-biotite-quartz-feldspar gneiss, medium-grained, schistose in part; interlayered with sphene-epidote-muscovite-quartz-feldspar-biotite schist, medium- to coarse-grained; garnets may be present, but are small and scarce. Hornblende gneiss/amphibolite lenses and layers (commonly boudinaged) are common. Contains, in many places, lenses and discontinuous layers of unfoliated granite on a scale of feet and ten's of feet. Concordant and discordant quartz veins are common. Pegmatitic layers and coarse pegmatites up to 60 inches thick are abundant and characteristic; shear foliation in the gneiss/schist wraps around the coarse pegmatites and small bodies of granite, which are generally not sheared.
>
> This rock mass is extremely contorted; foliations are quite variable over short distances, and are generally low-angle and undulatory. Random fractures are abundant; through-going joint sets are scarce and not well-developed.
>
> The zoned feldspar gneiss consists of an epidote-muscovite-biotite-quartz-feldspar gneiss, fine- to medium-grained, with disseminated very coarse zoned feldspar crystals; very feldspathic overall; deep weathering is characteristic.
>
> The Brevard Zone black mylonite is generally composed of biotite, quartz, and feldspar. This unit is typically extremely fine-grained and weakly foliated. Where the foliation is better developed, the rock is shown to be very contorted. In most outcrops, the black mylonite is dark gray to black and locally contains thin light colored layers of white mylonite (see rock unit 2B description). Weathering of this unit generally yields a reddish brown to red, uniform fine clayey residuum.
>
> The Brevard Zone white mylonite is interpreted to be sheared granite. This mylonitized granite is composed of muscovite, quartz, and feldspar; much of the feldspar is pink and coarse-grained. Shearing was pervasive and produced a well-developed shear foliation. Reduction in grain size was not as extreme as in Rock Unit 2A. Weathering of this unit generally yields a white to tan, uniform fine clayey residuum.

At the time of writing, rock mass conditions encountered during construction of the Quarry shafts and TBM tunnel are consistent with the information as provided in the preliminary geologic report. Foliation is quite contorted over the scale of the excavation and degrees of schistosity vary across the excavation.

QUARRY DESIGN

Quarry Highwall Evaluation

During an earlier phase of the project, the DWM conducted a study of the quarry highwall stability (*2*). The objective of this phase was to determine if there were any significant stability issues that would jeopardize the use of the quarry as a water storage facility.

The evaluation of the highwall was focused on the long-term stability of the highwalls during operation of the Quarry as a reservoir. As discussed in a following section, highwall stability during construction is managed by the Contractor responsible for the tunnel and shaft construction.

The main items included as part of the highwall evaluation included;

- Review of geological data collected during design and construction of a tunnel located approximately 700 feet east of the quarry,
- Review of exploration drilling data provided by the previous quarry operator and discussion with the previous quarry operator's staff regarding quarry highwall stability,
- Field geologic mapping in the quarry and around the top of the quarry, and
- Photo-geologic mapping of portions of the quarry highwalls.

Due to the height of the quarry walls, and limited access to the quarry walls, photo-geologic mapping was used to collect structural data of the discontinuities exposed in the quarry highwalls. Model processing and mapping were performed using Sirovision, a rock slope modeling and photo-geologic mapping computer program developed by the Commonwealth Scientific and Industrial Research Organization (CSIRO) Mining and Exploration Group based in Brisbane, Australia.

The structural and photo-geologic mapping found that the general dip of foliation ranges from approximately horizontal to approximately 20° and the dip direction generally ranges from southwest to east. Foliation undulates throughout the quarry at a scale of tens of feet between crests on the foliation surfaces, and locally may dip up to 25° in any direction at any particular location. Foliation is reflected by the central pole clusters shown on the stereonet plots on Figure 4.

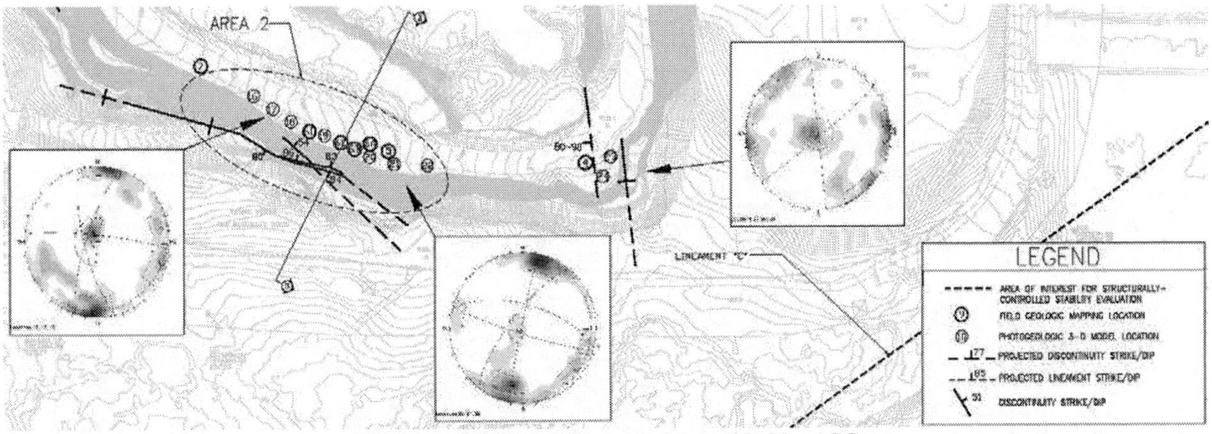

Figure 4. Portion of the analysis provided by ASG.

Projections of individual fractures and stereonet plots of great circles representing fracture sets are shown on Figure 4. As can be seen from this figure, fractures observed in the highwall generally tend to dip at angles greater than 60° (high angle fractures). The foliation fractures tend to be rough and undulating, and tight or closed with no alteration or infilling. Foliation fractures tend to have low persistence relative to the scale of the highwall (trace lengths were observed to be generally less than 30 feet), and the spacing between foliation fractures is irregular, but generally greater than 2 feet. The high angle joints were typically rough and planar, stepped, or undulating; fresh to slightly weathered, with no infilling. High angle joints tend to be moderately widely to extremely widely spaced (from 2 feet to more than 20 feet apart).

The highwall evaluation did not identify any large scale features that would prevent the quarry from operating as intended. Localized areas with potential for rock falls were identified. These areas included zones with blast damage to the quarry walls and zones of localized jointing. The project design included methods to control rockfalls near the tunnel portal during operation, as well as during construction, which are described in following sections.

Tunnel Portal Stabilization

The contract stipulated that while final design of the drape was specified, safety during construction was the responsibility of the tunneling contractor. Therefore, substantial scaling program was undertaken by contractor, ATJV, to provide safe egress and ingress to the quarry bottom and TBM location. Scaling around the quarry rim took place from April through August 2016. While scaling of the quarry could last indefinitely, following initial inspection, ATJV implemented a scaling protocol that requires visits quarterly to inspect the rockmass and quarry rim. An outcome of this plan is that ATJV and their subcontractor conduct daily and periodic inspections of the highwall around the perimeter of the quarry that has resulted in additional scaling.

To secure the approximate 300-foot-tall rock face above the tunnel portal at the base of the quarry, a designed stabilization system was included in the contract documents. The system covers the full depth of the quarry over a width of approximately 400 foot centered over the TBM tunnel portal. The general area of stabilization is shown in Figure 6.

Figure 6. General area of tunnel portal stabilization area and the two canopies.

Although the stabilization system was designed as part of the "permanent works," ATJV came up with an innovative way to combine the permanent stabilization system with supplemental rockfall measures so that the overall system could function as both temporary and permanent works.

The stabilization system consists of TECCO 3 mm mesh from Geobrugg and rock dowels in the locations that are identified as locations of potential rock wedge failures. Canopies were installed as additional protection to workers at the two portals in the quarry as shown on Figure 6. The canopies are designed to catch any rocks that may come loose and fall behind the drape above the portals. At the portals for both the tunnel and 636' adit, 20 foot long spiles are installed along the crown to stabilize more fractured ground.

TUNNEL DESIGN

The tunnel is about 24,000 fteet long and 250 feet to 450 feet below ground surface. It is sloping up from the quarry to the drop/construction shaft at CWTP with a grade of 0.2% and will be partially concrete lined with an internal (lined) diameter of 10 feet. The service life of the final lining system is designed to be 100 years.

Tunnel Initial Ground Support

The design provided for a two-pass tunnel support system, which is common for Atlanta area tunnels. Excavation ground support will be installed immediately following the TBM excavation to stabilize the tunnel and provide a safe work area. The ground was categorized into three ground types (Types A, B and C) for support based on rock mass properties with three excavation ground support types installed, respectively. Type A support consists of two 5-ft long double corrosion protection dowels as both excavation support and permanent support, since most of Type A ground is not anticipated to be concrete lined. Type B support consists of four 5-ft long friction dowels with welded wire mesh, and Type C support consists of steel ribs with welded wire mesh as lagging. Both Type B and Type C ground will be concrete lined.

Tunnel Permanent Lining

Following completion of TBM tunnel mining, both Type B and C ground will be lined, while most of the Type A ground is anticipated to remain unlined. The minimum lining thickness is designed to be 12 inches, not only for sustaining the design loads but for facilitating quality concrete placement. The double corrosion protection dowels installed in unlined tunnel sections of Type A ground is considered as part of the permanent support system and will support the ground during the tunnel service life.

As the tunnel is part of the water storage facility, the permanent lining system not only needs to support all the external loading, including rock load and groundwater pressure, but also to sustain the internal water pressure, which is about 300 ft. head. Under certain conditions the internal pressure could result in tension loads in the concrete lining; as such, reinforcement is designed for the lining in Type B and C ground since such ground is expected to provide less constraint than Type A ground. The transient pressures during filling the tunnel are also considered in the lining design.

SHAFT DESIGN

As aforementioned, the system consists of 10 shafts with different sizes, depths, and construction techniques. Pump station shafts at the quarry and the drop/construction shaft at the CWTP will be built with conventional drill-and-blast methods from the top down. The tangential drop shaft and riser shaft at the quarry will be raise-bored from the bottom up. The five pump station shafts at HWTP will be drilled from the surface with blind boring techniques. To the authors' knowledge, the five 9.5 ft. diameter blind bore shafts at HWTP will be the largest and deepest shafts in the Piedmont geology to use this technique.

Blind Bore Shafts at HWTP

As shown on Figure 8 on the following page, the five pump station shafts will be constructed using blind bore techniques since surface blasting is prohibited at HWTP due to the existing adjacent reservoirs. Upon completion of the five 11-ft diameter steel casing installations in overburden, drilling of the 9.5-ft diameter 400-ft deep blind bore shafts will start from the surface into rock through the steel casings.

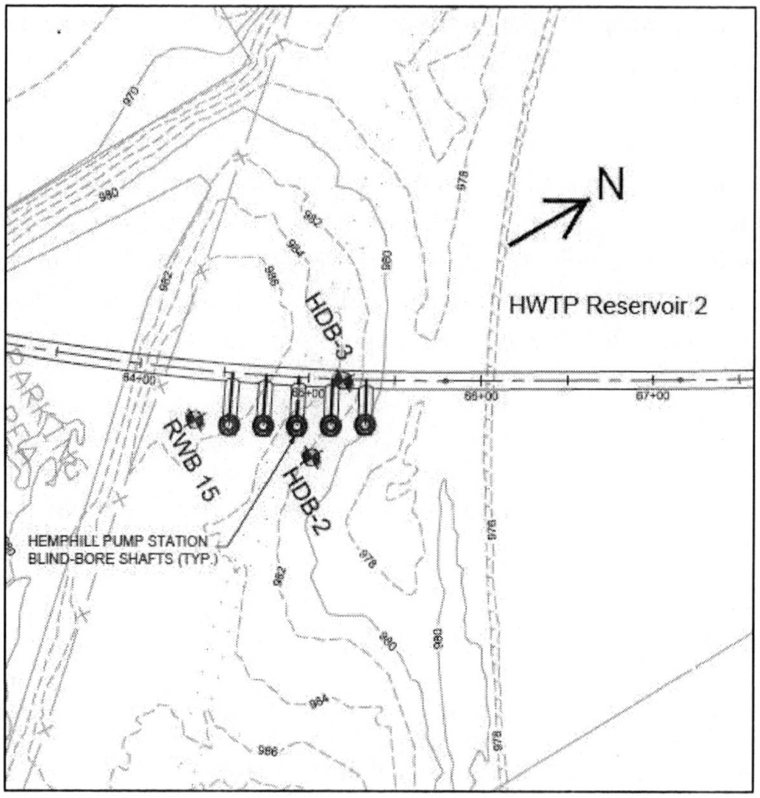

Figure 8. Site layout at Hemphill showing blind bore shafts and adits connecting to the main tunnel.

Two blind bore rigs will be mobilized to meet schedule requirements. Each rig has a rotary table that provides the torque or turning action for the reamer. Throughout the entire shaft development, both the shaft and the hollow drill string are filled with water to create two independent columns of water. The water column inside the drill string is made much lighter by injecting compressed air. The heavier water column inside the shaft thus pushes down and across the bottom of the shaft. The water is then forced through a small opening on the reamer body and displaces the lighter water in the drill string to create upward flow or reverse circulation. The reverse circulation generates tremendous vacuum at the reamer opening and removes the cuttings from the face. Maintaining a constant water level in the shaft during the entire drilling operation is critical. In addition to cutting removal, the water also provides outward pressure on the shaft wall to improve the shaft stability. The returned water from the shaft is collected in an adjacent settling pond, and the water can be re-circulated to the drilling operation after the cuttings have been settled out.

In order to meet the verticality tolerance, a pilot hole is required for each shaft. The pilot hole will be directionally advanced utilizing an optical technique that allows continuous monitoring for deviation. Once completed, an optical survey will be performed to verify that the pilot holes meet the required verticality.

Upon completion of the blind bore drilling, a 76" ID steel pipe with 1-inch wall thickness will be lowered into the shaft and grouted in place in the wet. The steel pipe is provided in 40-ft long sections that will be welded together. All welds will be ultrasonically tested. After the shaft construction, the vertical turbine type pumps will be installed inside the steel casings.

Pre-Excavation Grouting

The Hemphill Site includes the construction of a 136 million gallons per day (136 MGD) firm capacity raw water pump station (Hemphill Pump Station or HPS), consisting of 5 pumps. The 5 pumps are each housed in a shaft, all of which are located less than 100 feet from Raw Water Reservoir 2 at the HWTP (refer to Figure 8). The construction of these shafts poses a significant risk to the unlined reservoir. As such, a shaft pre-excavation grouting program was designed for the soil to rock transition zone and rock zone to greatly reduce the chance of communication between the reservoir and the 5 pump station shafts during construction.

During the geotechnical investigation for the project, the Hemphill site was scrutinized for two reasons. First, the City indicated that all risk associated with inadvertent dewatering of the Hemphill Reservoir due to construction of any aspect of the project was to be kept to an absolute minimum. Second, given the results from the initial borings and subsequent borings, poor ground conditions were identified within the limits of excavation. These ground conditions required mitigation to facilitate excavation with the blind bore shaft sinking technique.

The most practical mitigation method was determined to be pre-excavation grouting of the area. The pre-excavation grouting program addressed these concerns by mitigating risk for the reservoir through consolidation of the rock mass to lower permeability of the rock mass and reduce the potential for loss of drilling fluids during blind bore operations. During design, a third risk was identified that is also addressed through the pre-excavation grouting program. This is the potential for catastrophic fluid loss during blind bore shaft sinking after the tunnel passes through the area, thus flooding the tunnel excavation.

As design of the HPS was fluid and changed during the course of the project, the grouting program evolved as well from preliminary layouts addressing conventional shaft configurations, shifting to the present blind bore shaft configuration. Conventional grouting layouts for shafts were not considered, and a design more typical of underground chambers was implemented. This was due to needing an increased area of reduced rock mass permeability for protection of the reservoir.

As noted, the need for protection of the tunnel from potential flooding during blind bore shaft sinking also factored into this decision. Initially, all the pre-excavation grout holes were planned to be vertical with primary holes on 16-foot centers, as well as the secondary grout holes. This resulted in a battered spacing of 8 ft. between the primary and secondary grout holes.

Additional borings, HDB-2 and HDB-3, were drilled at the site in January 2016 while site design was underway and the initial pre-excavation grouting program had already been designed.
Results of borehole geophysics from the additional borings were received a week before the Hemphill pricing set of Contract Documents was to be released. Analysis of the geophysical data indicated two primary joint sets that were steeply dipping (>75°) as shown on Figure 9.

Geophysical data also indicated numerous fractures within the three identified joint sets, which contained apertures ranging from 0.25in. to 5in. While open fractures within the foliation joint set were not considered an issue with vertical grout holes, potentially missing open fractures within the two high angle joint sets was judged to be a risk to both the reservoir and the blind bore shaft sinking operation. Consequently, the grout hole orientation was changed from vertical to inclined 10° off vertical at a bearing of 260° (refer to Figure 10).

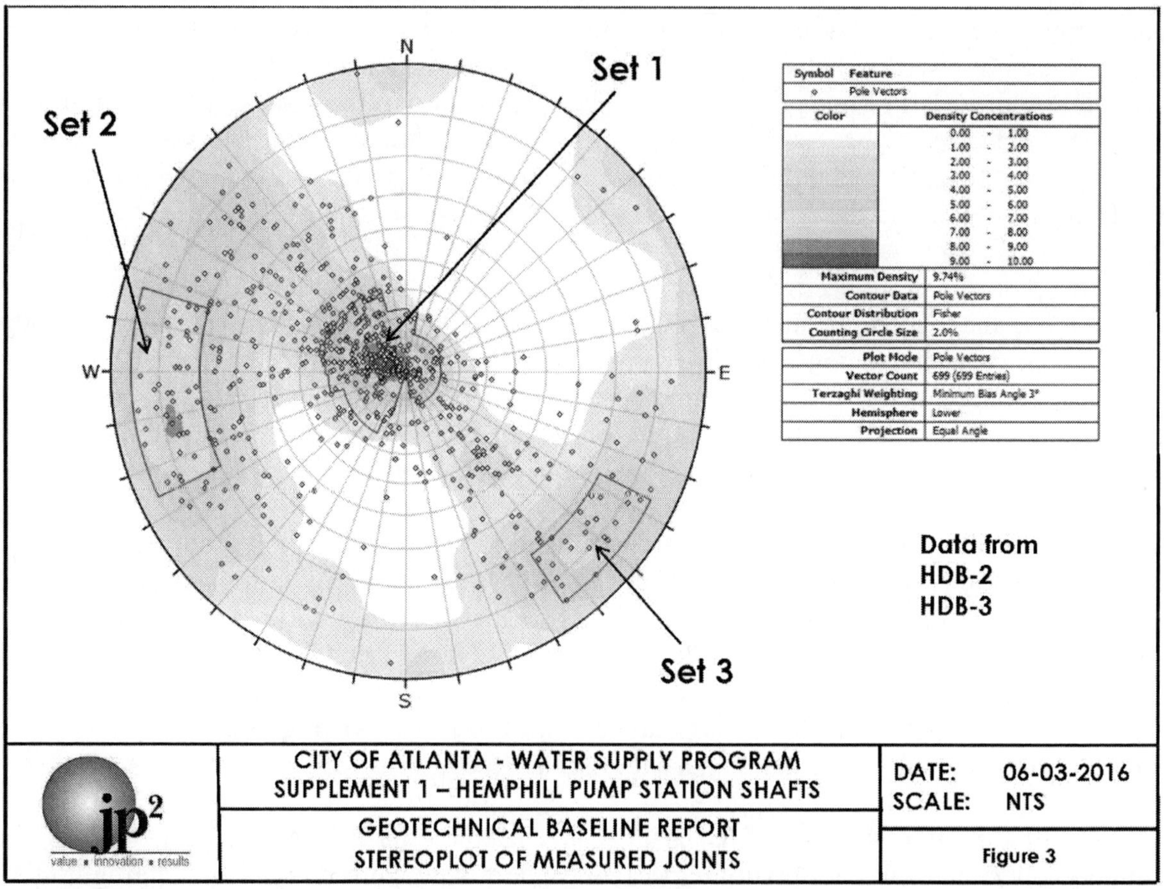

Figure 9. Stereoplot from the Geotechnical Baseline Report of geophysical information showing the high angle joint sets.

This orientation allows for a higher potential for intersecting all the identified features as indicated from the geotechnical investigation and analysis (while staying within the footprint of the surface site), thus reducing the potential for the identified risks to occur.

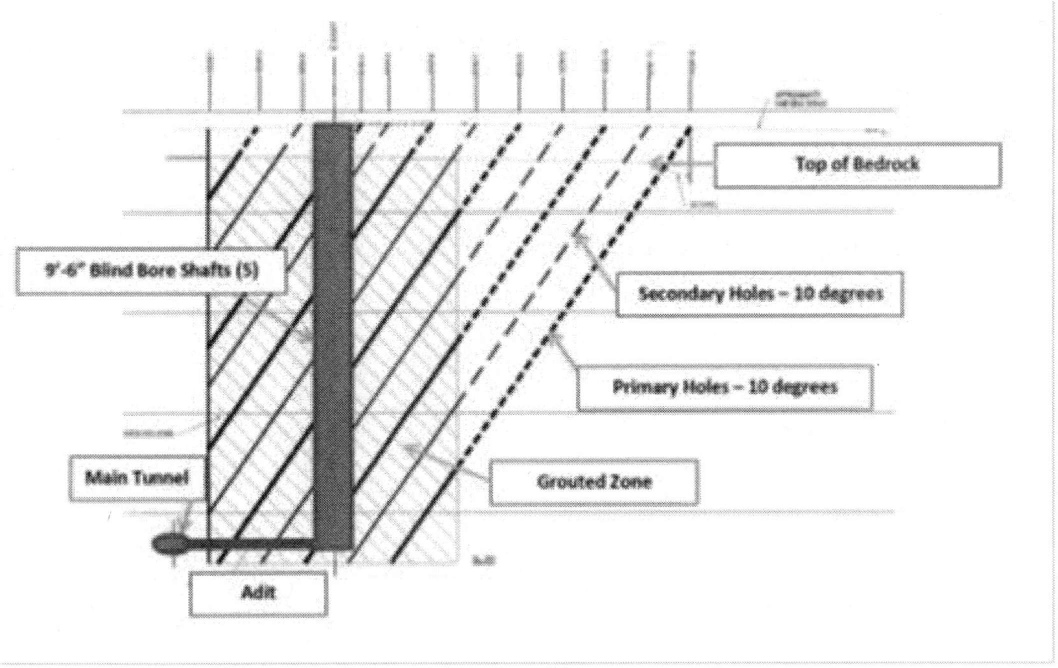

Figure 10. Cross-section view of the pre-excavation grout holes.

A significant variable in pre-excavation grouting programs is the grouting shut-off criterion. For this project the shut-off criterion is defined as a grout injection rate of 1/4 gallon per minute or less, as measured each minute for five consecutive minutes at 100% of the required grouting pressure and constant grout consistency.

CONCLUSION

The City of Atlanta Water Supply Program is a large, multi-faceted construction project that incorporates many "firsts," including the deepest of all the Metro-Atlanta tunnels. An evolving design allowed for portions of the project to be under construction while other elements were still under design. The WSP tunnel project incorporated many criteria into the design including pre-excavation grouting, tunnel lining analysis, blind bore shaft design, as well as a substantial quarry highwall stabilization program. These design elements are all in place to secure the City of Atlanta's drinking water supply. As the largest re-purposed quarry in North America, the City of Atlanta is once again leading the way.

References

1. **PetroLogic Solutions**, *Detailed Geologic Mapping Along The Proposed City Of Atlanta Raw Water Tunnel Alignment, Fulton County, Georgia*, Atlanta, Georgia, 2014

2. **Atlanta Service Group**, *Geotechnical Investigation and Evaluation – Technical Memorandum*, Atlanta, Georgia, 2007

An Example of Risk-Based Geotechnical Asset Management

Scott. A. Anderson
BGC Engineering, Inc.
710 10th Street, Suite 170
Golden, CO 80401
(720)289-9430
scanderson@bgcengineering.ca

Mark J. Vessely
Shannon & Wilson, Inc.
1321 Bannock Street, Suite 200
Denver, CO 80204
mjv@shanwil.com

Ty Ortiz
Colorado Department of Transportation
4670 Holly Street
Denver, CO 80216
Ty.ortiz@state.co.us

Prepared for the 68th Highway Geology Symposium, July, 2017

Disclaimer

Statements and views presented in this paper are strictly those of the author(s), and do not necessarily reflect positions held by their affiliations, the Highway Geology Symposium (HGS), or others acknowledged above. The mention of trade names for commercial products does not imply the approval or endorsement by HGS.

ABSTRACT

The Colorado Department of Transportation (CDOT) has recently implemented a Risk-Based Transportation Asset Management Plan (RB TAMP) that incorporates geotechnical assets and hazards. CDOT's RB TAMP includes an ancillary wall structures program that includes all earth retaining structures, and a geohazards management program which is used to manage multiple hazards related to slopes, embankments, and roadway subgrade. The RB TAMP states multiple performance goals to be achieved, including safety, infrastructure condition, reliability, congestion, and maintenance, and the state will measure and report progress in these areas. Natural hazards, physical failures, external agency impacts and operational risks are risk types that present threats to CDOT's achievement of their goals. The way these risks act on assets to impact performance goals can be visualized in a cubic form, and this allows for recognition of how many elements of risk there are, for making explicit decisions on which risks to address and how, and for communicating these decisions to others. Risk analysis at CDOT includes both qualitative and quantitative approaches in accordance with data availability. The quantitative estimate of risk is expressed in terms of exposure cost for all assets, risk types and performance goals and then used by CDOT subject matter experts for project selection and planning. The estimated risk exposures are also categorized into Level of Risk grades that are used to concisely communicate risk levels to executive management and to compare the long-term performance risks between asset types under different funding scenarios in the RB TAMP.

INTRODUCTION

Geotechnical engineers have a long history of recognizing risk management as a part of their responsibilities. Casagrande wrote about calculated risk for highway embankment construction in his 1964 Terzaghi Lecture (1965) and many more authors have contributed to this discussion in the years since. Pierson et al. (1990) published important work for the transportation sector using risk as a basis for prioritizing decision on state highway rockfall sites. Those authors presented a rockfall hazard rating system (RHRS) that has since been adopted, and customized by many states, including Colorado, and for other uses as well. Though the word hazard is in the title, the characteristic assessed is one of risk (including likelihood and consequence), and the resulting decisions are risk-based.

The Colorado Department of Transportation (CDOT) has more recently implemented a risk-based transportation asset management plan which it uses to help make funding and project planning decisions. CDOT's plan is consistent with the requirements of MAP-21 and the FAST federal authorizations that require all states to develop such a plan. CDOT's plan addresses risks related to geohazards such as rockfall, and geotechnical assets such as slopes and retaining walls, and are thus related to the work that evolved from Pierson (1990). This is because the potential consequence of a geohazard or poor performance of a geotechnical asset can threaten CDOT's plan to achieve its mission to "provide the best multi-modal transportation system for Colorado that most effectively and safely moves people, goods, and information".

CDOT is pioneering the way geohazards are being measured with respect to risk and the way risks from geohazards are compared directly with risks from other assets and used to prioritize spending and plan projects. Risk is a word and concept that can mean different things and be used in different ways, and CDOT's RB TAMP considers risk in a more expanded way than Pierson (1990). CDOT's RB TAMP includes an ancillary wall structures program that includes all earth retaining structures, and a geohazards management program which is used to manage geohazards related to slopes, embankments, and roadway subgrade. The RB TAMP explicitly states multiple performance goals and addresses risk types that present threats to CDOT's successful achievement of established targets for their goals. The way these multiple risks act on multiple assets to impact multiple performance goals can be visualized in a cubic form, called a risk cube. The visualization allows for recognition of how many elements of risk there are; for making explicit decisions on which risks to address and how, and for communicating these decisions to others, which CDOT has done.

BACKGROUND

Hazards and Assets

The CDOT RB-TAMP recognizes 11 asset groups (pavements, bridges, maintenance [traffic and safety devices], buildings, intelligent transportation systems

[ITS], fleet/road equipment, tunnels, culverts, geohazards, retaining walls, and traffic signals). These are things that CDOT has purchased or built (with plans and specifications). It is a little different with respect to the geohazard asset group, but CDOT maintains the right of way and has built a highway corridor that is underlain by subgrade and bounded by slopes. The vast majority of the subgrade miles and the slopes are unaffected, but the rest of them are where geohazards are realized, so effectively CDOT owns the hazards and corresponding threats. The geohazards recognized by CDOT in this way are specifically defined as follows:

- debris flows
- drainage/seepage features
- embankment distress
- landslides
- rockfall sites
- rockslides
- sinkholes
- subgrade distress

It is possible to look at the 8 geohazards independently and to manage them directly. It is also possible to condense them into three categories of slopes, embankments and subgrade, as is useful for some of the discussion in this paper. When grouped in this way, it is more intuitive that slopes, embankments and subgrade are assets much in the same way as bridges and pavement. They are not just important; they are mandatory components of a highway that forms a transportation corridor. Thus, slopes, embankments and subgrade could be added to CDOTs list of 11 asset groups (in lieu of geohazards). These 3 asset types, along with retaining walls (already in the CDOT list) have elsewhere been identified as independent geotechnical assets (Anderson, 2016) because they are comprised of soil and rock or have performance governed largely by soil or rock, and they are independent of other asset classes typically considered by highway transportation agencies. In summary, CDOT has a RB TAMP to manage several different assets with respect to risk, and four of them are geotechnical.

Performance Goals

Performance goals have been set at the federal level in highway transportation through the MAP-21 and FAST act legislation and these goals have been adopted by states in a context that is meaningful to the state. CDOT has the following primary goals set by state policy directive and identified in their RB-TAMP:

1. Safety – Reduce traffic fatalities and serious injuries and work toward zero deaths for all users.
2. Infrastructure Condition – Preserve the transportation infrastructure condition to ensure safety and mobility at a least life-cycle cost.
3. System Performance – Improve system reliability and reduce congestion, primarily through operational improvements and secondarily through the addition of capacity.
4. Maintenance – Annually maintain CDOT's roadways and facilities to minimize the

need for replacement or rehabilitation.

CDOT has other goals beyond these, such as freight movement and environmental sustainability that also align with MAP-21. CDOT treats these as "Planning Principles" and in that way incorporates them with the achievement of other goals. In summary, CDOT is now very explicit in trying to achieve targets for multiple performance goals simultaneously. Other agencies are too. The path to doing this includes management of multiple assets and making decisions based on opportunities and threats, which is where risk-based management is engaged.

Sources of Risk

Risk comes from multiple sources and there are many ways of categorizing risk, but for the purposes here a categorization made by the American Association of State Highway and Transportation Officials (AASHTO) is particularly useful (AASHTO, 2011). AASHTO defines four types of risk that are all relevant for geotechnical assets:

1. Natural hazards
2. External agency impacts
3. Physical failure
4. Operational risk

The risk from natural hazards can be viewed as something originating beyond the ability to control and it can be mitigated primarily by actions that prepare for it and prepare for recovery from it. In other words, the risk can be reduced by actions that reduce its consequence and not its likelihood. Some examples are extreme events, such as earthquake hazards, and large, rare floods, as well as regionally pervasive geohazards such as swelling soil. Landslides such as the tragic 2014 landslide in Oso Washington that took 43 lives and buried SR 530 are another example of a natural hazard. An important recognition of this is that natural hazards will still occur and pose some level of risk, even with the best application of risk-based asset management. Geohazards are a subset of natural hazards.

In contrast to the natural hazard risk type, the physical failure type of risk is that which happens through an ongoing process of deterioration, much like pavement or bridge decks deteriorate (e.g. Galehouse et al., 2003). The shape of a deterioration function for geotechnical assets is not well known because these assets haven't had much study in this regard. Experience does show, however, that many geotechnical assets deteriorate and do so at an increasing rate if actions aren't taken to preserve them. Examples are the maintenance of surface and subsurface drainage on slopes or earth retaining structures, the maintenance of vegetation and riprap, and facing materials, and the maintenance of rock slopes and rockfall mitigation equipment through scaling and repair of improvements. These preservation actions and others serve to reduce the consequence and/or likelihood of the physical failure risks and their efficacy in doing so depends in part on their timing, which is also within control of CDOT. This risk type is, therefore, very effectively addressed with transportation asset management practices.

The operational risk type and external agency impacts risk type apply differently and are not as well managed by using transportation asset management principles as they are addressed using other means. Operational risk is the risk related to business decisions and whether the owner agency makes good or poor decisions related to its ability to get the right people delivering the right project at the right time and to accidents from the public use of the system. External agency impacts are risks related to what is delivered to the agency. These include the quality or price variance of materials and design or construction services purchased. These risks are managed to an acceptable level by business practices, for example implementing quality assurance to manage external impacts, and business planning, personnel practices, and training/education for operational risk.

Risk Cube

Each asset group (walls, slopes, etc.) can be viewed as a path through which each risk type acts to threaten achievement of the performance goals. Risk of any type and consequence (which can be measured uniquely in the context of each goal) can be mitigated by taking actions that impact the asset (path). Thus, CDOT owns an element of risk related to each combination of asset, goal and risk type, and it can be mitigated by actions on the assets. Considering the assets, goals, and risk types defined by CDOT and summarized in the previous sections, this elemental risk matrix is a 3-D form that can be viewed as a "risk cube" as shown in Figure 1 (Anderson, 2016).

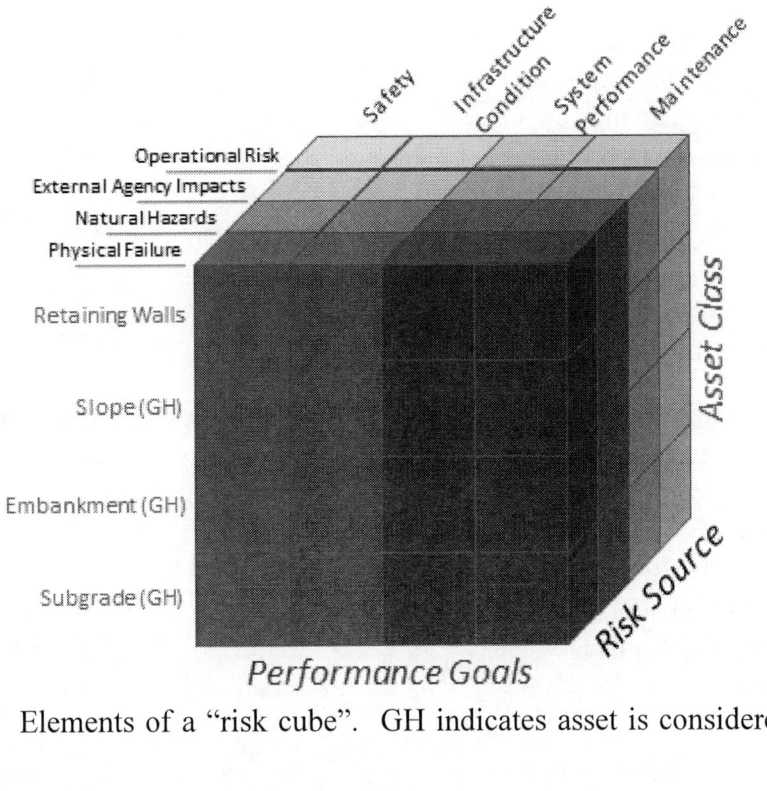

Figure 1. Elements of a "risk cube". GH indicates asset is considered only through

impact of geohazard in CDOT's RB TAMP.

A way to look at this figure is to view each element as a place holder for the calculated or estimated risk of a specific type acting through a given asset type and having a consequence related to a specific performance goal. Thus, this cube shows 64 elements of risk. The magnitude of risk in some of these elements will be far greater than in others and many elements, rows, columns, or even planes of risk elements can be recognized as secondary based on inspection or preliminary analysis. For other elements of risk it will be important to make more careful assessments or analysis of risk, and to consider actions and the desired timing of actions that will reduce those risks.

As demonstrated in the following sections, the risk cube helps to communicate what risks are high, what are being addressed by certain actions, and what risks remain. The risk cube visualization applies at any scale: individual asset, corridor segment, corridor, or highway system. Retaining walls are considered by CDOT as individual assets, and for the geohazards program all geohazards are lumped together for a segment of a corridor. As can be imagined, the risk mitigation strategies envisioned for an element of risk will depend on the scale being considered.

GEOTECHNICAL RISK CONSIDERATION BY CDOT

Risk-based management of geotechnical assets (geohazards and retaining walls) involves evaluating a range in potential consequences that align with pre-established department performance goals (in parentheses) as follows:

- condition deterioration to the specific asset (infrastructure condition and maintenance);
- public safety (safety);
- traveler delay, congestion, and mobility impacts (system performance);
- department maintenance expenses for asset repair (maintenance);
- environmental resource damage (other);
- economic loss (system performance); and
- private property damage (system performance and economic vitality).

During an initial phase of risk assessment, retaining wall assets were determined to have greater impacts to mobility and asset (infrastructure) condition than to other goals. These impacts could be evaluated based on traveler delay and department maintenance expenses, respectively. By using maintenance expense as a measure of consequence for the condition goals, it is effectively rolled into these goals and not considered separately (as in bullet four, above). For other geotechnical assets, similar conclusions were made, and in addition it was possible to assess the risk to public safety. Other performance areas such as environmental resource damage could be evaluated; however, these impacts were generally minor, found to be duplicative to other performance areas, or not seen as reliable for incorporation into the asset management plans.

Retaining Wall Management Program

The retaining wall management program (CDOT, 2016) consists of over 3,000 walls and is based on the National Bridge Inventory (NBI) ratings and the element level rating required for all bridges. The NBI has been established to help ensure safety by tracking the condition of various visible elements through time. The idea is that a bridge element which is deteriorating will be detected and can be addressed before there is a safety consequence. Because these data are available, they are now being used for more than safety: they are being used for bridge performance management, and that is the same approach that CDOT's risk-based wall management program is using.

Though CDOT's wall management program is an integral part of their RB TAMP, it addresses only two elements of geotechnical risk, as shown in Figure 2. Ten elements of risk in the upper plane for retaining walls are not addressed by the RB TAMP. Note that Figure 2 uses CDOT terminology for performance goals: safety, maintenance and mobility. Recognition that only two of twelve possible risk elements are considered by the RB TAMP is important. CDOT can consider if the other elements are significant and if they are managed in other ways, or if they should be incorporated in the RB TAMP.

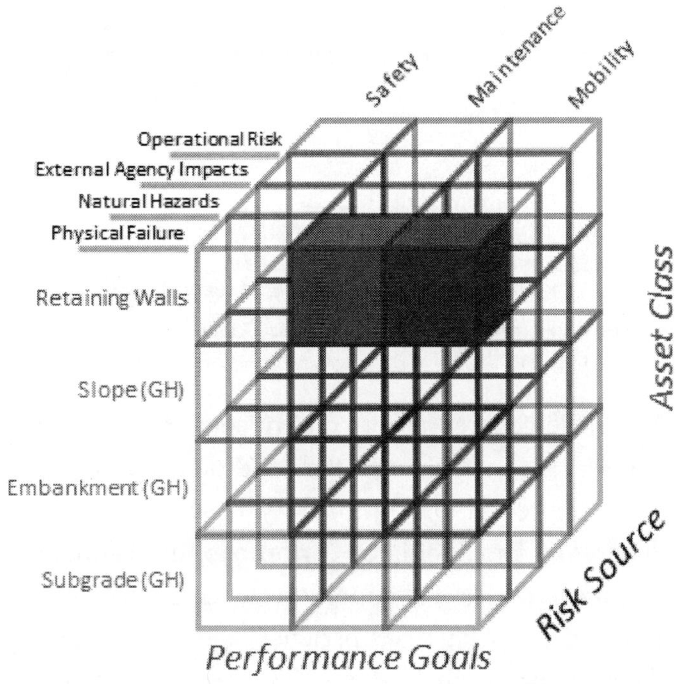

Figure 2. Elements of risk in the CDOT wall management plan.

CDOT started with an initial phase of risk assessment to establish a priority for doing more labor and data intensive assessments of all walls. The Tier 1 assessment did not distinguish the two risk cells shown in Figure 2; however, the second tier of assessment, which is still going on, does. The Tier 2 assessment incorporates measurable data collected during inspections of wall and structure element conditions to estimate the risk exposure, develop performance goals and metrics, and support decisions for long-range planning. The measurable parameters used in the maintenance and mobility risk calculations are outlined in the following sections.

Maintenance Goal Risk

Each wall asset is composed of elements that are defined as visible features such as facing, coping, and drainage components. The maintenance risk exposure is determined based on a weighted repair cost that considers the quantity and category of these structure elements (primary or secondary) and the field assessed condition state of the element. The parameters used in the calculation are shown in Table 1. Primary elements, which are structural in nature, are assumed to have a greater priority for repair than secondary elements, which tend to be cosmetic or ancillary. Structural elements are thus weighted more heavily in the determination of risk cost due to the potential for the financial consequences to be recognized by CDOT for these elements over the wall life cycle. Similarly, defects in elements categorized in good or fair condition typically have a lower priority for repairs than those in severe condition and are, accordingly, weighted less heavily.

Table 1. Retaining wall maintenance risk calculation parameters

Performance Goal	**Factor**	**Parameter**
Maintenance		
	Consequence	Quantity of Elements
		~Unit Costs
	Likelihood	Condition State
		Element Type
		Element Category (primary or secondary)

~Data compiled based on inspector experience and with CDOT input.

The unit costs to repair defects represent consequence in the determination of maintenance risk cost. The element category and the condition state score are used as surrogates for likelihood and represent the probability of repairs being made and maintenance costs being incurred. The likelihood (or probability) estimates presented in Table 2 for various element categories and condition states are based on input from CDOT and consultant staff and reflect past experience and professional judgment. These can be interpreted as annual probabilities. The resulting maintenance risk exposure is calculated as the sum of the product of each element cost (unit cost x quantity) and likelihood that a direct maintenance cost for that element would be incurred (values from Table 2 based on element type and condition state for that element).

Table 2. Probability Values for Risk Exposure Calculation

Likelihood of Incurring Maintenance Cost		
Condition State	**Primary Elements**	**Secondary Elements**
1 (best)	0%	0%
2	11%	7%
3	59%	37%
4 (worst)	98%	66%

Mobility Goal Risk

User costs represent an estimate of the consequence to mobility in the determination of mobility risk. User costs are indirect costs for closure or delay and they are calculated for both the roadway in front and the roadway carried, as applicable in scenarios of tiered roadways. The parameters used to calculate user cost are shown in Table 3. Geometric parameters such as height and closeness to the road indicate how big the impact will be to the road and the average annual daily traffic (AADT) reflects how many users experience that impact.

The likelihood of an event is determined by the condition of the wall as dictated by the lowest inspection score received for the items of main structure condition, foundation condition, or scour critical condition. These items are adapted from the NBI. The state has been collecting this type of information on bridges for many years, but using these condition scores to predict likelihood of an impact to the mobility performance goal is a new idea. The likelihood of risk exposure based on the condition score is based on input from CDOT and contractor inspection staff and reflects experience and professional judgment, and is presented in Table 4.

Based on this process, the final mobility risk calculation, representing the product of consequence and likelihood, can be calculated as follows:

$$\text{User Costs} = \frac{\text{Delay Time}}{3600} \times \frac{(\text{AADT Actual} - \text{AADT Delay}) / 2}{24} \times \text{User Value} \times \text{Occupancy Rate}$$

$$\text{Mobility Risk Exposure} = \text{User Costs} \times \text{Wall Condition}$$

Table 3. Retaining wall mobility risk calculation parameters

Performance Goal	Factor	Parameter
Mobility		
	Consequence	Avg. Wall Height
		Avg. Distance from Road in Front
		Avg. Distance from Road Carried
		AADT
		^Delay Time, 2 hours
		*User Value, $30.50
		*Occupancy Rate, 1.67
		*ADT Delay, 33% of Actual ADT
	Likelihood	Main Structure Condition
		Foundation Condition
		Scour Critical Condition

^Assumed value based on likely time of delay from an urgent adverse event, similar to delay associated with over-height bridge strikes.
*Per AASHTO 2010.

Table 4. Estimated experience-based correlation of wall item condition and likelihood of risk exposure.

Wall Item Condition	Likelihood
9 (best)	2%
8	
7	
6	5%
5	
4	26%
3	
2	78%
1	
0 (worst)	

Through these approaches a financial risk exposure is calculated for both shaded cells shown in Figure 2, and these individual risk costs are summed to arrive at a total risk exposure for each wall. As the sum of estimated risk exposures for retaining walls are determined, CDOT can evaluate the data for deterioration trends related to wall and element types, age, and location. Further, CDOT anticipates recognizing cost savings through the bundling of wall rehabilitation projects to address similar performance issues, such as repair of common drainage and wall facing systems for multiple walls. Additionally, CDOT can better evaluate the long-term performance of decisions made during design, such as the trade off in asset management performance between different wall systems.

Geohazard Management Plan (GMP)

The CDOT Geohazard Management Plan (GMP) consists of over 1,600 highway segments with an identified threat that has been documented through a prior geologic event. About half of these sites consist of rockfall locations previously identified by the CDOT Rockfall Hazard Rating System and the remaining locations consist of geohazards such as unstable soil and rock slopes or subgrades. The geohazards program comprises the three remaining independent geotechnical assets: embankments, slopes, and subgrades, and calculates risk for CDOT's three performance goals, as defined for walls. For each performance goal the risk estimate is based on the associated threat such as traveler injury from a geohazard event, highway closure, or direct maintenance costs to the department. The condition of the geohazard asset is determined based on the number of recorded events, which is then converted to an annual probability that there would be a consequence to one of the performance areas. For example, a rockfall geohazard location may experience three events in a year; however, not all events will result in an impact to the traveler safety, mobility, or maintenance direct expenses. CDOT subject matter expertise was used to estimate the annual probability for a performance impact based on the number of events. Further, the estimation of safety risk exposure includes a vulnerability value to account for the likelihood that not all accidents attributed to a geohazard location will result in an injury.

Different levels of consequence are assigned to each geohazard based on historical ranges of impact. Estimated costs associated with each level of safety and mobility consequence were then assigned based on internal department economic studies that are being developed for broad planning purposes. The cost consequence associated with threats to maintenance was based on the judgment of the ability of maintenance budgets to accommodate unplanned expenses. As this is a new process that relies on historical and current data as well as evolving estimates of consequence, the input values may change as the plan evolves. However, in the interest of initiating risk management CDOT is moving forward with a plan that can be adjusted as the confidence in data and means improves. The process for the assessment of geohazard risk exposure is presented in Figure 3.

CONSEQUENCE

Safety Threat

- No Reported Accidents........ 0
- 1 or more Accidents........ $3,500 ea.
- Injury........ $91,600
- Fatality........ $6,297,000

Mobility Threat

Road Closure Time = full closure (hrs) + (partial closure (hrs) x 0.5)

- Negligible: No closure........ 0
- Minor: Less than 1 hour of closure........ $10.50
- Major: 1 to 24 hours of closure........ $252.00
- Critical: 1 to 5 days of closure........ $1,008.00
- Catastrophic: More than 5 days of closure..... $2,520.00

Maintenance Threat

- Incidental: < $25,000........ $5,000
- Minor: $25,000 - $100,000........ $50,000
- Major: $100,000 - $500,000........ $200,000
- Critical: > $500,000........ $1,000,000

LIKELIHOOD

Number of Events	Probability
0	0
1	10%
2	20%
3	63%
>3	86%

RISK

Safety Risk = Safety Threat * Likelihood * P(s:h)* 0.3 (*vulnerability*)

P(s:h) = [(ADT/24) * (Length/5280)]/speed limit

Length = [(Off Peak Truck Traffic * 45 ft *truck*) + ((100-Off Pk Truck Traffic) * 15 ft *car*)]/100

Mobility Risk = Mobility Threat * Likelihood * (ADT/24)

Maintenance Risk = Maint Threat * Likelihood

Total Risk = Safety Risk + Mobility Risk + Maintenance Risk

Figure 3: Calculation of Total Annual Geohazard Risk Exposure

CDOT's GMP is focused on physical failure as a risk type, though it also considers natural hazards because events tied to natural hazards are not distinguished in the data. In other words, there has been no attempt to retroactively assign a risk type to the 1600 highway segments that have been identified. The inclusion of two risk sources for the GMP differs from the wall program which considers only physical failure. Thus, the calculated risk exposure from Figure 3 represents the both the natural hazard and physical failure elements of risk and the elements of natural hazard risk type are included with those of the physical failure type in Figure 4 because of their implicit role in the past data collection and valuation. A possible future activity for CDOT is to separate these risk types. The data being collected now as part of this program will be helpful for doing so.

While the historical distinction between physical failure and natural hazards has not routinely occurred some interesting observations are possible. For one, expectations for performance levels and funding sources after extreme event natural hazards are different. Damage is expected when extreme events have a recurrence interval greater than the design life of a structure (including obsolete structures), for example, and federal Emergency Relief funds can be available after a government declaration of disaster. Further, recent work by CDOT has estimated the broader economic consequences from both natural hazards and physical failure of geohazards. Through this work, CDOT is able to demonstrate to external stakeholders the benefits that CDOT delivers through their response to natural hazards that originate from outside the CDOT right-of-way. In the future, it may be possible to obtain additional funding contributions from stakeholders should there be a strong desire to improve the performance of a corridor exposed to natural hazards.

Additionally, CDOT and other state transportation departments frequently assume the responsibility of both risk sources because they are most capable of quickly responding with resources for repair and construction. As an example, the U.S. Forest Service or a state/federal land management agency does not have missions that involve maintaining the economic vitality of a region through good performance of transportation corridors. Should a hazard originate from property managed by these other agencies, they are typically not able to respond in a manner that would rapidly restore the affected transportation corridor. If a distinction between risk types is made in the future, it may help CDOT quantify this value.

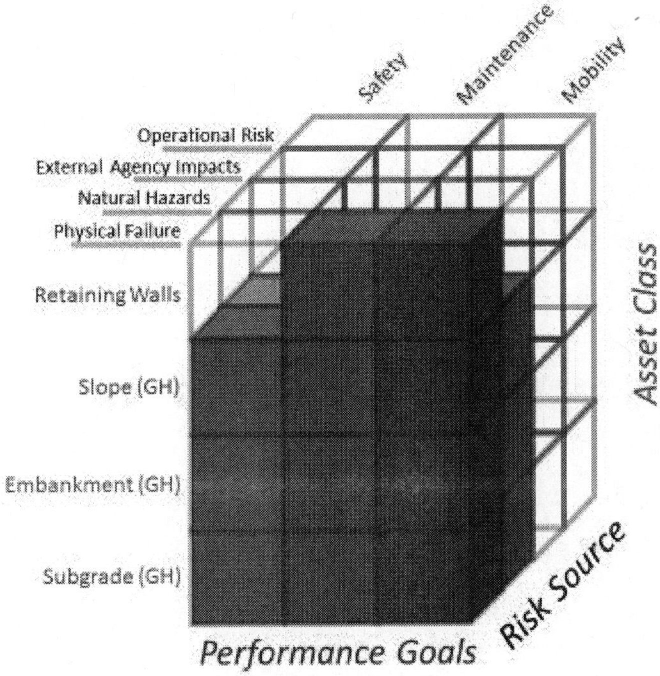

Figure 4. Elements of risk in the CDOT wall and geohazard management plans.

The total risk exposure calculation (Figure 3) allows CDOT to develop projects and maintenance plans following a risk management approach that is in alignment with the department performance goals. Additionally, when viewing the risk exposure by highway segment in a geographic information system (GIS) or other mapping environment, it is possible to identify geographic concentrations of risk. This allows CDOT to define management corridors that can be prioritized for mitigation based on the potential for greater levels of risk reduction in a concentrated area. This approach results in a more rapid and measureable improvement in system performance because an entire corridor is improved in a shorter duration, versus dispersing projects among several corridors without significant reductions in risk in those corridors. In fact, this is a key anticipated outcome of risk-based asset management.

Additionally, the benefit of routine maintenance activities is able to be better qualified in terms of multiple performance goals rather than treated as an isolated nuisance cost. For example, CDOT recently mitigated a rockslide feature through a planned removal project for a relatively direct low cost. A photograph of the site immediately following the slope hazard reduction work is presented in Figure 5 and illustrates the potential threat to safety, mobility, and condition of the system (the "maintenance" goal). By proactively performing this work, CDOT was able to control the safety threat at the site while also minimizing the consequences to mobility and department maintenance expenses. When the cost of the mitigation is compared with the reductions in risk exposure in terms of safety, mobility, and department maintenance expenses, actions like this can demonstrate a favorable benefit:cost ratio. In other words, and using the risk cube for visualization, the action, timed as it was, mitigated risk in the three elements that are the intersection of the geohazards on slopes (Slope (GH)) plane and the physical failure plane, which is the risk source here.

Figure 5. Rock Slope Risk Exposure Reduction on US 24 near Minturn, Colorado.

Other Risk Elements

Despite the very proactive approach taken to manage risk by CDOT, it can be seen from Figure 4 that there are many elements of risk not addressed by the wall and geohazard programs. The risk cube helps communicate this point so that people coming from all perspectives can see what other risks are present. CDOT can then make informed decisions to accept the risk in other elements, as secondary, low-level risks or they can complete other programs to mitigate them. For example, CDOTs asset management plan does not address operational risk and external agency risk because these are mitigated by business practices, which CDOT does undertake and continuously evaluates in other processes.

LEVEL OF RISK

The measurement and reporting of asset risk to executive and planning professionals in a transportation department needs to be conveyed in a simplified manner. For this to occur within CDOT, the concept of a level of risk (LOR) grade was established to communicate the qualitative categorization of the risk exposure. The LOR concept was modified based on other related categorical measures, such as level of service, which are commonly used within a DOT to communicate performance to executives and public.

For wall and geotechnical assets, a monetized risk exposure was estimated for each of the performance goals, such as safety, mobility, and asset condition (planes in the risk cube). The individual risk costs are then aggregated to define the LOR as follows.

- A – less than $1,000 risk exposure
- B – $1,000 to $5,000 risk exposure
- C – $5,000 to $50,000 risk exposure
- D – $50,000 to $100,000 risk exposure
- F – greater than $100,000 risk exposure

The LOR category values were selected based on the assumed tolerance for differing economic consequence levels for the annual performance of walls and geohazard assets. The underlying exposure in the LOR categories is intended to be an estimate of the economic consequence, considering both direct and indirect costs, associated with ownership and maintenance of the assets. Initially, the categories for the GMP were established on the basis of a relative, non-monetized risk score, similar to the Tier 1 process used for the wall asset group. However, the goal of an asset management plan to develop financial and investment strategies that are measurable and the use of the estimated risk cost enables this to occur. Further, there is a real cost for delaying risk management as demonstrated by an internal study commissioned by CDOT on the economic consequences from geologic hazard events within right-of-way (Vessely et al., 2017). Vesseley et al. provide an estimate of the direct and indirect costs associated with ownership and maintenance of the GMP asset group and also provide a means to compare actual economic impacts with the total estimated risk exposure in the GMP. As a result, the emergence of these CDOT risk based plans are supported by executive management based on the goal to quickly achieve the benefits from risk reduction rather than waiting on others to develop a process.

The comparison of LOR between walls and geohazards has not yet occurred in the asset management process at CDOT, but the option does exist. A productive future process will be for the wall and geohazard asset groups to overlay LOR values for the respective assets and identify locations where combined investment strategies will demonstrate improvement to both asset groups, resulting in a more favorable cost to benefit ratio.

While LOR is used for measurement and reporting to department executives, the underlying data are available to subject matter experts for project planning and development. For example, if the department has a mandate to develop projects that improve traveler safety or mobility, the data can be de-aggregated consider only those goals. This would eventually allow the department to examine the risk exposure associated with each individual cell within the risk cube.

CONCLUSIONS

The management of risk in geotechnical engineering is well-established. These risk-based concepts and processes can be adapted within the transportation sector for the management of threats originating from constructed assets and natural hazards, as demonstrated by the inclusion of retaining walls, slopes, embankments, and roadway subgrade in a risk-based transportation asset management plan.

CDOT's RB TAMP considers natural hazard and physical failure risk types for geohazards and the physical failure risk type for walls. Geohazard risk calculations include some component of deterioration (physical failure) and some component of resilience to natural hazards. Bridge management approaches, and therefore wall management approaches that mimic them, are based on observing things deteriorate, not ensuring resilience to extreme events, so the wall management program does not address risks from natural hazards. The elements of risk that are not captured here are omitted explicitly because either they were judged to be a second or third order contribution to risk, and not where management should be directed, or they are addressed by programs other than RB-AMP. The cubic form of the "risk cube" helps convey this clearly.

REFERENCES

American Association of State Highway and Transportation Officials (AASHTO). *AASHTO LRFD bridge design specifications*: U.S. customary units (7th ed. with 2015 and 2016 interim revisions): Washington, D. C., AASHTO, 2016, 2 v.

American Association of State Highway and Transportation Officials (AASHTO). *Transportation Asset Management Guide*: A Focus on Implementation, 2011.

Anderson, S.A. "*Communicating multiobjective risk: a new geotechnical need for transportation.*" Transportation Research Record No. 2580, 2016, pp. 80-88.

Casagrande, A. *Role of the 'Calculated Risk' in Earthwork and Foundation Engineering*, Journal of the Soil Mechanics and Foundation Division, ASCE, Vol. 91,

No. SM4, July, 1965, pp. 1-40.

Colorado Department of Transportation (CDOT). *Retaining and Noise Wall Inspection and Asset Management Manual, Version 1.0*, April 2016. Available at: https://www.codot.gov/library/bridge/retaining-and-noise-wall-inspection-and-asset-management-manual/retaining-and-noise-wall-inspection-and-asset.pdf

Galehouse, L., J. S. Moulthrop, and R. G. Hicks. *Principles of Pavement Preservation. TR News, No. 228*, Transportation Research Board of the National Academies, Washington, D.C., 2003, pp. 4-15.

Pierson, L.A.; Davis, S.A; and Van Vickle, R. *Rockfall hazard rating system: implementation manual.* Washington, D. C., Federal Highway Administration, Report no. FHWA-OR-EG-90-01, 1990, 172 p.

Vessely, M., Richrath, S., Weldemicael, E., *Economic Impacts from Geologic Hazard Events on Colorado Department of Transportation Right-of-Way*. Paper ID 17-02718, Transportation Research Board, 2017 Annual Meeting Compendium of Papers, 2017.

Montana's Rock Slope Asset Management Program (RAMP)

Darren Beckstrand, C.E.G., Aine Mines, P.G., Brent Black, C.E.G.
Landslide Technology
10250 SW Greenburg Road Suite 111
Portland, OR 97223
(503) 452-1200
darrenb@LandslideTechnology.com, ainem@LandlideTechnology.com,
brentb@LandslideTechnology.com

Jeff Jackson, P.E., Bret Boundy, P.G., Scott Helm, P.G.
Montana Department of Transportation
2701 Prospect Avenue
Helena, MT
(503) 452-1200
jejackson@mt.gov, bboundy@mt.gov, shelm@mt.gov

David A. Stanley
DASC Consulting
5391 Bellaire Drive
Bellingham, WA 98226
(907) 267-9416
dastanley44@gmail.com

Paul D. Thompson
Consultant
17035 NE 28th Place
Bellevue, WA 98008
(425) 224-5443
pdt@pdth.com

Prepared for the 68th Highway Geology Symposium, May, 2017

ABSTRACT

The Montana Department of Transportation (MDT) is currently reassessing their rockfall hazard evaluation process. MDT implemented the Rockfall Hazard Rating System (RHRS) in 2005, where over 2,600 rock slopes were evaluated as A, B, or C sites, followed up by detailed ratings at 869 preliminary 'A' sites. An RHRS detailed rating score of 350 points established the cutoff score, resulting in 368 'A' sites on MDT's highway system. After a decade of using the RHRS, MDT sought to update their rockfall database, gain additional insight on data use to help guide decision making, and build upon recent developments in Transportation Asset Management that include geotechnical assets.

The new MDT Rock Slope Asset Management Program (RAMP) includes a number of new enhancements. RHRS score components recombine to create sub-scores to isolate specific evaluation attributes. The slope's Condition is calculated as a function of rockfall history and ditch effectiveness and scored using a 100 (good, like new condition) to 0 (poor or failed condition) linear score. Five Condition State categorizations facilitate deterioration modelling and risk analysis. Evaluation of rockfall event records permitted estimation of rockfall event likelihoods based on slope dimensions and condition for use in risk calculations. Programmatic cost estimates to improve the slope, also based on size and condition, allow rapid network-wide estimation of improvement costs. Performance Measures and Decision Support Tools help guide the planning process. Tools that leverage MDT's cloud-based GIS services permit collection of rockfall events and maintenance activities across multiple computing platforms.

INTRODUCTION

The Montana Department of Transportation (MDT) implemented the Rockfall Hazard Rating System (RHRS) (1) between 2003 and 2005 (2) to obtain further information on the state's rock slopes and the hazards posed. This initial implementation resulted in a review of 2,653 rockfall sites, detailed ratings at 869 of those sites, and a final 368 'A' sites spread throughout the state (Figure 1).

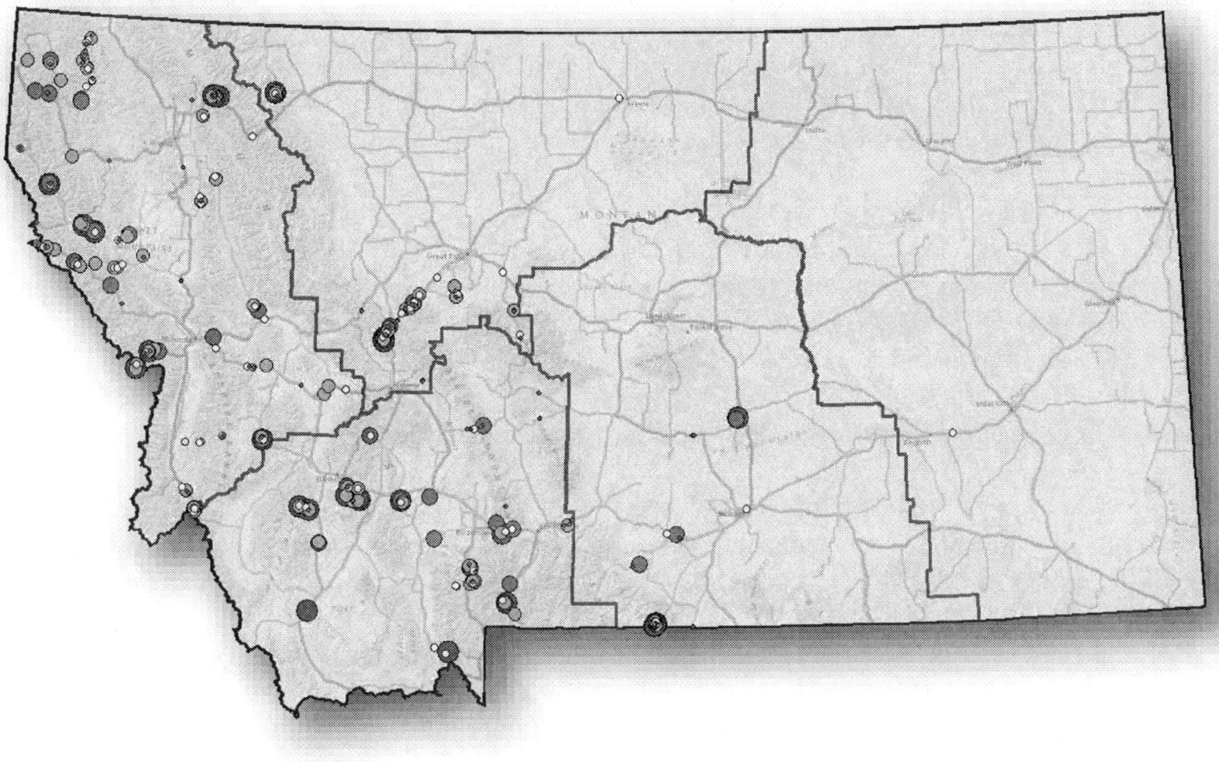

Figure 1: Map of 368 'A' sites, with the five districts outlined and highway functional classification shown.

MDT implemented the RHRS ratings in an informal process, reviewing ratings and comparing them to event occurrences, maintenance needs, and rockfall mitigation project selection in the decade since completion. MDT found the RHRS to be a valuable tool providing relative rankings between sites, but a combination of changed sites, a desire for additional tools to aid in project selection, and a need for incorporating principles of Transportation Asset Management (TAM) prompted the Department to request an update to assess its rockfall hazard process, which begun in 2015.

The research project was subdivided into eight tasks, beginning with a literature review followed by a series of field evaluations of proposed new approaches to assessing rock slopes, and finalized with critical site determination, benefit/cost analysis, and reports. This paper summarizes the key changes to its process, usability, performance measures, and the decision support tools to help guide the planning and programming process.

REVIEW OF ROCKFALL EVALUATION SYSTEMS

Based on the literature review, the most widely used rockfall ranking and management systems in North America are variations or modifications of the Rockfall Hazard Rating System (1). Other similar hazard rating systems, such as those for landslides for the Ohio Department of Transportation, use a similar exponential scoring system as found in the RHRS (3). The DOTs of New York (4), Ohio (3), Utah (5), Washington (6), Alaska (7), Tennessee (8) (9), and Missouri (10), among others, are all examples of agencies that, along with MDT, have utilized RHRS-based systems for ranking and evaluating rock slopes, some with more extensive alterations than others. In a 2008 survey, 25 U.S. State or Canadian Provincial transportation agencies utilize a management system to track rock slope data and most of these (88%) are based on the RHRS (11). Most of these agencies have made modifications to the RHRS to meet departmental goals and objectives, such as Montana's relatively minor modification for climatic criteria.

There have been two primary modifications of the rockfall assessment systems in recent years. The first comes from the province of Ontario, Canada. In their Ontario Rockfall Hazard Rating System (RHRON), the rating categories are subdivided and grouped into four Factors to approximate 1) magnitude, 2) instability, 3) reach, and 4) consequences and evaluated on a 0 (good) to 9 (bad) scale (12). This system uses categories from the RHRS and adds additional lab testing or estimations to further assess certain rock characteristics.

The second set of modifications is the result of ongoing research into developing concepts of geotechnical asset management (GAM) by the Alaska Department of Transportation (7). This research project's purpose to develop a comprehensive plan to manage geotechnical assets, focused on rock slopes, unstable soil slopes and embankments, retaining walls, and material sources. The research has included development of a GAM Plan, inventorying assets, developing rating systems, conducting field ratings, developing condition states, deterioration curves, programmatic cost estimations, and modelling funding scenarios on maintaining these assets. The nearly completed project will be a comprehensive asset management program for geotechnical assets compatible with Alaska's Transportation Asset Management (TAM) plan. This project has demonstrated how to adjust RHRS-like inventory and rating programs into TAM-compatible systems based on condition states, which can be utilized for deterioration modeling and life cycle cost analyses to support efficient and cost-effective management.

The advent of readily available and inexpensive GPS-capable mobile computing platforms in the past five years has made inventory, mapping, and analysis more accessible to geotechnical personnel and planners (13). Utilizing these platforms would modernize the IT interface and make the use of the data less challenging and more intuitive and therefore more useful at various Department technical and managerial levels.

Major developments in the field of laser scanning and photogrammetry have occurred and are becoming more widespread in the last 10 years (14). The use of aerial and ground-based laser scanning has made landform interpretation and monitoring much more accessible and accurate. Advancements in photogrammetry now make it possible to remotely measure joint orientations for engineering analysis, zoom in with great detail for visual inspections, and generate detailed surface models for change detection and volume calculations for quantitative analysis of rockfall activity. Figure 2 exhibits an example of photogrammetric techniques in northwestern Montana.

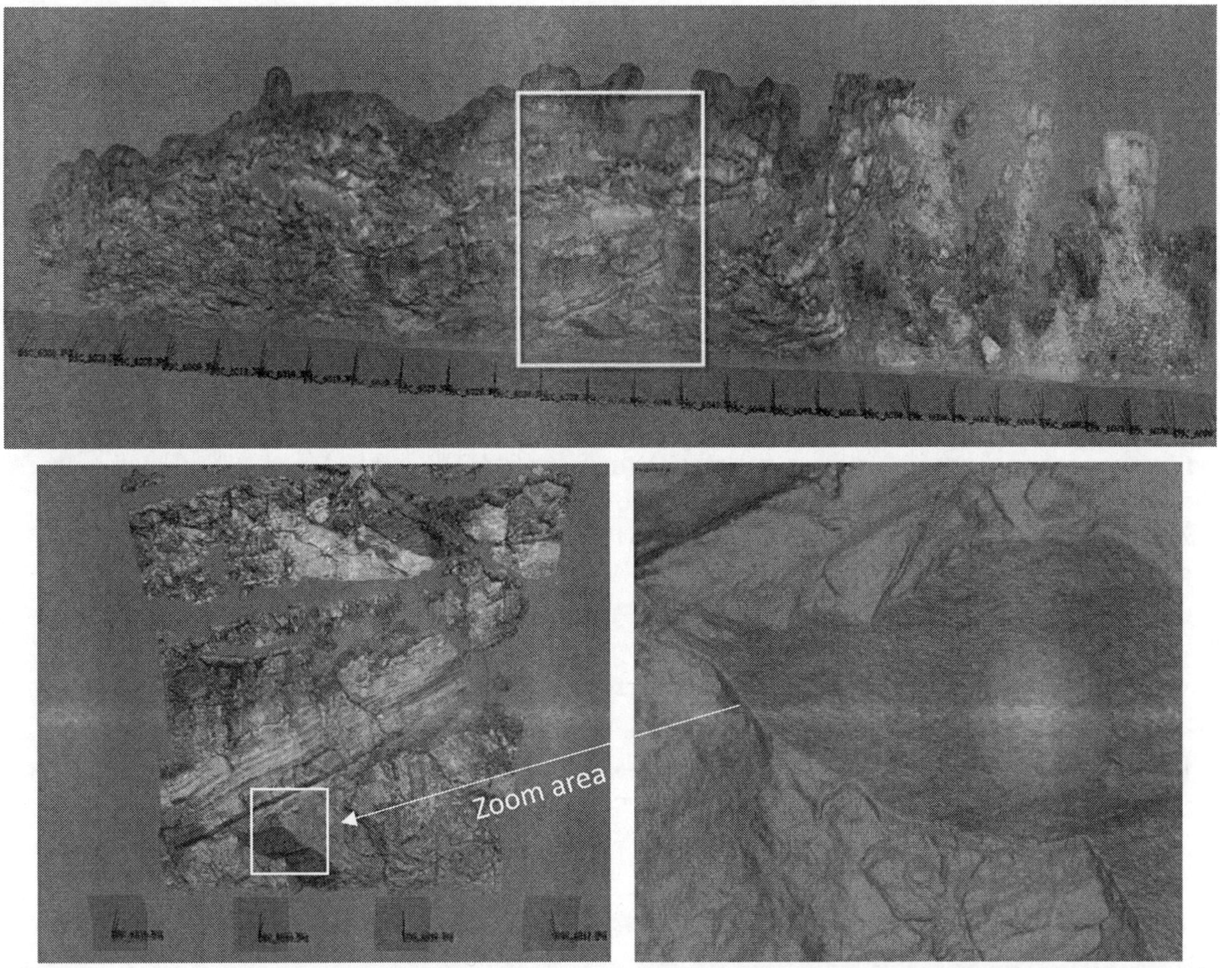

Figure 2: US 2 Badrock Canyon Model. Photo model on the left and a detailed solid model on the right, zoomed in from the upper photo model at the top.

The maturation of rockfall hazard management programs through alignment with asset management systems has been partially driven by the 2012 Moving Ahead for Progress in the 21st Century Act (MAP-21) (15) and partially by increasing agency awareness of advances in management and technology. Through these advances, the process to inventory and assess slope condition and risks were incorporated into MDT's improved Rockfall Hazard Process. A modernized rockfall management system should meet the goals of MTD's developing asset management program and improve safety, mobility and efficiency for the road system. The MAP-21 law requires a streamlined and performance-based and risk-based transportation program for bridges and pavements but also encourages similar management practices for other types of transportation assets, such as rock slopes. The goals of this current project align with both the objectives of federal mandates and with MDT's goals and objectives.

PROPOSED ASSESSMENT SYSTEM

MDT's existing performance and asset management programs, including the Performance Programming Process (P3) Program and the Transportation Asset Management Plan (TAMP), create the link between agency goals, objectives and policies, and successful operation of the RAMP program. Performance management is a means for transportation agencies to measure progress towards agency goals and is an integral part of the RAMP's future compliance with the TAM programs required under federal law (15). Note that only pavements and bridges are required under this code, but that inclusion of other assets, such as rock slopes, into their TAMP is encouraged. Performance management is the tool commonly used by transportation agencies to measure progress toward federal and state goals and objectives. Within this toolkit, performance measures are indicators of work performed and results achieved (16).

In addition to technical management of MDT's numerous rock slopes, RAMP provides support for management decision-making and allocating funding for the design, construction, maintenance, and eventual replacement/reconstruction of MDT's rock slopes. Combined with additional deterioration analysis and life cycle cost analysis, MDT would have all the information it needs in order to include the RAMP in MDT's TAMP.

The roadmap laid out below outlines the steps for creation of the RAMP and using its data to measure performance of MDT's rock slope assets. These steps are useful at the Executive, Planning, and Technical levels. Figure 3 contains a flow chart of the various steps in the process, which are discussed fully in project documentation, available at http://www.mdt.mt.gov/research/projects/geotech/rockfall.shtml.

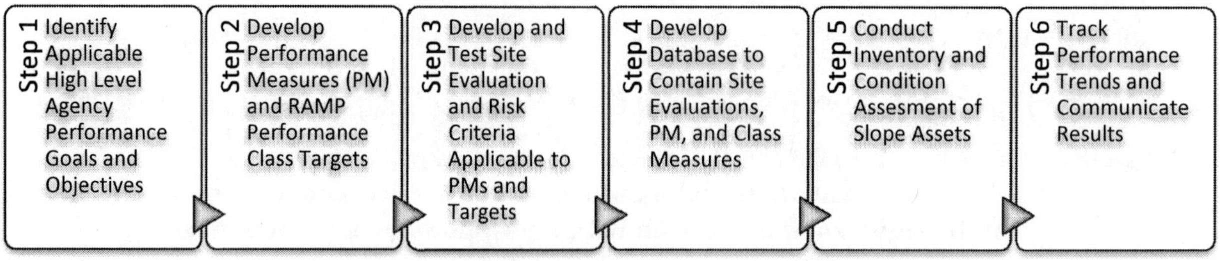

Figure 3: Flow Chart for RAMP Process.

RAMP Performance Management is largely based on slope performance, condition, and risk measures, with performance targets expressed as RAMP Performance Classes. The RAMP Performance Classes are similar to the winter condition Levels of Service currently in use by MDT in that they categorize performance targets and expectations. These specific targets are a necessary precursor for the performance measures and decision support tools developed for the RAMP. The process also includes development of detailed decision support tools for condition and performance of rock slopes statewide and for individual assets (specific slopes) as set out below.

Generalized, Statewide RAMP Performance Measure

Recent research (17) and proposed federal regulations recommend categorizing condition assessments into Good/Fair/Poor divisions, in addition to the purely numerical rankings like those generated by the above scoring and rating methods. In their current form, Good/Fair/Poor divisions are intended to improve an agency's ability to assess the overall health of the highway infrastructure and serve two primary objectives:

- Define a consistent and reliable method of assessing infrastructure health; and
- To develop tools to provide FHWA and State DOT personnel ready access to key information that will allow for a better and more complete view of infrastructure health.

To meet these objectives, the research focused on the development of an approach for categorizing assets as Good, Fair, or Poor, which can be used consistently across all asset classes. Asset performance in this context is based on condition information. Utilizing this guidance for rock slopes with a focus on rock slope performance yields the classification narratives in Table 1.

Table 1: Rock Slope Good/Fair/Poor Classification

Classification	Description
Good	Rock slopes and appurtenant rockfall mitigation elements are free of significant defects and are of a condition that does not adversely affect good performance. Preventive maintenance such as regular ditch cleaning keeps the slopes and mitigation elements in good condition. There is a low likelihood of adverse effect on users.
Fair	Rock slopes exhibit minor deterioration with occasional rockfall that does not frequently interfere with operation of the roadway or create significant delays to users. Rock slope maintenance may include some scaling, or more frequent ditch cleanout. Rockfall mitigation elements exhibit some deterioration or damage, but continue to function adequately without significant maintenance effort. Rockfall fences and drapes may require replacement of small amounts of damaged fence panels, braking elements and cables. Roadside barriers may require repair or replacement of a small percentage of barrier.
Poor	Rock slopes and mitigation elements exhibit advanced deterioration and damage. Individual slopes in a District, or groups of slopes along a corridor may have deteriorated to a level that requires an unacceptable amount of maintenance and repair costs for slopes and rockfall mitigation. Some slopes may have failed catastrophically, requiring major cleanup efforts and reconstruction projects with associated impacts on users, including detours and delays.

Consistent with many bridge and pavement TAM programs, the primary statewide Performance Measure is proposed to be the fraction of rock slopes in Good, Fair, or Poor condition. Relative amounts would be measured as a function of approximated surface area of rock cut.

In this application, performance measures will track how well the agency is managing and improving its rock slopes over time. Using data from slope rating and maintenance activities, an agency can track the condition of its slopes by periodically re-rating the slopes. MDT should also track the frequency of repairs, road closures, and the user costs related to adverse geotechnical events.

Asset Level Condition Assessments and Ratings

MDT internally developed three modified rating methods and requested that Landslide Technology (LT) test them using the existing 2004 data. All three methods seek to give more weight to factors that may be under-valued in the original RHRS rating system, but they would not alter or replace the rating categories currently used in the MDT RHRS program. Figure 4 exhibits a map of the various evaluation criteria and their distribution along a northern Montana corridor.

RHRS Total Score

Scores from both the 2004 and 2015/2016 ratings, without alteration of the RHRS system, were compiled and compared. Original 2004 rating data was entered into an Excel sheet and served as the basis for this comparison as well as all the other rating approaches evaluated.

MDT Methods

MDT geotechnical personnel familiar with the RHRS requested an evaluation of the additional various combinations of RHRS criteria, summarized below.

MDT Rating Method 1 Rating Method 1 assessed a rock slope site's ditch catchment effectiveness, potential traffic impacts, failure potential (expressed as geologic character scores of the RHRS), and rockfall history. Each category has a maximum possible score of 100 points, and the total possible score for a site under Method 1 is 400 points.

MDT Rating Method 2 Rating Method 2 assessed a rock slope's ditch effectiveness, potential traffic impacts, immediate hazard, failure potential, scale of the potential threat, and rockfall history. Each category has a maximum possible score of 100 points, and the total possible score for a site under Method 2 is 600 points.

MDT Rating Method 3 Unlike Rating Methods 1 and 2, Rating Method 3 generates three distinct sub scores – slope rating, vehicular risk, and impact to traffic. The slope rating score comprises ditch effectiveness, potential for failure, and rockfall history. The ditch effectiveness and rockfall history scores are obtained directly from the RHRS rating categories, while the potential for failure is derived using the same equation applied in Method #1. The maximum possible Slope Rating Score in Method #3 is 300 points.

Condition Indexes and States

The Alaska Department of Transportation (AKDOT) has endeavored to develop the nation's first Geotechnical Asset Management (GAM) program and is nearing completion (7). This program incorporates its previously existing Unstable Slope Management Program (USMP), which was developed to assess soil and rock slopes. Like MDT's original RHRS, this component of AKDOT's program uses rating categories with exponential scoring systems. Both states' rating systems are based on the RHRS, though the Alaska rating system includes a few additional categories to capture its extreme climate challenges.

The *Condition Index* is a linear continuum from 100 (ideal condition) to 0 (a failed condition) that is evenly divided into five Condition States. It is a combination of the potential for a rockfall event and the ability of the roadside ditch to contain the rockfall event and prevent it from reaching the roadway. The RHRS "Ditch Effectiveness" and "Rockfall History" categories provide these components. The Condition Index is useful to illustrate potentially minor changes between slopes. For instance, an Index score of 100 indicates a totally effective ditch and low rockfall activity. An Index score of 87 can indicate a less effective ditch but an equally low rockfall activity.

Five *Condition States* are subdivisions of the Condition Index, divided by 20-point spreads of the Index. This subdivision facilitates additional modelling, such for deterioration, likelihoods of adverse events, recommended management actions, or fiscal modelling. The two hypothetical sites used in the preceding example are both Condition State 1, 'Good' slopes, but one could be less capable of keeping the very infrequent rockfall from reaching the road. It is likely that neither site would warrant mitigation and therefore it is reasonable to be within the same Condition State 1 classification.

The three-category *Good/Fair/Poor* criteria, described previously, are further subdivisions of the Condition States to facilitate compatibility with other TAM directives. This continual subdivision may appear needlessly complex; however, it serves the purpose to communicate intricacies familiar to geotechnical personnel into easily understood assessments for management, the public, and legislators. The relationship between these categorizations presented in Table 2.

Table 2: Rock Slope Condition Category Descriptions

Cond. Index Range	Cond. State	G/F/P Descriptor	Description
100 - 80	1	Good	Rock slope produces little to no rockfall and no history of rock reaching the road. Little to no maintenance needs to be performed due to rockfall activity. Rockfall mitigation measures, if present, are in new or like new condition.
80 - 60	2	Fair	Rock slope produces occasional rockfall that may rarely reach the road. Some maintenance needs to be performed on a scheduled basis due to rockfall activity to address safety. Mitigation measures, if present, are in generally good condition, with only surficial rust or minor apparent damage.
60 - 40	3	Fair	Rock slope produces many rockfalls with rock occasionally reaching the road. Maintenance is required bi-annually or annually to maintain safety. Mitigation measures, if present, appear to have more significant corrosion or damage to minor elements. Preventative maintenance or replacement of minor mitigation components is warranted.
40 – 20	4	Poor	Rock slope produces constant rockfall with rocks frequently reaching the road. Maintenance is required annually or more often to maintain ditch performance. Much of the required maintenance response is unscheduled. Mitigation measures, if present, are generally ineffective due to significant damage to major components or apparent deep corrosion.
20 – 0	5	Poor	Rock slope produces constant rockfall and nearly all rockfall reaches the road. Virtually no rockfall catchment exists or is effective. Maintenance must respond to rockfalls regularly, possibly daily during adverse weather. If present, nearly all mitigation measures are ineffectual either due to deferred maintenance, significant damage, or obvious deep corrosion.

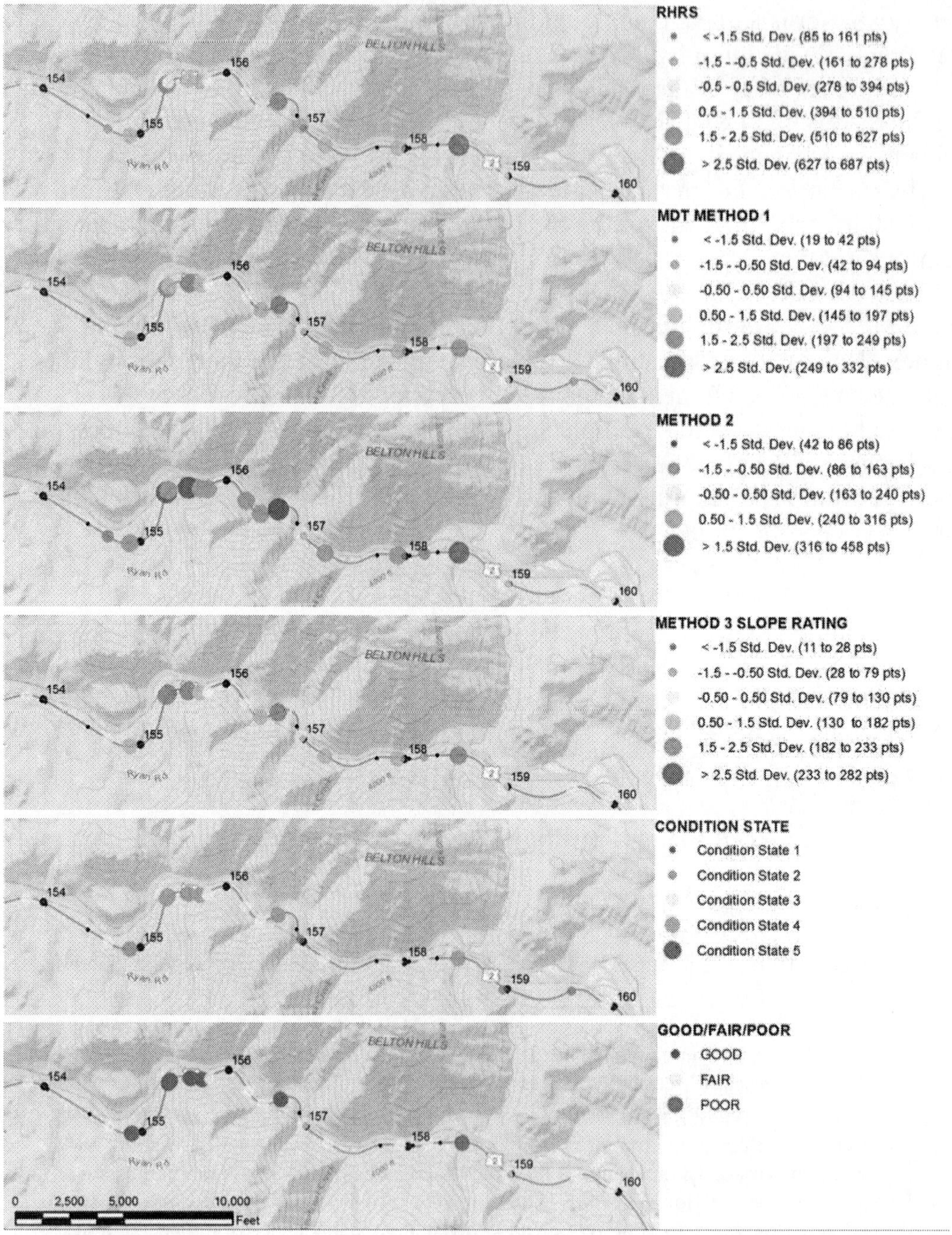

Figure 4: Comparison of methods for Hwy 2, MP 154 to 160, east of West Glacier.

SAMPLE EVALUTAION CRITERIA RESULTS

Thirteen mitigated rock slopes were visited within the Missoula District, shown in Figure 5. Table 3 contains the RHRS rerates, test-rating approaches, and sample user cost risk calculations for the evaluated sites within the Missoula District. This exercise evaluated the various approaches in their ability to quantify constructed improvements to the slopes, indicated by percent changes in the scores before and after mitigation.

The slopes include four sites on Interstate 90 (MP 6.5, 22.5, 24.0, and 24.5) that have been mitigated in response to three road-closing events where significant quantities of rock debris entered the roadway. These four events have all occurred since 2012. These events forced MDT into an emergency response with consequences to public safety, mobility, and public perception. The response necessitated the closure of the westbound lanes and the diversion of all traffic onto eastbound lanes for a number of months. A similar reactionary response was needed when a rock block larger than 10 feet in size failed on a planar feature near Lolo Pass, west of Missoula (C000093E, MP 18.11). This event affected traffic for over one week and required a specialty contractor to break-up and remove the rock.

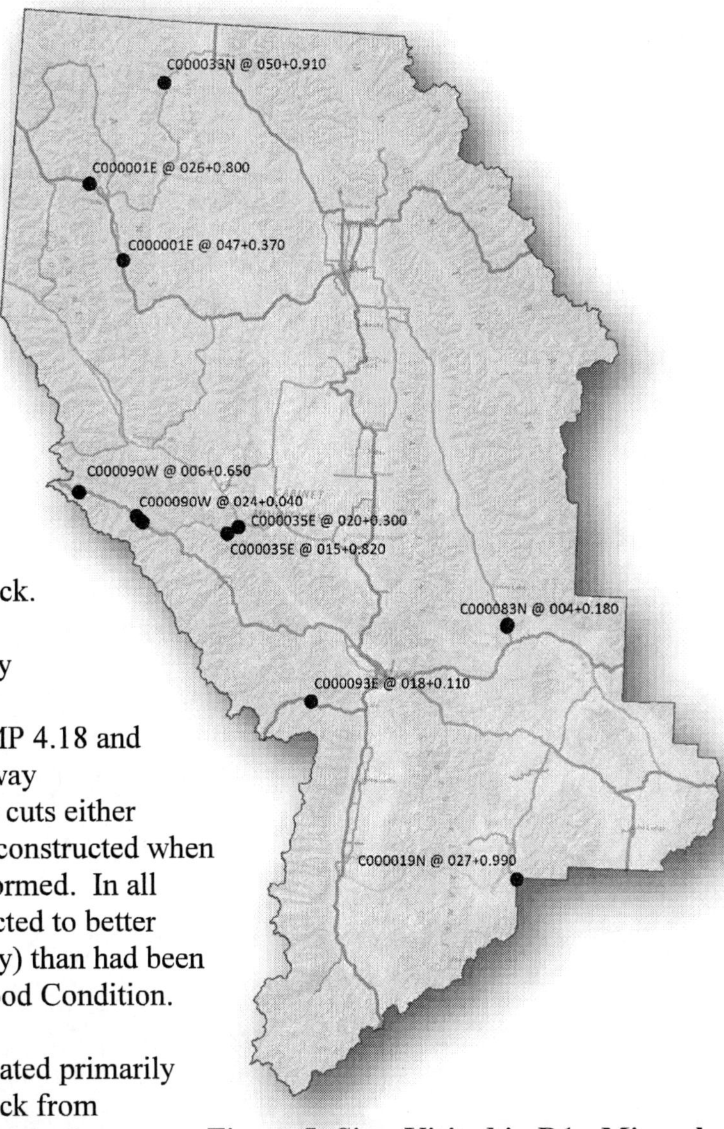

Figure 5: Sites Visited in D1 - Missoula

Three slopes at two locales (Libby Creek South, C000001E, MP 47.37 and Clearwater Junction North C000083N, MP 4.18 and 4.63) were reconstructed as part of highway improvement projects. Previously, these cuts either were small "B" rated slopes or were not constructed when the 2004 rating reconnaissance was performed. In all three cases, the new slopes were constructed to better condition (ditch effectiveness and activity) than had been present prior or were constructed in a Good Condition.

Two of the slopes had been mitigated primarily to reduce rockfall activity and prevent rock from entering the roadway, the Libby Wedge and Flint Creek (C000001E, MP 47.37 and C000019, MP 27.99, respectively). Mitigation measures included scaling, blast scaling, rock bolting and dowels, shotcrete, and barrier fences. Maintenance personnel have reported significant decreases in rockfall activity at both sites, though some deterioration of mitigation measures has occurred and will eventually result in increased rockfall activity.

The two sites located between St. Regis and MT200 (C000035E at MP 15.82 and 20.30) are included as examples of slopes that may have worsened in the years following rating, one of which may be included as part of an annual monitoring survey.

Table 3: Missoula District Re-rates and Test Approach Results.

Feature, Highway, Corridor & Mile Post	RHRS and % change	MDT #1 and % change	MDT #2 and % change	#3 Slope rating & % change	#3 Vehicle Risk and % change	#3 Impact and % change	Mob. & safety risk cost of 30 yr loss*	Condition Index & % change**
Libby Wedge Hwy 2, C000001E 26.90-27.02	499 / 354 -29%	196 / 115 -41%	302 / 169 -44%	171 / 92 -46%	19 / 19 0%	25 / 22 -10%	$734 / 367; -50%	43 / 75 +74%
Libby Ck. S. Hwy 2, C000001E 47.37-47.60	-- / 296 NA	-- / 85 NA	-- / 169 NA	-- / 76 NA	-- / 97 NA	-- / 9 NA	-- / $20 NA	-- / 75 NA
Hwy 135 C000035E 20.3	423 / 338 -20%	139 / 61 -56%	244 / 145 -41%	127 / 51 -60%	29 / 102 250%	12 / 10 -20%	$91 / 20 -79%	53 / 88 +66%
I-90 C000090W 6.5	-- / 361 NA	-- / 108 NA	-- / 142 NA	-- / 52 NA	-- / 19 NA	-- / 56 NA	-- / 17,047 NA	-- / 88 NA
I-90 C000090W 22.36-22.45	379 / 310 -18%	151 / 94 -38%	212 / 155 -27%	92 / 35 -62%	75 / 86 +15%	59 / 59 0%	$16,090 / 11,745 -27%	50 / 92 +84%
I-90 C000090W 24.04-24.19	551 / 432 -22%	176 / 127 -27%	314 / 210 -33%	117 / 72 -38%	107 / 88 -18%	59 / 56 -5%	$24,214 / 15,341 -27%	53 / 78 +47%
I-90 C000090W 24.59-24.72	564 / 406 -28%	217 / 113 -48%	342 / 201 -41%	158 / 57 -64%	89 / 107 +20%	59 / 56 -5%	$24,215 / 13,864 -43%	43 / 80 +86%
Clearwater Jct. Hwy 83 C000083N 4.18-4.22	-- / 190 NA	-- / 46 NA	-- / 116 NA	-- / 26 NA	-- / 116 NA	-- / 20 NA	-- / $47 --	-- / 92 NA
Clearwater Jct. Hwy 83 C000083 4.66-4.72	118 / 111 -6%	59 / 44 -25%	89 / 68 -23%	42 / 25 -41%	44 / 21 -53%	17 / 20 +14%	$37 / 55 +48%	63 / 100 +59%
Lolo Pass Hwy 12 C000093E 18.11-18.20	564 / 429 -24%	124 / 92 -26%	282 / 230 -18%	112 / 85 -24%	127 / 127 0%	12 / 7 -42%	$155 / 66 -58%	69 / 63 -9%
Flint Ck. Hwy 1 C000019N 27.99-28.44	683 / 539 -21%	269 / 126 -53%	427 / 285 -33%	261 / 121 -54%	132 / 132 0%	8 / 5 -33%	$1,670 / 230 -86%	16 / 63 +294%

* in thousands.
** Note that positive percent increases denote an improvement for Condition assessments.

RISK ANALYSIS

An annualized rate for adverse events was estimated for each RAMP site in the survey. For each Condition State group, the annual likelihood of an event somewhere in the Missoula District (D1) was calculated by summing all estimated slope face areas that had exhibited an adverse event. Each Condition State sum was compared to the total inventoried square footage in each condition state in the district, to generate a likelihood per square foot based on slope condition. This permits the estimation of event likelihood based on slope condition and its size. Utilizing this approach rather than evaluating each site only based on its condition, it will follow that a large Condition State 3 slope may pose a greater risk than a small Condition State 4 slope.

Likelihood values of Condition State and average annual likelihood of a service disruption is shown in Table 4. A service disruption is defined as a road closure or traffic slowdown, and some rockfall events may trigger both. A reported 1 in 10 to 1 in 20 (5 to 10%) of road closing rockfall events result in an accident of some kind.

Table 4: Condition and likelihood of service disruption.

Condition State (CS)	Annualized Risk of Service Disruption per sq ft of rock face.	Example annual likelihoods for a semi-triangular 500-ft long by 75-ft high slope
1 (Good)	1.2E-08	0.03%
2 (Fair)	4.8E-08	0.12%
3 (Fair)	3.9E-07	0.95%
4 (Poor)	1.3E-06	3.17%
5 (Poor)	2.0E-06	4.88%

Event consequences were estimated for user mobility and safety using approaches recommended by AASHTO for estimating roadway user benefits (18). Standard AASHTO-recommended values for occupancy and time value were factored into detour lengths in the event of road closures. Average daily traffic volumes from 2014 were provided by MDT. A consistent estimate of a six-hour closure (based on event records) and a slowdown of one week was factored into the risk models. Only on Interstate 90 near the Idaho border, where slope failures have led to traffic slowdowns lasting months, were longer slowdown estimates used.

Multiplying the annual likelihoods (slope condition likelihood factor x estimated area of face) by the consequence of a road closing event (temporary road closures, accidents, slowdowns, and detours) provides a total annual risk cost in US Dollars. The risks can then be grouped by corridor, road segment, or evaluated on a site-by-site basis. GIS techniques facilitate the analyses, as shown by Figure 6.

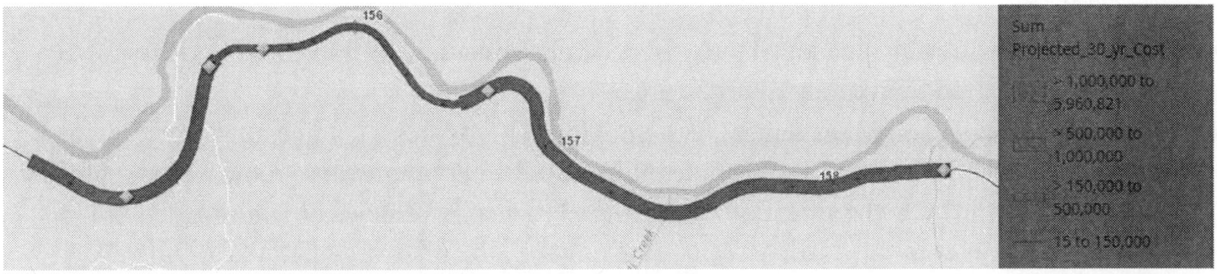

Figure 6: Example Risk Analysis Result on Highway 2 near West Glacier, Montana

PROGRAMMATIC COST ESTIMATES

In addition to identifying corridors with high risk and/or poor condition slopes, this project was able to incorporate Montana-specific data set analyses compiled for other GAM research projects (7). The MDT-specific 'Top 100' highest-scored sites and corresponding cost estimate datasets from the 2005 RHRS program (2) was reviewed and analyzed to determine relationships between slope condition and the costs to mitigate the sites. This work was incorporated into the Alaska research and also published separately (19).

Overhead costs, calculated as a percentage increase of the straightforward geotechnical elements (i.e. rock bolts, scaling, mesh, etc.), were calculated from a Washington DOT dataset of

89 unstable slope sites. Results indicate a reasonable, programmatic escalation factor for costs that encompass PS&E, traffic control, mobilization, and other ancillary design and construction activities was 105%. This factor was compared to a 2015 MDT rockfall mitigation project and compared favorably to this estimate, with a 109% factor based on the Engineer's estimate and 70% based on the low bid (19).

This information is applied to the statewide dataset for determination of cost and benefits, project selection, life cycle cost estimates, and investment models.

Table 5: Average cost to improve a rock slope by a given number of Condition States, with average overhead costs incorporated.

Number of Condition States Improved by Mitigation Activities	Average Mitigation Costs per sq. ft. of Rock Slope Face	
	Geotechnical Element Cost	Incorporating Overhead Costs (105%)
1	$3.56	$7.30
2	$7.12	$14.60
3	$10.68	$21.90
4	$14.24	$29.20

DECISION SUPPORT TOOLS

Use of RHRS/RAMP data to guide decision-making is an under-explored topic for practitioners. Performance goals and targets relative to statewide Performance Measures gauge how well a department is managing their assets, but tools to guide geotechnical personnel on which sites and corridors to add to a file of candidate investments have been lacking. While MDT does not have a generalized agency-wide performance classification scheme to guide the RAMP, there are examples in other MDT agency programs. These include:

- Statewide: Winter Maintenance Standards – Six classifications of Levels of Service (LOS) based on AADT and proximity to urban areas.
- Statewide: Congestion Management System (CoMS) provides a "congestion index" with key performance indicators for Interstate, NHS and Primary highways. CoMS also includes a five-level A - E LOS classification scale. Level A means vehicles are unimpeded in their ability to maneuver in the traffic stream. For Level E, the roadway operates at full capacity with few usable gaps in the traffic stream.
- There are also local/regional classification examples, such as the 2007 "MDT TRED (Transportation Regional Economic Development) – Theodore Roosevelt Expressway Working Paper #5 on Level of Service and Safety" which has a six-level classification scheme.

RAMP Performance Classes

For the above examples, the performance classes are effectively based on goals related to the mobility of the road user. Some classes are indicative of little to no mobility, such as winter pass road closures (Level 5). Others indicate the public's ability to drive at their desired speed and limited time waiting to pass slow moving vehicles.

The five-tier classification scale is typical of many transportation agencies and sets the targets for the quality of road service to users. As with the winter LOSs above, MDT can vary its goals for rock slope performance rather than using a standardized approach that treats each rock slope and corridor identically. A five-tier performance classification scheme for the RAMP that focuses on slope condition and likelihood of road closing events is proposed and should guide MDT on how and where to implement decision support tools.

Table 6 contains proposed route/segment performance goals and the associated Performance Measures based on the roadway's Functional Classification. The performance goals and percentage targets would be applied to these routes and where no rock slopes exist, the default RAMP performance would be 'A', as shown in Table 7.

This approach recognizes that some routes and highway systems are higher priority than others. Follow-up inventory and condition surveys can be prioritized based on functional classification or other metrics, such as the AADT.

Table 6: Proposed RAMP Performance Classification Scheme

RAMP Perf. Class	Road Segment Performance Classification, Likelihood, and Associated Condition Targets
A	Very high performance level. Rock slopes pose a very low likelihood (<0.25% annual likelihood per centerline mile) of user delays. Condition target: >80% of rock slope area (square-foot basis) in GOOD condition and <2% in POOR.
B	High level. Rock slopes pose a low likelihood of user delays (<0.5% annual likelihood). Condition target: >70% of rock slopes in GOOD condition and <5% in POOR.
C	Minimum acceptable level. Rock slopes pose a moderate likelihood of user delays (<1% annual likelihood). Condition target: >50% of rock slopes in GOOD condition and <10% in POOR.
D	Unacceptable level. Rock slopes pose a high likelihood of user delays (<3% annual likelihood). Condition target: <50% of rock slopes in GOOD condition and <10% in POOR.
F	Failing level. Rock slopes pose an unacceptably high likelihood of user delays (>3% likelihood). Condition target: >50% of rock slopes in FAIR condition and >10% in POOR.

Table 7: Functional Classification and Performance Targets

Roadway Functional Classification	Example	Target RAMP Class	Min. RAMP Class
Principal Arterial – Interstate	I-90, I-15	A	B
Principal Arterial – Non-Interstate	US 191 Belgrade to W. Yellowstone	B	B
Minor Arterial	MT 56 Troy to Noxon, Beartooth Pass	B	C
Major Collector	Rt 279 Helena to MT 200	B	C
Minor Collector (all Off System, not part of original RHRS)	Stampede Pass Road Dillion to Rt 357	C	C

Sample Decision Support Tool

The public has certain expectations for roadway performance, such as paved roads will generally be open for travel (with seasonal exceptions); road-closing events are cleared as quickly as possible; and traffic-slowing events are addressed daily (i.e., a rock on the road requiring evasive maneuvering to avoid will be moved off the roadway and into the ditch as needed).

In the following table, the proposed scores are used to determine unacceptable conditions for the various RAMP Classes. Depending on the target performance class, poor condition slopes are addressed by percentile analyses, with higher expectations on Class A routes (interstates). These scores could be adjusted to reflect different percentiles or raw scores, or to ensure that rock slopes meet certain minimum criteria (i.e., Poor condition slopes are not tolerated).

Table 8: Proposed Minimum Tolerable Slope Conditions.

DST Objective – Improve system-wide rock slope conditions

Maintain slope condition to applicable service levels statewide, as measured by service disrupting events (road closure or slowdown). The goals in this table correspond to the RAMP Class Targets in Table 7.

RAMP Class Target A (Interstates): Roads will require only application of routine maintenance to remain open. Sites are selected for mitigation based on slope condition. Consider sites scoring in the worst 15th percentile in the various rating schemes for mitigation. These scoring cutoffs are:

- Condition Index/Condition State: <35/Poor (4/5)
- Total RHRS Score: >450
- Method 1: >175
- Method 2: > 280
- Method 3 Slope Rating: >160

RAMP Class Target B (Arterials and Major Collectors): Road closing events occur on an annual or biannual basis. Consider sites scoring in the worst 10th percentile on the various rating schemes for mitigation. These cutoffs are:

- Condition Index/Condition State: <30/Poor (4/5)
- Total RHRS Score: >485
- Method 1: >190
- Method 2: > 305
- Method 3 Slope Rating: >175

RAMP Class Target C (Minor Collectors and off-system routes): Road closing events may occur multiple times yearly, seasonally concentrated. Consider sites scoring in the worst 5th percentile in the various rating schemes for mitigation.

- Condition Index/Condition State: <25/Poor (5)
- Total RHRS Score: >550
- Method 1: >215
- Method 2: > 345
- Method 3 Slope Rating: >200

ONLINE SPATIAL RAMP DATABASE

The former RHRS utilized a table-based enterprise Oracle system. Users would search for sites based on highway, milepost, or ID number. No map interface was available, limiting its functionality for professionals that are most comfortable with spatially distributed data. The newest RAMP database interface is utilizing MDT's preexisting ArcGIS Online (AGOL) subscription service. This service will facilitate active use of the extensive RAMP database and ensures that the Department can effectively track and manage their assets and properly incorporate system benefits. Housing the data on this site permits interactive use of the data, both online and on desktop GIS platforms. Maps are interactive with zoom, filter, and other query capabilities and do not require extensive GIS knowledge to create custom maps. Figure 7 features sample images of a preliminary online map interface exhibiting various RAMP classifications and results.

One of the critical steps of maintaining and improving MDT's RAMP system over time is tracking and recording geotechnical failures, their mobility impacts, and the direct maintenance costs to the Department. Similar to work performed for AKDOT, a tool was created to track adverse events and rock slope related maintenance activities. The tool requests information on event specifics such as date, type, and size; and follows up with the consequences of the failure such as accidents, closure duration, department resources responding to the event, and the approximate cost to respond.

a) b) c)

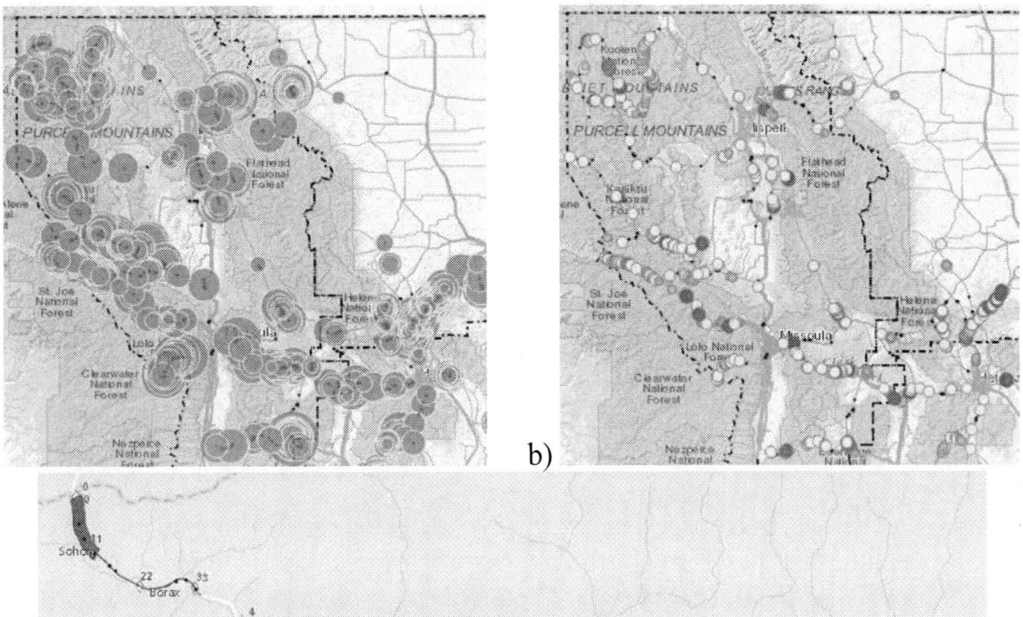

Figure 7: Screens of NW Montana of a Conceptual RAMP Cloud-Based GIS Interface. A) Total RHRS Score, B) Good/Fair/Poor Classifications, and C) Risk at I-90 near Idaho.

CONCLUSION

The Montana Department of Transportation has been on the forefront with addressing rockfall hazards on over 50 slopes over the past decade. The revised RAMP process will provide MDT geotechnical personnel, planners, and TAM managers with the tools necessary to reduce risk and prioritize sites and corridors in a proactive, transparent, and state-of-the-art manner.

By implementing these tools, MDT is improving their ability to achieve state and national objectives of safety, mobility, and resilience in a manner that would not be possible by focusing on the federally mandated asset classes of pavements and bridges alone. Recognition that rock slopes provide critical function to the Department and road user and whose performance can be routinely monitored and managed to minimize the risk of service disruptions and unforeseen costs will assist MDT with allocating scarce funds for the greatest benefit in the coming decades.

REFERENCES

1. **Pierson, Lawrence and Van Vickle, Robert.** *Rockfall Hazard Rating Program - Participants' Manual.* Washington DC : FHWA, 1993. FHWA-SA-93-057.

2. **Pierson, Lawrence, Beckstrand, Darren and Black, Brent.** *Rockfall Hazard Classficaton and Mitigaton System.* 2005. Report No. FHWA/MT-05-011/8174.

3. **Liang, Robert.** *Landslide Hazard Rating Matrix and Database.* Columbus, Ohio : Ohio Department of Transportation, 2007. p. 131. FHWA/OH-2007/18.

4. **Hadjin, Douglas.** *New York State Department of Transportation Rock Slope Rating Procedure and Rockfall Assessment.* Washington D.C. : Transportation Research Board, 2002. TRB 02-3978.

5. **Pack , Robert, et al.** *UDOT Rockfall Hazard Rating System: Final Report and User's Manual.* Salt Lake City : Utah Department of Transportation, 2006. UT-06.07.

6. **Washington Department of Transportation.** *WSDOT's Unstable Slope Management Program.* 2010.

7. *Alaska's Geotechnical Asset Managment Program.* **Beckstrand, Darren, et al.** Roanoke, VA : s.n., 2017. Proceedings of the Third North American Symposia on Landslides.

8. **Rose, Brett.** *Tennessee Rockfall Management System.* Blacksburg, VA : Virginia Polytechnic Institute and State University, 2005.

9. **Bateman, Vannessa.** *Development of a Database to Manage Rockfall Hazard.* Washington, D.C. : Transportation Research Board, 2003. 03-001624.

10. *A Risk-Consequence Hazard Rating System for Missouri Highway Rock Cuts.* **Youssef, Ahmed, Maerz, Norbert and Fritz, Mike.** Burlington, VT : 54th Highway Geology Symposium, 2003. pp. 175-195.

11. **Pierson, Lawrence.** Rockfall Hazard Rating Systems. [book auth.] A. Keith Turner and L. Robert Schuster. *Rockfall Characterization and Control.* Washington DC : Transportation Research Board of the National Acadamies, 2012, pp. 56-71.

12. **Franklin, John, et al.** *RHRON: Ontario Rockfall Hazard Rating System - Field Procedures Manual.* Downsview, ON : Ministry of Transportation, Materials Engineering and Research Office Report, 2012. MERO-043 , ISBN 978-1-4435-9148-5.

13. *Communicating Geotechnial Risk and Complexity.* **Beckstrand, Darren, et al.** Kona, Hawaii : Association of Environmental and Engineering Geologists, 2016.

14. **Lato, M. J., Gauthier, D. and Hutchinson, J.** Selecting the optimal 3D remote sensing technology for the mapping, monitoring and management of steep rock slopes along transportation corridors. *Transportation Research Board 2015 Annual Meeting.* 2015.

15. **23 U.S. Code § 119.** *National highway performance program - Moving Ahead for Progress (MAP21).*

16. **NCHRP.** *Performance Measures and Targets for Transportation Asset Management.* Washington DC : Transportation Research Board, 2006. NCHRP Report 551.

17. **Guerre, Joseph, et al.** *Improving FHWA's Ability to Assess Highway Infrastructure Health Pilot Study Report.* Washington DC : Federal Hihgway Administration, 2012. FHWA-HIF-12-049.

18. **AASHTO.** *User and Non-User Benefit Analysis for Highways.* s.l. : American Association of State Highway and Transportation Officials, 2010.

19. *Development of Mitigation Cost Estimates for Unstable Soil and Rock Slopes Based on Slope Condition.* **Beckstrand, Darren, et al.** s.l. : Transportation Research Board, 2016. The 2016 Annual Meeting Compendium of Papers.

Foothills Parkway:

Micropiles Support Closing the Missing Link

Anthony Sak, P.E.
Hayward Baker, Inc.
515 Nine North Court
Alpharetta, Georgia 30004
(770)-442-1801
asak@haywardbaker.com

Prepared for the 68th Highway Geology Symposium, July, 2017

Acknowledgements

The author would like to thank the Jesus Treto, P.E. (Project Manager) and John Wolosick, P.E., D. GE, F.ASCE (Director), both of Hayward Baker, Inc. for their contributions on this project and the preparation of this paper.

Disclaimer

Statements and views presented in this paper are strictly those of the author(s), and do not necessarily reflect positions held by their affiliations, the Highway Geology Symposium (HGS), or others acknowledged above. The mention of trade names for commercial products does not imply the approval or endorsement by HGS.

ABSTRACT

When completed, the Foothills Parkway (Parkway) in east Tennessee will skirt the northern boundary of the Great Smoky Mountains National Park, offering panoramic views of America's most visited park. At a proposed length of approximately 72 miles, to date only 22 miles are open to the public, approximately 6 miles on the east end and approximately 17 miles on the west end. Another 16 miles of the Parkway is expected to open in 2017 between Wears Valley and Walland, TN. All but 1.6 miles of this section of the Parkway were constructed in the 1980's.

The last 4,000 feet of the Missing Link are currently under construction. Five bridges are needed to traverse the rugged terrain. Foundations for these bridges bear directly on rock or micropiles bonded into rock. This paper will provide some history of the Parkway, the site and construction conditions, and the micropile foundations needed to support closing the Missing Link.

INTRODUCTION

The Foothills Parkway (Parkway) is a scenic roadway located just north of and outside the boundary of the Great Smoky Mountains National Park, Figure 1. Several other parkways are in the Southern US, including the Blue Ridge Parkway and the Natchez-Trace Parkway.

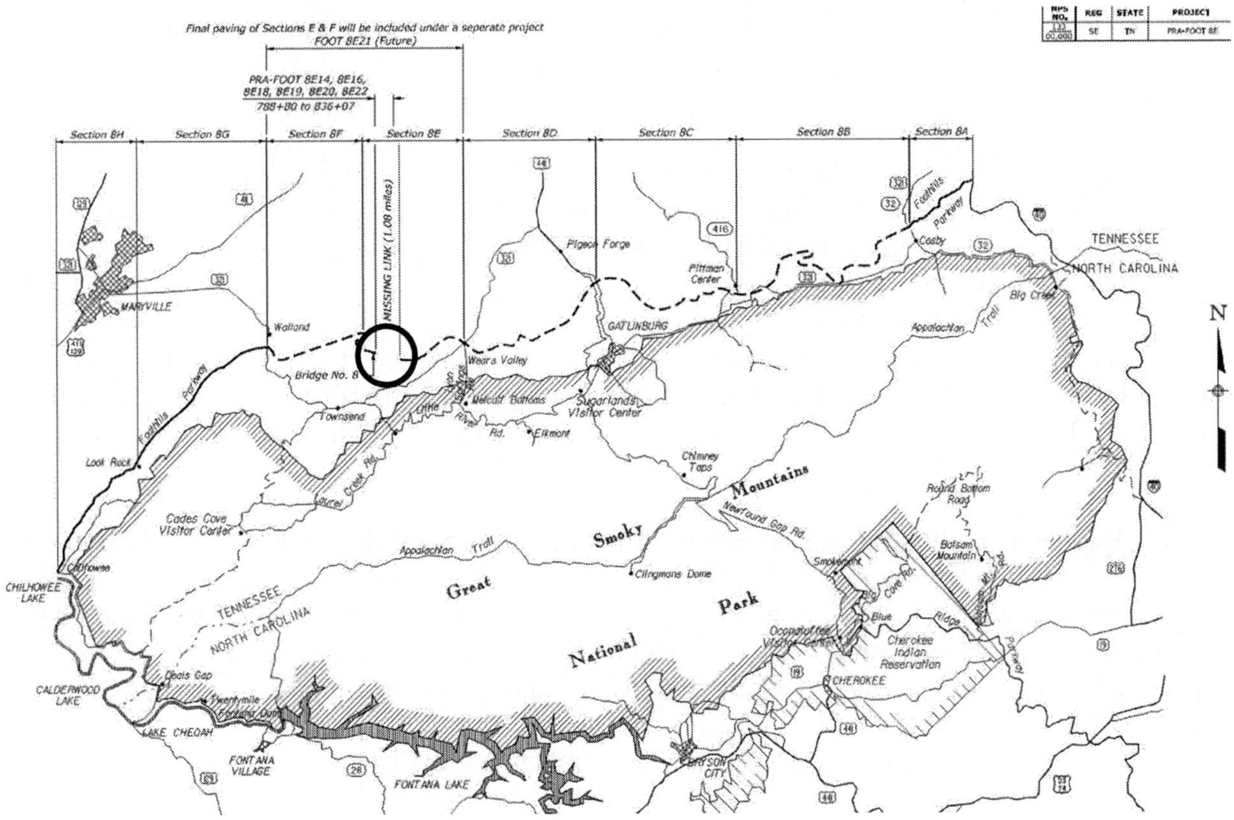

Figure 1 – Foothills Parkway and Missing Link Location

Construction of the Parkway has been an off and on process since the 1960's, when the first sections opened. The latest section of Parkway under construction, between the towns of Wears Valley and Walland, Tennessee, is slated for initial completion in 2017. Construction of this section initially began in the early 1980's and proceeded until environmental and geotechnical issues resulted in the termination of the construction contract. This termination led to a gap in completed alignment that became known as the "Missing Link".

In response to the construction issues, the Missing Link alignment was changed, incorporating a series of 10 bridges to aid in traversing a rugged landscape. The last 4,000 feet of the Missing Link is currently under construction, including 5 of the 10 required bridges. Foundations for these bridges are supported either on spread footings bearing on or micropiles bonded in bedrock.

PARKWAY HISTORY (PAST AND FUTURE)

The Parkway was envisioned by Frank Maloney. Mr. Maloney wanted scenic access to the National Park from the Tennessee side after Congress authorized the Blue Ridge Parkway extending from the North Carolina side of the Smoky Mountain National Park to Virginia's Shenandoah National Park (Los Angeles Times, 2000). Congress authorized acquisition of the right-of-way in 1944 (National Park Service, 2017). Envisioned as a 72 mile stretch of 2-lane highway providing beautiful vistas of Great Smoky Mountains National Park, the Parkway was divided in to 8 segments, labeled A through H from east to west.

Initial construction of the Parkway began in the 1960's with approximate 6 mile Segment A on the east and a total of approximately 17 miles of both Segments G and H on the west. Construction of another 16 miles (total) of Segments E and F began in 1984 and 1985, respectively. Much of these 2 Segments were completed to subgrade elevation when failures of large mechanically stabilized earth (MSE) walls supporting large highway embankments occurred during construction (Lee et. al. 1994). These conditions along with environmental impacts resulting from erosion and sedimentation issues of acid runoff from pyritic rock led to the termination of the construction contract for Segment E in 1989. Thus, the Missing Link was born, measuring in at 1.65 miles.

Beginning in the 1990's, the Federal Highway Administration (FHWA), on behalf of the National Park Service, redesigned the Missing Link. This design intended to minimize surface disturbance and environmental impacts, as well as previous erosion issues and grade separation requirements of the MSE walls. This was accomplished by incorporating a series of 10 bridges (1 through 10) between at-grade and cut sections of roadway alignment.

Between 2001 and 2012, Bridges 1 and 2 and 8 through 10 were completed under several contracts, including design-build. The remaining 4,000 feet of the Missing Link is currently under construction in a design-build delivery contract to Lane Construction Co., Inc. (Contractor) / HDR Engineering, Inc. (Designer) valued at $46.2 million. At completion of the current contract, the Missing Link will be completed, however, Segments E and F will not yet be open to the motoring public. Funding for final paving, signage, and safety measures has been allocated for completion of this work by the end of 2018. Segments B through D, totaling the final 30 miles of the Parkway, await funding.

PROJECT DETAILS

Each of the 5 bridges are post-tensioned, cast-in-place concrete superstructure on cast-in-place concrete substructure of one or two spans. Other bridge characteristics are shown in Table 1, including foundation type.

Table 1 – Bridge Characteristics			
Bridge	Length (feet)	Spans	Foundation Type
3	158.5	1	Micropiles – Both abutments
4	290	2	Micropiles - Abutment 1 and center pier Spread footing – Abutment 2
5	336.5	2	Spread Footing – Both abutments and center pier
6	215	1	Spread footing – Both abutments
7	215	1	Micropiles - Both abutments

Site Conditions

The site was generally forested with mature deciduous trees along the alignment. A pioneer road generally parallels and is offset uphill from the Parkway alignment. The pioneer road remained from previous geotechnical and site work in the Missing Link. As shown in Figure 2, the Parkway alignment is on the south side of a topographic ridge.

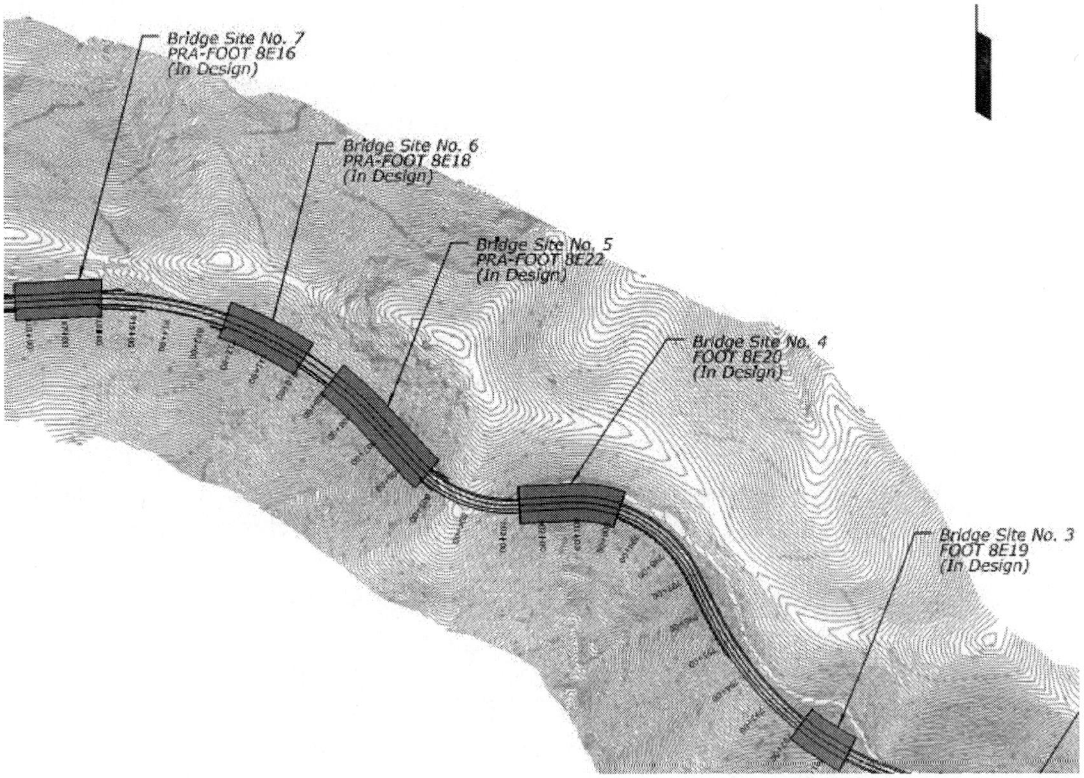

Figure 3 – Foothills Parkway Alignment in Project Area

The cross slope (perpendicular to bridge centerline) elevation change from right to left was typically around 30 feet, giving a nearly 1 horizontal to 1 vertical slope at each abutment and pier. For instance, Figure 3 shows the cross slope of the ground surface at Bridge 3. In the longitudinal direction, the elevation difference was between 20 and 50 feet from bridge abutment to mid-span.

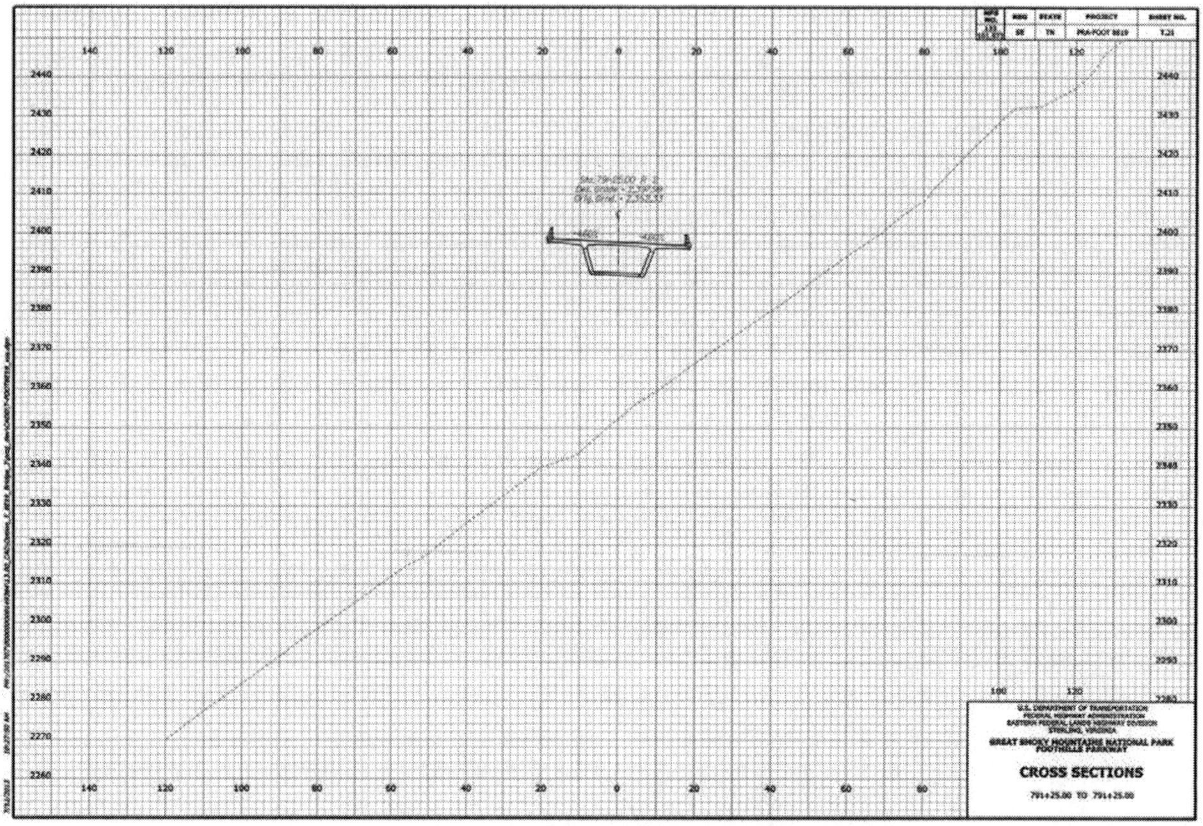

Figure 3 – Cross Section of Bridge 3

The project site is in the foothills of the Great Smoky Mountains in eastern Tennessee within the Blue Ridge Physiographic Province, a region characterized by moderate to high topographic relief formed primarily by uplift of the Great Smoky Thrust Sheet. The site is underlain by Precambrian rock of the Ocoee Series which has been subjected to degrees of folding and metamorphism into metaconglomerate, metasandstone, metashale, and phyllite (HDR, 2011-2013).

Overburden in the project area is composed of residual materials (defined as standard penetration test [SPT] N-value less than 50 blows per 6 inches) generally less than 10 feet in thickness. These materials generally consist of residual sand and gravel with some silt and clay fines. Underlying the residual soils are highly-weathered rock (material that can be drilled and has an SPT N-value greater than 50 blows per 6 inches) to fractured to competent bedrock. The depth to competent bedrock (core recovery greater than 85% and rock quality designation greater than 50%) varies between 0 and 55 feet below ground surface (HDR, 2011-2013).

MICROPILE FOUNDATIONS

As shown in Table 1, two bridges were founded on micropiles, one bridge was founded on a combination of micropiles and spread footings, and the remaining 2 bridges were founded on spread footings only. Where competent rock was not available at design foundation grade for spread footing construction, micropiles were the only deep foundation type specified, as weathered rock of variable quality and site topographic conditions were such that driven piles and drilled shafts would be impractical deep foundation options. These site conditions were ideal for micropile installation.

Micropile Technology

Micropiles were developed in post-World War II Italy for foundation construction and underpinning of historic structures. Introduced to the United States in the 1970's, they again found their use in underpinning applications. As their use grew in the United States, the FHWA began evaluating micropiles as a foundation option for highway construction. A 1997 "State of the Practice Review" was published by the FHWA followed by an "Implementation Manual" in 2000. The Implementation Manual was updated in 2005 (Sabatini et. al. 2005). The design for micropile foundations in transportation projects is outlined in the AASHTO LRFD Bridge Design Specifications.

A micropile is a small diameter (less than 12-inch), drilled and steel-reinforced grouted element with working (factored) load resistance in competent bedrock of 250 tons or more. The micropile transfers axial load through a cased, unbonded zone (like a ground anchor) into a bond zone bearing layer. Due to their small diameter, the end bearing resistance is typically ignored, thus, the axial load is transferred through grout-to-ground bond stress in the bond zone material.

In drilling the borehole for a micropile using one of a variety of drilling methods, a steel casing is typically advanced through the unbonded zone. This steel casing is typically high-strength (minimum 80 ksi) steel originally developed for petroleum wells. Steel reinforcement bars may be grouted inside the casing. The grout used for micropile construction is a low water to cement ratio (typically 0.45 to 1, by weight) to minimize bleed water and grout shrinkage. Figure 4 shows a typical micropile installation process.

As micropiles were originally used for building applications, particularly underpinning, the drilling equipment has become specialized, including, small, low headroom (8 feet or less) equipment. As mentioned above, the geotechnical conditions were ideal for micropiles. However, the site conditions were ideal for small, powerful micropile drilling equipment.

The grout-to-ground bond stress is assumed from typical values (see FHWA 2005) or from local experience during the design phase. In construction, load testing on sacrificial micropiles confirms the design value and proof tests on selected production piles allow for confirmation of the construction techniques.

Installation of micropiles is typically performed by specialty geotechnical contractors. Design details are typically left to the specialty contractor, who uses their expertise and familiarity with drilling techniques to optimize a design that meets the project needs.

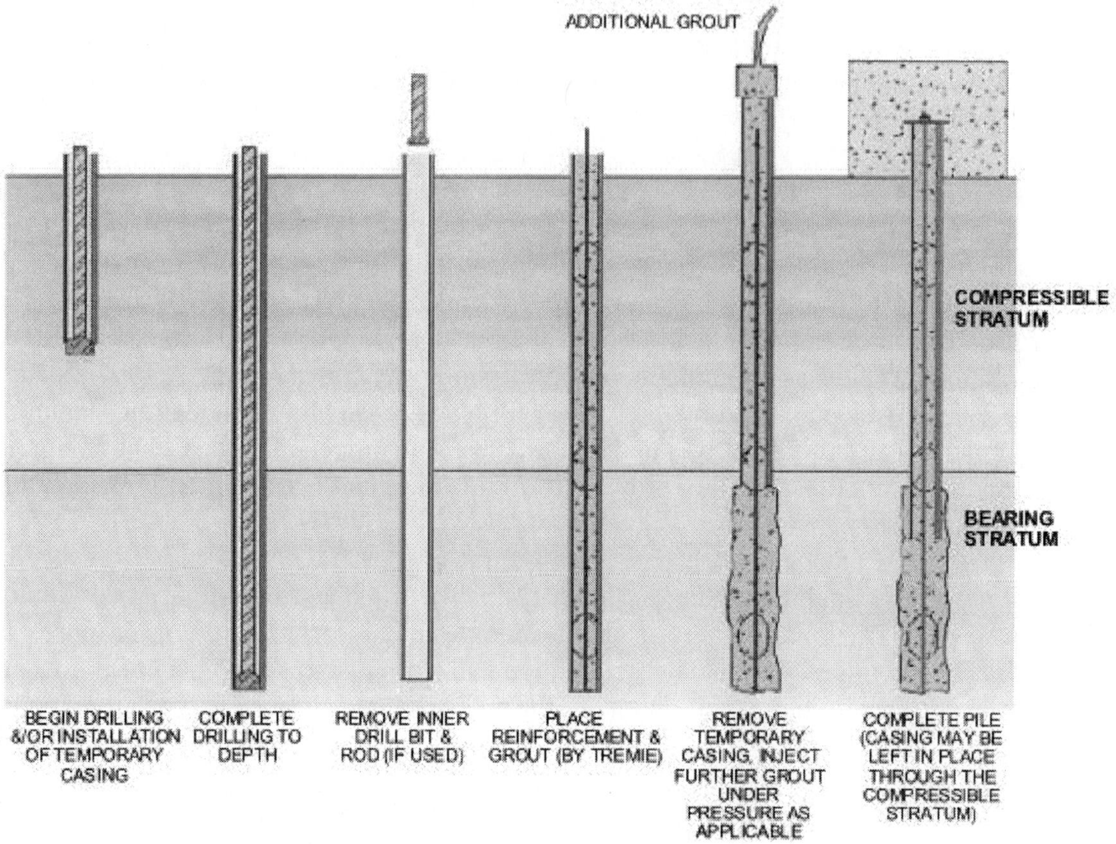

Figure 4 – Micropile Installation (Sabatini et. al. 2005)

Parkway Bridge Micropiles

For the Parkway, the design grout-to-ground bond stress for competent rock (assumed for sandstone) was 20 ksf (139 psi), within the typical range given in AASHTO. Other design details, such as hole diameter, casing size, etc., were also specified. However, the specialty contractor was free to provide an alternate, value-engineered (VE) design for the project. Figure 5 shows the project design vs. the value-engineered design micropile details. Table 2 provides a comparison between the design and VE micropile details. The VE bond stress was based on the same 20 ksf bond stress.

Figure 6 is a photograph of micropile installation at a bridge abutment.

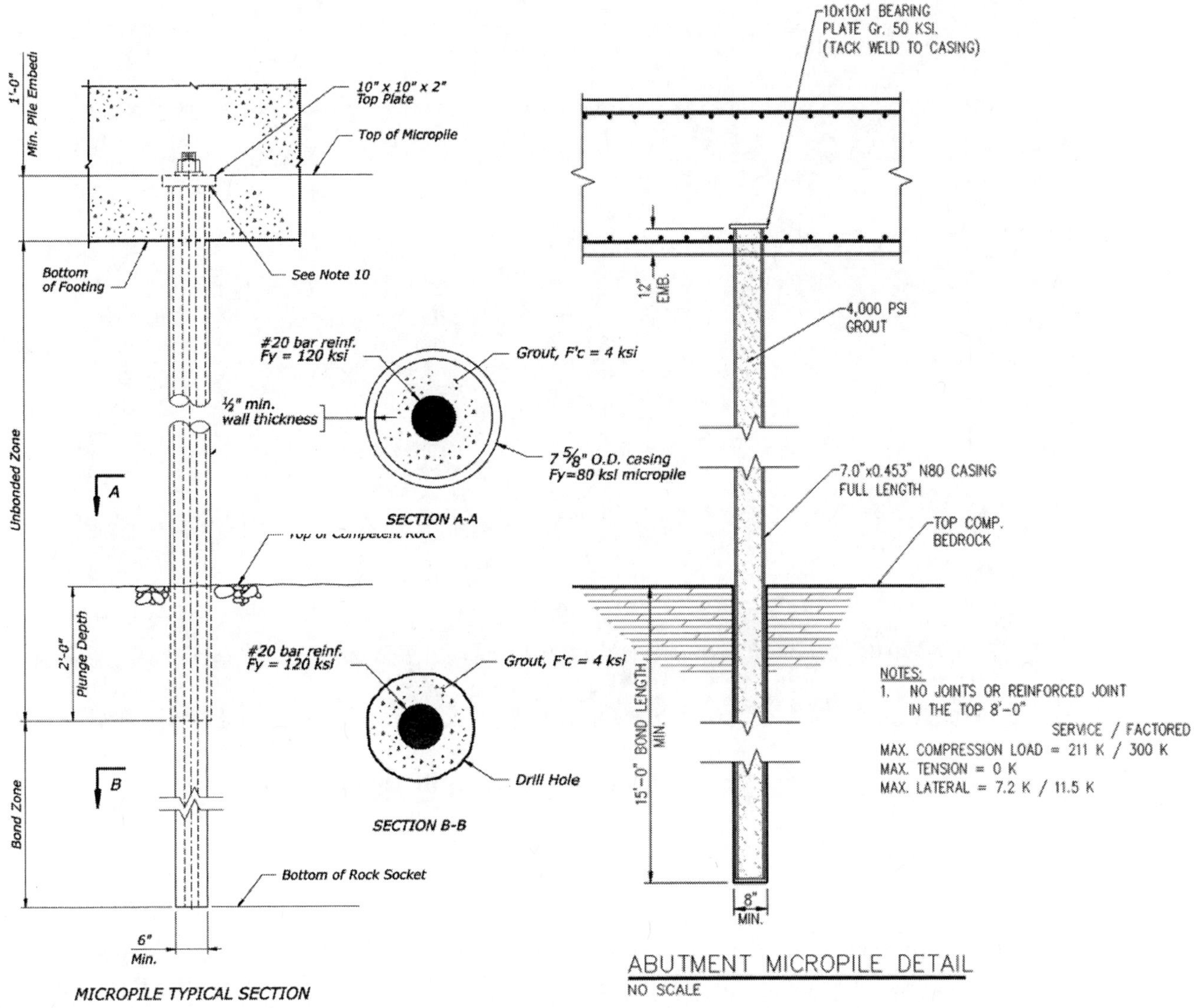

Figure 5 – Design vs. Value-Engineered Micropile Details

Table 2 – Micropile Design Comparison

	Design	Value-Engineered
Borehole Diameter	8 inch (unbonded) 6 inch (bond)	8 inch
Bond Zone Length	19 feet	15 feet
Casing Size	7-5/8 inch 0.498-inch wall	7 inch 0.453-inch wall
Steel Bar Reinforcement	#20, f_y 120 ksi, full length	NA

Figure 6 – Bridge Abutment Micropile Installation

Falsework Micropiles

As mentioned above, the bridge superstructure consisted of post-tensioned, cast-in-place concrete. While structure design accounted for load combinations on the completed structure, it was left to the contractor to deal with the temporary construction conditions. For the Parkway, the temporary construction condition required temporary falsework on which the superstructure formwork was set. The falsework needed foundations to carry the weight of the superstructure until post-tensioning was completed, when the bridge would be self-supporting according to the bridge design. Again, micropiles and micropile drilling equipment were ideal for providing the foundations for the falsework.

To accomplish this, temporary drilling platforms were set at each foundation location. A mini-drill rig was then lifted to each platform, to drill the 4 micropiles per foundation. Figure 7 shows setting a drilling rig on a platform. After completion of the micropiles, cast-in-place concrete pile caps were constructed, followed by the falsework columns and beams. Figure 8 shows completed falsework in place. Once the superstructure was complete, all falsework was removed.

Figure 7 – Setting Drilling Rig on Temporary Platform

Figure 8 – Temporary Falsework

Load Testing

As described above, load testing was conducted to confirm the design bond stress. Load test procedures were the same as those for other deep foundation types. For the Parkway, compression verification testing, in accordance with ASTM D1143, was conducted on pre-production, sacrificial micropiles at each bridge. Figure 9 shows a verification load test setup at a bridge abutment. Four reaction micropiles were used to resist the compression load. The verification test load was typically about 400 kips.

Figure 9 – Compression Load Test Setup

Proof tests were conducted on production piles. As micropiles develop their resistance through grout-to-ground bond, the behavior of micropiles in compression and tension are essentially the same. Therefore, proof tests can be efficiently conducted in tension, in accordance with ASTM D3689, which eliminates the need for reaction micropiles. Tension proof load testing was approved by the contracting officer on the Parkway. Proof tests were conducted to 160% of the design load.

CONCLUSIONS

With completion of the last 4,000 feet, including 5 bridges, the Missing Link of the Foothills Parkway will be completed. After two decades of delays brought about by geotechnical and environmental issues the motoring public will be able to travel on an additional 16 miles of Parkway between Wears Valley and Walland. These 5 bridges are founded on spread footings bearing on, or micropiles bonded in bedrock.

While micropiles were the only deep foundation specified, this technology was perfectly suited to the site and foundation conditions of the Parkway, for both the permanent bridge foundations and the temporary falsework needed to construct the bridge superstructure.

REFERENCES:

HDR Engineering, Inc., *Foothills Parkway Final Geotechnical Engineering Reports.* Multiple dates, 2011 to 2013

Lee, K., C.J.F.P. Jones, W.R. Sullivan, and W. Trolinger. Failure and Deformation of Four Reinforced Soil Walls in Eastern Tennessee. *Geotechnique* vol. 44, No. 3, 1994, pp. 397-426.

Los Angeles Times. *It's All Uphill for Tennessee Parkway Construction Project.* http://articles.latimes.com/2000/aug/06/news/mn-65305. Published August 6, 2000. Accessed April 21, 2017.

National Park Service. *Completion of the Foothills Parkway.* U.S. Department of the Interior Federal Lands Transportation Program Fact Sheet. www.nps.gov/transportation/factsheets/Foothills_Parkway.pdf. Dated February 2012. Accessed April 21, 2017.

Sabatini, P.J, B. Tanyu, T. Armour, P. Groneck, and J. Keeley. *Micropile Design and Construction.* Report No. FHWA-NHI-05-039. FHWA, U.S Department of Transportation, 2005.

Stub Pier Stabilization Performance

Swaminathan Srinivasan, P.E., M. ASCE
Terracon Consultants, Inc.
611 Lunken Park Drive
Cincinnati, OH 45226
(513) 321-5816, ext. 356
ssrinivasan@terracon.com

Jess A. Schroeder, P.E., M. ASCE
Terracon Consultants, Inc.
611 Lunken Park Drive
Cincinnati, OH 45226
(513) 321-5816, ext. 356
jaschroeder@terracon.com

Prepared for the 68th Highway Geology Symposium, May, 2017

Acknowledgements

The authors would like to than the following individuals for their contributions in the work described:

Joseph Smithson, P.E. - Ohio Department of Transportation
David W. Westendorf, P.E. - Terracon Consultants, Inc.
Andrew Bodocsi, PhD, P.E. – Terracon Consultants, Inc.

Disclaimer

Statements and views presented in this paper are strictly those of the author(s), and do not necessarily reflect positions held by their affiliations, the Highway Geology Symposium (HGS), or others acknowledged above. The mention of trade names for commercial products does not imply the approval or endorsement by HGS.

ABSTRACT

Landslide activity along U.S. 50 in Cincinnati, Ohio has caused roadway damage for decades. After a necessary closure of three lanes due to slope movements, emergency stabilization measures were undertaken to protect the roadway by providing a short-term solution necessitated by ODOT budget constraints.

The landslide shear plane occurred on a sloping bedrock surface as much as 50 feet below grade. Drilled shafts were installed 40 feet downslope of the roadway shoulder. The shafts were heavily reinforced across the shear plane but steel reinforcing did not extend the full length of the shafts and was stopped well short of the ground surface. The goal was to provide shear resistance across the failure plane, forcing the theoretical failure surface higher into the overburden soils. These "Stub Piers" were installed and found to meet all of the project goals.

The stub piers and surrounding ground were instrumented with inclinometers, strain gages and earth pressure cells. Data collected 11 years after construction has shown this option has provided much more than a short-term solution to the problem and offered an attractive alternate to conventional deep shafts or tieback drilled shaft options. This paper provides an in-depth long-term performance evaluation of the "stub piers". Instrumentation data is reviewed in this paper, as well as geotechnical design assumptions used for input in the original LPILE analyses.

INTRODUCTION

Landslide activity has occurred along U.S. Rt. 50 in western Cincinnati, Ohio for many decades. The site is located between North Bend and Addyston, OH, on the right descending (cutting) bank of the Ohio River, at about river mile 485.25. Road distress caused by slope movements required periodic repairs over recent decades. The railroad tracks downslope of the roadway has also shown signs of horizontal displacement and periodic repair. Visual evidence in 2005 suggested the shear plane extended below the US 50 roadway and extended out into the Ohio River.

In brief review, the road elevation at the time of our original site study was at about 508 to 516 ft., increasing in an east-northeast direction. A weed and brush-covered slope extended southwest and downward toward the Ohio River at about 3H:1V. Above US 50, the hillside continued to rise more than 100 feet. On the downhill side of US 50, grade sloped down about 15 to 20 feet in elevation to a railroad right-of-way, occurring at about elevation 490 ft. The riverbank then sloped down at about 2.5H:1V to 3H:1V to the water's edge. Normal pool elevation of this section of the Ohio River is 455 ft.

In 2005, Terracon Consultants was retained by the Ohio Department of Transportation (ODOT) to perform a geotechnical study that included 17 test borings and inclinometer monitoring at 4 locations. After only a few weeks of monitoring, the inclinometer casings sheared off about 50 feet below grade, near the soil / bedrock interface. Soon after, the roadway distress worsened, causing ODOT to close 3 of the 4 lanes to traffic and reroute traffic onto the remaining lane and shoulder. Terracon was asked to develop a stabilization design under emergency repair conditions. However, both limited budget and time-related constraints necessitated a direction by ODOT that the solution be at least "pseudo" short-term (3 to 5 years).

Figure 1 - Roadway failure in 2005.

The landslide displayed deep-seated movement extending down to the top of bedrock that lay about 40 to 50 feet below present grade.

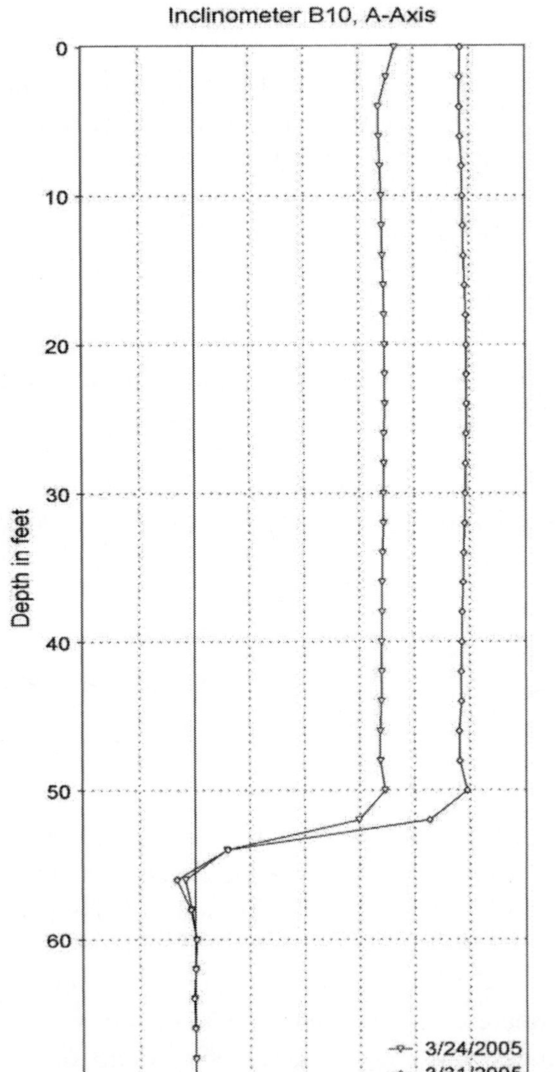

FIG. 2 - Typical inclinometer results showing horizontal displacement <u>before</u> construction.

The shear plane is clearly evident near the deep soil / bedrock interface. Some of these original inclinometer casings sheared off within a few weeks of installation.

The toe of the landslide most likely extended out into the Ohio River. The use of a toe berm or MSE-type retaining wall was not considered practical or feasible for remediation due to the ODOT right-of-way limitations and also because such a repair method would add unwanted load and driving forces to the landslide mass and probably accelerate slope movements.

The most appropriate and effective long-term remedial measure appeared to be construction of a drilled pier wall containing multiple rows of tieback anchors. The anchor installation would likely involve substantial excavation for equipment access to anchor elevations. While effective, this method would involve significant cost and time to construct. ODOT did not currently have sufficient budget for such a repair and needed to reopen the roadway as soon as possible. Instead, ODOT requested a recommendation from Terracon for a "temporary" repair. The primary goal was to allow US 50 to be reopened and remain open for some period of time (3 to 5 years). This time period would allow for plans to proceed with a more permanent solution and build sufficient budget.

Due to the significant depth to bedrock and deep shear plane, the use of "stub piers" was proposed by Terracon as the "pseudo-temporary" repair. The pier reinforcement was designed to act as "shear elements." Details are presented in the following paragraphs, as well as instrumentation results.

GEOLOGIC SETTING

The overburden profile consists of cohesive embankment fill, alluvium, colluvium, and residuum. Fill ranges from 10 to 25 feet deep and is underlain by alluvium that is interbedded and sometimes lying atop colluvium. Residuum is also present in some areas at a thickness of about 3 feet.

Bedrock lies between 31 and 50 feet deep and consists of gray shale and limestone. However, about 3 feet of brown weathered shale with limestone occurs in some locations above the gray shale. The horizontally-bedded shale and limestone belongs to the Kope Formation (Ordovician System) and includes shale that rates as very soft to soft in terms of bedrock hardness. There are numerous documented landslides in this local geologic setting. Shale comprises about 90% of the Kope's mass. Very hard limestone makes up the remainder, occurring in layers up to about 3 inches thick. Refer to Figure 1 for a general subsurface profile illustration.

The Ohio River in this area has a normal pool elevation of 455 feet and official flood elevation of 485 feet. The 100-year flood elevation is 501 feet while the highest recorded river level in Cincinnati occurred during the 1937 flood at elevation 512 feet. With the U.S. 50 roadway elevation at 508 to 516 feet and the railroad at 490 feet, at least the lower portions of this slope are subject to periodic flooding and river drawdown conditions. These periodic events worsen the overall slope instability.

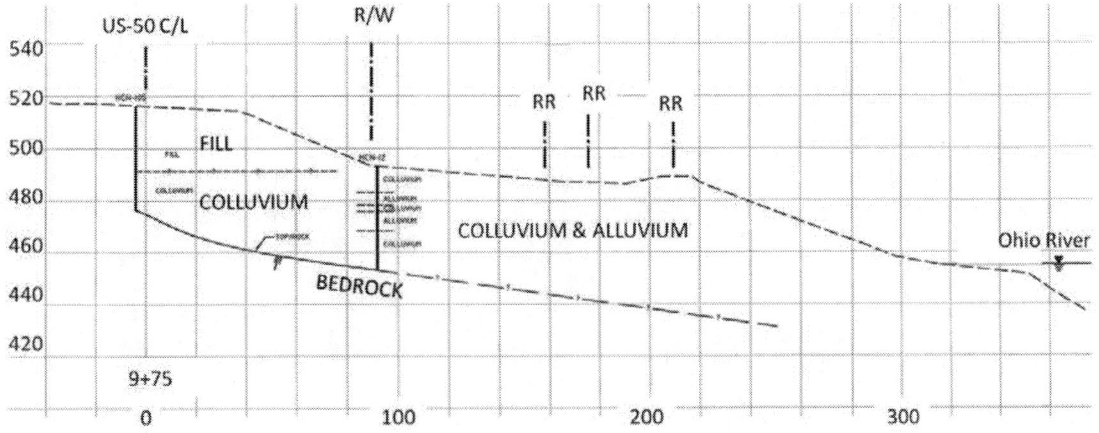

FIG. 1 - Typical Subsurface Profile

STUB PIER DESIGN APPROACH

The prescribed repair method included a single row of straight-sided drilled piers socketed into bedrock. Due to the thickness of overburden, typical local practice would include tieback anchor system with the rock-socketed drilled piers as a more permanent repair method. However, it was assumed that the steel-reinforced portion of the concrete piers only extended part of the way upward through the overburden soils. These "stub piers" were assumed to be closely spaced where soil arching was counted on to make the piers behave as a continuous wall. The piers would therefore force a theoretical shear plane upward from the top of bedrock to above the steel-reinforced portion of the piers.

The stub piers would be located within the right-of-way about 40 feet downslope of the roadway shoulder. The innovative and cost-effective aspect of this scheme involved the steel-reinforcing length. Only the zone near the deep shear plane would be heavily reinforced, thus creating shear pin-type support. The structural steel would be terminated as much as 35 feet below the ground surface.

The writers were unaware of any local experience with this solution. They were also unable to locate any case histories in published literature that would potentially validate this approach and provide useful information on its design.

From an analytical point-of-view, the "pseudo" short-term solution criterion was quantified by slope stability analyses. Laboratory tests were conducted and soil parameters were then adjusted slightly by back-calculating for a failed slope condition (safety factor of 1.0) and observed shear plane depths. Then, the shear plane was forced upward to the planned top-of-steel elevation of the stub piers. The theoretical safety factor increased from the original 1.0 failure condition to about 1.2 (see Figure 2). ODOT was conferred with and agreed with this potential improvement, as a short-term solution. Otherwise, a more long-term solution would result in a theoretical safety factor of perhaps 1.4 or higher.

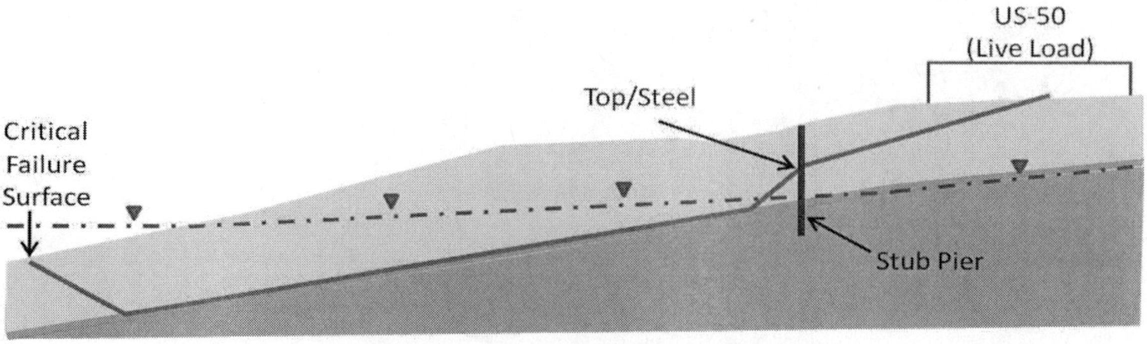

FIG. 2 - Slope stability schematic.

Stub pier design details were then developed. The lateral earth pressure was estimated assuming triangular pressure distribution from the ground level to the shear plane. This resulted in a trapezoidal-shaped earth pressure diagram acting on the reinforced portion of the piers. For potential arching effects above the steel, it was assumed that the contributing pressure extended to one pier diameter above the top-of-steel. This estimated earth pressure was also checked using slope stability analysis to compute the resisting pressure required to generate a safety factor of 1.2. Refer to Figure 3 for schematics of the assumed earth pressure diagram.

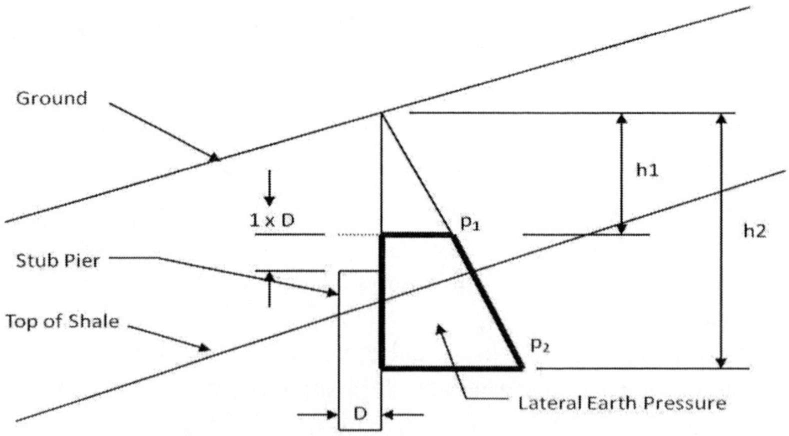

FIG. 3 - Earth Pressure Schematic

Stub pier design was developed using the LPILE computer program. The drilled shafts included 30 and 36-inch diameter units and were socketed 10 to 15-ft. into gray unweathered shale bedrock. The steel reinforcement within the drilled shafts consisted of rolled steel sections that included HP14X73, W18X119, and W24X117. In some cases, additional bending resistance was deemed necessary and developed by welding a steel plate to the uphill face of the beam. The steel beams extended to the bottom of the hole; however, they were limited in length and only extended about 20-ft. above the top-of-rock. Therefore, steel beam lengths ranged from 30 to 35-ft. and stopped well short of the ground surface. The top-of-steel was essentially determined to be the top-of-shaft, thereby assuming that shear failure of the slope could occur at the top-of-steel.

The shaft opening above the steel beam was backfilled with either unreinforced structural concrete or a flowable fill product, as determined by ODOT in the field. The final design included a row of 154 shafts installed on 5-ft. centers.

CONSTRUCTION

The 154 Stub Piers were installed from July to September 2005 under an emergency repair contract. The roadway was repaved on October 6 and 7, 2005, adding upwards of 2 feet of new asphalt in some areas to relevel the road. Traffic was reopened on October 7, 2005.

ODOT indicated the cost for stub pier installation was about $500,000.00 (in 2005 dollars). This cost included drilling, reinforcing, and backfilling 154 stub piers. As-built quantities included 8386 feet of shaft drilling, 1485 cu. yds. of concrete backfill, 553 cu. yds. of flowable fill backfill, and 273 tons of structural steel beams plus stiffening plates.

INSTRUMENTATION

ODOT approved a Terracon-recommended instrumentation program, which began shortly after construction was underway. Locations for instrumentation devices were selected for their critical locations, as well as to coordinate with the contractor's activities and schedule. The instrumentation program consisted of the following:

1. Five Inclinometers were installed within selected Stub Piers.

2. Four Inclinometers were installed upslope of selected Stub Piers.

3. Two Inclinometers were installed about 10 feet downslope of selected Stub Piers.

4. Three Push-In Earth Pressure Cells (Geokon Model 4830) were installed within boreholes located about 8 to 10 feet upslope of selected Stub Piers. These devices were installed about 40 to 45 ft. below grade with the intent of being just above the bedrock surface (close to the interpreted shear plane).

5. Six strain gages were installed in each of two piers (four on the tension side and two on the compression side). The strain gages (Geocon Model 4000 Strain Gages, weldable mounting blocks, plucking coil and thermistor) were vibrating wire gages welded directly to the soldier pile. A thermistor is integrated into the strain gages to account for temperature induced strain. Individual pieces of angle iron were welded over the strain gages to prevent damage during concrete placement.

The strain gage cables were extended up the two respective stub piers to the ground surface. These cables, as well as the earth pressure cell cables, were routed laterally to a terminal box installed on a post embedded within the top of a nearby Stub Pier.

INSTUMENTATION DATA REVIEW

In 11 years since construction was completed, some instrumentation cables were missing and some inclinometer casings were damaged. However, access was still available to some inclinometer casings and cables to allow for periodic readings. Recent readings were taken in February 2017. Analyses were performed to compare apparent earth pressures and displacements with original design predictions. See Figure 4 for measured horizontal deflections at two inclinometers installed within stub piers.

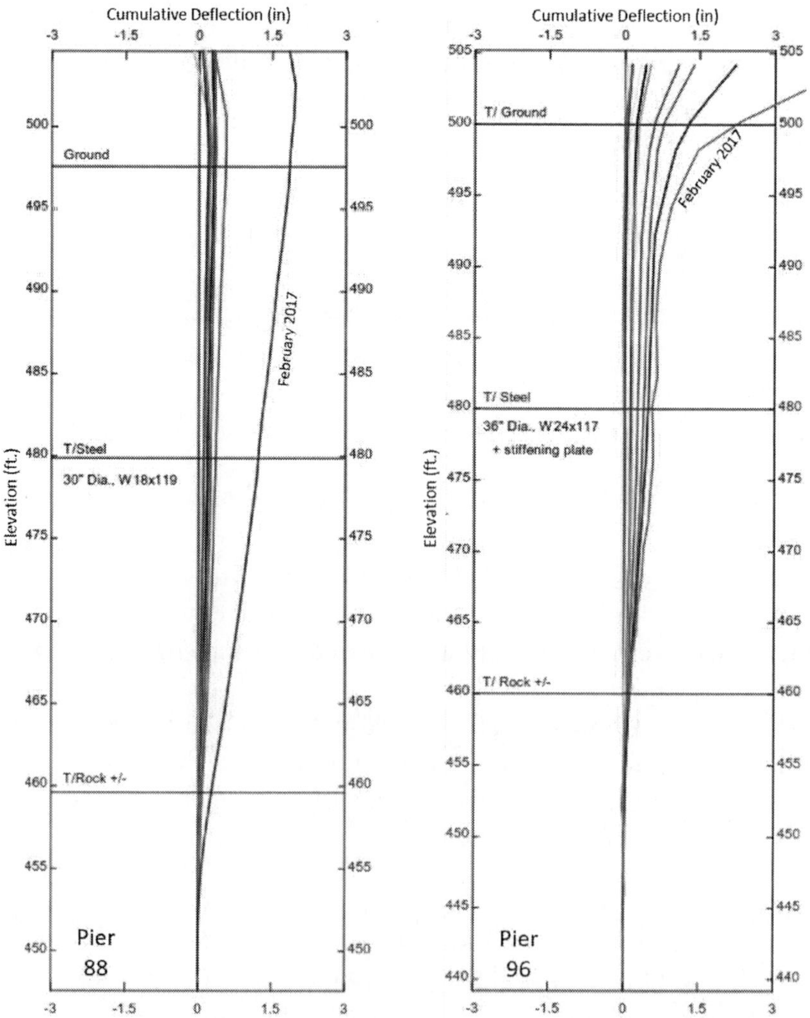

Figure 4 – Inclinometer Readings at Piers 88 and 96, 2005 to 2017

Analysis Procedure

1. The original 2005 design analysis was based on the following:
   a. Slope stability analyses modelled the slope failure and generated back-calculated soil parameters shown below:
      i. Total unit weight: 128 pcf
      ii. Effective Friction Angle (Ø) = 20 degrees
      iii. Effective cohesion = 200 psf

   (Similar values were generated by laboratory triaxial compression tests performed on undisturbed soil samples.)

   The bottommost three feet of the overburden soil stratum was assumed to contain the shear plane and an estimated Ø value of 12 degrees.

   b. Lateral earth pressure was assumed to be triangular distribution extending from the ground surface to the top of rock, as illustrated on Figure 3. The parameters included unit weight of 128 pcf and angle of internal friction, Ø of 20 degrees. It was further assumed that the ground sloped at about 19 degrees (or about 2.9H:1V) and no wall friction was included. An active earth pressure coefficient of 0.76 was computed.

   c. Earth pressure distribution on the stub pier was assumed to be the trapezoidal portion of the triangular distribution, as it only acted on the embedded steel beam part of the stub pier. Additionally, it was assumed there was a vertical arching effect whereby the active pressure within a zone of one pier diameter above the top of the steel added to the stub pier lateral pressure as a concentrated load at the top of steel (Figure 3).

2. In 2017, the analysis began in a similar manner as described above, but as-built conditions were assumed at Stub Piers 88 and 96. Each pier was socketed into bedrock. Other as-built conditions for two of the studied piers are listed below:

Stub Pier 88

- 30" diameter shaft
- 38.5 ft. from ground surface to bedrock.
- 18.5 ft. from ground surface to top-of-steel
- Steel beam is W18x119 (Moment of Inertia = 2190 in.^{4})
- Inclinometer PCI-6 was installed inside the pier (attached to the inside web and flange faces of the steel beam).

Stub Pier 96

- 36" diameter shaft
- 39.5 ft. from ground surface to bedrock
- 19.5' from ground surface to top-of-steel
- Steel beam is W24x117 plus a stiffener plate. Hand calculations indicated the Moment of Inertia for the stiffened W24x117 section was about 5727 in.^{4}
- Inclinometer PCI-7 was installed inside the pier (attached to the inside web and flange faces of the steel beam).

Applying the lateral earth pressure and concentrated horizontal load on Stub Piers 88 and 96 as described above, LPILE analyses were performed. The lateral deflection at the top-of-steel was computed to be about 6.0 and 3.6 inches, for Stub Piers 88 and 96, respectively.

3. Inclinometer casings were still accessible at Stub Piers 88 and 96 in February 2017. Field readings indicated horizontal deflections at the top-of-steel elevation to be about 1.2 and 0.6 inches, for Piers 88 and 96, respectively. Therefore, the theoretical analyses highly over-predicted the actual deflections (see summary on Table 1).

Table 1 - Maximum Horizontal Deflection at Top-of-Steel

Stub Pier	Theoretical Deflection, in.	Measured Deflection, in.
88	6.0	1.2
96	3.6	0.6

Predicted and measured horizontal deflections about 1.5 feet above the bedrock surface (near the original, true shear plane), compared closely, as shown in Table 2 below.

Table 2 - Maximum Horizontal Deflection about 1.5 feet above Bedrock

Stub Pier	Theoretical Deflection, in.	Measured Deflection, in.
88	0.70	0.7
96	0.55	0.6

4. LPILE was again utilized to determine what approximate lateral earth pressure and concentrated horizontal load at the top-of-steel would be needed to generate field-measured defections. Results at both stub pier locations 88 and 96 obtained

reasonably good correlation with field-measured lateral deflection at about 50 to 55 percent of the original assumed lateral load intensity.

5. As a check to the results of values computed in Step 4, field-measured earth pressures were obtained, as cables for the vibrating wire "push-in" earth pressure cells were still accessible in 2017. Three such devices had been installed about 8 to 10 feet directly upslope of selected piers and close to the base of the soil overburden profile. One of those three devices was directly upslope of Stub Pier 96. The measured earth pressure was 1130 psf. Comparisons are tabulated below for Stub Pier 96 (see Table 2).

Table 3 - Maximum Lateral Earth Pressure Near Top-of-Bedrock at Stub Pier 96

Stage	Maximum Horiz. Earth Pressure, psf
Original Design, 2005	3890
Estimated using LPILE and matching with measured deflection	2000
Earth Pressure Cell field measurement	1130

The measured earth pressure value shown above is well below the back-calculated value of 2000 psf. We suspect the earth pressure monitoring device may have rotated before being seated at the bottom of the borehole and therefore would not have its sensor properly oriented perpendicular to the slope forces. Another possible explanation for readings below expected values is if the gauge were installed slightly deeper than the bottom of the colluvial soil stratum (and perhaps not wholly within the shear zone).

6. Comparisons were made between the maximum bending moments and average earth pressures between the original theoretical design analyses and those estimated from measured strain gage data. Results of those comparisons are described below.

Strain gages were installed on Stub Pier steel at Piers 96 and 110. Calculations were conducted in 2011, about 5 to 6 years after construction. The measured or "apparent" strain was converted to bending strain by subtracting the calculated compressive strain due to the weight of the pier above (carried by steel and concrete) from the measured apparent strain. The bending stress and bending moment were then computed from the bending strain value at each strain gage location. The computed bending moments based on these measured strains were well below theoretical estimates generated from the original LPILE analyses. For example, bending moments on the tension side of the steel were computed from strain gage readings to be approximately 25 percent of the LPILE results at the time. Additionally, the strain gage data generated bending moments significantly higher on the tension side than the compression side of the steel. One potential explanation could be that the concrete contribution in resisting bending is neglected in the analysis.

One significant inconsistency in the strain gage data occurs when earth pressures were back-calculated from the computed bending moments. These computed earth pressures were a fraction of those generated by earth pressure theory and are also well below those measured in the three earth pressure cells. There was no clear explanation for these results.

DISCUSSION

1. The data and analysis about 11 years after construction of the stub pier wall has confirmed the landslide shear plane has been successfully cut off and stub pier performance appears to be much better and more of a long-term solution than the original goal of creating a "pseudo-temporary" solution.

2. The owner (ODOT) realized a successful repair solution because the repair was designed and constructed quickly, where the 154 stub piers were installed and the roadway repaved in under 3 months. The costs were significantly less than the alternative of a tieback-anchored drilled pier arrangement. A tieback approach would likely have involved excavating and installing multiple rows of tiebacks due to the deep shear plane. Excavated soil materials would have had to be removed from the site to avoid stockpile loads on the slope, only to be returned later for burying the deeper tieback heads. A much longer construction period would have been required at significant inconvenience to roadway users. A tieback anchor and drilled pier approach cost was estimated to be about 3 to 4 times the cost of the constructed stub pier approach.

3. Movement continues downslope of the stub piers. Eleven years after construction, inclinometer measurements at two locations just downslope of the stub piers indicated about 1.5 to 2 inches of creep movement has continued downslope and away from the stub piers. Those movements have possibly worsened in recent years, as evidenced by a new soldier pile wall that was installed by the CSX railroad, just downslope of existing railroad lines and about 80 to 100 feet down slope of the stub piers. The writers have not been successful in gathering background information on that soldier pile wall (see Figure 5).

Figure 5 – Soldier pile wall installed downslope of the railroad.

4. Various field readings collected on inclinometers installed upslope of the stub piers, within the stub piers, and downslope of the stub piers have generally indicated annual horizontal movement rates as listed below:

Table 4 - Approximate Annual Rate of Measured Maximum Horizontal Movement

Location	Annual Rate, in./yr.	Remarks
Upslope of Piers(1)	0.3	13 ft. upslope of Pier 88
Within the Piers	0.1 to 0.2	Measured at top-of-steel reinforcement
Downslope of Piers(1)	0.15 to 0.2	11 ft. downslope of Piers 88 and 110

(1) Measured at the ground surface.

5. The field data collected 11 years after stub pier installation suggests original design assumptions were conservative. That degree of conservatism has been estimated in terms of applied lateral earth pressure.

   Instrumented Stub Pier 96 is used here for evaluation. If the apparent maximum lateral earth pressure acting on the stub pier is about 2000 psf (as described above), the back-calculated active earth pressure coefficient, K_a, would be about 0.4, as compared with the original design value in 2005 of 0.76.

$$p_a = \text{unit weight} \times \text{height} \times K_a,$$

$$K_a = p_a / (\text{unit weight} \times \text{height}) = 2000 / (128 \times 39.5 \text{ ft.}) = 0.4$$

   The active earth pressure coefficient of 0.4 would back-calculate a friction angle (phi) of about 32 degrees accounting for sloping ground and no wall friction. In the writer's opinion, this Ø angle is too high to represent the overburden soils and landslide geometry at this site. If the sloping ground is ignored, the computed Ø would be about 25 degrees, which is more reasonable.

   Another simplified way to look at this would be to neglect the sloping ground effect and calculate the maximum active earth pressure using the effective strength parameters determined in the original 2005 slope stability analyses. This approach uses a combination of cohesion and Ø.

$$p_a = \text{unit weight} \times \text{height} \times (\tan^2(45\text{-}\varnothing/2)) - 2 \times \text{cohesion} \times (\tan(45\text{-}\varnothing/2)$$

$$= 128 \times 39.5 \times \tan^2(45\text{-}(20/2)) - 2\times 200 \times \tan(45\text{-}(20/2) = 2199 \text{ psf}$$

This shows a computed earth pressure about 10% higher than the target value of 2000 psf. Adjusting to the target value, a combination of about Ø = 22 degrees and c = 200 psf would compare closely, giving an active earth pressure of about 2000 psf. These parameters are more realistic, in the writer's opinion.

As mentioned, the original 2005 design accounted for sloping ground in developing an active earth pressure coefficient. The subsequent field monitoring and analyses suggest that sloping ground should be neglected. The Ohio Department of Transportation's Geotechnical Bulletin GB 7, which was published in 2014, also suggests neglecting sloping ground.

6. Using Pier 96 as an example, this analysis has shown that if the original design had been based upon a maximum active earth pressure (at the top-of-bedrock)

equal to 2000 psf (plus the concentrated load at the top-of-steel to account for vertical arching effects, as described earlier), LPILE analysis would generate a theoretical maximum horizontal deflection of about 0.6 inches at the top-of-steel, using the W 24X117 steel section (plus stiffening plate) that was installed.

The writers would caution taking the approach further by allowing a much greater theoretical deflection, such as 2 inches (a value suggested by ODOT for roadside pier walls) to determine a required steel section. Using LPILE, this example would generate a required steel section of W18 x 35, which is significantly lighter than the constructed version and one that the writers do not necessarily agree with. Keep in mind that a 2-inch horizontal deflection at top-of-steel for these stub pier units would translate into a much larger displacement at the ground surface and may not be advisable. Also, horizontal deflections have continued to increase at a rate approaching 0.2 inches per year over the past 11 years and will likely continue to do so. Selection of steel section based upon a theoretical horizontal deflection should anticipate long-term creep increases and anticipated movements at the ground surface, where features like pavement stability, guardrail support, etc. might be affected.

LESSONS LEARNED AND CONCLUSIONS

- The Stub Pier approach works for deep shear planes.
- This approach is not suitable for all settings, as shallow landslide potential after construction must be quantified.
- Quantifying shallow landslide potential (by slope stability analysis) appears to be a valid basis for evaluating "longevity" of the system.
- There is potential for significant cost savings over more traditional solutions and allows for relatively quick installation.
- Stub Pier installation can be done with minimal specialty materials or equipment.
- The original design assumption for active lateral earth pressure distribution and intensity for this project appears to be conservative. For example, the original prediction for lateral displacement at the top-of-steel section was about 3 to 6 inches. Measured values after 11 years are about 0.6 to 1.2 inches at the top of the steel beam. A design active earth pressure coefficient also appears to be conservative, where active earth pressures appear to be about 50 to 55 percent of the original predictions. Predictions of active earth pressure of this lower magnitude were better determined using a combination "c" and "Ø" with no wall friction and a horizontal ground surface.

- Selection of steel reinforcing based upon a theoretical horizontal deflection should anticipate long-term creep increases and anticipated movements at the ground surface, where features like pavement stability, guardrail support, etc. might be affected.

- A well-designed instrumentation program is needed and recommended to validate design assumptions and to monitor performance.

- Considering the complex soil-structure interaction and sensitivity of design active earth pressures to deformation, a numerical analysis approach would be beneficial.

Finally, the stub pier approach at this site appears to be functioning well 11 years after construction and may provide many more years of support. Therefore, the original goal of providing a "short-term" solution appears to have been met and exceeded.

REFERENCES

- Department of Navy, Naval Facilities Engineering Command, Foundations and Earth Structures, Design Manual 7.2, NAVFAC DM-7.2, May 1982.

- Ohio Department of Transportation Geotechnical Bulletin GB 7, "Drilled Shaft Landslide Stabilization Design," November 20, 2014.

- Computer Software
  1. W. M. Isenhower, PhD, P.E. and Shin-Tower Wang, PhD, P.E., Technical Manual for LPile, Version 2012-06, "A Program for the Analysis of Deep Foundations under Lateral Loading," ENSOFT, INC., November 2012.

Information Modeling workflows for using geotechnical data in civil engineering

BIM and subsurface data

Nicolas Loubier
Bentley Systems
685 Stockton Drive
Exton, PA 19341
(800)-236-8539
Nicolas.Loubier@bentley.com

Katie Aguilar
Bentley Systems
685 Stockton Drive
Exton, PA 19341
Katie.Aguilar@bentley.com

Prepared for the 68th Highway Geology Symposium, July, 2017

ABSTRACT

Information Modeling Systems

Applying Information Modeling principles to geotechnical engineering as an integral part of Transportation Asset Management Plans

Title: Information Modeling workflows for using geotechnical data in civil engineering

Many organizations that rely on subsurface information fail to integrate this information in to an information model for lack of tools to easily transfer and integrate the data to the model.

This disconnect is caused primarily by the fact that the geotechnical industry is still a report driven industry. This means that geotechnical data often ends up isolated and not part of the information modeling warehouse. Moreover, this data is usually available only a transactional basis (when a report is finalized and handed over to owner operator), rather than being a constant data flow or "plugged in" data source.

This causes huge challenges in collaboration between disciplines and sometimes a level of mistrust to use and analyze data for multiple purposes.

In order to bridge this gap in the geotechnical area, two actions must be taken: working with digital data, and allowing the geotechnical information to be displayed in proper context

- Digital data : geotechnical data should be stored in systems that are open for multiple uses rather than "just" the generation of a paper based report; more importantly, this system should support interoperability standards (which are just emerging for now) so that data can flow from one system to another one.
- Context: Because of the Information Modeling framework, data from multiple disciplines and systems can be brought together in order to provide context for the information produced by all disciplines. It allows geotechnical data to be viewed in conjunction with transportation assets, and aid in long-term management strategies organization wide.

This paper will focus on the context aspect and how geotechnical data can be used in the context of civil workflows to help in project analysis, options development, and maintenance programs of transportation assets.

INTRODUCTION

The traditional geotechnical workflow is report driven: get the logs out to the clients for interpretation and design. This can be internal or external users and clients. There is little consideration to the use and reuse of the data in various applications and other reports for actual consumption. As projects become bigger, longer in duration, and more time critical for information, the concept that the log is an end-product must change.

The geotechnical site investigation is the foundation for all subsequent site work and design. And yet little consideration to time spent entering and re-entering geotechnical data in various summaries is given. How can a wide variety of end users consume the data and be confident of the data when not from a PDF sheet? Why is the PDF our most important document?

It is not just the geotechnical engineer, geologist, or geoenvironmental engineer consuming the data for design and analysis. To which we need to simplify. The information is used across the broad scope of civil engineering from preliminary site design through the final design; from construction: site clearing (topsoil removal), subgrade stabilization, haul road design, site drainage plans, excavation shoring, foundation design, retaining walls, landscaping and vegetation, etc; to operations.

If an issue develops in any one of these stages, it is often the geotechnical logs and reports that are the first documents examined for fault with the question: what was overlooked?

This paper explores the context aspect of data, and how geotechnical data can be used in the context of civil workflows to help in project analysis, options development, and maintenance programs of transportation assets.

The geotechnical report is not an end-product, but an integral part of the BIM process.

Traditional Geotechnical State of Practice

Traditionally, in geotechnical engineering the single source of truth is the geotechnical log. On it is everything from sample data, lithology descriptions (often from sample descriptions), laboratory data, water level data, field test results, well installations, inclinometer installations, and environmental measurements. Yes, there are a gathering of lab reports, but often primary reports from the lab are on the log report as well.

Geotechnical engineers will use that data to create cross sections, various plots, spread sheet summaries, and sometimes a 3D model to get an idea of what is going on across the site to make design assumptions for everything from road surface design to retaining wall design, to bridge and sign foundations, and more.

To use the data in other applications, often this means manually entering data from a PDF or paper log and a series of laboratory reports. And these additional data entries are not typically 'smart' meaning they are data points or drawing elements (depending on the application), and no supporting data is attached to it. For that an engineer must refer back to the logs.

Without considering geotechnical data as a resource to be managed, geotechnical engineers often are isolated: internally they

1) spend hours creating a 'final report'
2) Spend hours searching for data used on previous investigations that can be helpful on a current investigation
3) Have data for a single project scattered across excel sheets, word documents, and even PDFs printed out or saved on a computer.
4) Must spend hours to check calculations and validate analyses.
5) Enter data multiple times for different reasons: reports, summaries, and analyses
6) and need to re-enter data manually to use other design and evaluation applications.

This isolation prevents workflow from occurring between disciplines. And causing additional work down the design chain as well. Geotechnical design is slowed as engineers enter data in various sources to understand site conditions, and create and validates design assumptions. Downstream design and construction is waiting for the final reports and information.

Geotechnical Data Management

Instead of reading and manually entering data in various applications for review and validation of site design assumptions, the data should be entered in to, and managed from, a single source of truth: a database. It can then be used in other applications by geotechnical engineers, and others downstream needed the data for their use including interpretation and design.

A database format allows this repeated, multi-discipline use. The geotechnical log is not the end-product to be passed on to be used further downstream within geotechnical engineering and beyond to other civil disciplines. Rather, the log is the starting point for geotechnical site evaluation, and general civil design.

With geotechnical data in a database, a wide world of data transfer, and site evaluation and design is open and available instantaneously.

Creating a Geotechnical Model

A geotechnical model created across large areas is typically generated with relatively distinct point data information from the boreholes, meaning large assumptions and extrapolations are made in developing any mode, 2D (i.e. fences) or 3D.

Soil borings are taken along the project, but at predetermined intervals based on spacing rather than actual ground conditions. For example, a typical roadway project may require soil borings every 150 - 500 feet (give or take, depending on the project) just along the centerline of construction, or a soil boring at a foundation location.

Often these models, if created, are made with drawing elements only. And there is no associated data. Again, the model is the end. Instead of being a part of a continuing workflow for a site.

Geotechnical Data in the Context of BIM

Many engineers want to see subsurface information in to BIM (Building Information Modeling) for design and maintenance of sites. For BIM to be useful, these subsurface elements must be smart, and not simply drawing elements. In other words, data must be attached to the elements, and not just what the symbols represent. But the information associated with that symbol, be it a lithology description, sample length and/or blow counts, water level description, and laboratory data.

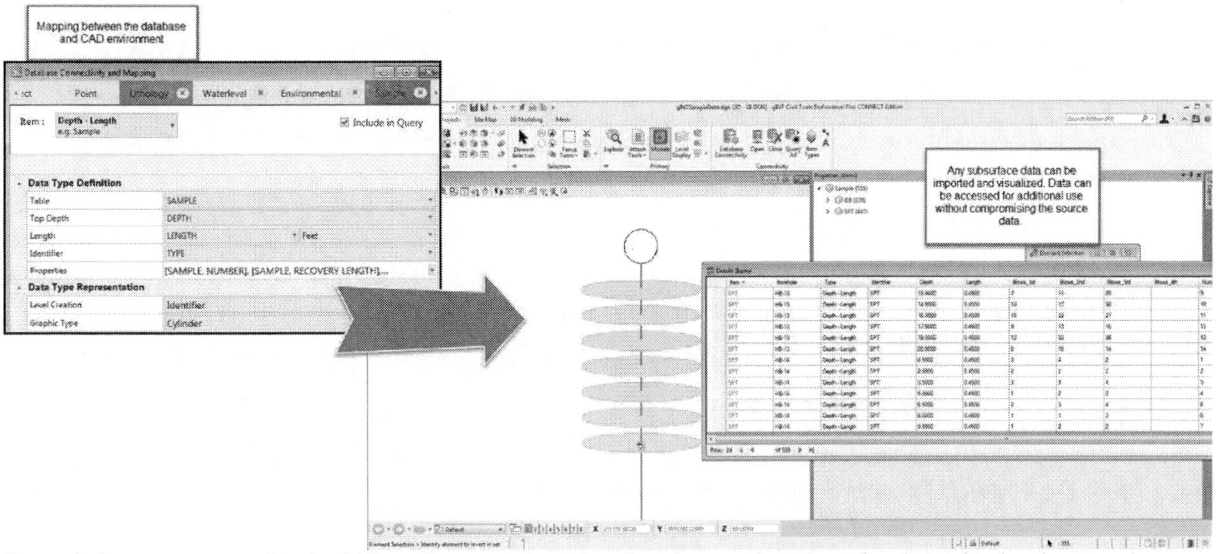

Figure 1: Data mapping and in the CAD environment. Any subsurface data can be imported and visualized and reviewed

Many organizations that rely on this subsurface information fail to integrate this information in to a BIM model for lack of tools to easily transfer and integrate the data to the model.

Applications are now available that are enable geotechnical data to be integrated in to a BIM model.

With their geotechnical data in a BIM model, geotechnical engineers, geologists, and geoenvironmental engineers can see data in the context of site design and available GIS data and:

- Create terrains (surfaces) composed of elements like boundaries, break lines, spots …Can be updated
- Generate triangles and contours on demand. Symbolization is easy to update for creating easily interpreted reports
- Review and refine subsurface interpretation in multiple passes
- Create dynamic profiles
- Visualize boreholes plus surfaces
- View civil projects
- Quickly interpret subsurface conditions
- Use editing tools to work with terrain and real "DTMs"
- Visualize all geotechnical data in 3D
- Overlay site design plans to provide context view
- QA for boreholes elevation
- Data QA/analysis by combining with other data (point clouds, …)
- Access data within the model for confirmation of design assumptions and necessary information without returning to the source

In addition, there are several output format options that enable the geoprofessional to share their data without compromising their single source of truth. Additional end users have the ability to view and further use the data in the context of the project as designed in a CAD environment.

Output options include Microstation (.dgn), imodel, and smart (3D) PDFs with layers. These outputs allows users to share data, and downstream users use and consume data without altering the source (the database) without having to re-enter data.

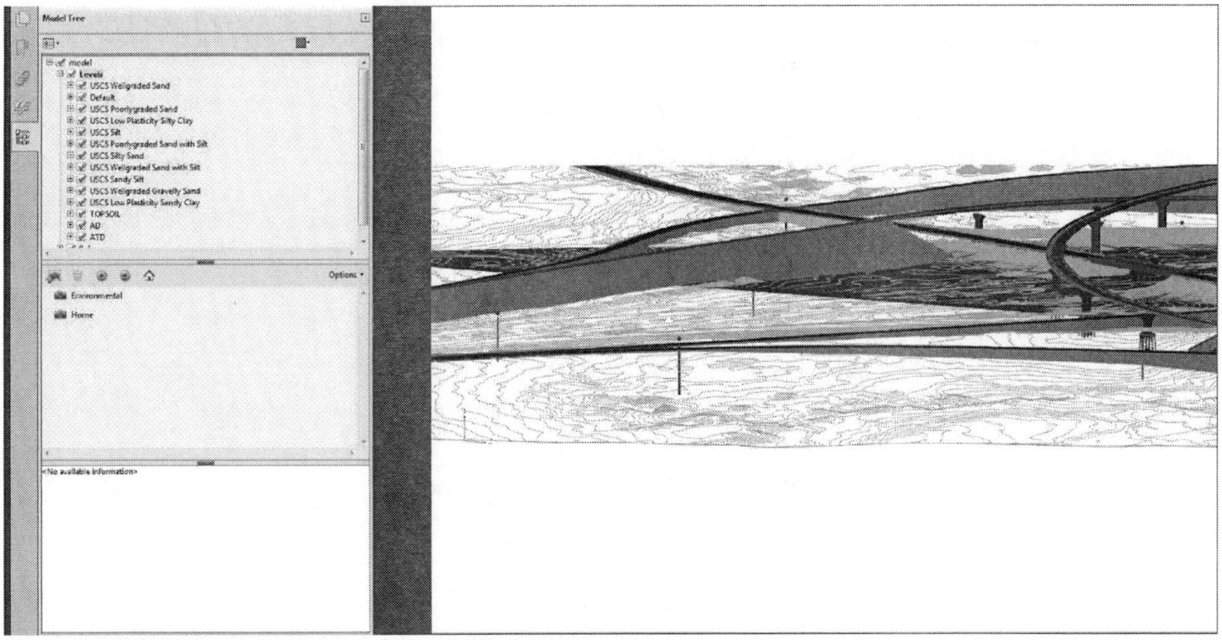

Figure 2: 3D PDF. Levels can be turned on and off.

Conclusion

Begin with the end in mind. Is the log the end? No. It is the beginning of a long process to evaluate site conditions to make design assumptions. This is not just for the geoprofessional, but also for many site civil design applications.

Geotechnical data is the foundation for all subsequent design work. And must be a part of the integrated building information modeling network.

The civil industry is advancing, and projects becoming bigger and more complex. The integration of geotechnical data in to BIM models is inevitable, and necessary. Tools are now available to integrate the data for design, construction, and operations.

Road Widening in Creep-Prone Bituminous Sandstone

Gresham D. Eckrich, P.E., C.E.G.
Yeh and Associates, Inc.
391 Front Street, Suite D
Grover Beach, California 93449
(805) 481-9590
geckrich@yeh-eng.com

Jerko Kocijan, Ph.D., P.E., G.E.
Fugro Consultants, Inc.
4802 McGrath Street, Suite 100
Ventura, California 93003
(805) 650-7000
jkocijan@fugro.com

Prepared for the 68th Highway Geology Symposium, May, 2017

Acknowledgements
County of San Luis Obispo Public Works Department
Mike Britton, Mark Davis, Pete Newel

Disclaimer

Statements and views presented in this paper are strictly those of the author(s), and do not necessarily reflect positions held by their affiliations, the Highway Geology Symposium (HGS), or others acknowledged above. The mention of trade names for commercial products does not imply the approval or endorsement by HGS.

ABSTRACT

Road Widening in Creep-Prone Bituminous Sandstone

Gresham D. Eckrich, P.E., C.E.G.
Yeh and Associates, Inc.
391 Front Street, Suite D
Grover Beach, California 93449
(805) 481-9590
geckrich@yeh-eng.com

Jerko Kocijan, Ph.D., P.E., G.E.
Fugro Consultants, Inc.
4802 McGrath Street, Suite 100
Ventura, California 93003
(805) 650-7000
jkocijan@fugro.com

Price Canyon Road is a two-lane route that runs roughly north-south between the City of Pismo Beach and the town of Edna on Highway 227 in San Luis Obispo County, California. The roadway passes through the Arroyo Grande Oil Field that is developed on both sides of the roadway and currently operated by Sentinel Peak.

The project consists of widening the existing roadway for Class II bike lanes. Three relatively deep through-cuts are located along the roadway alignment. The existing cut slopes are inclined at approximately 1H:1V (horizontal:vertical) and are up to approximately 58 feet high. In areas where the project is constrained by right-of-way and environmentally sensitive areas, a rock dowel wall was constructed to retain cut slopes that varied in height from approximately 5-1/2 to 16 feet.

The Edna Member of the Pismo Formation is extensively exposed in road cuts along Price Canyon Road, and a subunit of the Edna Member is characterized as a bituminous sandstone that includes active tar seeps. Unconfined compressive strength test results indicate the heavily bituminous sandstone unit can be prone to long-term creep under certain environmental conditions. Test results clearly indicate that the likelihood of material creep and eventual failure increases as the temperature of the material increases. Consequently, the design approach accounted for a potential strength reduction due to temperature increases, as well as long-term creep considerations.

INTRODUCTION

The County of San Luis Obispo (County) constructed Price Canyon Road in the 1960's between the City of Pismo Beach and the unincorporated town of Edna to connect the Highway 227 and Highway 101 corridors in San Luis Obispo County, California. The road was constructed with two 12-foot lanes, with about 2- to 3-foot shoulders.

In the early 2000's, the County initiated a project to widen the existing roadway for Class II bike lanes. The roadway will be widened in both directions by approximately 8 feet. The project was phased to expand the roadway in segments, moving from north to south. Construction of Phase 1 was completed in 2012. Studies for the Phase 2 alignment, which is the focus of this paper, were performed between 2013 and 2015, with construction starting in May 2016. Figure 1 shows the Phase 1 alignment in green, the Phase 2 alignment in yellow, and the proposed widening dimensions for a typical roadway section.

Topography along the Phase 2 alignment is generally defined by grass-covered rolling hills that are occupied by cattle ranches, vineyards, and the active Arroyo Grande Oil Field. Because of varying topography, the original roadway construction required numerous cut and fill slopes. The roadway widening of about 8 feet in both directions necessitated additional cuts and fills, including three through-cut slopes with total heights up to approximately 58 feet and inclinations at about 1H:1V. A retaining wall up to 16-feet in height was designed to accommodate the widening along an approximately 600-foot length of the Phase 2 alignment.

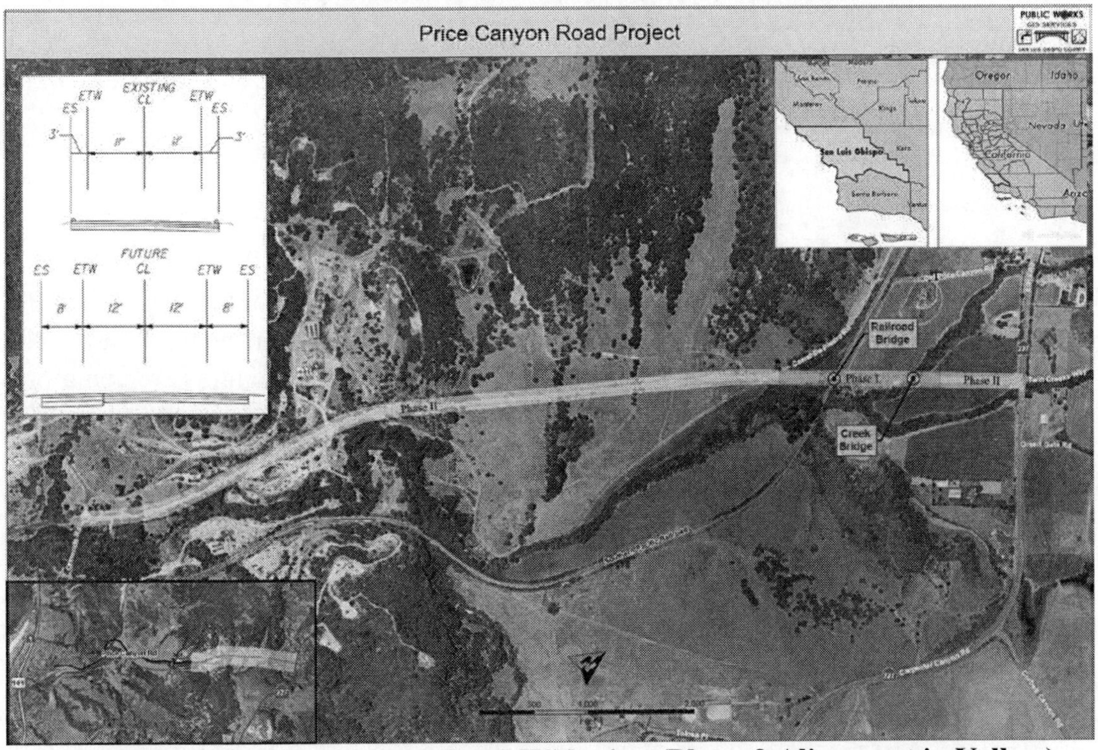

Figure 1. Price Canyon Road Widening (Phase 2 Alignment in Yellow)

An assessment of surficial and deep-seated slope stability in the native bituminous sandstone presented a unique challenge due to the material's multi-dimensional strength properties, particularly the material's temperature-controlled strength loss and propensity to creep. The geotechnical engineering lexicon defines "creep" as slow, progressive shear failure while an earth material is subjected to constant stress. This paper describes how these challenges were incorporated into the design of the proposed retaining wall.

GEOLOGY

The site is within the Coast Ranges geologic and geomorphic province. Hall (1973) maps the geology along the project alignment as surficial sediments of alluvium that overlie sedimentary bedrock of the Tertiary-age Pismo Formation. The alluvial sediments are deposited within southeast-trending drainages that cross beneath Price Canyon Road and flow to Pismo Creek.

The Edna Member of the Pismo Formation is exposed in road cuts along the Phase 2 alignment. Hall (1973) described the Edna Member in the site vicinity with the following subunits: Tmpe, Tmpe2, Tmpe3 and Tmpec. Tmpe is bituminous sandstone that includes active tar seeps (Figure 2). Tmpe2 and Tmpe3 are non-bituminous sandstone subunits composed mostly of fine and coarse sand, respectively, with varying degrees of cementation, and Tmpec is described by Hall (1973) as a pebbly conglomerate unit.

Figure 2. Bituminous Sandstone with Active Tar Seeps

Antonellini et al. (1999) attributed the bitumen (tar) to structural traps within the sandstone unit, formed by deformation bands and small-offset bedrock faults within the sandstone. Sharp boundaries observed between the bituminous and non-bituminous sandstone units are commonly associated with sets of deformation bands, in which sandstone grains are crushed and re-crystallized, forming a permeability barrier. Multiple subunits are locally interbedded along the project alignment and were mapped as part of the field exploration program (Figure 3). The contact between bituminous units (labeled as $Tmpe_{HB}$) and non-bituminous units ($Tmpe_{NB}$) was distinct in places on the existing, vegetated slope. Where the subunits were difficult to distinguish on the basis of field observations, the slope was mapped as interbedded sandstone ($Tmpe_{INT}$).

The geologic structure in the site vicinity consists of northwest-southeast trending faults and folds, mainly associated with the Edna Fault Zone and the Pismo syncline (Hall, 1973). Regional mapping by Hall shows the bedding dips 37 to 44 degrees, with predominantly southwest and northwest dip directions resulting from the folded synclinal structures. Measurements of bedding planes along the Phase 2 alignment indicate a dip magnitude ranging from about 35 to 49 degrees and a predominantly southwest dip direction.

FIELD EXPLORATION

The field exploration program included hollow-stem auger borings along Price Canyon Road to depths of 9 to 40-1/2 feet. Block samples were collected by hand from various outcrops exposed on the existing rock cuts along Price Canyon Road during the field exploration program. Geologic mapping consisted of noting selected geologic features observed at the site such as rock types at outcroppings, springs, and measuring predominant discontinuity orientations.

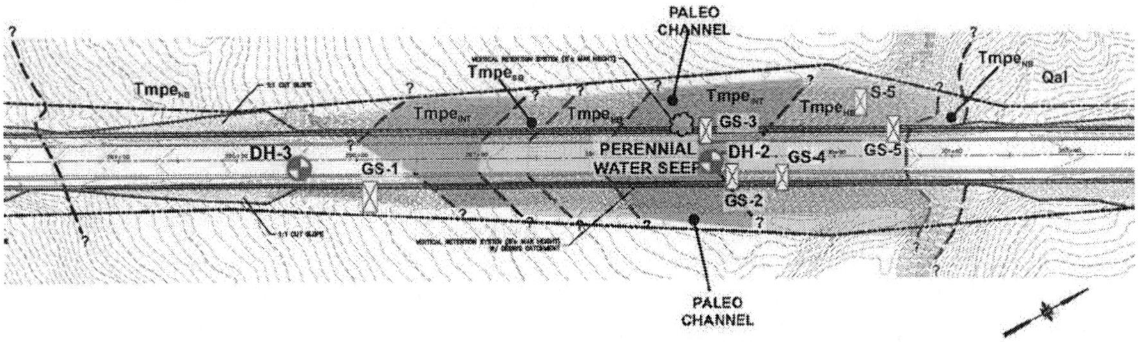

Figure 3. Field Exploration Map with Mapped Edna Member Subunits

To better understand the temperature dependent behavior of the bituminous sandstone, the field exploration program also included installation of three thermistors. Thermistors were installed at depths of 1-inch, 6-inches, and 13-inches to monitor the temperature variations with depth from the slope surface. The thermistors were placed in a single 3/8-inch pilot hole and temperature-sealed with insulating foam.

SLOPE STABILITY ASSESSMENT

The retaining wall design included an atypical slope stability assessment that considered the complex strength properties of bituminous material. The heavily bituminous unit ($Tmpe_{HB}$) is effectively an "oil sand" with a bitumen content typically exceeding 10-percent by dry weight. For comparison, the typical bitumen content of dense-graded hot mix asphalt is about 5-percent by dry weight. Engineering parameters for the heavily bituminous sandstone unit were developed on the basis of laboratory test results and back-analysis of existing slope conditions. Due to the creep potential of the heavily bituminous sandstone unit, the design considered the strength behavior of the unit for a range of loading and environmental conditions.

Laboratory Testing – Short-Term Strength

A number of unconfined compressive strength (UCS) tests were performed on bituminous samples recovered from the slope surface and carved to meet UCS test shape requirements. The results indicated that sampled bituminous materials have a peak UCS strength of 18 to 94 kips per square foot (ksf), which corresponds to a peak shear strength of about 9 to 47 ksf. These values indicate that under short-term loading conditions (e.g., seismic loading, temporary excavation), bituminous materials are anticipated to provide significant resistance to short-term loads when insufficient time is available for creep to develop. Therefore, a short-term shear strength of 5,000 pounds per square foot (psf) was selected for use in the analyses of transient seismic loading.

Laboratory Testing - Creep Under Long-Term Loading

The UCS tests performed on bituminous materials also indicate the heavily bituminous sandstone unit is prone to long-term creep. When subjected to stresses for a prolonged period of time, the bituminous material can exhibit progressive creep and eventual failure at stresses significantly below the peak strength measured during the short-duration UCS tests (less than 1 ksf in some of the UCS tests performed). Heavily bituminous samples failed under constant test loads that were applied for durations ranging from less than 5 minutes to more than 40 days. Laboratory tests indicate that the potential for creep increases as the temperature of the material increases, however, insufficient data was available to develop a rigorous correlation between creep, temperature, and induced stresses.

To account for the creep potential of heavily bituminous sandstone slopes, the design considered the stable existing condition (i.e., height and inclination of the existing, non-creeping cut slopes) as a baseline to estimate slope stress levels that are unlikely to induce long-term creep. Because the existing slopes have not exhibited evidence of global instability for 50+ years, it can be reasonably assumed that the current stress levels in the slope are below the stresses that result in progressive creep of the material, and the factor of safety is 1 or greater. Global stability back-analysis was performed to estimate the stress level that is unlikely to induce long-term progressive creep of existing heavily bituminous sandstone slopes. For a calculated factor of safety of 1, shear stress levels in the slope were estimated to be on the order of about 900 psf (Table 1).

Sandstone Unit	Unit Weight (pcf)	Long-Term Stress without Progressive Creep (psf)	Short-Term, Seismic Shear Strength Parameters		Ultimate Bond Strength (psf)
			Cohesion (psf)	Friction Angle (degrees)	
Heavily-bituminous	125	900[1]	5,000[2]	--	900[3]

Notes:

[1] Back-calculated parameter based on a slope stability assessment of existing slopes as the shear stress level along the failure surface with the lowest calculated factor of safety.

[2] Based on peak strength in unconfined compressive strength testing of bituminous materials.

[3] Based on the shear stress level that is expected to limit the potential for progressive long term creep.

Table 1. Engineering Parameters for Retaining Wall Analyses

THERMAL VARIATION ZONE

Existing slopes in the heavily bituminous unit are prone to surficial instability, particularly during periods of hot weather. This is evident from field observations, where sloughing of the heavily bituminous materials seems to be limited to zones with a thickness of about 6 to 12 inches (Figure 4).

Figure 4. Typical Surficial Failure on Bituminous Sandstone Slopes. Depth of Failure Surface is Approximately 8 Inches Below Slope Face.

Thermistor data indicate thermal differences within about 13 inches of the slope surface can vary up to approximately 34 degrees (F) daily and approximately 90 degrees (F) annually (Figure 5). High temperatures and thermal variations likely contribute to weathering of the heavily bituminous unit that is common during hot summer months, as temperature variations (volume change associated with heating and cooling) can promote development and propagation of fractures.

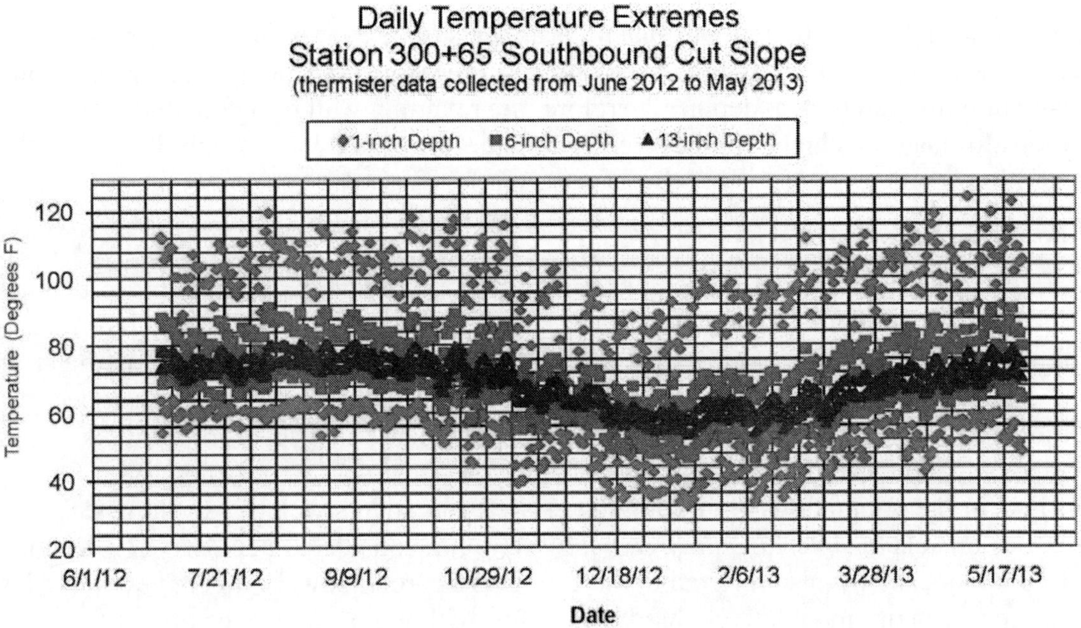

Figure 5. Daily Temperature Extremes Within a Bituminous Sandstone Slope

A simple extrapolation of the annual extreme temperatures indicates the zone of significant thermal variation extends to a depth of about 3 to 3-1/2 feet (36 to 42 inches) below the rock slope surface (Figure 6). Based on the same extrapolation, the temperature of rock at depth appears to be at about 60 to 62 degrees (F). Although the interpreted zone of thermal variation extends to a depth of about 3 to 3-1/2 feet, major thermal variations are expected to occur at lesser depths. The peak temperature measured at 1-inch depth was about 125 degrees (F) – approximately 65 degrees higher than rock at depth, while at 13 inches depth, the peak temperature measured was about 80 to 82 degrees, only 20 degrees higher than rock at depth.

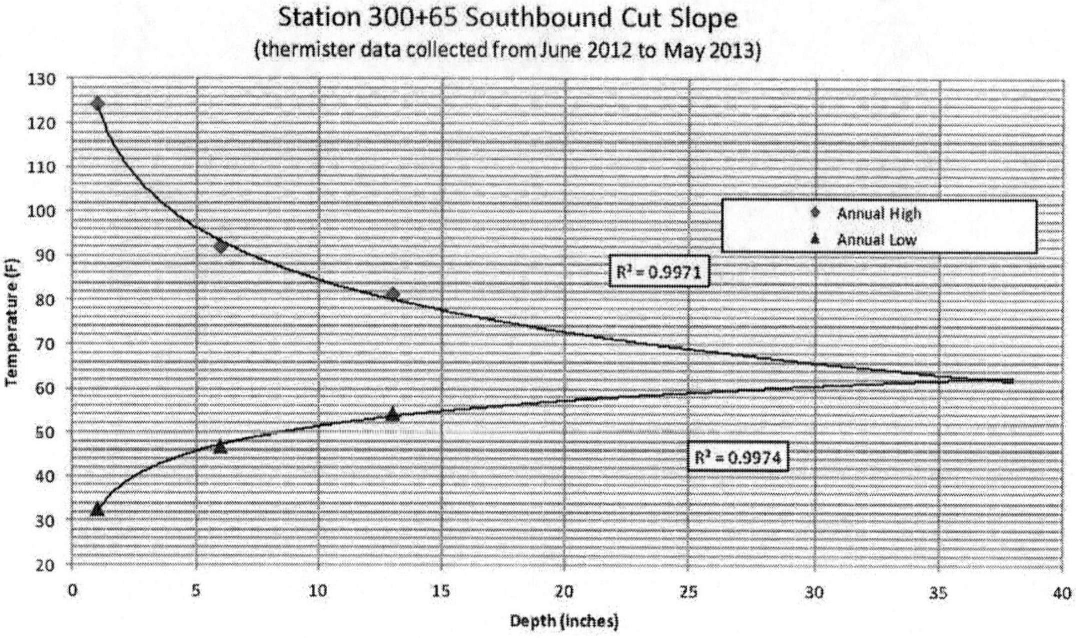

Figure 6. Annual Temperature Extremes within Bituminous Sandstone

Because progressive creep in bituminous materials appears to be exacerbated by increased temperatures, the shallow zone with high peak temperatures were considered more susceptible to creep-related failures than rock at depth. Therefore, the retaining wall design included considerations for materials highly prone to creep and the associated loss of stability within the thermal variation zone.

ROCK DOWEL WALL DESIGN

The retaining wall design included rock dowel reinforcement for stabilization of the vertical cut slopes. Rock dowels are defined herein as passive elements that rely on mobilization of the tensile strength of the steel reinforcement, as well as the shear strength of surrounding material, at relatively small wall displacements. In other applications, rock dowels may be designed to improve the interface shear capacity and resist lateral movement of adjacent rock blocks. Geologic mapping indicated that discontinuities within the rock mass were generally healed/infilled, or the orientations of discontinuities were not adverse relative to slope stability. Therefore, rock dowels *for interface shear capacity* were not considered necessary for design. The construction documents also referred to the elements as "rock dowels" to convey to bidders the potential for encountering relatively hard rock zones within the wall excavation.

Global Stability

The assessment of global stability included an evaluation of critical failure planes passing behind the zone penetrated by rock dowels. By extending rock dowels deeper into the slope, the critical failure planes were extended deeper and the associated stresses along the critical failure surface were lowered to the target stress levels.

To select the minimum rock dowel lengths, we assessed the state of stresses along the critical failure surface for the proposed slope configurations at different typical sections (cut height and rock dowel length) to confirm that the calculated stress levels were at or below the stress levels estimated from back-analyses of existing, stable slopes. Following this method, the design assumed that a calculated factor of safety of 1 or greater indicates the new wall and slope configurations will perform similar to existing slopes, as the new stress levels will be at or below stress levels that existed before construction (i.e., stress levels that the slopes have maintained for 50+ years).

Internal Wall Stability

To evaluate anticipated kinematic loading mechanisms and possible failure scenarios within the rock dowel wall, design analyses considered the temperature-controlled behavior of the heavily bituminous materials near the slope surface (Figure 7). Because of its higher creep potential due to exposure to higher temperatures, the surficial thermal variation zone could have a significant impact on the loads applied to the wall facing and rock dowels. By accounting for the thermal-variation zone, the design considered the potential for flexural and shear failure within the facing in addition to pullout of the rock dowels.

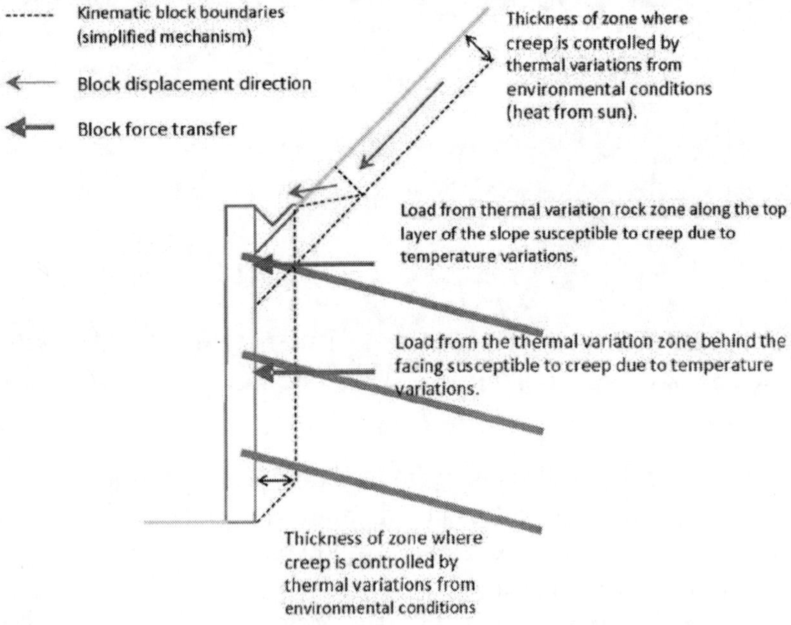

Figure 7. Rock Dowel Wall Surficial Loading Mechanisms

Ultimate Bond Strength of Rock Dowels

For rock dowels in the heavily bituminous sandstone unit, the design considered the possibility that the bond between the grouted rock dowel and heavily bituminous material may be prone to long-term creep. Therefore, the ultimate bond strength was limited by the shear stress level (900 psf) estimated from back-analyses of existing, stable slopes. In addition, the ultimate bond strength was adjusted by a reduction factor of 0.5 (i.e., factor of safety of 2), and the bond strength was neglected within a 3-foot depth from the slope surface to account for the thermal variation zone discussed above. The effective bond length (i.e., total bond length minus the 3-foot thermal variation zone) is depicted in Figure 7 above.

CONSTRUCTION

The general construction contractor, Whitaker Construction (Whitaker), started excavating cut slopes in August 2016. The geotechnical report recommended a staged, top-down construction with maximum exposed heights of 6 feet. However, the specifications allowed greater heights, if approved on the basis of stability testing. Following stability testing of 30-foot sections, Whitaker excavated vertical cut slopes with a maximum height of 15 feet, to avoid the need for top-down construction and provide the contractors with more flexibility during construction of the wall. Days after excavating the entire wall length (Figure 8), temperatures at the site exceeded 100 degrees. Whitaker monitored the slope face and did not observe evidence of instability.

Figure 8. Excavation of The Wall Alignment (Facing South). Note Non-Bituminous Sandstone in Foreground. Heavily-Bituminous Sandstone in Background.

The wall construction contractor, DrillTech, excavated dowel locations and performed verification tests in late September 2016. As noted above, design loads for dowels in heavily bituminous material corresponded to a bond stress of 900 psf. To evaluate creep behavior of the heavily bituminous material, the specifications required supplemental verification testing with dowels loaded to 300-percent of the design load for a 24-hour test period. Unfortunately, temperatures measurements were not taken with each test displacement measurement, however, daily temperatures during the testing period ranged from about 50 to 87 degrees (F). During one supplemental verification test, DrillTech mistakenly increased the load to almost 400-percent of the design load for a 2-hour period, however, field representatives did not observe significant movement of the dowel.

Verification test results indicated the heavily bituminous material is not prone to long-term creep under the design loads, and the factor of safety against pullout is likely at least 3. Following successful proof testing on approximately 5-percent of production dowels, DrillTech placed geocomposite strip drains between rock dowels, which were designed with a diamond configuration on the slope face. Drainage strips were connected to 2-inch PVC pipes at the base of the wall and outlet via weepholes at regular intervals along the wall face. Structural reinforcement consists of welded wire fabric and No. 5 rebar mesh placed prior to application of the temporary and permanent shotcrete layers, respectively.

In general, construction of the wall within the heavily bituminous material was successful and the bulk of construction was completed in November 2016. Whitaker added architectural treatment/facing in February 2017 and the wall has performed well through the first 6 months of service (Figure 9).

Figure 9. Constructed Rock Dowel Wall with Architectural Treatment (facing north)

SUMMARY AND CONCLUSIONS

The design of a retaining wall in relatively complex geologic material required a customized design approach, which included both temperature- and shear stress-related creep considerations. The design evaluated global slope stability and kinematic block loading to assess rock dowel lengths required to maintain shear stress levels within the heavily bituminous sandstone at or below the pre-existing stress levels. The excavation and stand-up time of vertical cut slopes validated design assumptions about the heavily bituminous material's short-term strength, and verification test results indicate the heavily bituminous material's rock dowel bond strength, over a 24-hour period with fluctuating temperatures, is greater than the long-term strength assumed.

During design and construction, communication between the County, project designers, and contractors helped the project participants understand and appreciate the complexity and challenges associated with bituminous material. The County will continue to monitor the rock dowel wall and may eventually encounter similar material in future phases of the Price Canyon Road Widening project.

Design of projects in bituminous (oil sand) materials should recognize the potential for creep and if necessary, adapt an investigation plan and laboratory testing program to develop rigorous correlations between creep, temperature, and induced stresses. It is anticipated that heavy infrastructure work in or near oil fields would especially benefit from a better understanding of bituminous materials' multi-dimensional strength properties.

REFERENCES

- Hall (1973), Geology of the Arroyo Grande 15' Quadrangle, San Luis Obispo County, California, Scale: California Division of Mines and Geology Map Sheet 24, 1:48000.
- Antonellini et al. (1999), "Outcrop-Aided Characterization of a Faulted Hydrocarbon Reservoir: Arroyo Grande Oil Field, California, USA". Department of Geological and Environmental Sciences, Stanford University, Stanford, California.

REPURPOSING AN AGGREGATE QUARRY FOR RAW WATER STORAGE IN METRO-ATLANTA & THE ORIGIN OF BARITE AT THE EMERSON MINE, CARTERSVILLE MINING DISTRICT

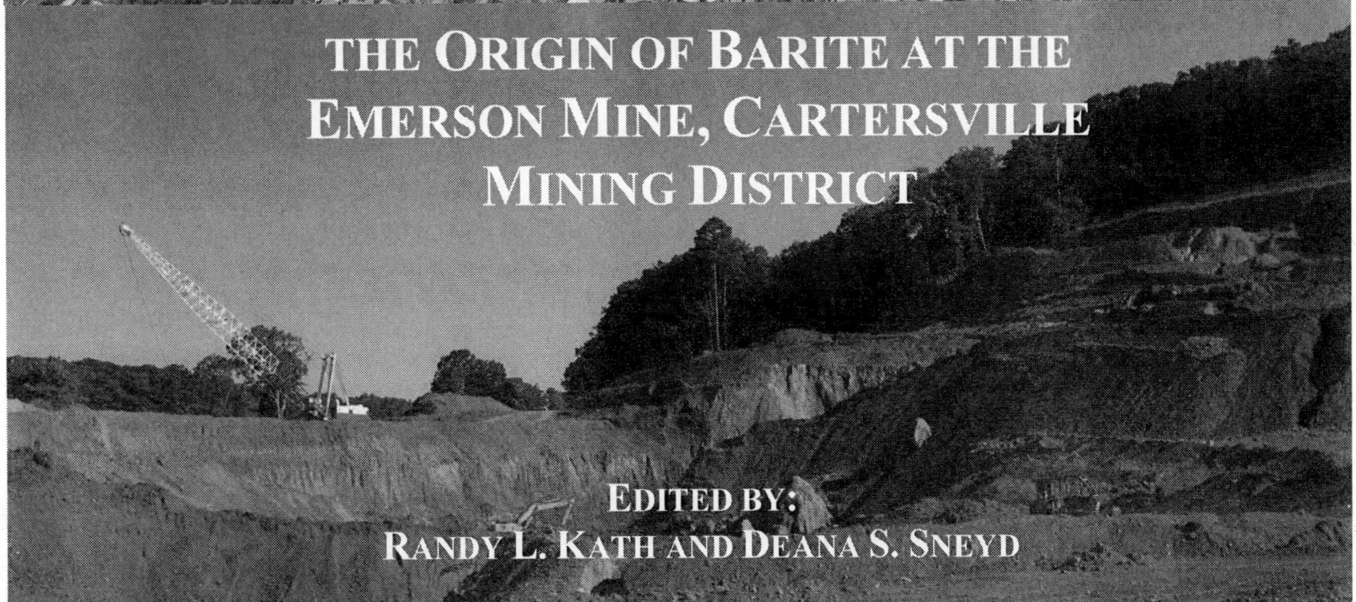

EDITED BY:
RANDY L. KATH AND DEANA S. SNEYD

68TH ANNUAL FIELD TRIP
OF THE HIGHWAY GEOLOGY SYMPOSIUM

OFFICERS OF THE HIGHWAY GEOLOGY SYMPOSIUM

2017

Chairman ----------------------- Ken Ashton
West Virginia Geological Survey
P.O. Box 879
Morgantown, WV 26507

Vice Chairman----------------- Krystle Pelham
New Hampshire DOT
P.O. Box 483
5 Hazen Drive
Concord, NH 03302

Secretary ----------------------- Bill Webster
CalTrans
5900 Folsom Blvd.
Sacramento, CA 95819

Treasurer ---------------------- John Pilipchuk
NCDOT Geotechnical Engineering Unit
1589 Mail Service Center
Raleigh, NC 27699

2017 FIELD TRIP

Field Trip Organizers and Editors:
Randy Kath, University of West Georgia
Deana Sneyd, Petrologic Solutions, Inc.

Field Trip Leaders:
Randy Kath, University of West Georgia
Adam Bedell, Stantec
Wayne Warburton, Stantec
David Scarpato, Scarptec, Inc.
Stanley Bearden, New Riverside Ochre

Special Acknowledgements:
Brian Jones, City of Atlanta- Dept of Watershed Management
Joseph Stika, Buzzi Unicem USA
Rusty Simmons, U.S. Army Corp of Engineers
Stanley Bearden, New Riverside Ochre

REPURPOSING AN AGGREGATE QUARRY FOR RAW WATER STORAGE IN METRO-ATLANTA AND THE ORIGIN OF BARITE AT THE EMERSON MINE, CARTERSVILLE MINING DISTRICT

EDITED BY:
RANDY L. KATH AND DEANA S. SNEYD

68TH ANNUAL FIELD TRIP OF THE HIGHWAY GEOLOGY SYMPOSIUM
MARIETTA, GEORGIA, MAY 1-4, 2017

HIGHWAY GEOLOGY SYMPOSIUM GUIDEBOOKS
MAY 2017

FIELD TRIP OVERVIEW

Field Trip Logistics

Welcome to the 68th Highway Geology Symposium in Marietta, GA. Marietta is located within Cobb County, one of 20 counties that comprise Metropolitan Atlanta. The metro area covers 8,376 square miles, is comparable in size to the state of Massachusetts, and encompasses three of the four physiographic provinces within Georgia.

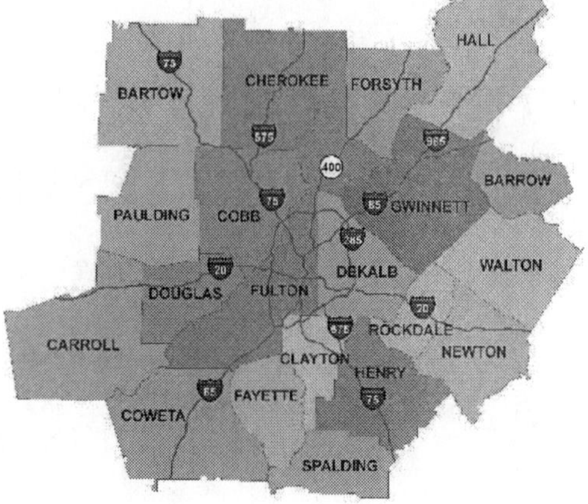

We will explore sites of geologic, historical and cultural interest in the Piedmont, Blue Ridge and Valley Ridge Physiographic Provinces, all occurring within the metro Atlanta area. The field trip will begin at the Hilton Atlanta Marietta Hotel and Conference Center, which is located in the Piedmont Physiographic Province, and travel south to the Bellwood Quarry water-supply project near downtown Atlanta. This project involves excavation of two vertical shafts and a five-mile long TBM tunnel that will convey raw water from the Chattahoochee River to the quarry which will be used as an emergency 30-day backup water supply for Atlanta. The Bellwood Quarry produced aggregate for road construction for over 100 years, ceasing production in 2006, and has more recently been used as the location for film and TV shoots, including *"The Hunger Games"* and *"The Walking Dead."* Construction activities related to the water-supply project required significant slope stabilization measures within the quarry which will be observed during the field trip along with an overview presentation of the project. From there we will visit an exposure of the Clairmont Mélange behind the architecturally historic Buzzi Unicem Cement Terminal, which will provide an example of the lithology and structure encountered along the reach of the tunnel. Lunch will be at the Sweetwater Brewery, named after Sweetwater Creek, a tributary to the Chattahoochee River. Lunch will be catered by Dickey's Barbeque, a variety of beer can be sampled, and a tour of the facility will be offered for those interested. We will travel north to the Blue Ridge Province following lunch, to view the site of Georgia's "Iron Empire" which was destroyed by Sherman in 1864 on his march to Atlanta. The National Historic Register Site, known as Cooper's Iron Works and the only remnant of the 1830's town of Etowah, is located here. The base of Allatoona Dam and powerhouse, constructed by the USACE in the 1940's, can also be observed at this stop. We will then cross the Cartersville-Great Smoky Fault and go into the Valley and Ridge where we will visit the New Riverside Ochre barite mine, which hosts up to 130-foot high cuts in soil. Figure 1 is a map of the route and showing stop locations.

Field Trip Itinerary

Wednesday May 3, 2017

6:30 – 7:15 am	Buses arrive at the Hilton Atl Marietta Hotel and Conf Center for boarding
7:15 am	Buses leave the Hilton Atlanta Marietta Hotel and Conference Center
8:00 – 10:45 am	Stop 1: Bellwood Quarry Project
11:15 – 11:45 am	Stop 2: Buzzi Unicem USA Cement Distribution Terminal
12:00 – 1:15 pm	Lunch: Sweetwater Brewery
2:00 – 3:00 pm	Stop 3: Cooper's Furnace Park and Allatoona Dam
3:15 – 4:30 pm	Stop 4: New Riverside Ochre's Emerson Barite Mine
4:30 – 5:00 pm	Return to Hilton Atlanta Marietta Hotel and Conference Center

Field trip sponsors include: ***Geobrugg*** for Lunch; ***Golder Associates Inc***, for field trip beverages.

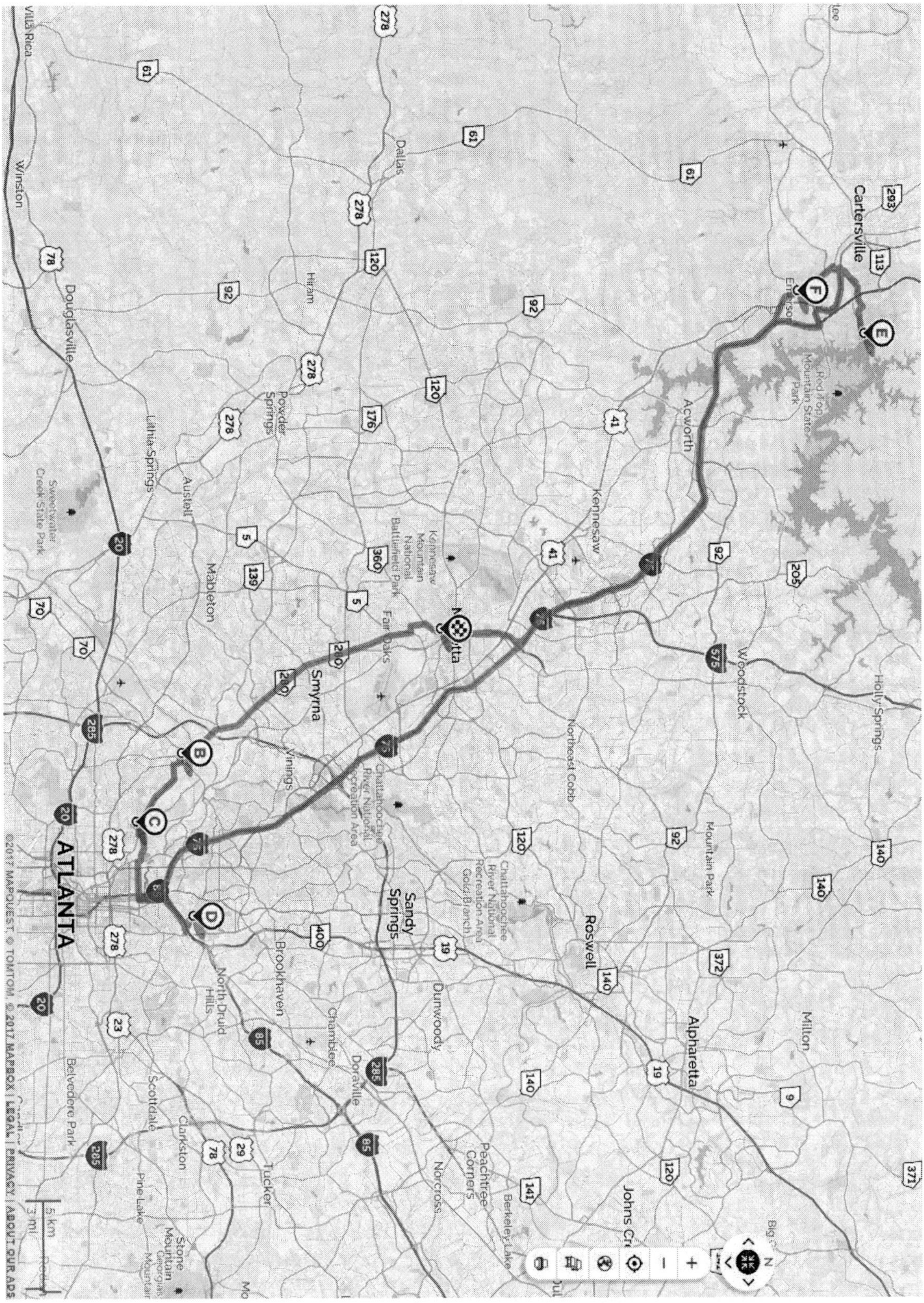

Figure 1. Field trip route map.

FIELD TRIP PHYSIOGRAPHY

The field trip route shown on Figure 1 will transect three physiographic provinces in the state. The conference headquarters and Stops 1 and 2 (Bellwood Quarry and Buzzi Unicem) are within the Piedmont Physiographic Province, Figure 2. The Piedmont Province is characterized by Neoproterozoic to Ordovician metasedimentary and metavolcanic rocks associated with the Iapetus and Rheic Ocean basins that were docked onto the Laurentian (North America) margin during the Late Paleozoic.

After lunch, the field trip route will proceed north on I-75 out of Atlanta toward Cartersville, Georgia. On I-75 near Emerson, Georgia, I-75 crosses the nearly east-west trending Emerson-Talladega Fault which is the boundary between the Piedmont and Valley and Ridge. The Valley and Ridge Province is characterized by Cambrian to Pennsylvanian sedimentary rocks that were deposited on the Laurentian platform. These sedimentary rocks were sourced from the Laurentian interior.

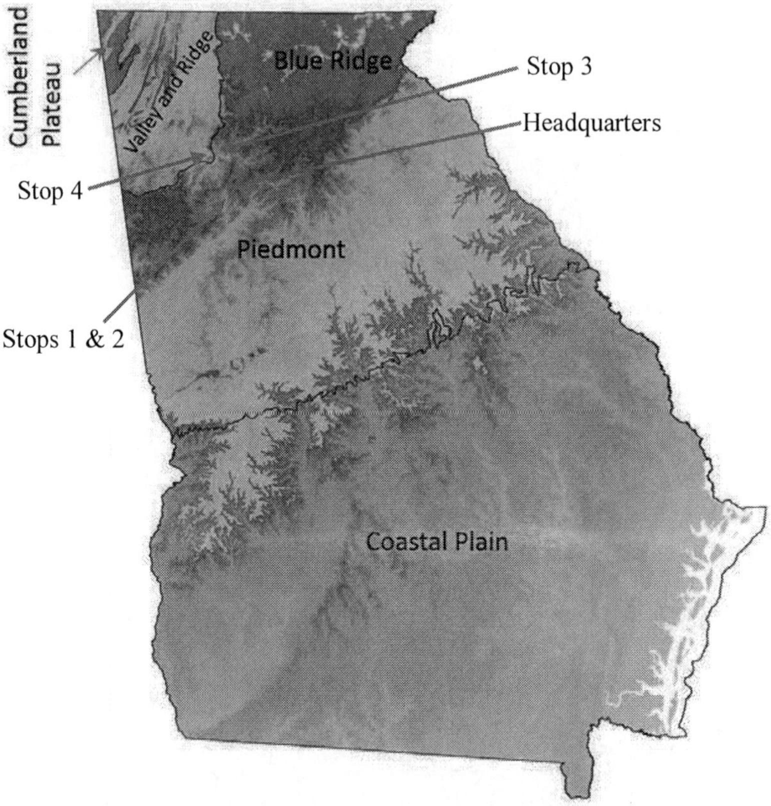

Figure 2. Generalized physiographic map of Georgia showing geologic provinces.

After entering the Valley and Ridge Province, the field trip route turns east toward Stop 3, Cooper's Furnace. Just west of Cooper's Furnace, Old River Road crosses the nearly north-south trending Cartersville-Great Smoky Fault which separates the Valley and Ridge from the Blue Ridge Province. The Blue Ridge Province contains metamorphic rocks that are associated with the Laurentian Margin. This province contains Neoproterozoic to Cambrian to Middle Ordovician (?) rifted margin clastic (with minor carbonates) sedimentary rocks (Ocoee Supergroup) that grade into drift-sequence sedimentary rocks (Chilhowee Group, Shady Dolomite, and Rome Formation). This rift- to drift-sequence is associated with the Neoproterozoic break up of Rodinia.

After leaving Cooper's Furnace, Old River Road again crosses the Cartersville-Great Smoky Fault and the field trip returns to the Valley and Ridge. At the Emerson Mine (Stop 4) we will be in the lower part of the Cambrian stratigraphy at the Chilhowee Group/Shady Dolomite contact. The lower part of the Shady Dolomite produces economic quantities of ochre, umber and barite.

General History of Metropolitan Atlanta: Hub of the South

The City of Atlanta's rich and diverse history began as a transportation hub in 1836, when the State of Georgia decided to connect the Midwest with the Southeastern United States by rail. Atlanta was selected as the line's terminus, with a stone pillar marking the Atlanta Zero Mile Post for the Western and Atlantic Railroad being placed near Forsyth St. in downtown Atlanta. For the next 20 years, rail lines converged in Atlanta from four different directions, promoting significant growth and development and confirming Atlanta's role as a rail hub for the entire South. Although the original name for the city was Terminus, it eventually was renamed Atlanta, which was shortened from the proposed name of Atlantica-Pacifica, and was incorporated in 1847.

Although Atlanta is known as a center of black wealth, political power and culture, being the cradle of the Civil Rights Movement, and home to Dr. Martin Luther King, Jr., slavery constituted the main reason for African American residency beginning in 1823. Development of cotton plantations around antebellum Atlanta in the early 1840's required intensive labor, the needs of which were met by slaves transported to the area from port cities such as Savannah and Charleston. As the railroads converged in Atlanta, population grew from 30 total residents in 1842 to over 2,500 residents in 1850, 25% of which were black slaves.

Simultaneous with the growth in agriculture, a booming mining industry began about 40 miles northwest of Atlanta in an area known as the Cartersville Mining District. Mining of brown iron ore began in the early 1840's, largely in support of railroad construction. **Field trip stop #3** shows the remains of Etowah Iron Works, the mining and manufacturing empire built by Mark Anthony Cooper around 1845. Cooper's efforts are considered to have influenced the alignment of the Western and Atlantic Railroad to pass within a few miles of Etowah. A mixture of slave and hired black labor produced up to 12 to 15 tons of iron per day, manufacturing nails, bolts, hollow ore, railroad iron, pots and pans. In 1861, Cooper gained a contract between Etowah Iron Works and the Confederacy, serving as a foundry for the manufacture of cannons and other Civil War munitions.

During the Civil War, Atlanta served as a critical railroad and military supply center. The most decisive military incursion into the Deep South occurred when William Tecumseh Sherman's army marched south from Chattanooga, Tennessee to Atlanta. During May 1864, the main portion of Sherman's army forded the Etowah River and burned Etowah Iron Works and the town of Cartersville. The Battle of Kennesaw Mountain, a critical battle that was fought north of Atlanta near the HGS headquarters in Marietta, ensued on June 27, 1864. Other notable battles included the Battle of Peachtree Creek, the Battle of Atlanta, and the Battle of Ezra Church. Sherman severed the last supply line to Atlanta at the Battle of Jonesboro, and on November 11, 1864, Sherman's troops burned Atlanta's assets and buildings to the ground, only allowing churches and hospitals to be preserved. It was during and after this time frame that Margaret Mitchell penned, "*Gone with the Wind*"; the iconic Southern plantation fiction novel set in Atlanta.

Following the Civil War and Reconstruction, Atlanta emerged from the ashes as it was gradually rebuilt, giving rise to the city's symbol: the phoenix. Growth and development related to post-war construction projects created new jobs, promoting a boom in employment and population, allowing Atlanta to become the industrial and commercial hub of the South. One of the more notable businesses that developed during this time frame was based on the creation of Coca Cola in 1886 by Atlanta pharmacist Dr. John S. Pemberton.

Despite infrastructure setbacks during the Civil War, iron mining in the Cartersville District continued to thrive, and in 1877, mining in the district continued to expand with the extraction of ochre. Ochre was used as a natural iron pigment, and is also used to color other construction materials such as concrete, masonry cement, bricks, and ceramic tile, to name a few. Barite mining in the District, which will be observed at **field trip Stop #4**, began in 1894. Production of barite increased sharply in 1916 and has been the principal mining industry in the district since that time.

Figure 3. Historic photograph of Bellwood Quarry. (Courtesy of Georgia State University Special Collections)

The Bellwood Quarry, **field trip Stop #1**, began producing aggregate in support of road pavement at the end of the 19th century. The quarry was run by the City of Atlanta and also served as a prison labor camp, Figure 3. During this time frame, prison labor camps in the South were viewed as an extension of slavery and exemplified the racial tensions that permeated politics of the south after the Civil War. Prisoners at the Bellwood Convict Labor Camp were chained together, hence the term "chain gang", and were subjected to whipping, malnutrition, and inadequate housing. Following removal of the prison labor system in the early 1950's, the Bellwood Quarry has been privately owned until 2006 when Atlanta purchased the quarry from Vulcan Materials for public use.

Transport by horse-drawn (1871) and then electric streetcars (1888) stimulated real estate development and continued growth of the City. The first north-south paved highway in the United States, the Dixie Highway, was constructed in 1915, passing through downtown Atlanta as it connected Canada to Florida, Figure 4. The Highway Department of Georgia, the predecessor to the Georgia Department of Transportation (GDOT), was created in 1916 to oversee construction contracts and coordinate with the Federal Government. As funding sources were developed, the State Highway Board (SHB) was developed to begin modernizing the transportation systems in Georgia, constructing four state highways running a total of 800 miles with 28 bridges.

Figure 4. Outline map of the Dixie Highway. (Image from Wordpress)

Transportation growth in the region continued with Delta Airlines moving its headquarters to Atlanta in 1941, and construction of Georgia's interstate system throughout the 1950's and 1960's. Major interstate highways that support the Metropolitan Atlanta area consist of: I-85, the northeast-southwest interstate that connects with I-75 through downtown Atlanta; I-285, which serves as a 63-mile circumferential loop around Atlanta; and I-20, the east-west interstate that bisects I-285. The Metropolitan Atlanta Rapid Transit Authority (MARTA), is the principal public transport operator in Metropolitan Atlanta, forming in 1971 strictly as a bus system. Construction of the heavy rail system began in 1975, with the first rail service being offered in 1979. MARTA's rail system currently has 47.6 miles of route and 38 rail station located on four service lines.

Continued development led to significant population growth, which has progressively increased stress on metro area infrastructure systems over the past 120 years. As a part of the municipal sanitation programs beginning around 1875, a system of sewage collection pipes were installed throughout Atlanta. The sewage collection system was designed to also collect surface water runoff. During dry weather and small storms, flow within the system was handled by the publically owned treatment works (POTW); however, during large storms, some of the combined storm water and sewage were discharged in the Chattahoochee River. Significant water pollution problems increasingly occurred during combined sewer overflow (CSO) events, which led to significant environmental and ecological impacts to the Chattahoochee downstream of Atlanta. Due to aging infrastructure and increasingly diminished capacity of pipelines, and motivated by development of the 1989 Environmental Protection Agency (EPA) CSO regulations, Atlanta embarked upon an aggressive 10-year capital improvement program in 1993, investing over 1-billion dollars on wastewater and sewer improvements. Several large-diameter tunnels have been excavated over the past 10 years to capture and store combined sewer overflows. This stored water is now conveyed to separate treatment systems designed to handle overflows.

At the same time that the municipal sanitation programs began, more reliable sources of water supply were needed. In the 1890's, water quality in Atlanta's first reservoir became increasingly polluted from contaminated source streams. The City constructed a new reservoir on Hemphill Avenue, where water diverted from the Chattahoochee River was treated at a new pumping station. The late Victorian architectural style of the Hemphill waterworks plant, as applied to an industrial complex, has resulted in the listing of this structure on the National Register of Historic Places. As Atlanta's water supply struggled to keep pace with industrial growth, additional reservoirs were constructed by the United States Army Corps of Engineers to provide hydroelectricity, navigation, flood control and water supply for Atlanta and surrounding areas. Allatoona dam and reservoir, **field trip Stop #3**, was completed on the Etowah River in 1949 and currently supplies most of the drinking water for three northern metro Atlanta counties.

The Buford Dam and Lake Lanier Reservoir project, completed in 1959, was framed by politicians to serve as a navigation, hydropower and flood control project, with water supply being incidental to the primary intended use. The dam was constructed across the Chattahoochee River approximately 45 miles northeast of Atlanta. The City has increasingly used water from this reservoir for water supply as population growth has skyrocketed in the metro region. The neighboring states of Alabama and Florida have complained since 1990 of impact to industry in their states due to diminished flow in the Chattahoochee downstream of Buford Dam. These complaints came to a head in 2007 when the region suffered a severe drought, further impacting downstream users, culminating in a tristate water war. In July 2009, a U.S. District judge ruled that Lake Lanier was never authorized for use as water supply for metro Atlanta and gave three years to stop withdraw from the lake except for two adjacent cities. Although this ruling was overturned in 2011, Florida has filed action against Georgia in the Supreme Court of the United States, which is currently pending. Atlanta's water needs continue to grow; fortunately, the Bellwood Quarry project will boost the city's backup supply from 3 to 30 days.

Along with the challenges of water supply and treatment, population growth and aging infrastructure have led to horrific traffic problems that regularly make national news. In 2009, metro Atlanta experienced a 500-year flood that washed out several bridges over the Chattahoochee River, effectively cutting off residents west of Atlanta from the city. More recently, in 2014 *snowpocalypse* incapacitated Atlanta, stranding 1000's of people on the highways, many without food and water, abandoning their cars and walking home, giving rise to the understanding of why the zombies take Atlanta first. And just last month, the failure of a major interstate bridge along I 85 severed a major artery through Atlanta, impacting both local and regional traffic flow. For those citizens who have no alternative but to navigate the congestion of Atlanta, somedays, particularly after yet another all-lanes blocked accident, it leaves many residents whistling to the John Prine tune,

"Blow up your TV, throw away your paper
Go to the country, build you a home
Plant a little garden, eat a lot of peaches
Try an find Jesus, on your own"

References used to develop this summary:

Wikipedia; Wordpress; Aggregate Research; EVHS Online; Artery.org; CleanwaterAtlanta.org

Borden, Skye, 2015, Thirsty City Politics, Greed, and the Making of Atlanta's Water Crisis: SUNY Press, 202 pp.

FIELD TRIP SPONSORS

City of Atlanta
Department of Watershed Management

Stantec

UNIVERSITY of West Georgia

Golder Associates

Buzzi Unicem

GEOBRUGG®
BRUGG

New Riverside Ochre
Since 1905

US Army Corps of Engineers®

GEOLOGIC CHARACTERIZATION ALONG THE CITY OF ATLANTA RAW WATER TUNNEL, FULTON COUNTY, GEORGIA

RANDY L. KATH
Department of Geology
University of West Georgia
Carrollton, Georgia 30118-3100

INTRODUCTION

Project Understanding-

The City of Atlanta Department of Watershed Management has initiated the Raw Water Delivery Project that will provide a reliable and secure raw water conveyance system to move water from the Chattahoochee River to the City's water treatment plants. The scope of this project includes constructing a new raw water conveyance system that will connect the Chattahoochee River to the Chattahoochee Water Treatment Plant, the Hemphill Water Treatment Plant, the Hemphill Complex Reservoirs, and the Bellwood Quarry.

The tunnel, referred to as the Raw Water Tunnel (RWT), will be about 5 miles long and have a finished diameter of 10-feet, excavated at depths ranging from about 200 feet to 400 feet below ground surface. The alignment of the RWT, shown on Figure 1, is located near the Chattahoochee River, Brevard Zone, and the Katy Creek Fault. The tunnel will cross numerous surface drainages and several faults/fault zones, which may present geotechnical challenges related to design and construction of the tunnel.

The City of Atlanta retained Stantec as their design engineer, who in turn has employed United Consulting (United) to support them on geotechnical aspects of the project. Petrologic Solutions Inc. (Petrologic) was retained by United to characterize the geologic and hydrogeologic conditions anticipated to be encountered along the tunnel alignment, as well as within the Bellwood Quarry, which will store raw water.

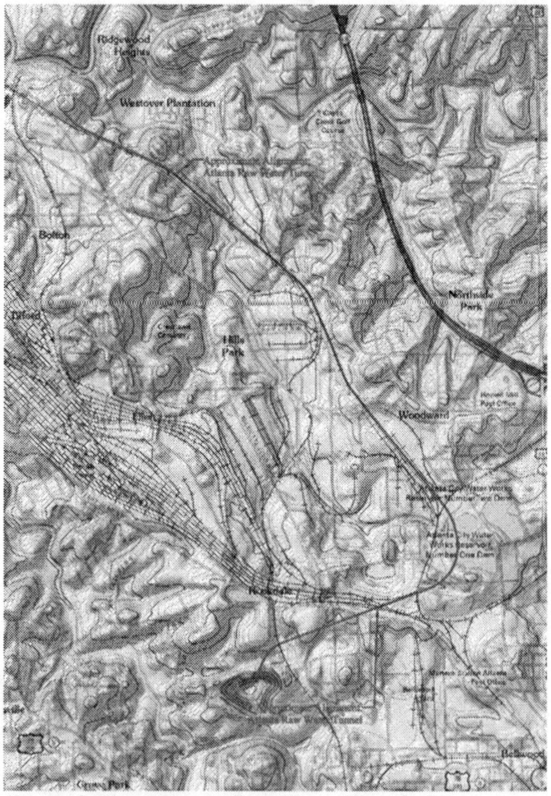

Figure 1. Approximate Raw Water Tunnel alignment.

Technical Approach-

For the current phase of work, Petrologic performed a desk study, which includes a literature review and topographic lineament analysis, detailed geologic mapping, and review of existing project-related documents. The objective of this work effort is to identify geologic and topographic features which can be used as an indication of geotechnical and hydrogeological conditions that might be encountered during construction of the tunnel, and that may impact use of the Bellwood Quarry for raw water storage.

Regional geologic conditions, based on review of published literature, are characterized in this report to provide context for the detailed geologic mapping conducted along the tunnel corridor. Information collected during the detailed mapping, which is the primary focus of this paper, includes: distribution and characterization of lithologic units and geologic structures; depth of physical and chemical weathering; and other geologic features considered to control ground conditions and groundwater movement. Topographic lineaments within a ~3-mile radius of the RWT were identified and characterized with respect to orientation and length as a part of the topographic lineament analysis. Results of this analysis are summarized and compared to structural discontinuities measured during field mapping and presented herein. A discussion related to the potential hydrogeologic impact these features may have on construction of the tunnel is also included at the end of this paper.

Information that we relied on for the literature review and detailed geologic mapping includes the following:

<u>Publicly Available Information</u>

- Topographic map of the Northwest Atlanta, 7.5-minute Quadrangle, Fulton County, Georgia (USGS, 2011).
- Published and unpublished geologic and hydrogeologic information available for the Site, specifically including:
  - unpublished mapping by Kath and Crawford
  - unpublished geologic map of the Northwest Atlanta quadrangle by Robert Dooley, Georgia Geological Survey
  - Detailed Geologic Mapping along the proposed Chattahoochee Interceptor Tunnel, Cobb County, Georgia: by Petrologic Solutions, Inc. (1998).
  - Geologic Map of the Atlanta, Georgia 30' by 60' Quadrangle (Higgins and others, 2003)
  - Geologic Map of the Brevard Fault Zone near Atlanta, Georgia Higgins 1976)
  - Geologic Map of the Brevard Fault Zone from Abanda, Alabama, to northeast of Atlanta, Georgia, by Kath and Crawford (2015)

<u>Information Provided by Stantec</u>

- Tunnel alignment map
- Preliminary Engineering Report, Bellwood Quarry Reservoir, by Atlanta Services Group (2009)

REGIONAL GEOLOGIC SETTING

Site Description and Physiography-

The approximate RWT alignment begins at the R. M. Clayton Wastewater Treatment Plant (WWTP) near the confluence of the Chattahoochee River and Peachtree Creek and generally trends southeastward toward the City of Atlanta Reservoirs (Figure 1). An unnamed tributary is shown to intersect the tunnel alignment along this section. The alignment trends southwestward from the Reservoirs, crossing beneath several rail lines within the Bellwood/Inman yard, and ends in the Bellwood Quarry.

The proposed RWT alignment is located in the Piedmont Physiographic Province, which is bounded to the southeast by the Fall Line and Coastal Plain Physiographic Province and to the northwest by the mountains of the Blue Ridge Physiographic Province. Because geologic characteristics are similar between the Piedmont and Blue Ridge, these Physiographic Provinces are considered to occur within the same geologic province.

Topography in this part of the Piedmont/Blue Ridge geologic province is characterized by gently rolling hills, deeply weathered bedrock, and a relative paucity of solid outcrop at ground surface. The rocks are deeply weathered due to the humid climate and bedrock is typically overlain by a variably thick blanket of residual soils and saprolite. Relief along the proposed RWT alignment is greater than 200 feet, with natural topographic lows of about 750 feet above mean sea level (ft. msl), where Peachtree Creek joins the Chattahoochee River, and 770 ft. msl where the unnamed tributary crosses the alignment. Topographic highs of greater than 970 ft. msl occur where the rail lines cross the alignment. The Bellwood Quarry excavated base grade is shown to occur at less than 650 ft. msl.

General Geology-

The Piedmont/Blue Ridge geologic province contains the oldest rocks in the Southeastern United States. Since their origin, some 276 to 1100 million years ago (Ma), these late Precambrian (Neoproterozoic) to late Paleozoic (Permian) rocks have undergone repeated cycles of igneous intrusions and extrusions, metamorphism, folding, faulting, shearing, and silicification. The latest regional metamorphism and associated deformation has been attributed to the collision of the North America plate with the Eurasian plate approximately 200 to 230 Ma. More recent deformation and emplacement of mafic dikes is associated with the rifting of the North American craton during the Mesozoic and Cenozoic Eras.

Much of the information provided in the following sections is taken from available literature (Higgins, 1968; McConnell and Abrams, 1984; Higgins and others, 1988; Crawford and Kath, 2001; Kath and Crawford, 2001; Higgins and others, 2003; Harden and others, 2013).

Regional Structure

The Brevard Zone (BZ) is a major regional zone of deformation in the Piedmont/Blue Ridge that extends from Alabama to Virginia. The BZ has been interpreted by many workers to represent various structural features, ranging from a nappe root zone, to a suture zone, to a terrain boundary. However, most agree that the BZ is a zone of intense shearing which reduced the grain size of the parent rocks forming a variety of tectonic rock types, including phyllonite, button schist, and mylonitic rocks. Generally, the BZ and associated shear fabric are subparallel to lithologic unit contacts. North of the proposed tunnel, Harden and others (2013) found a discordance of approximately 8 degrees between the Brevard shear fabric and lithologic contacts.

Lithologic contacts and major structural features in the BZ generally trend northeast-southwest. Structural features include folds, faults (thrust, oblique-slip, and strike-slip), foliation, shear foliation, joints, and other discontinuities. The alignment from the R.M Clayton Waste Water Treatment Plant (WWTP) to the Bellwood Quarry will cross various rock types and structures which are within and southeast of the BZ, Figure 2.

Typically up to four different joint sets formed in this area due to tectonic stresses imposed upon the bedrock. Dip joints form parallel to foliation/compositional layering dip direction and are typically perpendicular to fold axes, representing extension in the maximum principal stress direction or direction of compression. These joints are commonly near vertical. Strike joints develop parallel to the strike of foliation/compositional layering and fold axes, typically forming from tension along fold hinges. The dip direction and angle of these joints is orthogonal to the dip direction and angle of bedding. Oblique joints develop diagonal (± 30°) to the principal stress direction and represent conjugate sets formed from shear.

Regional Stratigraphy

Rock types within the BZ along the Chattahoochee River from the R.M. Clayton WWTP southeastward include: an unnamed Button Schist (POb), Mylonitic Granitoid (PSm), and Mylonitized Ben Hill (PCmb). Rocks within these various formations and groups have been intensely deformed, sheared, chemically altered, and are generally repeated because of movement along faults both within and outside of the BZ.

Rock units southeast of the BZ include: Clairmont Formation (OZcm) and Lithonia Gneiss (Dl). The general description of each unit is given below and the aerial distribution of each unit is shown in a generalized regional geologic map, Figure 2.

Button Schist- The button schist (POb) unit is characterized by gray to silvery, tan-weathering, feldspar-quartz-sericite button schist (Higgins, 1971). Typically, this schist contains a strong S-C mylonite texture. The S-C texture is responsible for the development of buttons. In many exposures, the schist is manganiferous and contains chlorite. Where shearing is more intense, the button schist becomes a phyllonite. This unit typically weathers to a red soil with abundant schist buttons.

Mylonitic Granitoid- The mylonitic granitoid (PSm) consists of light-gray to nearly white

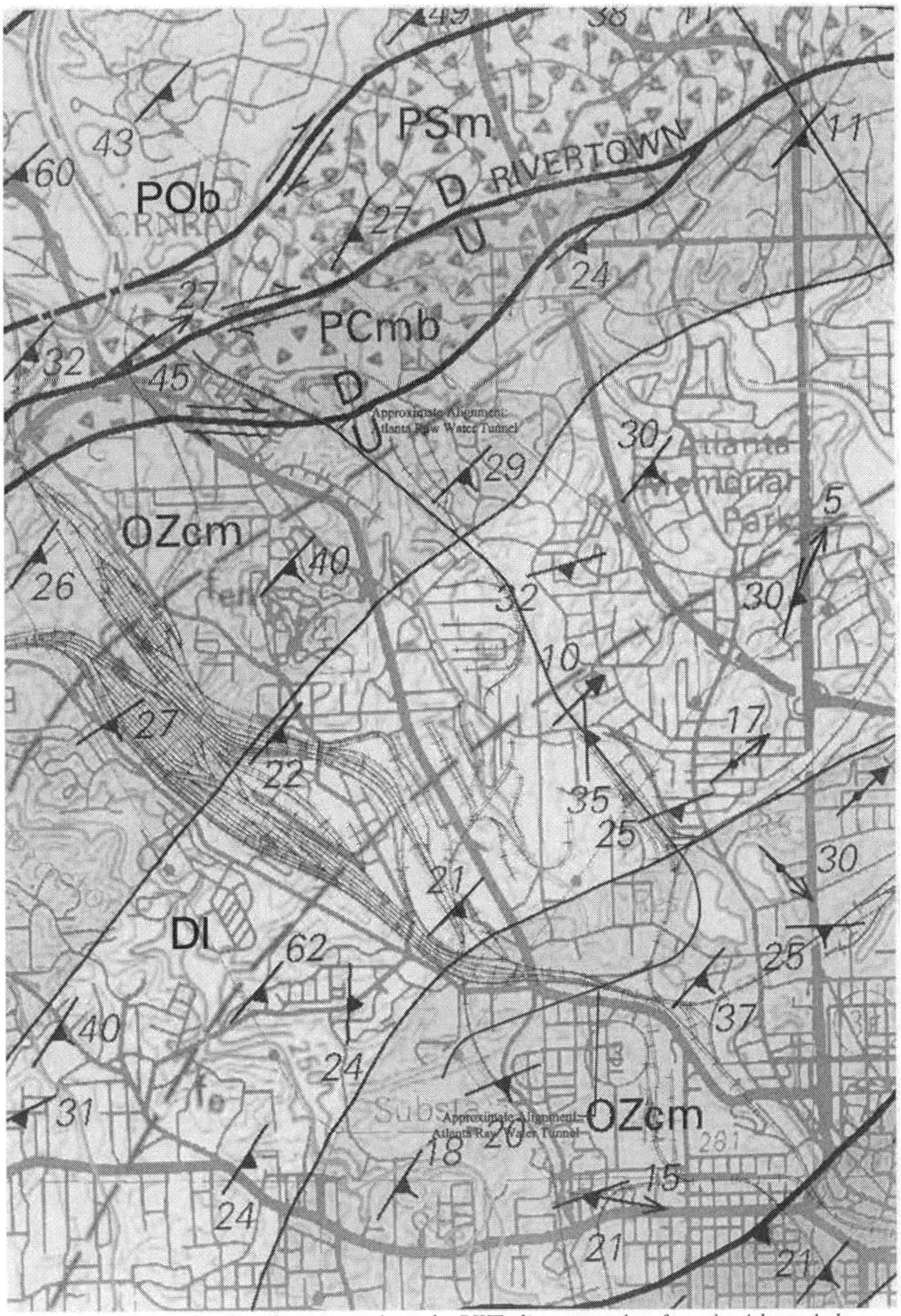

Figure 2. Generalized geologic map along the RWT alignment taken from the Atlanta 1-degree sheet by Higgins and others (2003). See text for description of individual units.

mylonite and/or mylonitic gneiss. The mylonite and mylonitic gneiss are thought to be derived from granite and/or granitic gneiss.

Mylonitized Ben Hill Granite- The Ben Hill granite is Permian-aged granite that intrudes the BZ southeast of the RWT. This unit consists of a generally fine-grained, light-gray to whit e mylonitic gneiss that contains a few scattered porphyroclasts of potassium feldspar. Based on the regional geologic mapping, the contact between mylonitized and unmylonitized Ben Hill has not been observed.

Clairmont Formation- The Clairmont Formation (OZcm) is a tectonic mélange which contains a wide variety of rock types. Generally clasts within the mélange are contained in a fine-grained biotite gneiss matrix. Higgins and others (2003) describe the Clairmont as a broken formation that contains a variety of exotic clasts. Clasts within the mélange include thinly layered amphibolite and hornblende gneiss, light-gray granofels, light- to medium-gray granitic gneiss, epidosite, meta-granite, quartzite, and ultramafic rocks.

Lithonia Gneiss- The Lithonia Gneiss is a complex of metagranites and granitic gneisses. The most common rock type in this complex is a light-gray to grayish-white, medium-grained, poorly foliated metagranite that is cut by numerous pegmatite and aplite dikes and sills of several generations. Higgins and others (2003) suggest that this rock type accounts for 40 to 50 percent of all rocks mapped as Lithonia Gneiss in the metro-Atlanta and metro-Griffin areas. The remainder of rocks mapped as Lithonia Gneiss are migmatitic gneiss that belong to the Mount Arabia Migmatite of Grant and others (1980). Grant and others (1980) describe this rock unit as a light-gray to whitish-gray, medium-grained, muscovite-biotite-microcline-oligoclase-quartz gneiss with a well-defined, contorted, generally 3-mm to 1-cm thick gneissic layering. Higgins and others (1988, 2003) suggest that the migmatitic gneiss is the dominant rock type of the Lithonia Gneiss near its margins, whereas the metagranite is more prevalent in the interior of the body.

Locally the Lithonia Gneiss contains small scattered xenoliths of amphibolite. Pavement-style outcrops are characteristic of both the migmatitic gneiss and metagranite. Where weathered, the Lithonia forms a light-whitish to yellow sandy soil.

Regional Hydrogeology

Groundwater in the Piedmont/Blue Ridge geologic province generally occurs in a series of discreet but locally interconnected aquifer systems. Groundwater is recharged through precipitation that is stored in high porosity, low permeability residual soils and saprolite that generally overlie bedrock. This groundwater storage recharges bedrock aquifer systems by moving through preferentially weathered discontinuities in the bedrock mass, such as foliation/compositional layering, joints, and faults. The occurrence and characteristics of discontinuities (size, orientation, dilation, infilling, spacing, and persistence) are dependent on the lithology of the rock and the type of stresses applied to them. These discontinuities are locally enlarged along individual planes as well as at the intersection of planes due to physical and chemical weathering, providing preferential pathways for enhanced groundwater flow.

Weathering generally increases the porosity and permeability of igneous and metamorphic rocks. However, some processes taking place in this zone, such as the growth of clay minerals, mineral deposition in fractures, and development of iron oxide "hardpan", can significantly decrease the permeability of the weathered zone.

At the interface between unweathered rock and weathered rock, there is commonly a "transition zone" where chemical weathering has changed the chemistry and created open spaces but not yet destroyed the rock's texture. This weathered rock, referred to as "saprolite", is generally more permeable than the overlying residuum, and the underlying fresh rock, and serves to concentrate ground water along a tabular zone of enhanced permeability. A thick (several 10's of feet) soil weathered zone above the saprolite will store ground water and allow it to move into the saprolite and fresh rock on a continuous basis; provided it can be recharged, and permeability has not been too severely impacted by the growth and concentration of clay minerals.

PROJECT GEOLOGIC SETTING AND STRUCTURAL ANALYSIS

General-

Detailed geologic mapping of the site was performed by Randy L. Kath, Ph.D., P.G. of Petrologic and Thomas J. Crawford, P.G., an independent geologist subcontracted to Petrologic, using the Northwest Atlanta Topographic Quadrangle map produced by the United States Geologic Survey (USGS) as a base. These maps are at 1:24,000 scale and are 7½-minute series topographic maps. The geologic map presented as Figure 3 is based on the detailed geologic mapping along and adjacent to the alignment corridor.

Figure 3. Detailed geologic map along the RWT alignment. See text for characteristics of each rock unit. OZbs- button schist, OZmw- Brevard Zone white mylonites, OZmb- Brevard Zone black mylonites, OZmbs- Brevard Zone black mylonites and button schist, OZzf- zoned feldspar gneiss, OZcm- Clairmont Mélange, OZcg- Clairmont Mélange gneiss.

Soil Development-

The parent rocks of soils in the region are comprised primarily of quartz, feldspars, muscovite mica, hornblende, biotite mica, and a wide variety of accessory minerals such as magnetite, garnet, epidote, and sphene. Because of the crystalline nature of the parent rock, chemical decomposition initially occurs along individual grain boundaries. The derived residual soil occupies the same general position as that previously occupied by grains in the original rock. As a result, partially weathered rock typically resembles the parent rock in appearance. However, strength and permeability characteristics are more similar to very dense silty sand or sand (SM or SP), depending on grain size of the parent rock. With further weathering, the individual crystals other than quartz and muscovite are altered and the mass typically becomes a micaceous silty sand (SM) or micaceous sandy silt (ML). In this stage, the original texture of the parent rock is still apparent, but the original crystalline structure is no longer preserved. Depending on the composition of the original rock, muscovite flakes, rather than quartz grains, may comprise the majority of the sand-size particles. The weathered rock resulting from this stage of weathering is termed saprolite, a soft, earthy, clay-rich, thoroughly decomposed rock formed in-place by chemical weathering of igneous and metamorphic rocks.

In the most advanced stages of chemical weathering, the material is changed into a red or reddish-brown silty clay (CL or CH) or clayey silt (ML or MH) with a sandy fraction directly related to the quartz content of the parent rock. In this weathered stage, the banding and crystalline structure of the parent rocks have been obliterated. This material is referred to as residuum or residual soils.

The rock types likely to be encountered along the tunnel alignment proposed in this area will generally consist of phyllonite, button schist, mylonitic granite (white mylonite), mylonitic biotite gneiss (black mylonite), biotite gneiss, granitic biotite gneiss, muscovite/biotite schist, amphibolite/hornblende gneiss, ultramafic rocks, and granite. When unweathered, these rocks are strong and of high quality for tunneling. However, the strength and quality of the rock and the mass permeability are strongly related to the degree of weathering. Thus, an evaluation of the degree of weathering at tunnel-level is important for assessing potential tunnel excavation methods, potential problems, and support requirements.

Because these rock types have different mineralogy, texture, and chemistry, they will weather differently. In general, the overall degree of weathering, from least weathered to most weathered, is: granite, black mylonite, button schist, phyllonite, muscovite/biotite schist, ultramafic, amphibolite/hornblende gneiss, biotite gneiss, and white mylonite. However, because of structural attitudes, zones of intensely-weathered rock may be present at depth, underlying units that are very resistant to weathering.

The overall depth of weathering in the Atlanta area is generally about 20 to 60 feet; however, the depth of weathering along discontinuities and/or very feldspathic rock units may extend to depths greater than 100 feet. Because of such variations in rock types and structure, the depth of weathering can vary significantly over short horizontal distances. Detailed knowledge of the geologic setting is essential for successful planning and implementation of a tunnel project.

Lithologic/Rock Units-

For purposes of description, major lithologic units were identified along the alignment and assigned a relative number. Rock Unit 1 is at the northern end of the alignment (R.M. Clayton intake shaft), and numbers are in sequence southeastward along the alignment, with Rock Unit 4 being at the southern end (Bellwood Quarry; see Figure 3). In the following descriptions, the mineral components of rock units is listed in order of increasing abundance and a general description of observed and anticipated weathering conditions is presented for each of the lithologic units.

Rock Unit 1- Phyllonite, Button Schist and Mylonitic Gneiss- The phyllonite, button schist and mylonitic biotite gneiss are interlayered on a scale of inches, feet, and 10's of feet. The phyllonite is composed of sericite, quartz, and feldspar, extremely fine-grained, with a well-developed, anastomosing, shear foliation. The button schist is composed primarily of fine sericite, muscovite, quartz, and feldspar; with medium- to coarse-grained muscovite forming distinctive "eyes" or "buttons." The development of the buttons indicate that there are two well-developed foliations that intersect at a low angle. The mylonitic biotite gneiss is composed primarily of biotite, quartz, and feldspar, very fine-grained; with a well-developed shear foliation.

Even though there are considerable differences in the mineralogy of these interlayered lithologies, the overall fine grain size seems to be the dominant control on weathering, and results in generally moderate and uniform depths of weathering.

Rock Unit 2A- Brevard Zone Black Mylonite- The BZ black mylonites (Figure 4) is generally composed of biotite, quartz, and feldspar. This unit is typically extremely fine-grained and weakly foliated. Where the foliation is better developed, the rock is shown to be very contorted. In most outcrops, the black mylonite is dark-gray to black and locally contains thin light colored layers of white mylonite (see rock unit 2B description). Weathering of this unit generally yields a reddish brown to red, uniform fine clayey residuum.

Figure 4. Brevard Zone black mylonite.

The uniform fine grain size, uniform composition, and poorly developed contorted foliation all inhibit weathering; however, the abundance of feldspar enhances weathering. Combined, these characteristics result in a generally shallow and uniform depth of weathering.

Rock Unit 2B- Brevard Zone White Mylonite- The BZ white mylonite is interpreted to be sheared granite, Figure 5. This mylonitized granite is composed of muscovite, quartz, and feldspar; much of the feldspar is pink and coarse-grained. Shearing was pervasive and produced a well-developed shear foliation. Reduction in grain size was not as extreme as in Rock Unit 2A. Weathering of this unit generally yields a white to tan, uniform fine clayey residuum.

Figure 5. Characteristic Brevard Zone white mylonite.

Where shear foliation is absent or poorly developed, this rock unit is massive, with few discontinuities, and shallow weathering. The development of a shear foliation in parts of the granite has provided discontinuities which weakened the rock and allowed more rapid weathering, resulting in tabular zones of deeper, more intense weathering.

Rock Unit 3- Zoned Feldspar Gneiss- The zoned feldspar gneiss consists of an epidote-muscovite-biotite-quartz-feldspar gneiss, fine- to medium-grained, with disseminated very coarse zoned feldspar crystals, Figure 6. Rock Unit 3 is massive and uniform, generally with moderate to deep weathering. Along shear zones and joints, weathering proceeds to greater depth.

Figure 6. Zoned feldspar gneiss.

Rock Unit 4- Clairmont Mélange-

Main Mass of the Clairmont (contorted unit): The majority of the contorted unit consists of a sphene-epidote-muscovite-biotite-quartz-feldspar gneiss, medium-grained, schistose in part; interlayered with sphene-epidote-muscovite-quartz-feldspar-biotite schist, medium- to coarse-grained; garnets may be present, but are small and scarce. Hornblende gneiss/amphibolite lenses and layers (commonly boudinaged) are common. Contains, in many places, lenses and discontinuous layers of unfoliated granite on a scale of 1ft to 20ft. Concordant and discordant quartz veins are common. Pegmatitic layers and coarse pegmatites up to 5ft thick are abundant and characteristic; shear foliation in the gneiss/schist wraps around the coarse pegmatites and small bodies of granite, which are generally not sheared.

This rock mass is extremely contorted; foliations are quite variable over short distances, and are generally low-angle and undulatory. Random fractures are abundant; through-going joint sets are scarce and not well-developed.

The following subunits occur as layers, lenses, and pods, within the main mass of the contorted unit, and, in places, as small mappable bodies.

Granite subunit- The granite subunit is a biotite-muscovite-quartz-feldspar granite (biotite and muscovite generally about equal), medium- to coarse-grained. This unit is typically unsheared and unfoliated. However, locally moderate to intense shearing has caused growth of muscovite along shear planes, producing a schistose texture. In places, this unit contains coarse feldspar crystals similar to the Ben Hill Granite. Locally, small inclusions (xenoliths?) of the main mass of the contorted unit are present within the granite.

Saprolite is very light-colored; soil is pale tan to light-gray, to medium-red. Joint sets are well-developed.

An "exotic blocks" within these granite bodies locally occur. These blocks contain biotite-quartz-feldspar gneiss interlayered on an inch-scale with chlorite-hornblende-quartz-feldspar gneiss, both very-fine- to fine-grained; both contain coarse hornblende crystals and disseminated tiny garnets.

Perry Boulevard subunit: The Perry Boulevard subunit is comprised of garnet (minor)-muscovite-biotite-quartz-feldspar gneiss, very fine- to fine-grained; garnets are tiny, pink, and scarce. This subunit has a uniform texture and is generally weakly foliated. Joint sets are very well-developed; blocky weathering is characteristic. Locally contains inclusions (xenoliths?) of the main mass of the contorted unit.

Amphibolite/Hornblende Gneiss subunit- The amphibolite/hornblende gneiss subunit occurs as pods, lenses, and layers of amphibolite and hornblende gneiss that occur sporadically throughout the main mass of the Contorted Unit and their abundance is characteristic of that Unit. Textures range from fine-grained, and thinly laminated to coarse-grained irregular masses. Generally, these are too small to be mapped.

Structure-

The structures within and adjacent to the BZ are complex and have been debated in the geologic literature for many years. Crawford and Kath (2001) provide a literature summary of the BZ. Two prominent features of the BZ are a granulation of the rocks and shear-induced foliation (shear foliation). Generally, the shear foliation trends northeasterly and dips at moderate angles to the southeast. This foliation nearly parallels geologic contacts on a local scale; however, on a regional scale, the foliation may be slightly oblique to geologic contacts (Harden and others, 2013).

Foliation

One of the most prominent features of the Brevard Zone is the presence of a well-developed shear foliation. The shear foliation is generally parallel to compositional layering or transposed compositional layering. Equal-area stereonet analyses of the foliation measurements for the entire Northwest Atlanta quadrangle have a maximum pole concentration representing a foliation of N46E, dipping 33 degrees to the southeast (Figure 7a). Within a 2-mile corridor of the tunnel alignment (2-mile on each side of the alignment), the maximum concentration of poles to foliation planes illustrated in Figure 7a, represents foliation planes trending N41E, inclined 17 degrees to the southeast.

The flattening of the averaged dip of the foliation in the vicinity of the tunnel versus the average dip for the entire quadrangle is due to the overall gentle dip observed and measured in the

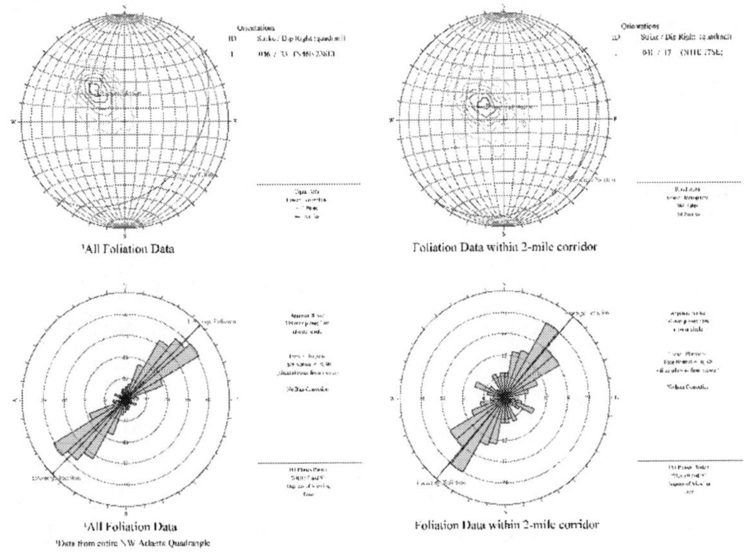

Figure 7a. Equal-area stereonet and rose diagrams of measured foliation from all mapping and measured foliation within a 2-mile corridor.

contorted unit, which underlies a major part of the tunnel alignment.

Because of the structural differences between the rocks that have been influenced by BZ shearing and the contorted unit, measured foliation within these two units were plotted separately (Figure 7b). Equal-area stereonet analyses of the foliation measurements for rocks that have been influenced by BZ shearing (Units 1, 2A, 2B, and 3), within the 2-mile corridor, have a maximum pole concentration representing a foliation of N41E, dipping 39 degrees to the southeast (Figure 7b). Analyses of the foliation measurements for the rocks within the contorted unit (Unit 4), within the 2-mile corridor, have a maximum pole concentration representing a foliation of N26E, dipping 13 degrees to the southeast (Figure 7b). These differences in foliation attitude must be taken into consideration when assessing tunnel stability.

Faults

Igneous and metamorphic rocks in the Piedmont/Blue Ridge have been extensively faulted. There are faults coincident with lithologic contacts, faults which cut across lithologic units, and faults within single mappable units. The major criteria for recognizing faulting in the southern Piedmont/Blue Ridge are: discontinuity of lithologic units; omission or repetition of lithologic units in a sequence; and the presence of shear textures, mylonite, or breccia.

Major faults within the BZ are interpreted to be sub-parallel or parallel to foliation (Figures 2 and 3). Between R.M. Clayton WWTP and the Bellwood Quarry three of these major faults are crossed by the tunnel alignment (refer to Figures 2 and 3). The northern-most major fault is approximately parallel to the northern bank of the Chattahoochee River near the confluence of Peachtree Creek. This fault separates the phyllonite and button schist lithologies (Unit 1) from the BZ mylonites (Units 2A and 2B). This unnamed fault is expected to be encountered by the proposed tunnel or within a drop shaft constructed near the confluence.

Within the BZ mylonites (Units 2A and 2B), the tunnel will cross a series of high-angle strike-

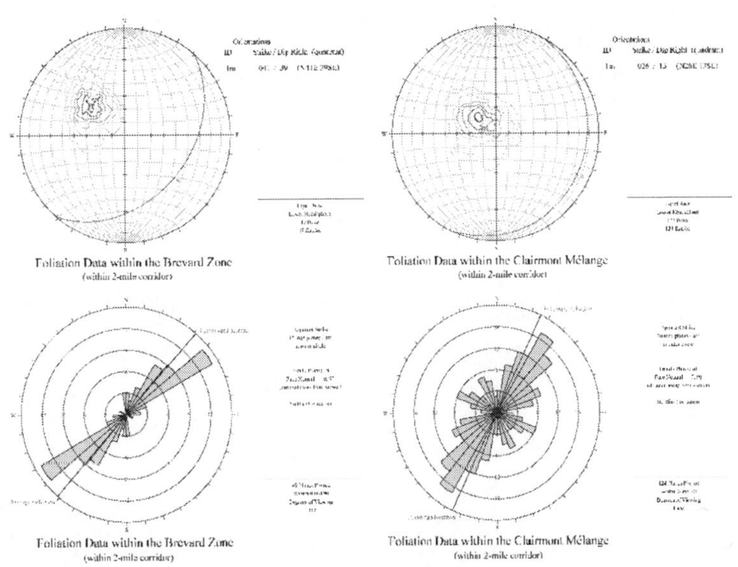

Figure 7b. Equal-area stereonets and rose diagrams of measured foliation from the Brevard Zone and Clairmont Mélange.

slip faults that generally separate the black and white mylonite units. These faults are part of the Rivertown Fault system as mapped by Higgins and others (1988 and 2006, shown on Figure 2). Traversing further southeastward, the tunnel will cross another strike-slip fault that separates the BZ mylonites from the Zoned Feldspar Gneiss (Unit 3). Based on the nearly straight outcrop trace of this fault, we interpret it to also represent a high-angle fault. The last mapped fault to be transected by the tunnel is a thrust fault (Katy Creek Fault) that separates the contorted unit, to the south, from the Zoned Feldspar Gneiss, to the north. The more sinuous outcrop trace of this fault suggests that it dips at moderate angles to the southeast.

Most of the faults in the southern Piedmont/Blue Ridge occurred at great depths, under high confining pressures and elevated temperatures. Consequently, brittle deformation was minimal and/or was healed during the tectonic processes, and resulted in little, if any, increase in porosity or permeability. Deformation within the BZ was dominated by high-pressure crushing and shearing; shear foliations and faults were produced as a result of these stresses. In addition, the crushing reduced the grain size of the rocks, which generally reduces permeability. Further, silica-rich metamorphic fluids associated with this crushing tended to heal fractures that were generated. Because of this healing, the permeability along the zones of intense shearing and silicification is expected to be very low and the rocks along these zones are expected to be strong and of high quality for tunneling.

Joints

Because the evaluation of joints is visual and judgmental, an effort is made for consistency in describing the relative frequency of occurrence using the following designations: Abundant (A); Common (C); and Scarce (S). These designations are relative to one another but are used consistently in descriptions made throughout the study area. An effort is made to record all of the different joint sets and, if an exposure is large, several same (or similar) joints may be recorded at the same Map Station. This deliberate method of visual evaluation in the field is more scientifically relevant and efficient than saturation-measurement of joints.

Joints within the BZ are common and persistent in most of the rock types. The joints are generally spaced on the order of a few inches to a few feet; however, there are more massive parts of various rock units which have a wider joint spacing.

Three major joint sets and one minor joint set were recorded during the detailed geologic mapping. Equal-area stereonet analysis of all joints measured in all lithologies is presented in Figure 8.

The three major joint sets are (quadrant and azimuth, right hand rule):

1) N19W 85NE (341/85),
2) N64E 85NW (244/85), and
3) N25E 83NW (205/83).

Locally, some of the joints contain clay infilling; however, most of the joints do not contain any infilling in surface exposures. The plane-surface morphology of each joint was noted in the field descriptions. Most of the joints are planar and smooth with little to no evidence of high fluid flow.

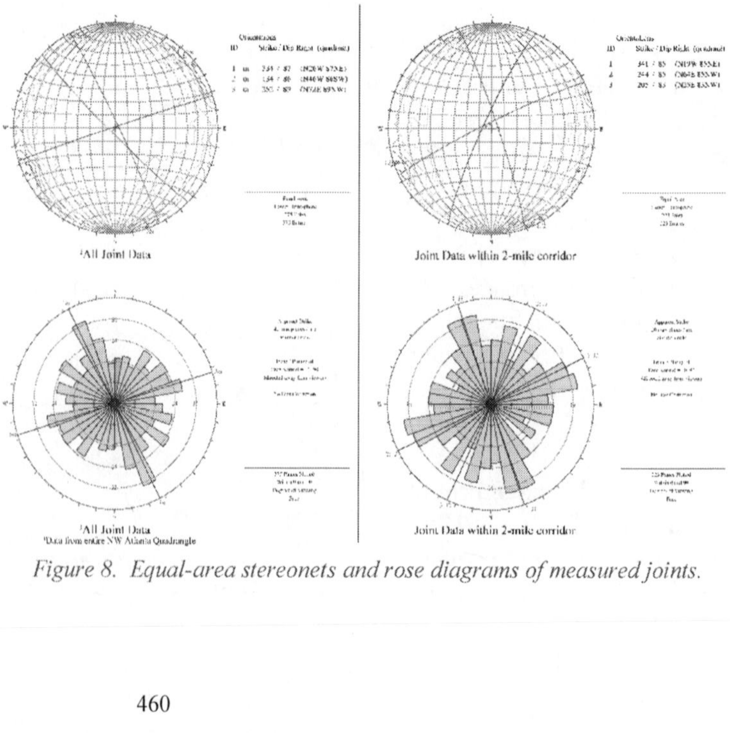

Figure 8. Equal-area stereonets and rose diagrams of measured joints.

LINEAMENT ANALYSIS

Subsurface geologic discontinuities such as lithologic contacts between resistant or non-resistant units, fracture zones, jointing, shear planes, and faults often have ground surface expressions that can be identified through analysis of photographic and topographic images. The discontinuities expressed as lineaments at ground surface commonly have enhanced porosity and permeability in the rock mass due to differential weathering. Groundwater in igneous and metamorphic rocks generally moves along discontinuities in the bedrock, enhancing the differential weathering processes.

Because discontinuity zones are typically less resistant to weathering, they are often expressed as natural topographic lows, such as straight stream valley segments, swales, aligned depressions and gaps in ridges or as linear tonal or vegetative alignments due to variations in soil thickness and moisture. These surface manifestations are referred to as fracture traces or lineaments and were identified for this project by remote-sensing techniques using topographic maps, aerial photographs, and shaded relief maps generated from 10-meter digital elevation model (DEM) data.

Discussion of Lineaments-

Lineament analyses were conducted on US Geological Survey (USGS) topographic maps, USGS Digital Elevation Models (DEM), and USGS low-altitude aerial photographs (verified with National High Altitude Photography Program (NHAP) high-altitude aerial photographs). Linear features or linear groups of features were identified and traced on digital overlays of the maps. Lineaments arise from a number of sources. Many lineaments observed on the small scale imagery or maps are related to fence, property, and section lines. However, many lineaments are related to local and regional geologic anomalies. Rectilinear segments of streams may be associated with local weakness in the underlying bedrock related to persistent joint sets. Faults tend to be long linear features that are often difficult to detect at ground surface, but generally form photographic and topographic lineaments.

Based on a total of 452 lineaments identified on the topographic maps, aerial photographs, and DEM, four major groups of lineament orientations were identified within the proposed tunnel corridor by the lineament analyses:

- L_1: N40 to 80E
- L_2: N30 to 60W
- L_3: N0 to 10E
- L_4: N0 to 20W

Structural weaknesses in rocks are reflected by the fractures formed, which subsequently can be weathered to form lineaments. These fractures are caused by application of directional stresses to the rock body. Generally, the stress is due to regional tectonics and/or unloading due to weathering and erosion. If one assumes a principal stress direction (σ_1), then the other two stress directions (σ_2 and σ_3) can be determined. Given σ_1, σ_2 and σ_3, one can predict a theoretical fracture or joint pattern in the rock body.

Along the proposed tunnel corridor, the principal stress direction has been estimated to be N40W and S40E based on regional tectonics, local structures, and geologic maps. Generally, the principal stress direction is perpendicular to regional foliation. Based on detailed mapping along the tunnel alignment, regional foliation strikes approximately N40 to 50E and dips at moderate angles to the southeast. However, variations in foliation were observed associated with open-style and ptygmatic folding.

Calculated joint (fracture) patterns that would be expected:

- J_1: ~N50E
- J_2: ~N40W
- J_3: ~N10W
- J_4: ~N70W

Comprehensive description of methodology and results of the lineament analysis have been presented in Petrologic's Report on "Lineament Analysis along the proposed City of Atlanta Raw Water Tunnel, Fulton County, Georgia" (2014).

Discontinuity Mapping and Lineament Analysis Correlation-

Lineaments identified are considered to be the ground-surface expression of preferential weathering related to discontinuities in rock. Based on this evaluation, the project area appears to be characterized by several persistent lineament sets whose orientations are consistent with the structural stresses experienced in this area (i.e., L_1 is related in orientation to J_1; L_2 is related in orientation to J_2; L_4 is related in orientation to J_3).

The orientation of these discontinuities forms a classic joint pattern that develops in rock formations in the Appalachians due to compressional stress. Because lineament orientations correlate with known regional tectonic fabrics, it is likely that most are true manifestations of subsurface fracture zones or low-resistance stratigraphic layers within the rock formations underlying the tunnel alignment.

HYDROGEOLOGIC CONDITIONS

General-

Concepts of groundwater movement in igneous and metamorphic rocks in areas with a subtropical climate, such as that of the southeastern United States, have evolved over many decades. Because of the dearth of research directed toward an understanding of the variables involved, much of the data set concerning the hydrogeology of igneous and metamorphic rocks is empirical data generated by groundwater exploration and development (Crawford and Kath, 2003).

Some of the concepts derived from these empirical observations and from limited applied research, have been presented and discussed in various papers dealing with the hydrogeology of igneous and metamorphic rocks in the Georgia Piedmont/Blue Ridge. Many of these concepts have a direct bearing on the geographic area involved in this project. However, 35 years of practice in the exploration for, and the development and management of, groundwater in the southeastern Piedmont/Blue Ridge has provided the major basis for interpretations of data and formulation of concepts presented here. Recent U.S. Geological Survey projects in the Lawrenceville area, Gwinnett County, Georgia, have provided insights into the hydrogeology of igneous and metamorphic rocks which have been incorporated into interpretations and concepts. The Groundwater Research Station at the University of West Georgia, Carrollton, has provided information useful to the objectives of this study.

Igneous and metamorphic rocks have, in many places, very diverse properties that change over short distances both vertically and horizontally. Because of this, groundwater movement is often most influenced by the relative properties of various rock units or discontinuities rather than by absolute properties of a particular rock unit or discontinuity. This relationship greatly complicates attempts to understand the hydrogeology of igneous and metamorphic rocks, and emphasizes the need for a strong database where it is necessary to make predictions concerning groundwater in these rocks.

Controls of Groundwater Movement in Igneous and Metamorphic Rocks-

Rock Type

As in any study of shallow subsurface earth processes, the study of groundwater in igneous and metamorphic rocks requires knowledge of the rock types involved. Metamorphic rocks and intrusive igneous rocks have very little primary porosity/permeability. Secondary porosity and permeability develop as these rocks are subjected to tectonic stresses and weathering stresses.

Because different rock types will react differently to the same stresses, it is important in any study area to determine:

1. the aerial distribution of each rock type;
2. projections of these into the shallow subsurface;
3. the major minerals and general compositional percentages;
4. grain size distribution; and
5. textures.

Each of these has a direct bearing on the rock's reaction to tectonic stress, physical weathering stress, and chemical weathering stress.

Discontinuities

A "discontinuity" as the term is used here refers to any feature that interrupts the homogeneity of the rock. In igneous and metamorphic rocks, the most common discontinuities are: compositional layering, foliation, joints, faults, and irregular random fractures. Of these, only compositional layering and foliation can be primary features; they may also be secondary, as are all the others. Regardless of their origin, once formed, all of these discontinuities have the potential to enhance the porosity and permeability of the rock and provide pathways for groundwater movement.

Each discontinuity is a plane of “different” strength/weakness in relation to its bounding interfaces. As such, it will react differently to stress, whether tectonic or non-tectonic. Weathering, whether chemical or physical, will proceed along the discontinuities at a different rate than outside the discontinuities.

A determination of the presence of discontinuities, and an understanding of their nature, size, abundance, structural attitude, and degree to which they are interconnected are critical to a study of groundwater in any area of igneous and metamorphic rocks.

Topography

Topography is a major factor in determining the percentage of precipitation runoff, and its direction and velocity. As such, it exerts a major influence on infiltration and consequent chemical and physical weathering. Just as basically and by the same token, topography exerts control on ground-water recharge.

Depth of Weathering

Weathering generally increases the porosity and permeability of igneous and metamorphic rocks. However, some processes taking place in this zone, such as the growth of clay minerals, mineral deposition in fractures, and development of iron oxide “hardpan,” can significantly decrease the permeability of the weathered zone.

At the interface between unweathered-rock and weathered-rock, there is often a “transition zone” where chemical weathering has changed the chemistry and created open spaces but not yet destroyed the rock’s texture. This weathered rock, referred to as “saprolite,” is generally more permeable than either the underlying fresh rock or the overlying residuum, and serves to concentrate groundwater along a tabular zone, but often irregular and undulatory, of enhanced permeability. A thick (several 10’s of feet) weathered zone above the saprolite will store groundwater and allow it to move into the saprolite and fresh rock on a continual basis, provided it can be recharged and permeability has not been too severely limited by growth and concentration of clay minerals.

Nature and Extent of Recharge Area-

The amount and rate of recharge at any given point (well) is a function of the nature and extent of the recharge area (assuming a precipitation constant). The nature of the recharge area is evaluated in the manner already discussed, by determining the rock types and discontinuities, and relating these to topography and depth of weathering.

Alluvial and colluvial material in the recharge area will have characteristics different from residual material and saprolite, and need to be evaluated differently. This is discussed, briefly, further along in this report.

Enhancing Permeability-

The tectonic stresses which created fractures in igneous and metamorphic rocks were a major factor in developing secondary porosity/permeability, which enhances ground-water movement. A second category of permeability enhancement is compositional layering. Where it is well developed, compositional layering has created zones of weakness, which react to both physical and chemical stress, enhancing ground-water movement.

A third category of porosity/permeability enhancement, which may be critical in creating the setting for potential high-yield wells and ground-water movement, is the non-tectonic process of unloading. With compositional layering and fracture networks already in place, unloading through erosional development of broad valleys could be the “stress release” which causes opening of the fracture system, further weakens the compositional layering planes, and allows groundwater to move more freely and at greater depths. As the process develops, it feeds on its own success.

The “stress release” envisioned here is not a release of “built-in” stress such as might be associated with the uncovering of igneous plutons of deep-seated origin. Rather, it is more comparable to a “bulge” where rock expands upward and laterally due to removal of overlying rocks in restricted geographic areas (broad valleys), causing opening of previously developed discontinuities such as compositional layering and tectonically induced fractures (joints).

Joints

Rocks may react to stress by breaking. Where these breaks are planar and no movement has occurred parallel to the fracture surface, they are called joints. Joints commonly occur as numerous parallel breaks referred to as a joint set.

Joints enhance the permeability of rocks a little or a lot, depending on: the nature of the joint surface; the spacing (density); the width of the openings; the degree to which they are through-going; and the degree to which they are interconnected.

Joints were described and measured throughout the mapped area, and used in evaluating potential for fluid movement in each lithologic unit. Different rock types react differently even when subjected to the same stresses. This has a direct influence on the ground-water storage and transfer capabilities of the various lithologic units.

Faults/Shear Zones

Igneous and metamorphic rocks in the Piedmont/Blue Ridge have been extensively faulted and sheared. There are faults/shears coincident with lithologic contracts, faults/shears within single mappable units, and faults/shears that cut across lithologic units.

The major criteria for recognizing faulting in these rocks are: discontinuity of lithologic units; omission of lithologic units in a sequence; and the presence of shear textures and mylonites. The size and shape of the area mapped for this project does not allow a valid application of this approach to determine whether any given lithologic contract is also a fault contact.

However, such a determination is of little, if any, value to the purpose of this study. There is no evidence of low-confining-pressure brittle fault deformation along lithologic contracts in the study area. Of most importance hydrogeologically are the differences in the mineralogy and texture of adjacent rock units.

Most of the faults in this region occurred at great depths, under high-confining pressures and elevated temperatures. Consequently, brittle deformation was minimal and/or was healed during the tectonic processes, and produced little, if any, increase in porosity/permeability.

Many geologic maps show some lithologic contacts as faults. This does not mean that deformation associated with faulting has enhanced permeability along that contact. The faulting may have, in fact, decreased permeability of the rock. For example, shearing and mylonitization reduce particle size, and are often accompanied by silicification; both tend to decreased permeability. The value of such a fault from a groundwater perspective would be in whether or not it juxtaposed lithologic units with great differences in lithology and/or texture.

Lithologic Contacts

Lithologic contacts can exert considerable influence on ground-water movement. The magnitude of the influence is directly related to the differences in the units which are juxtaposed. Where similar lithologic units are in contact, the contact zone has little influence. Many mappable units have greater internal differences in lithology and texture than the differences across contacts. In such cases, the contact zone would not enhance groundwater movement.

Groundwater occurrence and movement in igneous and metamorphic rocks is generally controlled by discontinuities. These discontinuities include contacts between rocks of differing mineral composition, small-scale compositional layering, and fractures (such as joints, faults, and shears). The permeability of these discontinuities is increased by a decrease in confining pressure (unloading) that has resulted from weathering and erosion of the overlying material and by in-situ weathering along the discontinuities.

Groundwater occurs in the soil, the weathered portions of the rock (saprolite), and along discontinuities in the underlying fresh rock. The water table surface in the saprolite generally mimics the shape of the overlying topography, but with more gentle slopes. Groundwater flow through the saprolite is controlled by the fabric of the saprolite. Flow in the underlying rock is controlled by the occurrence, distribution, and interconnection of discontinuities. The quantities of groundwater generally available in the Brevard Zone are usually small due to the granulation and silicification of the rocks, and the resultant sealing of discontinuities.

While the groundwater quantities are generally limited, concentrations of discontinuities in restricted zones cause extensive and deep weathering. The resulting increases in permeability allow increased groundwater flow.

The remote sensing lineament analysis and detailed geologic mapping techniques employed by Petrologic for this project and summarized in this report can help identify these high-permeability zones.

REFERENCES CITED

Crawford, T.J., and Kath, R.L., 2001, The Brevard Zone: A Literature Review: in Kath, R.L., and Crawford, T.J., eds., Across the Brevard Zone: The Geology of the Chattahoochee Interceptor Tunnel, Cobb County, Georgia: Georgia Geological Society Guidebooks, vol. 22, No. 1, pp. 1-18.

Crawford, T.J. and Kath, R.L., 2003, Groundwater Exploration and Development in Igneous and Metamorphic Rocks of the southern Piedmont/Blue Ridge, in: Williams, L.J, (ed), 2003, Methods Used to Assess the Occurrence and Availability of Ground Water in Fractured Crystalline Bedrock: An Excursion into areas of Lithonia Gneiss in Eastern Metropolitan Atlanta, Georgia: Georgia Department of Natural Resources, Environmental Protection Division, Georgia Geologic Survey Guidebook 23.

Harden, H. M., Kath, R.L., Crawford, T.J., 2013, Geologic Map illustrating the Tectonostratigraphy and Structural Geology of a part of the Brevard Zone in Georgia: Results from detailed geologic mapping in the Chattahoochee River National Recreation Area, Sandy Springs, 7.5-Minute Quadrangle: Geological Society of America, Abstracts with Programs, vol. 45, No. 7, p. 379.

Higgins, M.W., 1968, Geologic Map of the Brevard Fault Zone near Atlanta, Georgia, U.S. Geological Survey Miscellaneous Geologic Investigations Map I-511.

Higgins, M.S., Atkins, R.L., Crawford, T.J., Crawford, R.F.,III, Brooks, R., and Cook, R.B., 1988, The structure, stratigraphy, tectonostratigraphy, and evolution of the southernmost part of the Appalachian orogeny: U.S. Geological Survey Professional Paper 1475.

Higgins, M.W., Crawford, T.J., Atkins, R.L., and Crawford, R.F., III, 2003, Geologic Map of the Atlanta, Georgia 30' by 60' Quadrangle, U.S. Geological Survey Geologic Investigations Series Map I-2602.

Kath, R.L., and Crawford, T.J., 2001, Detailed Geologic Mapping along the Chattahoochee Tunnel, Cobb County, Georgia: in Kath, R.L., and Crawford, T.J., eds., Across the Brevard Zone: The Geology of the Chattahoochee Interceptor Tunnel, Cobb County, Georgia: Georgia Geological Society Guidebooks, vol. 22, No. 1, pp. 25-38.

Kath, R.L., 2014, Lineament Analysis along the proposed City of Atlanta Raw Water Tunnel, Fulton County, Georgia: Petrologic Solutions, Inc., pp. 42

McConnell, K.I., and Abrams, C.E., 1984, Geology of the Greater Atlanta Region, Georgia Geological Survey Bulletin 96, p. 121.

Terzaghi, R.D., 1965, Source of Error in Joint Surveys: Geotechnique, V15, No. 3, p. 287-304.

ATLANTA'S LATEST MEGA-TUNNEL

ADAM BEDELL, TAO JIANG, WAYNE WARBURTON, KONNOR HORTON, DON DEL NERO
Stantec Consulting
2310 Parklake Drive NE, Suite 400
Atlanta, Georgia 30345

BRIAN JONES
City of Atlanta
Department of Watershed Management
72 Marietta Street, 5th Floor
Atlanta, Georgia 30303

INTRODUCTION

The City of Atlanta is currently constructing an approximately 300-million dollar water supply system, known as the Water Supply Program (WSP), which includes the conversion of a century-old rock quarry into a 2.4 billion-gallon raw water storage facility, 4.5 miles of 400 ft.+ deep, 12 ft. diameter tunnel bore with ten shafts of various types. The tunnel is being mined from a portal at the base of the quarry and will connect the quarry to two water treatment plants and three pump stations. The project is being delivered using the Construction Manager at Risk (CMAR) model, which is an innovative contracting method that is fairly new to the tunnel industry. Multiple aspects of the project will be highlighted in this paper, including subsurface investigations, design elements, ground conditions, and tunnel lining.

PROJECT OVERVIEW

Resiliency is now in the lexicon of the global community and has become one of the primary goals of many urban centers. To that end, major urban centers like the City of Atlanta are building resiliency into their water, wastewater, and transportation infrastructure.

The City of Atlanta, like most municipalities, is estimated to have just a three-day backup supply of clean water and most of the world is experiencing some type of drought. Now more than ever, forward-thinking communities are seeking to build resiliency into their infrastructure. For Atlanta that meant the purchase of the Bellwood Quarry several years ago from Vulcan Materials Company with the intent to create a 2.4 billion-gallon raw water storage facility that would extend the City's back-up water supply to 30 days at full use, and around 90 days with emergency conservation measures. This is truly a "mega project" that involves getting water from the Chattahoochee River to the quarry, pumping it up multiple vertical shafts to two water treatment plants, and then pushing it into the City's water distribution system.

The Atlanta region is well known for its "firsts" in the world, and this project is no different. From the start of construction to project buildout, it will feature the:

- First blind bore shaft, over 400 feet deep in hard rock, in the Southeastern United States;
- Deepest tunnel in Georgia; and
- Largest quarry repurposed as a raw water storage facility in North America.

The current water supply program operated by the City's Department of Watershed Management (DWM) consists of four aged raw water pipelines, one of which dates to 1893. Based on previous assessments completed by the DWM, the entire water system is at, or will soon reach, its recommended useful life. As such, the City acquired the Bellwood Quarry in 2006 with the intention to create a raw water storage facility with a volume of approximately 2.4 billion

Figure 1. Bellwood Quarry prior to construction.

gallons to serve approximately 1.2 million people.

Using the quarry (Figure 1) as a water storage facility greatly enhances the reliability and security of the drinking water supply to the greater Atlanta metropolitan area. For many years it was used for mining granitic gneiss and crushed-stone aggregate production. The Quarry has nearly vertical sides and ground elevations around the rim ranging from approximately 850 feet to 970 feet above mean sea level based on the NAVD 88 vertical datum. Quarry floor elevations range from about 520 feet to 540 feet. The proposed full pool level for raw water storage is at an elevation of 840 feet.

The project will connect the quarry to the Hemphill Water Treatment Plant (HWTP), the Chattahoochee Water Treatment Plant (CWTP) and Chattahoochee River. Raw water will be supplied to the quarry storage facility from the Chattahoochee River. Stored raw water will be withdrawn from the quarry for treatment at the Hemphill and/or Chattahoochee water treatment plants, with treated water subsequently pumped to the City's treated water distribution system. This offline operating mode includes routine withdrawals and replenishments.

The project location is shown on Figure 2 which is generally in the Northwest part of downtown Atlanta, Georgia. The overall project has been divided into two phases. The Phase I project connects the Quarry and the HWTP, and the Phase II Extension project connects the HWTP to the CWTP and the Chattahoochee River.

The main features of the project include a TBM-excavated tunnel, seven pump station shafts, a drop shaft, a riser shaft, one combined drop and construction shaft, and a quarry highwall rockfall protection system to provide long-term protection of the tunnel inlet. Two of the seven pump station shafts will be constructed using conventional shaft excavation methods (including drill-and-blast in rock) while the remaining five will be excavated using blind bore methods. A 3-D rendering depicting the general

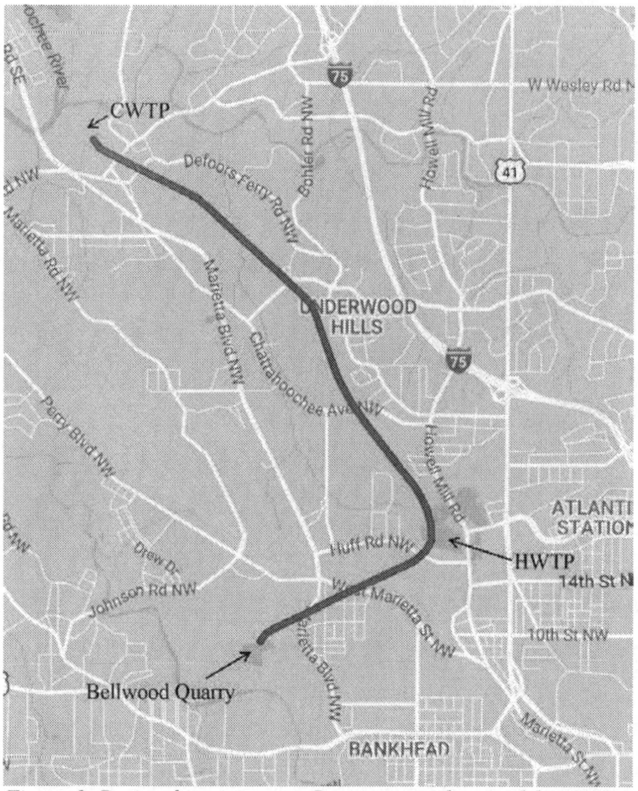

Figure 2. Project location map. Project is northwest of downtown Atlanta.

arrangement of shafts, tunnel and adits at the Quarry site is presented in Figure 3. These components, along with the other project components generally noted above include:

- A TBM tunnel that is approximately 24,000 feet long and partially concrete-lined with a finished diameter of 10 feet.
- A primary pump station shaft at the quarry that is approximately 250 feet deep with a finished diameter of 35 feet. The low level pump station shaft has a finished diameter of 20 feet and is approximately 340 feet deep. The primary and low level pump station shafts are connected to the tunnel and quarry via adits.
- A drop shaft at the quarry that is about 320 feet deep with a finished diameter of 25 feet above El. 805 feet and 4.5 feet below El. 805 feet. The drop shaft is connected to the Quarry low-level pump station shaft, the riser shaft, and the main tunnel through adits. The drop shaft provides a flow capacity of 90 million gallons per day.
- A riser shaft at the quarry that is about 320 feet deep with a finished diameter of 25 feet above El. 805 feet and 12 feet below El. 805 feet. The quarry riser shaft is connected to the quarry drop shaft and the main tunnel through adits. Five pump station shafts at the HWTP that are about 420 feet deep and 9.5 feet in diameter.
- Each of the five blind bored pump station shafts will have a 76-inch diameter grouted steel casing to house the pump, and are connected to the main tunnel by five, 8-foot diameter adits with lengths ranging from 20 feet to 30 feet.
- A construction/drop shaft at the CWTP site that is about 250 feet deep with a finished diameter of 30 feet.

The Construction Manager at Risk (CMAR) model was used as the contracting method, with the City selecting the joint venture PC Russell JV as the CMAR. Other important players include the Atkinson/Technique JV (ATJV) as the tunnel contractor and the joint venture design team of JP2. Stantec Consulting acted as the tunnel designer for JP2. At the time of this field trip, both pump station shafts at the quarry have been completed, along with the 636' adit and its breakout structure. The upper portions of the drop and riser shaft have been excavated and are being prepared for the start of raise bore operations and pilot holes for the five blind bore shafts are being drilled. The TBM tunnel is nearly 13% mined and the Peachtree construction/drop shaft is underway.

GEOLOGIC AND GEOMORPHIC CONDITIONS

The project is located in the Piedmont Physiographic Province. The geology of the Piedmont in the greater Atlanta area generally consists of medium-grade metamorphic rocks with granitic intrusions. These crystalline rocks are some of the oldest rocks in the Southeastern United States, ranging in age from some 275 million to over 1 billion years ago, with the youngest forming during the series of orogenic events that culminated in formation of the Appalachian Mountains. Since their origin, the rocks have undergone a complex history of

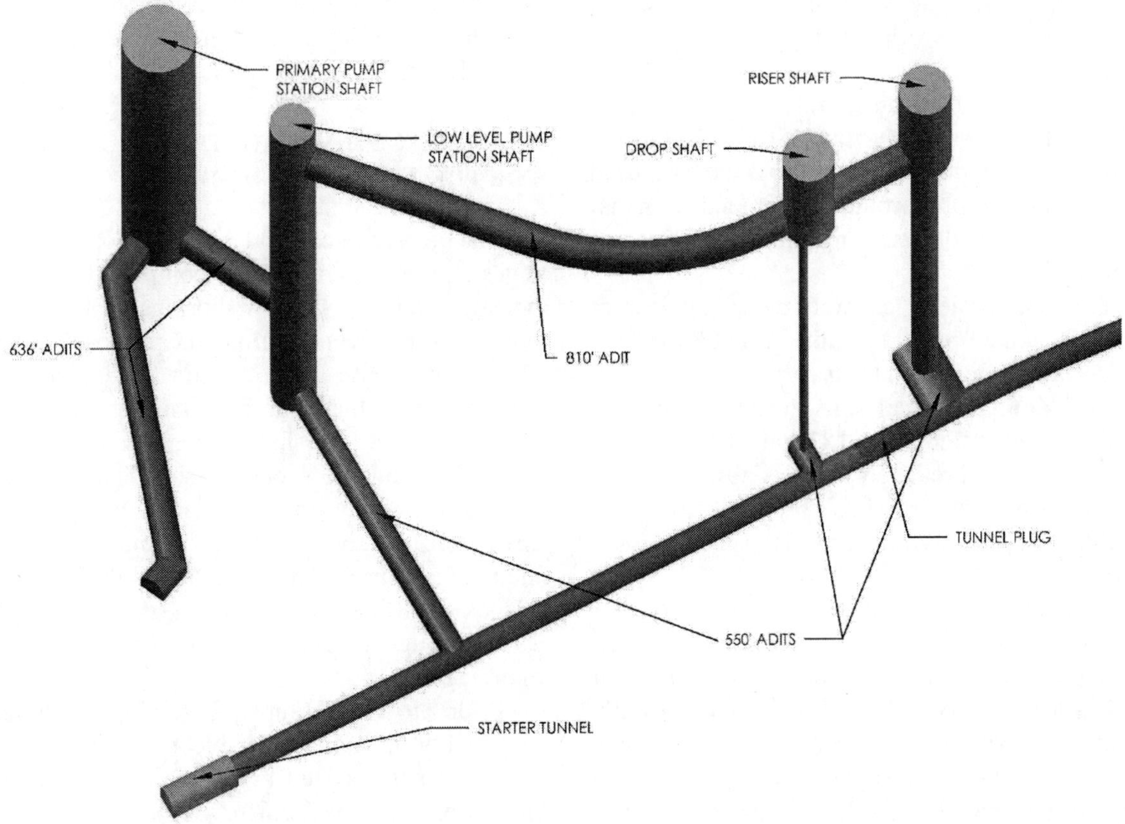

Figure 3. Arrangement of the structures at the quarry site.

metamorphism, weathering, and deformation. More specifically, the rocks in the greater Atlanta area have undergone episodes of both progressive and retrogressive metamorphism, with the peak regional metamorphism occurring in the Paleozoic Era, 360 to 380 million years ago.

As a result of this complex geologic history, structural features of the rocks include folds, fractures, and lineaments. The high pressures and temperatures at great depths resulted in a full range of deformational styles, ranging from medium-grade metamorphism, through fully-welded ductile shearing and mylonite formation, to brittle fracturing with rocks that commonly contain hydrothermally deposited minerals. At shallower depths, exfoliation fractures were formed in the rocks due to erosion of overburden and unloading. The exfoliation fractures occur mainly along the foliation "planes" of the rocks. The foliation "planes" tend to act as areas of weakness within the rock mass, and the exfoliation fractures tend to be open and act as conduits for water movements through the rock mass.

Lineaments, which are surface topographic expressions of underlying rock mass or crustal structure, occur throughout the Piedmont. The lineaments are often controlled by weathering associated with discontinuities in the bedrock. In many cases, the lineaments represent fracture zones in the underlying bedrock. At greater depth, the fracture zones are typically cemented with minerals. At shallower depths, erosion of these weathering minerals (primarily micas) often results in zones of broken, water-bearing rocks and topographic features such as valleys and draws.

A key characteristic of the Piedmont region is the mantle of residual soils, derived from weathering of the parent metamorphic rocks and localized granites in the area. These residual soils grade downward into the underlying unweathered bedrock. The humid climate promotes chemical weathering of the parent material. Degradation of the parent crystalline rock begins at the grain boundaries and progresses inward through the rock mass producing residual soil. The residual soil resembles the original rock in appearance, but its physical characteristics such as strength and permeability are more similar to a micaceous sandy silt (ML) or silty sand (SM). Within the Southeast United States, saprolite is the term used to describe a soft, thoroughly degraded rock that is clay rich, while retaining the original parent rock structure.

For this project, as well as a number of previous tunnel projects in the Atlanta area, the subsurface is divided into three zones:

Soil Zone. Residual soils in the project area are the result of continued chemical breakdown of saprolite. All relict structure is absent and the resulting soil mass is reddish-brown in color and is either a silty clay (CL or CH) or a clayey silt (ML or MH).

Transition Zone. The transition zone consists of partially weathered rock and highly fractured rock, underlying the overburden soils. The top of this zone occurs where rock and partially weathered rock begin to predominate over soils, and the bottom of this zone is defined where slightly weathered or fresh rock takes control of the rock mass.

Bedrock Zone. The bedrock zone lies below the transition zone. This zone is dominated by fresh rock and faintly weathered rock, with local occurrences of more weathered material typically along discontinuity planes.

Groundwater occurs in all three zones of the subsurface described above. The depth of the groundwater table varies significantly along the tunnel alignments, ranging from less than 10 feet to over 200 feet. The soil zone is generally considered to be a good producer of groundwater. The transition zone typically contains abundant open fractures and can become a major storage source for groundwater where its thickness is significant. The bedrock zone in the Piedmont generally has fewer open fractures with depth than the transition zone. However, large fractures with the ability of producing large volumes of water do exist in the bedrock. High-yield wells have been reported to produce sustained yields up to nearly 500 gallons per minute.

Potentiometric gradients may be steep in the Piedmont. Seasonal fluctuations in the water table are common in response to rainfall. Local observations of the water table rising and falling between 8 feet to 14 feet are common. Perennial streams are fed by bank seepage and upwelling groundwater along the course of their lengths.

SUBSURFACE INVESTIGATION PROGRAM FOR THE TUNNELS AND SHAFTS

The geotechnical and hydrogeological field investigations for the WSP comprised 25 deep borings and 30 shallow borings. The deep borings were advanced along the proposed tunnel alignment with the main purpose of characterizing the bedrock conditions near the tunnel horizon. The shallow borings were drilled at the locations of proposed shafts and surface structures with the primary purpose of characterizing the overburden soil conditions, including information on the transition from soil to rock. Drilling occurred in phases from August 2014 through August 2016 in concert with an evolving design.

Prior to initiation of the geotechnical investigation, readily available, relevant geologic data was summarized and reviewed, and some field work was performed. Ground conditions along previously constructed tunnels proximate to the WSP tunnel were also reviewed. In addition, data provided by geologic field mapping and other available background information were used to complete lineament and structural geologic analyses, see Kath (2017, this volume).

The geotechnical investigation was developed based on information contained in the background reports developed from the geologic mapping and associated investigative work. Triple-tube HQ coring was selected to obtain rock samples. In addition to coring, double-packer permeability testing was performed on most of the deep vertical boreholes. Once cores were extracted, they were logged and photographed.

Once drilling was complete, a suite of borehole geophysical tests was run in 21 of the deep borings. This provided the following information: optical and acoustic televiewer logs, full wave sonic logs, fluid temperature and conductivity logs, natural gamma logs, single point resistance logs, three-arm caliper logs, and EM flowmeter logs. These tests helped to further characterize the in-situ geologic conditions at depth while also providing hydrogeologic information and joint orientation data used to create stereoplots.

Following core analysis and geophysical testing, pumping test locations to determine overall hydrogeologic conditions were selected. The locations were selected based on the completed geologic mapping and proximity to identified geologic controls that were expected to influence groundwater movement once tunneling began. Of the three locations chosen, two yielded insufficient groundwater (as determined through air lift testing) to conduct the tests, and the pumping test holes were abandoned. Consequently, only a single pumping test was performed. It was run for 24 hours, and recovery was measured immediately following shutting off the pump.

The depth of the tunnel (greater than 400 ft. in areas) warranted in-situ stress testing. Agapito and Associates conducted the in-situ stress testing in three of the deep borings and attempted 12 tests, of which 7 were successful. They used the over-coring method as developed by Sigra, Pty of Brisbane, Australia. The purpose of this testing is to determine the magnitude and direction of the horizontal principle stresses. The results were factored into tunnel excavation support design.

Subsequent laboratory testing to determine the properties of the observed rock types was performed. These tests include unit weight, unconfined compressive strength, Cerchar abrasivity, Brazilian tensile strength, acoustic velocity, point load index strength, petrographic analyses, x-ray diffraction, and abrasivity/drillability tests.

Two of the three main project sites were scrutinized during the last phase of the geotechnical subsurface exploration program: The Peachtree Drop/Construction Shaft and the Hemphill sites. During the initial site investigation, deep boring RWB-15 at the Hemphill site encountered degraded rock conditions and borehole stability was a constant issue. During the evolving design, 3 additional borings were drilled to help characterize this site. These included permeability testing and borehole geophysics. Additional tests were run as RWB-15 was considered too risky to place any tooling in the borehole. During this time, while shaft configurations evolved, potential impacts to the existing HWTP reservoir were constantly evaluated.

Construction records for the R.M. Clayton Construction Shaft, built for the North Avenue tunnel as part of the West Area CSO Storage Tunnel were reviewed, as the Peachtree Drop/Construction Shaft is approximately 125 ft away. Construction photographs of the R.M. Clayton Construction Shaft depict deep weathering in the shaft. So, shallow borings were drilled around the perimeter of the Peachtree Drop/Construction shaft to determine the thicknesses of the subsurface zones. Typical of the Piedmont, depths to different subsurface zones may vary substantially over short distances.

Lithologies along the Tunnel Alignment-

The majority of the proposed tunnel alignment is located in the Clairmont Mélange, with the latter portions in a zoned feldspar gneiss followed by Brevard Zone black and white mylonites. The descriptive text that follows is taken from the Geologic Report (1) prepared by Petrologic Solutions as part of the preliminary geotechnical investigation (see Kath (2017, this volume). The order of the four geologic unit descriptions (Clairmont Mélange, Zoned Feldpsar Gneiss, Black Mylonite, and White Mylonite) are from the quarry to HWTP and then through to the CWTP .

- The majority of the Contorted Unit [of the Clairmont Mélange] consists of a sphene-epidote-muscovite-biotite-quartz-feldspar gneiss, medium-grained, schistose in part; interlayered with sphene-epidote-muscovite-quartz-feldspar-biotite schist, medium- to coarse-grained; garnets may be present, but are small and scarce. Hornblende gneiss/amphibolite lenses and layers (commonly boudinaged) are common. Contains, in many places, lenses and discontinuous layers of unfoliated granite on a scale of feet and ten's of feet. Concordant and discordant quartz veins are common. Pegmatitic layers and coarse pegmatites up to 60 inches thick are abundant and characteristic; shear foliation in the gneiss/schist wraps around the coarse pegmatites and small bodies of granite, which are generally not sheared.

This rock mass is extremely contorted; foliations are quite variable over short distances, and are generally low-angle and undulatory. Random fractures are abundant; through-going joint sets are scarce and not well-developed.

- o The zoned feldspar gneiss consists of an epidote-muscovite-biotite-quartz-feldspar gneiss, fine- to medium-grained, with disseminated very coarse zoned feldspar crystals; very feldspathic overall; deep weathering is characteristic.
- o The Brevard Zone black mylonite is generally composed of biotite, quartz, and feldspar. This unit is typically extremely fine-grained and weakly foliated. Where the foliation is better developed, the rock is shown to be very contorted. In most outcrops, the black mylonite is dark gray to black and locally contains thin light colored layers of white mylonite (see rock unit 2B description). Weathering of this unit generally yields a reddish brown to red, uniform fine clayey residuum.
- o The Brevard Zone white mylonite is interpreted to be sheared granite. This mylonitized granite is composed of muscovite, quartz, and feldspar; much of the feldspar is pink and coarse-grained. Shearing was pervasive and produced a well-developed shear foliation. Reduction in grain size was not as extreme as in Rock Unit 2A. Weathering of this unit generally yields a white to tan, uniform fine clayey residuum.

At the time of writing, rock mass conditions encountered during construction of the quarry shafts and TBM tunnel are consistent with the information as provided in the preliminary geologic report. Foliation is quite contorted over the scale of the excavation and degrees of schistosity vary across the excavation.

QUARRY DESIGN

Quarry Highwall Evaluation-

During an earlier phase of the project, the DWM conducted a study of the quarry highwall stability (2). The objective of this phase was to determine if there were any significant stability issues that would jeopardize the use of the quarry as a water storage facility.

The evaluation of the highwall was focused on the long-term stability of the highwalls during operation of the quarry as a reservoir. As discussed in a following section, highwall stability during construction is managed by the Contractor responsible for the tunnel and shaft construction.

The main items included as part of the highwall evaluation included;

- Review of geological data collected during design and construction of a tunnel located approximately 700 feet east of the quarry,
- Review of exploration drilling data provided by the previous quarry operator and discussion with the previous quarry operator's staff regarding quarry highwall stability,
- Field geologic mapping in the quarry and around the top of the quarry, and

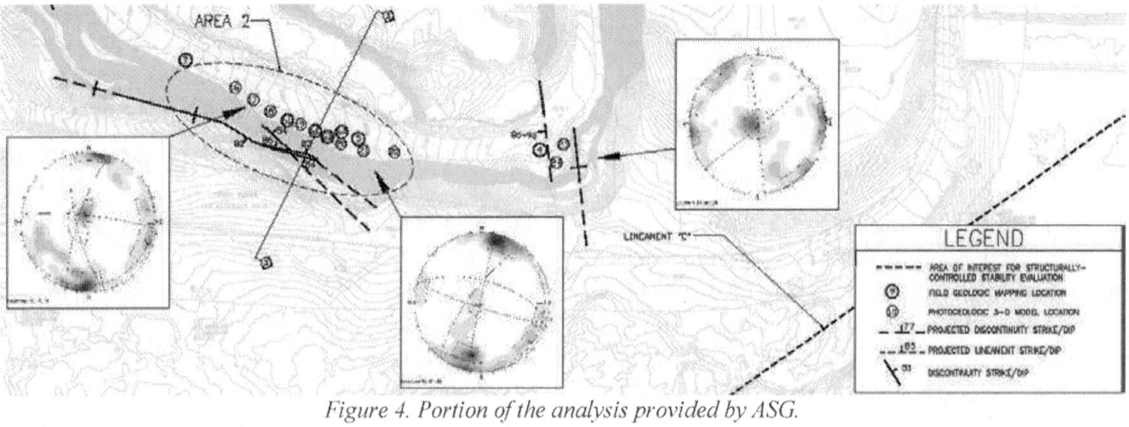

Figure 4. Portion of the analysis provided by ASG.

- Photo-geologic mapping of portions of the quarry highwalls.

Due to the height of the quarry walls, and limited access to the quarry walls, photo-geologic mapping was used to collect structural data of the discontinuities exposed in the quarry highwalls. Model processing and mapping were performed using Sirovision, a rock slope modeling and photo-geologic mapping computer program developed by the Commonwealth Scientific and Industrial Research Organization (CSIRO) Mining and Exploration Group based in Brisbane, Australia.

The structural and photo-geologic mapping found that the general dip of foliation ranges from approximately horizontal to approximately 20° and the dip direction generally ranges from southwest to east. Foliation undulates throughout the quarry at a scale of tens of feet between crests on the foliation surfaces, and locally may dip up to 25° in any direction at any particular location. Foliation is reflected by the central pole clusters shown on the stereonet plots on Figure 4.

Projections of individual fractures and stereonet plots of great circles representing fracture sets are shown on Figure 4. Fractures observed in the highwall generally tend to dip at angles greater than 60° (high angle fractures). The foliation fractures tend to be rough and undulating, and tight or closed with no alteration or infilling. Foliation fractures tend to have low persistence relative to the scale of the highwall (trace lengths were observed to be generally less than 30 feet), and the spacing between foliation fractures is irregular, but generally greater than 2 feet. The high-angle joints were typically rough and planar, stepped, or undulating; fresh to slightly weathered, with no infilling. High-angle joints tend to be moderately widely to extremely widely spaced (from 2 feet to more than 20 feet apart).

The highwall evaluation did not identify any large scale features that would prevent the quarry from operating as intended. Localized areas with potential for rock falls were identified. These areas included zones with blast damage to the quarry walls and zones of localized jointing. The project design included methods to control rockfalls near the tunnel portal during operation, as well as during construction, which are described in following sections.

Tunnel Portal Stabilization-

The contract stipulated that while final design of the drape was specified, safety during construction was the responsibility of the tunneling contractor. Therefore, substantial scaling program was undertaken by contractor, ATJV, to provide safe egress and ingress to the quarry bottom and TBM location. Scaling around the quarry rim took place from April through August 2016. While scaling of the quarry could last indefinitely, following initial inspection, ATJV implemented a scaling protocol that requires visits quarterly to inspect the rockmass and quarry rim. An outcome of this plan is that ATJV and their subcontractor conduct daily and periodic inspections of the highwall around the perimeter of the quarry that has resulted in additional scaling.

To secure the approximate 300-foot-tall rock face above the tunnel portal at the base of the quarry, a designed stabilization system was included in the contract documents. The system covers the full depth of the quarry over a width of approximately 400 foot centered over the TBM tunnel portal. The general area of stabilization is shown in Figure 5.

Although the stabilization system was designed as part of the "permanent works," ATJV came up with an innovative way to combine the permanent stabilization system with supplemental rockfall measures so that the overall system could function as both temporary and permanent works.

The stabilization system consists of TECCO 3 mm mesh from Geobrugg and rock dowels in the locations that are identified as locations of potential rock wedge failures. Canopies were installed as additional protection to workers at the two portals in the quarry as shown on Figure 5. The canopies are designed to catch any rocks that may come loose and fall behind the drape above the portals. At the portals for both the tunnel and 636' adit, 20 foot long spiles are installed along the crown to stabilize more fractured ground.

Figure 5. General area of tunnel portal stabilization area and the two canopies.

TUNNEL DESIGN

The tunnel is about 24,000 feet long and 250 feet to 450 feet below ground surface. It is sloping up from the quarry to the drop/construction shaft at CWTP with a grade of 0.2% and will be partially concrete lined with an internal (lined) diameter of 10 feet. The service life of the final lining system is designed to be 100 years.

Tunnel Initial Ground Support-

The design provided for a two-pass tunnel support system, which is common for Atlanta area tunnels. Excavation ground support will be installed immediately following the TBM excavation to stabilize the tunnel and provide a safe work area. The ground was categorized into three ground types (Types A, B and C) for support based on rock mass properties with three excavation ground support types installed, respectively. Type A support consists of two 5-ft long double corrosion protection dowels as both excavation support and permanent support, since most of Type A ground is not anticipated to be concrete lined. Type B support consists of four 5-ft long friction dowels with welded wire mesh, and Type C support consists of steel ribs with welded wire mesh as lagging. Both Type B and Type C ground will be concrete lined.

Tunnel Permanent Lining-

Following completion of TBM tunnel mining, both Type B and C ground will be lined, while most of the Type A ground is anticipated to remain unlined. The minimum lining thickness is designed to be 12 inches, not only for sustaining the design loads but for facilitating quality concrete placement. The double corrosion protection dowels installed in unlined tunnel sections of Type A ground is considered as part of the permanent support system and will support the ground during the tunnel service life.

As the tunnel is part of the water storage facility, the permanent lining system not only needs to support all the external loading, including rock load and groundwater pressure, but also to sustain the internal water pressure, which is about 300 ft. head. Under certain conditions the internal pressure could result in tension loads in the concrete lining; as such, reinforcement is designed for the lining in Type B and C ground since such ground is expected to provide less constraint than Type A ground. The transient pressures during filling the tunnel are also considered in the lining design.

SHAFT DESIGN

As aforementioned, the system consists of 10 shafts with different sizes, depths, and construction techniques. Pump station shafts at the quarry and the drop/construction shaft at the CWTP will be built with conventional drill-and-blast methods from the top down. The tangential drop shaft and riser shaft at the quarry will be raise-bored from the bottom up. The five pump station shafts at HWTP will be drilled from the surface with blind boring techniques. To the authors' knowledge, the five 9.5 ft. diameter blind bore shafts at HWTP will be the largest and deepest shafts in Piedmont-type geology to use this technique.

Blind Bore Shafts at Hemphill Water Treatment Plant-

As shown on Figure 6, the five pump station shafts will be constructed using blind bore techniques since surface blasting is prohibited at HWTP due to the existing adjacent reservoirs. Upon completion of the five 11-ft diameter steel casing installations in overburden, drilling of the 9.5-ft diameter 400-ft deep blind bore shafts will start from the surface into rock through the steel casings.

Two blind bore rigs will be mobilized to meet schedule requirements. Each rig has a rotary table that provides the torque or turning action for the reamer. Throughout the entire shaft development, both the shaft and the hollow drill string are filled with water to create two independent columns of water. The water column inside the drill string is made much lighter by injecting compressed air. The heavier water column inside the shaft thus pushes down and across the bottom of the shaft. The water is then forced through a small opening on the reamer body and displaces the lighter water in the drill string to create upward flow or reverse circulation. The reverse circulation generates tremendous vacuum at the reamer opening and removes the cuttings from the face. Maintaining a constant water level in the shaft during the entire drilling operation is critical. In addition to cutting removal, the water also provides outward pressure on the shaft wall to improve the shaft stability. The returned water from the shaft is collected in an adjacent settling pond, and the water can be re-circulated to the drilling operation after the cuttings have been settled out.

In order to meet the verticality tolerance, a pilot hole is required for each shaft. The pilot hole will be directionally advanced utilizing an optical technique that allows continuous monitoring for deviation. Once completed, an optical survey will be performed to verify that the pilot holes meet the required verticality.

Upon completion of the blind bore drilling, a 76" ID steel pipe with 1-inch wall thickness will be lowered into the shaft and grouted in place in the wet. The steel pipe is provided in 40-ft long sections that will be welded together. All welds will be ultrasonically tested. After the shaft construction, the vertical turbine type pumps will be installed inside the steel casings.

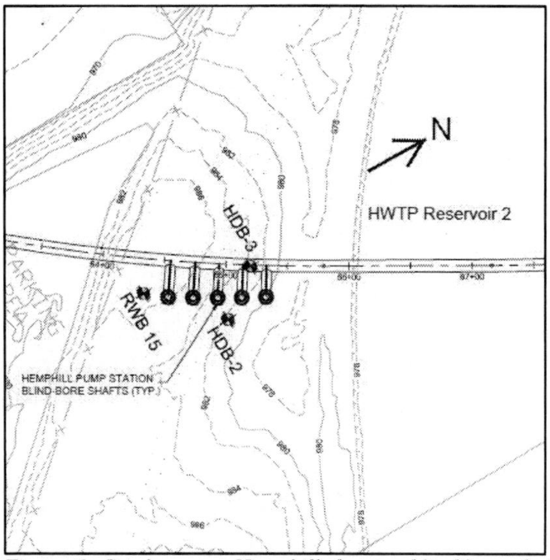

Figure 6. Site layout at Hemphill showing blind bore shafts and adits connecting to the main tunnel.

Pre-Excavation Grouting-

The Hemphill Site includes the construction of a 136 million gallons per day (136 MGD) firm capacity raw water pump station (Hemphill Pump Station or HPS), consisting of 5 pumps. The 5 pumps are each housed in a shaft, all of which are located less than 100 feet from Raw Water Reservoir #2 at the HWTP (see Figure 6). The construction of these shafts poses a significant risk to the unlined reservoir. As such, a shaft pre-excavation grouting program was designed for the soil to rock transition zone and rock zone to greatly reduce the chance of communication between the reservoir and the 5 pump station shafts during construction.

During the geotechnical investigation for the project, the Hemphill site was scrutinized for two reasons. First, the City indicated that all risk associated with inadvertent dewatering of the Hemphill Reservoir due to construction of any aspect of the project was to be kept to an absolute minimum. Second, given the results from the initial borings and subsequent borings, poor ground conditions were identified within the limits of excavation. These ground conditions required mitigation to facilitate excavation with the blind bore shaft sinking technique.

The most practical mitigation method was determined to be pre-excavation grouting of the area. The pre-excavation grouting program addressed these concerns by mitigating risk for the reservoir through consolidation of the rock mass to lower permeability of the rock mass and reduce the potential for loss of drilling fluids during blind bore operations. During design, a third risk was identified that is also addressed through the pre-excavation grouting program. This is the potential for catastrophic fluid loss during blind bore shaft sinking after the tunnel passes through the area, thus flooding the tunnel excavation.

As design of the HPS was fluid and changed during the course of the project, the grouting program evolved as well from preliminary layouts addressing conventional shaft configurations, shifting to the present blind bore shaft configuration. Conventional grouting layouts for shafts were not considered, and a design more typical of underground chambers was implemented. This was due to needing an increased area of reduced rock mass permeability for protection of the reservoir.

As noted, the need for protection of the tunnel from potential flooding during blind bore shaft sinking also factored into this decision. Initially,

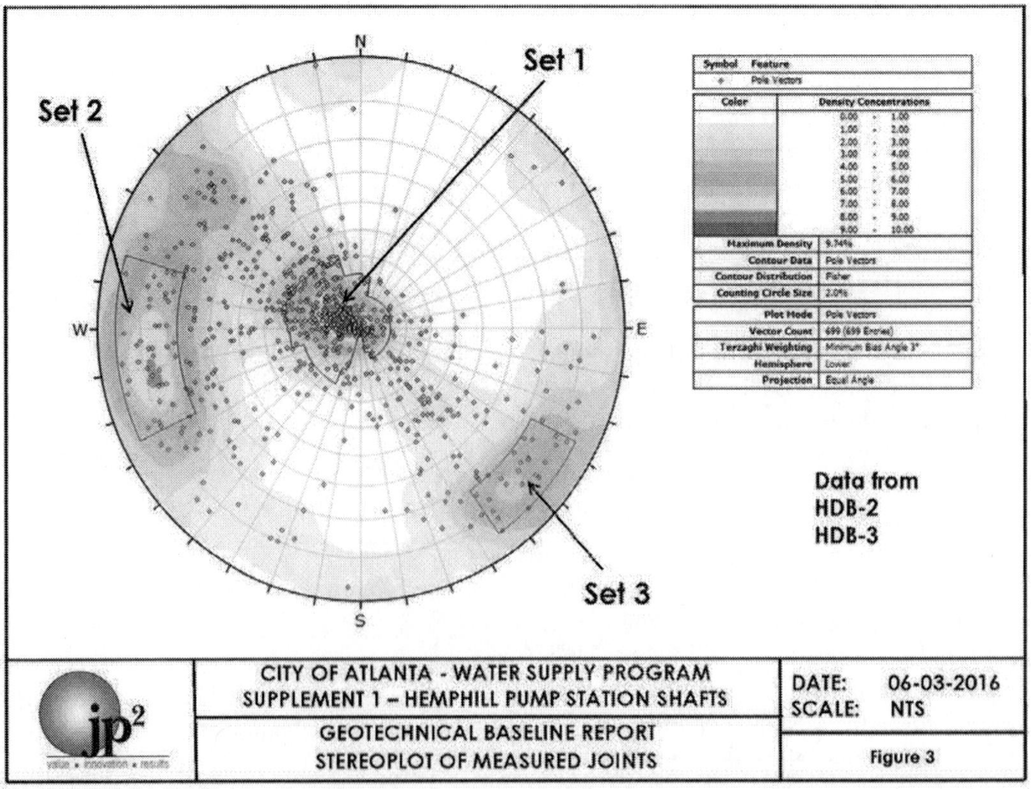

Figure 7. Stereoplot from the Geotechnical Baseline Report of geophysical information showing the high angle joint sets.

all the pre-excavation grout holes were planned to be vertical with primary holes on 16-foot centers, as well as the secondary grout holes. This resulted in a battered spacing of 8 ft. between the primary and secondary grout holes.

Additional borings, HDB-2 and HDB-3, were drilled at the site in January 2016 while site design was underway and the initial pre-excavation grouting program had already been designed.

Results of borehole geophysics from the additional borings were received a week before the Hemphill pricing set of Contract Documents was to be released. Analysis of the geophysical data indicated two primary joint sets that were steeply dipping (>75°) as shown on Figure 7.

Geophysical data also indicated numerous fractures within the three identified joint sets, which contained apertures ranging from 0.25in. to 5in. While open fractures within the foliation joint set were not considered an issue with vertical grout holes, potentially missing open fractures within the two high angle joint sets was judged to be a risk to both the reservoir and the blind bore shaft sinking operation. Consequently, the grout hole orientation was changed from vertical to inclined 10° off vertical at a bearing of 260° (see Figure 8).

This orientation allows for a higher potential for intersecting all the identified features as indicated from the geotechnical investigation and analysis (while staying within the footprint of the surface site), thus reducing the potential for the identified risks to occur.

A significant variable in pre-excavation grouting programs is the grouting shut-off criterion. For this project the shut-off criterion is defined as a grout injection rate of 1/4 gallon per minute or less, as measured each minute for five consecutive minutes at 100% of the required grouting pressure and constant grout consistency.

CONCLUSIONS

The City of Atlanta Water Supply Program is a large, multi-faceted construction project that incorporates many "firsts," including the deepest of all the Metro-Atlanta tunnels. An evolving design allowed for portions of the project to be under construction while other elements were still under design. The WSP tunnel project incorporated many criteria into the design including pre-excavation grouting, tunnel lining analysis, blind bore shaft design, as well as a substantial quarry highwall stabilization program. These design elements are all in place to secure the City of Atlanta's drinking water supply. As the largest re-purposed quarry in North America, the City of Atlanta is once again leading the way.

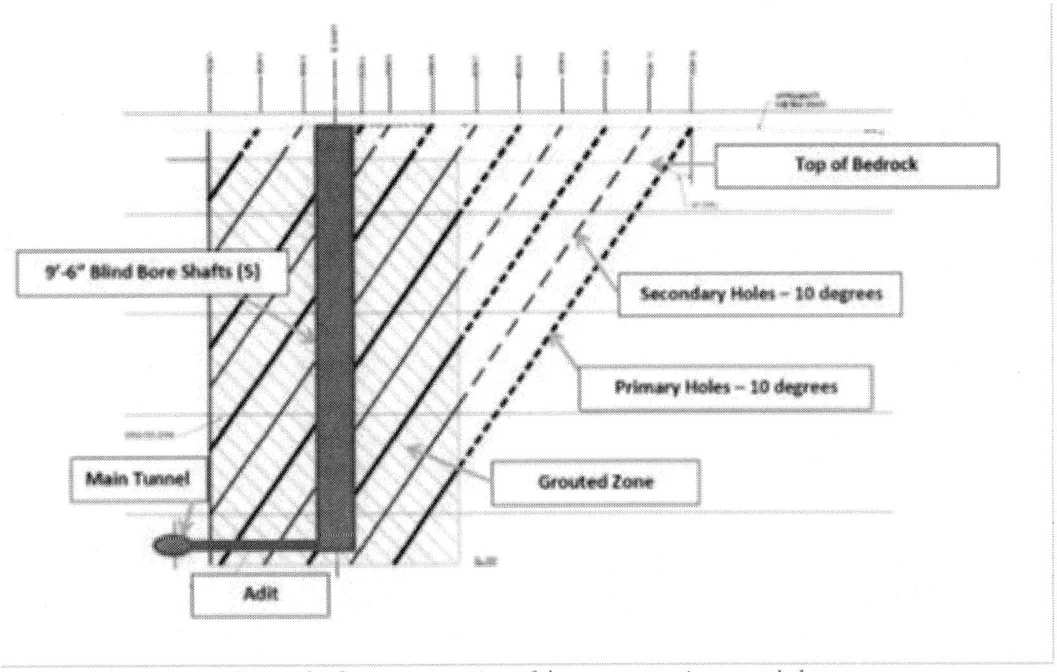

Figure 8. Cross-section view of the pre-excavation grout holes.

REFERENCES CITED

Petrologic Solutions, Detailed Geologic Mapping along the Proposed City Of Atlanta Raw Water Tunnel Alignment, Fulton County, Georgia, Atlanta, Georgia

Atlanta Service Group, Geotechnical Investigation and Evaluation – Technical Memorandum, Atlanta, Georgia, 2007

DESIGN AND CONSTRUCTION OF A TEMPORARY ROCKFALL MITIGATION SYSTEM AT THE BELLWOOD QUARRY RESERVOIR TUNNEL. PHASE I WATER SUPPLY PROGRAM, ATLANTA, GEORGIA

DAVID J. SCARPATO, P.E.
Scarptec Inc.
P.O. Box 326
Monument Beach, MA 02553

NICK H. STRATER, P.G.
Brierley Associates
167 South River Road #8
Bedford, NH 03110

BRANDON MANAHAN
Apex Rockfall Mitigation, LLC
2660 I Road
Grand Junction, CO 81506

ABSTRACT

The City of Atlanta is commissioning a new 1-mile-long, approx. 13-ft. diameter, lined, water conveyance tunnel as part of Phase 1 of the Water Supply Program. The tunnel will be excavated through bedrock with a TBM and will provide the City with potable water from the soon-to-be-filled Bellwood Quarry Reservoir. Construction of the tunnel and ancillary features was initiated in spring of 2016 and is expected to be complete in 2018. The previously mined Bellwood Quarry will serve as a reservoir to impound and distribute the water.

Prior mining activities have resulted in steep pit slopes, some as high as 350 ft., with an abundance of loose rock. In order to help maintain a safe and functional site for site access and tunneling, a temporary rockfall mitigation system was constructed (and is currently being maintained) above the main water supply tunnel and a secondary adit. Critical elements of the temporary system included post-scaling design and construction of draped netting, rock dowels, and two rockfall canopies. The draped netting and canopies were connected as part of a "slot" system, where falling rocks will be contained behind the drape and subsequently guided into (and arrested by) the canopy system.

This paper details the elements of the temporary rockfall mitigation system being utilized during tunnel construction, and will describe the challenges associated with installation of near-horizontal rockfall canopies at elevated, difficult access locations.

INTRODUCTION

The City of Atlanta is commissioning a new 1 mi.-long, approx. 13-ft. diameter, concrete lined, water conveyance tunnel as part of Phase 1 of the Water Supply Program. Subsequent phases of tunneling will result in the construction of another 4 mi. of tunnels to tie the underground water conveyance system together. Tunneling commenced from within the Bellwood Quarry Reservoir (currently drained), which is located approximately 2 miles from downtown Atlanta. The Phase 1 main tunnel consists of an initial short segment of drill and blast starter tunnel, with the remaining drive completed by means of a bedrock tunnel boring machine ("TBM").

Additionally, two pump station shafts were excavated within 200 ft. of the pit slope face, and a series of adits are being excavated that will connect the main tunnel to the pump station shafts. The previously-mined Bellwood Quarry will serve as a proposed 2.4 billion gallon reservoir with which to impound and distribute the water. The new tunnel system will provide the City with potable water from the soon-to-be-filled Bellwood Quarry Reservoir. The area surrounding the reservoir will be landscaped with walking trails, in what will eventually be designated as Westside Reservoir Park. Construction of the tunnel and ancillary features was initiated in spring of 2016 and is expected to be complete in late 2018.

The Bellwood Quarry has been the site of active rock extraction for over 100 years, providing a source of construction stone and aggregate. The principal lithology exposed in the quarry is biotite gneiss of the Clairmont Formation; however, the bedrock is also frequently referred to as granitic gneiss in public domain geologic literature. Prior drill and blast mining activities were utilized to develop the quarry, and have resulted in steep pit slopes, some locally as high as 350 ft. These exposed slopes have been subject to weathering processes, and as such, presented an abundance of loose rock. The exposed silica-rich bedrock is generally highly fractured, and very hard which can present a challenge to drilling operations.

In order to help maintain a safe and functional site during tunneling, a temporary rockfall mitigation system was constructed (and is currently being maintained) above the main water supply tunnel portal and secondary adit. Critical elements of the temporary system included scaling, post-scaling design and construction of draped netting, rock dowels, and two rockfall canopies. The overall site, rock slope and underground features are shown in Figure 1.

TEMPORARY ROCKFALL MITIGATION ELEMENTS

The temporary rockfall mitigation work at the site consisted of initial highwall scaling, followed by installation of a wire mesh rockfall drape, rock dowels, and two individual segments of rockfall canopy. In addition, system monitoring and maintenance efforts are also being conducted over the tunnel to maximize performance over the construction period. The temporary rockfall mitigation system was designed by Scarptec, Inc. (Scarptec) and Brierley Associates Corp. (Brierley), and was constructed by Apex Rockfall Mitigation, LLC (Apex). Periodic field engineering visits during installation of the temporary system were also completed by the design team. The underground workings are being constructed by Guy F. Atkinson Construction.

Highwall Scaling-

Initial rock slope scaling took place in the spring and early summer of 2016, prior to mobilization of tunneling equipment, with efforts being highly productive. Previous blasting activities and exposure to the forces of weathering resulted in an abundance of loose rock prior to construction activities. In order to

Figure 1. Northerly view of quarry highwall, tunnel and adit (on bench).

minimize the quantity of potentially unstable rock material, Apex completed scaling efforts using manual methods; (e.g., scaling bars, rope access techniques) and mechanical methods; (i.e., pneumatic air bags) which were employed using specialized rope access techniques.

Draped Rockfall Netting-

Draped steel netting was used for both temporary and permanent rockfall mitigation purposes, Figure 2. The temporary application was installed in June 2016 and was intended to mitigate rockfall potential during the 3-yr. period of tunnel construction. The permanent netting application, put forth by the tunnel designer and engineer-of-record (Stantec), considers mitigation of long-term rockfall occurrence to prevent large quantities of rock from clogging the tunnel entrance and impeding the flow of water. Transition from temporary to permanent protection systems will require a series of field-determined retrofits at the end of the tunnel construction period, and are described later in this paper.

Figure 2. Constructed rockfall drape.

Draped netting consists of galvanized G65/3mm Tecco® Mesh manufactured by Geobrugg supported at the crest of the slope by a series of 20-ft. long, ¾-in. dia. IWRC-EIP wire rope cable anchors and a top rope. The draped segment of slope in the vicinity of tunneling operations measures approx. 365-ft. in plan length along the slope crest by approx. 315-ft. in slope height. The top set cable anchors were subject to pull testing at both axial (i.e., vertical) and angled (45°) loading configurations in order to verify minimum load-carrying capacity.

In order to maximize rockfall capture, the temporary draped netting was locally tied into the canopy system. The intent of the connection between the canopy and drape was to create a "slot" with which falling rocks could be contained within the system and could not exit the limits of netting; in other words, the canopy formed the lower limit of the temporary drapery system.

Rock Dowels-

In order to design the temporary canopy-drape netting system, the Scarptec-Brierley design team needed to define the upper limits of rock block size and energy that could potentially compromise the system. Rock blocks greater than this critical size, conservatively assumed to be falling from near the slope crest, would require bolting if such blocks appeared to be loose based on field observations. Based on kinematic calculations and rockfall analyses of rock block free fall from 285 ft. in height, the critical rock block size that could exceed the maximum barrier deflection criterion of 28 ft. was estimated to be a cubic block measuring approx. 2.5-ft. (or the equivalent of 15-c.f.). Rock blocks greater than this size required rock reinforcement to arrest potential movement.

Passive rock dowels were chosen to reinforce potentially unstable rock blocks above both canopy systems due to their relative speed and ease of installation; however, to stiffen up the rock mass and pin down suspect key blocks without the benefit of tensioning requires that additional steel be installed. As such, the initial phase of rock reinforcement called for installation of 74 rock dowels that were marked-out in the field (Figure 3) and submitted to the Owner on plan sheets with calculations.

Figure 3. Rock dowel layout with paint using rope access.

Rock dowels were comprised of 1-¼-in. dia. grade 75 epoxy coated bars fabricated by Williams Form Engineering. Minimum embedment depths by location were provided to rock remediation technicians from Apex, who then drilled and installed the dowel bars using wagon drills. In two instances, temporary wire rope cable lashing was required as a precaution to stabilize rock blocks prior to drilling. Rock dowel lengths generally ranged from 10 to 20-ft. in total embedment length.

Rockfall Canopies-

Initially, a traditional barrier approach was considered whereby a barrier would be constructed along the crest of an intermediate bench slope; however, it quickly became apparent that vertical posts would not work for all locations given the complex geometry of the slope and need for access by tunneling personnel. Therefore, the design team opted for use of two rockfall canopy barrier arrangements, located above the tunnel and adit portals. Both canopies were adapted to the field conditions and would also not restrict construction access by the tunneling crews.

The temporary canopy barriers were constructed with GBE-1000-A rockfall barrier components from Geobrugg that includes segments of G65/4mm Tecco® mesh fabric spanning between the posts. Posts consist of 13.1-ft. long steel sections that are set at 25 ft. centers for a total of four posts with an effective length of 75 linear ft. above the main tunnel portal and adit (Figure 4).

Both canopies were connected to the draped netting as part of a "slot" system, so that falling rocks remain behind the drape and are subsequently guided into (and arrested by) the canopy system. To establish a "closed system", a cut line was established along the Tecco® drape and an additional segment of Tecco® mesh was connected between the drape cut line and the upper portion of the barrier post top cables (Figure 5).

Monitoring and Maintenance-

In order to maximize the reliability of the temporary rockfall mitigation system, the slope and constructed elements described herein are subject to periodic monitoring and maintenance efforts at the frequency of one visit every 6 months unless specific observations or events dictate more frequent monitoring. Geotechnical monitoring efforts generally consist of assessment of the following:

- the capacity and need for cleaning of rock debris within the canopies and drape;
- need for additional slope scaling;
- condition of canopy anchorage elements;
- need for additional spot rock dowels;
- condition of drape anchors;
- assessment of drape damage/over-stressing; and,
- canopy system tensioning and netting sag adjustments

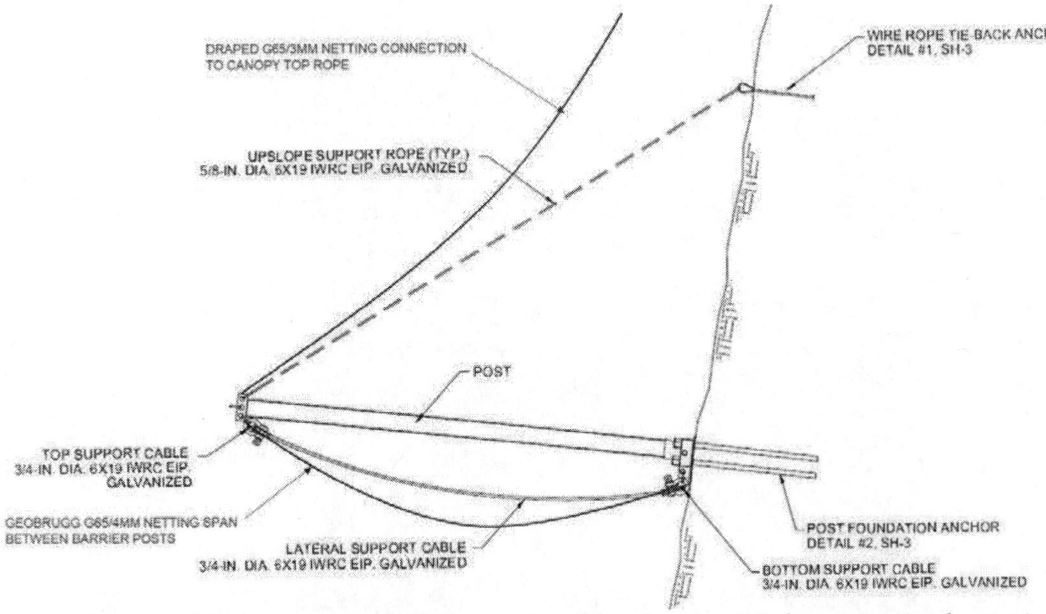

Figure 4. Canopy post section detail (Image adapted from Scarptec- Brierley construction drawings)

Figure 5. Canopy construction and drape tie-in.

Maintenance of the system over the tunneling construction period was (and continues to be) completed by Apex, based on monitoring visit observations. Small fragments of rock debris were removed from the tunnel canopy system in January of 2017 (Figure 6), and minor adjustments to the canopy system cabling were also completed.

TRANSITION TO PERMANENT CONDITION

The permanent condition will consider the effects of nearly full-submergence of the rock slope as the old quarry transitions to a long-term water supply reservoir for the City of Atlanta. Upon completion of the approx. 3-yr. construction period (temporary condition), the

Figure 6. Adit canopy rock debris subject to maintenance cleaning.

two rockfall canopy segments will be removed from service. Any interim connections between the drape and the canopies will be disassembled. The temporary rockfall drape will be converted to a permanent system through a series of minor repairs (if needed) and localized geometric reconfigurations which will be field-fit around the tunnel, adit, and any hard slope breaks. Within approx. 6 months of project completion, the temporary rockfall mitigation design team will consult with the tunnel designer regarding the transition from temporary to permanent system.

CONCLUDING REMARKS

The construction of temporary rockfall mitigation features during Phase 1 of the Water Supply Program are critical to site safety and will help provide for minimized down-time while tunneling continues from below. Design-during-construction efforts required the Apex-Brierley-Scarptec team to evaluate and adapt to field conditions "on-the-fly". Development of the canopy system concept initially posed some challenges given the complex slope geometry, height-related difficult access conditions and multitude of other construction priorities directly below the canopies. These initial challenges were overcome with solid field engineering input from the team during construction.

Although most surface and underground blasting is now complete, additional destabilizing forces from construction vibrations (e.g., TBM advancement), surface water and fracture-controlled drainage, and bedrock weathering may result in periodic rockfall at the site, all of which underscores the importance of this temporary rockfall mitigation system. To-date, the system has performed as intended and will be maintained as necessary to mitigate both the frequency and effects arising from potential rockfall events.

ACKNOWLEDGEMENTS

The author would like to thank the following individuals/entities for their contributions in the work described:

- Mr. Bob Lyne (GeoBrugg North America, LLC)
- Mrs. Lizan Gilbert (Guy F. Atkinson Construction)
- Mr. Brent Laminack (Guy F. Atkinson Construction)

REFERENCES CITED

Turner, A.K., Schuster, R.L., Rockfall – Characterization and Control, National Academy of Sciences, Transportation Research Board, Washington, D.C., 2012.

Geobrugg, Rockfall Product Range GBE-series 150303e.

Geobrugg, GS-1104us 140827 GBE-1000A System drawing, Layout1

Geobrugg, Tecco® Properties, Doc. No. 1.402.29.EN 1403

A REVIEW OF ORE DEPOSITS IN THE CARTERSVILLE DISTRICT, BARTOW COUNTY GEORGIA, INCLUDING RELATED DEPOSITS IN POLK AND FLOYD COUNTIES

TIMOTHY M. CHOWNS
Department of Geology
University of West Georgia
Carrollton, Georgia 30118-3100

RANDY L. KATH
Department of Geology
University of West Georgia
Carrollton, Georgia 30118-3100

INTRODUCTION

The Cartersville district is the oldest continuously active mining district in the Southeastern US (Bearden, 1990) with production of gold, graphite, barite, bauxite, iron and manganese oxides, umber and ocher. Related deposits occur elsewhere in Bartow, Polk and Floyd counties and although outside the main mining district, may throw light on their origin. The eastern and southern part of the district is part of the Blue Ridge physiographic province, while the northwestern part belongs to the Valley and Ridge (Figure 1). The division generally coincides with the Great Smoky (Cartersville) fault in the east and the Emerson (Talladega) fault to the south. Although complicated in detail, the boundary separates highly deformed metamorphic rocks in the east and south from sedimentary rocks, or sometimes low-grade metamorphic rocks to the northwest. Gold and graphite are restricted to the higher grade metamorphic rocks, while barite, bauxite, manganese, umber, and ocher occur within the sedimentary rocks. Iron ores are associated with both terrains. This review is mainly concerned with mineralization in the sedimentary rocks rather than with gold and graphite which are directly related to metamorphism.

Figure 1. Generalized mineral resources map of northwest Georgia. Modified from Smith, Green, Pickering, Auvil, and Furlow (1969).

General Geology

Figure 2 illustrates the stratigraphic section in the Cartersville area. Basement rocks including the Corbin Metagranite and Ocoee metasediments (Proterozoic) have been thrust over younger early Paleozoic sedimentary rocks and are therefore out of stratigraphic sequence. Northwest of the Emerson and Great Smoky faults, clastic rocks of the Chilhowee Group rest unconformably on concealed basement rocks. These are overlain in turn by the Shady Dolostone, sandstone and shale of the Rome Formation, shale and carbonates of the Conasauga Group and cherty carbonates of the Knox Group. Kesler (1950) differs from most workers in restricting the name Shady to the basal part of the older carbonate unit and assigning the upper part to the Rome. Here, we follow the more common practice of assigning the full thickness of carbonates to the Shady and drawing the base of the Rome at the base of clastic facies. We concur with Kesler (1950) that the Shady-Rome contact is diachronous. With the exception of Lower Ordovician beds in the upper part of the Knox Group all these strata are Cambrian in age.

The Role of Weathering, Hydrothermal Alteration and Primary Deposition in Formation of the Ores

A major problem encountered in the interpretation of the ore bodies in Bartow, Polk and Floyd counties is the high degree of weathering and the thickness of residuum and colluvium, which obscures fresh rock. With the exception of barite, all the ore deposits associated with the sedimentary section are oxides or hydroxides of relatively immobile elements (aluminum, iron and manganese). Ores of goethite-limonite, manganese oxide and bauxite occur in veins and pockets particularly over carbonate rocks. Goethite-limonite occurs in the residuum of both carbonates and adjacent metamorphic rocks. All deposits were mined from residuum or colluvium and both mineralogy and distribution suggest that weathering played a critical role in formation.

In northwest Georgia, mineralization occurs in two main carbonate units, the Shady Dolostone (Cambrian) and Knox Group (Cambro-Ordovician). The occurrence of barite and associated sulfides within massive dolostones of the Shady and Knox formations is similar to Mississippi Valley type (MVT) ore deposits described from the east side of the Appalachian basin from Alabama to Newfoundland and elsewhere (Kesler & van de Pluijm, 1990). Although mineralogy varies from district to district, characteristic ore minerals in this type deposit include galena, sphalerite, barite and/or fluorite associated with pyrite. Typically, these ores are associated with massive dolostones which are believed to have been invaded by basinal brines, perhaps associated with hydrocarbons, originating in the deeper parts of the Appalachian basin (now concealed). These carbonates served as paleoaquifers as a consequence of high porosity and permeability

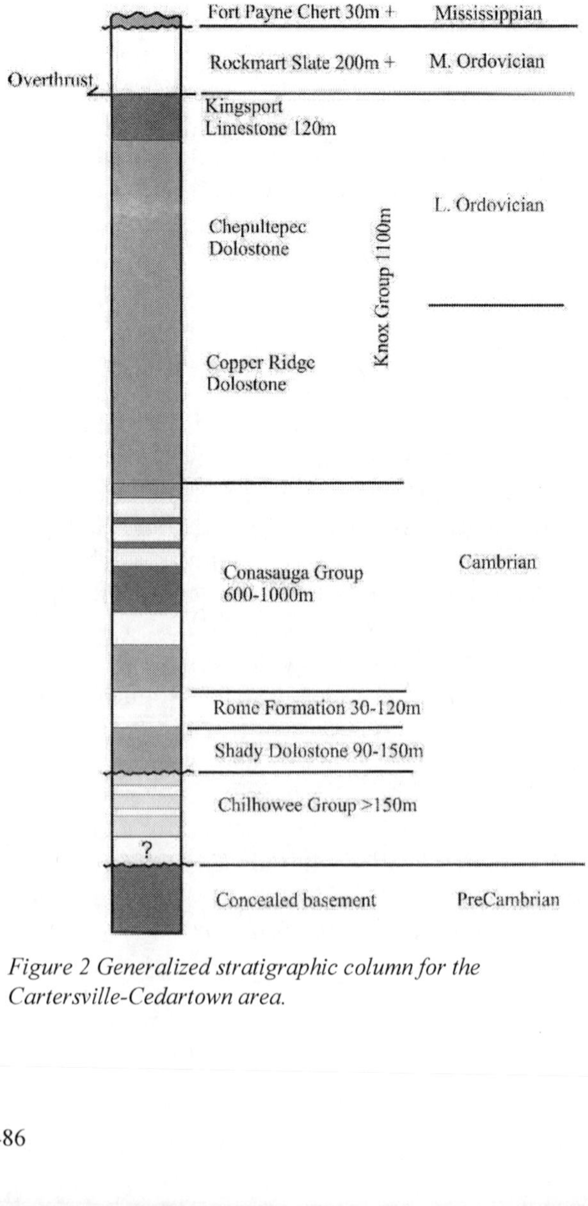

Figure 2 Generalized stratigraphic column for the Cartersville-Cedartown area.

created by dolomitization and the presence of paleokarst as well as their chemical reactivity to acidic brines. Collapse breccias are common, either as a result of preexisting karst or because of solution by hydrothermal fluids. The driving mechanism for the expulsion of these brines is thought to be tectonic loading during one or more of the Appalachian orogenies (Kesler and others, 1995, 1996, 1997; Leach and others, 2010).

The basal unit of the Shady Dolostone is particularly mineral-rich with deposits of iron and manganese oxy-hydroxides not only in Georgia but also in Tennessee and Virginia (King & Ferguson, 1960; Force, 1991). These deposits are closely related to an apparent disconformity between the Shady and Chilhowee Group. This together with their widespread distribution suggests a primary sedimentary source of iron and manganese.

Without weathering, none of these deposits would have been economic; but what are the relative roles of original synsedimentary mineralization, hydrothermal solutions, and weathering in creation of these deposits?

MINERALOGY

Barite-

Barium sulfate is the only non-oxide mineral mined in the Cartersville district. It occurs almost exclusively in residuum (and colluvium) from the Shady Dolostone and owes its abundance to its low solubility during weathering. Outcrops of unweathered dolostone are rare but occur as pinnacles exposed at the bottoms of many barite mines. Most significantly these exposed dolostones contain minor small veins and occasionally blocky masses of barite, but the average barite content is only about 2 percent by weight. Generally, barite crystals are white in hand specimen, clear and transparent in thin section with euhedral tabular rosettes (cockscombs) and appear to have grown within voids in the host rock. However, the occurrence of rare barite pseudomorphs after archaeocyathid fossils indicates that minor replacement occurred in the dolostones. Although, associated with sulfides, the majority of barite postdates the sulfides and is relatively free of sulfide inclusions.

Upon weathering, the dolostone is reduced to an ocherous silty clay residuum containing broken masses of barite. Some residuum appears to have been silicified (jasperoid) and contains angular fragments of barite. Weathering concentrates the insoluble barite and allows mining which would be uneconomic in the fresh rock. To produce a ton of barite requires about 4 to 20 cubic yards of residuum.

Although barite is mainly restricted to the Shady Dolostone, minor occurrences as veins have been noted in the Corbin and Ocoee metasediments close to major faults (Kesler, 1950). These have an important bearing on the age of emplacement.

Sulfides-

Sulfide minerals have been reported from both the Shady and Knox formations. In the Shady pyrite, sphalerite, galena, chalcopyrite, enargite and tennantite occur in small amounts associated with barite (Kesler, 1950). According to Kesler (1950), the sulfides occur mainly around the margins of the barite near the contact with host dolostone, indicating that sulfides precipitated before barite. Pyrite is also common in some chert and in the deeper parts of many of the brown iron ore mines.

Drilling within the Knox of the Cedartown area indicates breccias associated with Paleozoic karst (Chowns, 1993, 1994 unpublished). Pyrite is especially common as veins and cement but galena, sphalerite, quartz, calcite and dolomite are also present. Mineralization occurs particularly below the unconformity at the top of the Knox Group and within the Chepultepec Formation.

The karst at the top of the Knox Group is related to the unconformity at the base of the Middle Ordovician. Drill holes close to this unconformity in the vicinity of the old iron mines at Oremont, near Cedartown, Polk County reveal Paleozoic caverns filled by windblown sand and sandy shale that drape the limestone floor. Veins and breccias contain sulfides in the fresh rock and resemble rocks from the upper Knox of Tennessee (Hoagland and others, 1965; Misra and others, 1989).

Collapse breccias in the Chepultepec are associated with paleosols and also karst related. Again sulfides are present and the cavernous cherts and agates so characteristic of residuum from the Chepultepec are interpreted as siliceous flowstones (Chowns, 1994a).

Iron Oxides-

The principal iron mineral in the Cartersville district is goethite-limonite which occurs in veins and pockets ranging from massive to fibrous and botryoidal, indicating crystallization from groundwater, sometimes in open cavities. It seems to have two different origins, derived in one case from specular hematite and in the other from pyrite.

The basal ~10 m of the Shady Formation, immediately above the contact with the Chilhowee, is especially iron-rich and yields specular hematite in some mines (e.g. Roan Mine). According to Kesler (1950) hematite replaces skeletal debris and sometimes contains relics of ooidal texture reminiscent of sedimentary ironstones. The specular luster indicates low grade metamorphism.

Elsewhere in the Shady where faulting is suspected, pyrite is present at depth leading authors to interpret some of the ores as gossans formed by the weathering of original sulfides (Kesler, 1950: Hurst & Crawford, 1970). Similar gossans are also developed over the sulfides in the collapse breccias in the Knox Group around Cedartown.

Not all the iron deposits coincide with carbonate terrains. Outcrop belts of the Chilhowee and Ocoee groups are also mineralized as well as higher grade metamorphic rocks close to the Great Smoky and Emerson faults. Here goethite-limonite appears to have developed from sulfides deposited in fractured and faulted metamorphic rocks. However once again Tertiary weathering was critical in concentrating and enriching the gossans.

Manganese-

Manganese dioxide commonly accompanies goethite-limonite and is assumed to be a weathering residue. However, its primary source within the host rock is uncertain. In the Cartersville area it is particularly common in residuum from the Chilhowee and lower part of the Shady Dolostone (McCallie, 1926). Prismatic-tabular pyrolucite and massive cryptomelane have been identified by x-ray diffraction (Kesler, 1950) and occur as flattened, grape-like nodules (up to $0.5m^3$) with traces of original bedding. Manganese is also often associated with goethite in the Knox Group (Watson, 1904). Manganese is not known in solid solution in pyrite so it seems more likely the original source was within the hematite or perhaps the carbonate minerals of the Shady Dolostone. Analyses show a small percentage of manganese in most carbonates from the Shady (MnO, 0.01- 0.26 weight percent; Kesler, 1950). Force (1991) also reports small percentages of manganese (ave. 490 ppm) at various horizons within the Shady and Knox carbonates of Virginia.

Ocher and Umber-

Ocher and umber are mined principally from the basal iron rich-beds of the Shady Dolostone (Bearden, 1990, 2008). At first ocher was thought to be associated with quartzite in the Chilhowee Group (Watson, 1906), but Kesler (1950) suggested that the "quartzite" was really a peculiar weathered limonitic chert (jasperoid). Color varies from yellow-brown in ocher to purplish-brown in umber depending on the relative percentages of iron and manganese hydroxides (limonite and wad). Grain size is less than 10 μm. Fe_2O_3 ranges from 55-65 percent and MnO_2 from 0.5 percent in ocher to about 7 percent in umber (Fe 38-45 percent; Mn 0.3-4.4 percent). Based on values in unweathered dolostones given by Kesler (1950), this suggests an enrichment in Mn of around 6-80 times in ocher and umber, respectively, similar in magnitude to enrichment in Fe (65 times).

Silica-

Quartz is common in veins cutting most rock units close to the metamorphic front and is evidently related to deformation. However, carbonate rocks of the Knox and Shady formations are also commonly replaced by chert. Some chert has a vitreous luster and conchoidal fracture and occurs either as early diagenetic nodules or in association with paleosols and karst. In the Knox some agates appear to be siliceous speleothems. In some cases sulfide minerals are associated with this siliceous karst (Hoagland and others, 1965; Hoagland, 1976; Chowns, 1994).

In the Shady and Conasauga there is also a pervasive, diffuse silicification referred to as jasperoid. It has a rough, earthy luster, hackly fracture and ocherous color. This jasperoid occurs mainly in the residuum from carbonate rocks but not within the fresh rock. It is common in Shady residuum but also occasionally invades sandstones of the Chilhowee Group. In many cases it is brecciated or replaces brecciated dolostone with angular fragments of barite, vein quartz and older silicified dolostone. Where voids occur they are lined by minute euhedral quartz crystals. Kesler (1950) concluded that silicification was directly related to barite mineralization but its absence in fresh rock suggests this silicification occurred during weathering (King & Ferguson, 1960; Cressler and others, 1979)

Bauxite-

The first bauxite discovery in America dates to 1887 and was derived from residuum of the Knox Group near Hermitage, Floyd County adjacent to Bartow County (Watson, 1904). Subsequently, similar deposits were opened up south of Rome and also in the Rock Run district of Alabama (Cloud, 1967). All occur as pockets or pipes in the cherty residuum of the Knox dolostones. Some deposits are pisolitic or ooidal, others structureless. Many deposits are bedded and without the cherty debris characteristic of Knox residuum. This lack of normal cherty residuum was at first perplexing (Hayes, 1895; Watson, 1904) but is explained by the development of sinkhole lakes that concentrated the fine-grained suspension load derived by weathering. The role of sinkholes was made clear by drilling in the Rock Run area of Alabama (Cloud, 1967) and is clearly illustrated at the Gray fossil site within the Knox Group of northeast Tennessee (Zobaa et. al, 2011).

Fossil plant debris recovered from lignite in the Booger Hollow bauxite mine in Floyd County south of Rome, indicates a probable Paleocene or Eocene age for these deposits (White & Denson, 1966; Cloud, 1967).

PARAGENESIS

Synsedimentary Minerals-

The host rocks for the majority of orebodies in Bartow, Polk and Floyd counties are massive dolostones formed by the replacement of limestone. Both rock types are highly soluble and it is likely that the development of karst soon after deposition or in association with the unconformity at the base of the Middle Ordovician provided porosity and permeability for the passage of hydrothermal solutions. Carbonates commonly form solid solutions and small amounts of iron and manganese have been detected in both the Shady and Knox formations in Virginia (Force, 1991). Iron was particularly abundant at the base of the Shady, perhaps related to a disconformity. In places specular hematite has been recorded indicating an original ooidal sedimentary rock that underwent low-grade metamorphism (Kesler, 1950).

Mississippi Valley Type Deposits-

The presence of barite together with sulfides in the Shady and collapse breccias with sulfides in the Knox suggests they are Mississippi Valley type deposits and that these units acted as conduits for the migration of the connate water, thought to be the source of mineralization (Kesler and others, 1997). Fluid inclusion studies from barite in the Shady indicate emplacement temperatures between 126-297 °C, equivalent to burial depths of about 1.7 km. (Rife, 1971). Elsewhere it is not uncommon for barite to be associated with galena, sphalerite, pyrite and fluorite in Mississippi Valley type deposits. The widespread occurrence of Paleozoic karst in the Knox Group indicates that this may be the source of porosity and permeability in these massive carbonates. It is unknown whether the Shady was similarly karstic, but the occurrence of large masses of void filling barite would support this hypothesis. Alternatively, void space may have been created by the same hydrothermal solutions that precipitated the barite. Unfortunately very few unweathered outcrops survive and little drill core has been examined. Void space may also have been created by faulting and fracturing.

Since the paleokarst in the Knox Group is pre-Middle Ordovician and if the transporting brines were driven by tectonic loading the date of mineralization lies between Middle Ordovician and Permian. The occurrence of bedding-concordant detrital sphalerite in Early Ordovician karst (Hoagland, 1976) supports a relatively early age; prior to deformation and before occlusion of karst porosity. However, a combination of radiometric and paleomagnetic dating suggests that deposits from the Knox Group in east Tennessee are either middle or late Paleozoic (Acadian or Alleghenian) (Leach and others, 2001). Based on stratigraphic thicknesses in the southern part of the foreland basin two episodes of tectonic loading are identified during the Taconic and Alleghenian orogenies but not during the Acadian.

The ultimate source of mineralizing brines may be from connate water within the carbonate rocks or from adjacent clastic facies. Kessler and others (1988) and Saunders and Savrda (1993) have suggested Middle Ordovician black shale facies (Rockmart Slate) as a possible source of mineralized brines in eastern Tennessee and near Cedartown, Georgia.

Mineralization in the Chilhowee as well as metamorphic rocks east of the Great Smoky and Emerson faults is probably related to faulting and shearing. However, brines may also have been sourced through the Shady paleoaquifer, which underlies the Corbin Metagranite and Ocoee metasediments, in the foot wall of the Great Smoky fault (Kath, this guidebook). In this case hydrothermal solutions may have been newly expelled from the deeper parts of the Appalachian basin or locally remobilized from older deposits. Clearly the mineralization of the faults and fractures along the thrust front is related to Alleghenian events but mineralization in the paleokarst might be older.

Tertiary Weathering-

All the ore bodies in the Cartersville-Cedartown area are located in residuum (or colluvium), especially from massive carbonates. In places, residuum and overlying colluvium may be more than 75m thick and it is evident that the area has been subject to weathering over a long period (Reade and others, 1980). Judging by the high concentrations of iron and especially manganese in small deposits these elements were evidently mobilized and redistributed locally during weathering (Force, 1991).

A number of workers have commented on an apparent elevation control shown by the location of ore bodies. Bauxite and barite deposits are said to cluster between about 250-350 meters above mean sea level, perhaps related to the level of planation during the Tertiary (Watson, 1904). Whether, the kaolinitic clays and bauxites are remnants of more widespread deposits or were always isolated in sink holes is unknown. Plant fossils from lignite deposits associated with bauxite deposits indicate a Paleocene or Eocene age.

OUTSTANDING PROBLEMS

In writing this review we have glossed over many questions regarding the Cartersville mining district. Some of these questions concern the chemistry of the hydrothermal solutions and details of the driving mechanism that are hotly debated and outside the scope of this paper, but some are directly related to the Cartersville area.

Because of a lack of fresh exposures the relative importance of hematite versus pyrite as the source of goethite-limonite is still unclear. While hematite is favored as a source in the basal Shady, pyrite is preferred in the Knox and in faulted Chilhowee. How much iron was present in the original host rock and how much delivered by hydrothermal solutions? Is all the pyrite hydrothermal? Similarly, what is the original source of manganese?

The Mississippi Valley deposit model provides an attractive explanation for the migration of sulfides and barite into the Cartersville district but fails to explain the detailed distribution. Why is barite restricted to certain zones within the Shady while pyrite is most abundant in the underlying Chilhowee? If both were carried by the same hydrothermal solutions, why are they not more intimately associated? Why is most barite free of sulfide contaminants? On the other hand if they arrived separately and at different times, why was one host rock favored over another and how was porosity generated and retained?

Of the brecciation observed in the carbonates and especially the Shady Dolostone how much is related to Paleozoic karst, how much to solution by hydrothermal action and how much to solution-collapse during weathering?

How many episodes of silicification are involved? We have suggested three but without seeing more fresh rock and examining thin sections it is difficult to know. In particular, more work needs to done to characterize the jasperoid, prove it is a product of weathering, and distinguish it from older chert, either hydrothermal or early diagenetic.

A major problem remains attempting to establish original mineralogy and texture through the veil of Tertiary weathering. Based on the relative percentages of iron in fresh rock and ocher (0.6 : 40 percent respectively) weathering has reduced the original volume of rock by around 65 times and the resulting residuum has been let down upon a highly irregular karst surface, perhaps tower karst. We would certainly benefit from a suite of core from fresh rock before the last mines close in the Cartersville area.

CONCLUSIONS

1) Primary ore minerals in the Cartersville district include hematite, barite, and various sulfides. With the exception of hematite, which is probably of sedimentary origin all these original minerals are hydrothermal, probably Mississippi Valley Type deposits.
2) Massive dolostones in the Shady and Knox formations are the principal host rocks with porosity and permeability probably supplied by karst formed soon after deposition. Evidence of paleokarst is best seen in the Knox Group around Cedartown.
3) Although mineralization is most prevalent in paleoaquifers, it is not restricted to them. Sulfide, especially pyrite is widespread in the Chilhowee Group as well as in metamorphic rocks in the hanging wall of the Great Smoky and Emerson faults. In this case porosity seems to be related to faults and fractures.
4) Based on the Mississippi Valley Type model, the most likely source of barite and sulfides is from brines expelled from the Appalachian basin, and the most likely driving force tectonic loading. Two episodes of tectonic loading stand out; during the Late Ordovician and subsequently when the Great Smoky and Emerson thrust sheets were emplaced in the late Paleozoic. The mineralization associated with faults and fractures along the metamorphic front must be related to the latter, but karst mineralization is probably earlier.
5) Variations in mineralogy between different host formations and districts may be related to variation in source and possibly age. Some early Paleozoic ore minerals located in karst may have been remobilized and migrated into faults and fractures during late Paleozoic deformation.
6) Most of the ores mined in the Cartersville district are weathering residues derived from the primary ores during the Tertiary. This includes goethite gossans derived from the sulfides and residual barite. Some iron and manganese were likely present in sedimentary ironstones or in solid solution within the carbonate host and were concentrated by intense weathering.
7) Three episodes of silicification are recognized. The first is early diagenetic and responsible for the discrete chert nodules in the Shady Dolostone. A second generation of silica is associated with the emplacement of the sulfides and therefore hydrothermal. Finally, ocherous residuum and collapse breccias were replaced by jasperoid generated by weathering.
8) Tertiary weathering in the Knox produced sink holes that were infilled with kaolinitic clay residues that were sometimes converted to bauxite. Where Tertiary karst intersected mineralized Paleozoic karst, limonite and manganese accompanied kaolinite and bauxite.
9) If the residual deposits associated with the Shady around Carterville are related to the kaolinite-bauxite deposits, which occur over the Knox Group in Bartow, Polk and Floyd counties, they also owe their significance as ore bodies to Paleocene-Eocene weathering.

REFERENCES CITED

Bearden, S. D., 1990, Occurrence and exploitation of ore deposits in the Shady Dolomite, Cartersville Mining District, Georgia; p. 9-12, in Proceedings 24th Forum on Industrial Minerals, South Carolina Geological Survey,155 p.

Bearden, S.D., 2008, Ochre deposits of the Cartersville Mining District; Georgia Geological Society Guidebook v. 28 p. 55-57.

Chowns, T. M., 1993, Depositional environments and diagenetic history of the upper Knox Group in core holes from near Cedartown, Polk County, Georgia; unpublished report prepared for the Georgia Geologic Survey.

Chowns, T. M., 1994, Stratigraphy of the Knox Group in coreholes from near Cedartown, Polk County, Georgia; unpublished report prepared for the Georgia Geologic Survey.

Chowns, T. M., 1994a, Siliceous karst, paleosols and speleothems from the Knox Group (Cambro- Ordovician), Georgia; Geol. Soc. America, Abstracts with Programs, v. 26, p. 7-8.

Cloud, P.E., 1967, Geology and bauxite deposits of the Rock Run and Goshen Valley areas northeast Alabama; U.S. Geological Survey Bull 1199-N, 74 p.

Cressler, C.W., Blanchard, H.E., Hester, W.G., 1979, Geohydrology of Bartow, Cherokee and Forsyth counties: Georgia Geologic Survey, Information Circular 50, 45 p.

Force, E.R., 1991, Manganese contents of some Lower Paleozoic carbonate rocks of Virginia; U.S. Geological Survey Bull 1919-A, 9 p.

Haseltine, R.H., 1924, Iron ore deposits of Georgia: Georgia Geological Survey Bull. 41, 218 p.

Hayes, C.W., 1895, Bauxite; U.S. Geological Survey 16th Annual Report (1893-4), Part III, p. 547-597.

Hoagland, A.D., Hill, W.T., Fulweiler, R.E., 1975, Genesis of the Ordovician zinc deposits in east Tennessee; Economic Geology, v. 60, p.693-714.

Hoagland, A.D., 1976, Appalachian zinc-lead deposits; in Wolfe. K.H., ed. Handbook of stratabound and stratiform ore deposits; v. 6. p. 495-534.

Hurst, V.J, and Crawford, T.J., 1970, Sulfide Deposits in the Coosa Valley Area, Georgia: Economic Development Administration Technical Assistance project, U.S. Department of Commerce, Washington D.C., in cooperation with the Coosa Valley Area Planning and Development Commission, Rome Georgia.

Kesler, S.E., Appold, M.S., Martini, A.M., Walter, L.M., Huston, T.J., Kyle, J.R., 1995, Na-Cl-Br systematics of mineralizing brines in Mississippi Valley-type deposits; Geology, v. 23, p. 641-644.

Kesler, S.E., Martini, A.M., Appold, M.S., Walter, L.M., Huston, T.J., Furman, F.C., 1996, Na-Cl-Br systematics of fluid inclusions from Mississippi Valley type deposits, Appalachian Basin: Constraints on solute origin and migration paths; Geochim. et Cosmochim. Acta v. 60, p. 225-233.

Kesler, S.E., Vennemann, T.W., Frederickson, C., Breithaupt, A., Vazquez, R., Furman, F.C., 1997, Hydrogen and oxygen isotope evidence for origin of MVT-forming brines, southern Appalachians; Geochim. et Cosmochim. Acta v. 61, p. 1513-1523.

Kesler, S.E., van der Pluijm, B.A., 1990, Timing of Mississippi Valley-type mineralization: Relation to Appalachian orogenic events; Geology, v. 18, p.115-1118.

Kesler, T.L., 1950, Geology and mineral deposits of the Cartersville District; U.S. Geological Survey Professional Paper, 224, 97 p.

King, P.B., Ferguson, H.W., 1960, Geology of northeasternmost Tennessee; U.S. Geological Survey Professional Paper, 311, 136 p.

Leach, D.L., Bradley, D., Lewchuk, M.T., Symons, D.T.A., de Marsily, G., Brannon, J., 2001, Mississippi Valley-type lead-zinc deposits through geologic time: implications from recent age-dating research; Mineralium Deposita, v. 36, p.711-740.

Leach, D.L., Taylor, R.D., Fey, D.L., Diehl, S.F., Saltus, R.W., 2010, A deposit model for Mississippi Valley Type lead-zinc ores; U.S. Geological Survey Sci. Inv. Report 2010-5070-A, 52 p.

McCallie, S.W., 1900, A preliminary report on a part of the iron ores of Georgia (Polk, Bartow and Floyd counties); Georgia Geological Survey Bull. 10A, 190 p.

McCallie, S.W., 1926, A preliminary report on the mineral resources of Georgia; Georgia Geological Survey Bull. 23, 164 p.

Misra, K.C., Kopp, O.C., Paris, T.A., Linkous, M.A., 1989, Mississippi Valley-type fluorite-barite-sphalerite mineralization in the

Sweetwater district, Tennessee; Carbonates and Evaporites, v 4, p.211-230

Reade, E.H., Power, W.R., Keeners, M.J., White, D.H., 1980, Barite deposits in the Cartersville district, Bartow County, Georgia; in Excursions in Southeastern Geology (Geol. Soc. America Ann. Mtg.) v.2, p.379-384; American Geological Institute

Rife, D.L., 1971, Barite fluid inclusion geothermometry, Cartersville mining district, Georgia, northwest Georgia; Econ. Geology v. 66, p.1164-1167.

Watson, T.L., 1904, A preliminary report on the bauxite deposits of Georgia; Georgia Geological Survey Bull. 11, 169 p.

Watson, T.L., 1906, A preliminary report on the ocher deposits of Georgia; Georgia Geological Survey Bull. 13, 81 p.

White, W. S., Denson, N.M., 1966, Bauxite deposits of northwest Georgia; U.S. Geological Survey Bull 1199-M, 42 p.

Zobaa, M.K., Zavada, M.S., Whitelaw, M.J., Shunk, A.J., Oboh-Ikuenobe, 2011, Palynology and palynofacies analysis of the Gray fossil site, eastern Tennessee: Their role in understanding the basin-fill history; Palaeogeography, Palaeoclimatology, Palaeoecology v. 308, p. 433-444.

68TH GEOLOGY HIGHWAY SYMPOSIUM ANNUAL FIELD TRIP: ROAD LOG

KATH, R.L.[1] AND SNEYD, D.S.[2]

[1]*Department of Geology*
University of West Georgia
Carrollton, Georgia 30118-3100

[2]*Petrologic Solutions, Inc.*
3997 Oak Hill Road
Douglasville, Georgia 30135

Wednesday, May 3, 2017:

MILEAGE	*DESCRIPTION*
0.0/0.01	Begin at driveway of Hilton Atlanta / Marietta Hotel & Conference Center, 500 Powder Springs Street, Marietta, Georgia 30064. Turn right on Powder Springs Street.
0.30/0.30	Turn left onto Garrison Road SE
0.80/0.50	Turn right onto Pearl Street.
0.90/0.10	Turn right onto Georgia-280 S/S Cobb Drive. Continue to follow Georgia-280 S.
11.1/10.2	Turn left onto Peyton Road NW.
11.3/0.20	Turn right onto Hollywood Road NW.
11.5/0.20	Turn left onto Perry Boulevard NW. *Norfolk-Southern Inman Yard and CSX Tilford Yard are located to the left. Most of the major rail traffic in and out of the Atlanta area hits these rail yards, which are among the biggest rail yards of both companies in Georgia.*
13.9/2.40	Continue straight onto West Marietta Street NW
14.2/0.30	Turn right onto Lois Street NW (Partial restricted usage road)
14.4/0.20	Turn right onto Lois Street NW (Partial restricted usage road). Arrive at Bellwood Quarry, City of Atlanta Raw Water Tunnel project. Stop buses and unload.

STOP 1: BELLWOOD QUARRY, CITY OF ATLANTA RAW WATER PROJECT

Stop Leaders: Randy L. Kath, Adam Bedell, Wayne Warburton, and David Scarpeto

At this stop we will be dividing the group into three manageable sizes. Assuming we will have about 140 participants on the field trip, we will allow ~70 participants to descend into the quarry to look at the geology, raw water tunnel, slope stabilization measures, and associated infrastructure. The remainder of the group will be split into two groups, one group will assemble in the conference room

1 Mileage in the small font represents distance from the last stop.

for a presentation by Stantec Consulting regarding the overall project. The other group will be having light refreshments. After the Stantec's presentation is complete, the refreshment group will then proceed to the conference room to hear the project presentation. After these two groups are finished, the group that descended into the quarry should be returning and we will then swap.

Figure 1. Oblique aerial photograph of Bellwood Quarry.

The tunnel and shafts are excavate primarily in the Clairmont Mélange. This geologic unit is a tectonic mélange which contains a wide variety of rock types. Generally clasts within the Mélange are contained in a fine-grained biotite gneiss matrix. Higgins and others (2003) describe the Clairmont as a broken formation that contains a variety of exotic clasts. Clasts within the mélange include thinly layered amphibolite and hornblende gneiss, light-gray granofels, light- to medium-gray granitic gneiss, epidosite, meta-granite, quartzite, and ultramafic rocks.

Based on detailed geologic mapping, the only consistent part of the Clairmont is its inconsistency. This unit is highly contorted into ptygmatic-style folds, Figure 2. In the quarry, there are several lithologies exposed that are characteristic of the Clairmont. The majority of the Clairmont consists of an epidote-muscovite-biotite-quartz-feldspar gneiss, fine- to medium-grained, schistose in part. This gneiss can be either light- or dark-gray depending on the biotite concentration. Overall, the gneiss is equigranular and weakly foliated (granofelsic). Interlayered with the gneiss is an epidote-muscovite-quartz-feldspar-biotite schist, medium- to coarse-grained; garnets may be present, but are small and scarce, and well-banded migmatitic gneiss. Hornblende gneiss/amphibolite lenses and layers (commonly boudinaged) are locally present. In many places, lenses and discontinuous layers of unfoliated granite on a scale of 1ft to 20ft are intruded parallel to the low-angle foliation. Concordant and discordant quartz veins are common. Pegmatitic layers and coarse pegmatites up to 5ft thick are abundant and characteristic; shear foliation in the gneiss/schist wraps around the coarse pegmatites and small bodies of granite, which are generally not sheared.

The Clairmont is extremely contorted; foliations are quite variable over short distances, and are generally low-angle and undulatory (see Kath this volume). Random fractures are abundant; through-going joint sets are scarce and not well-developed.

Figure 2. Ptygmatic-style folding in the Clairmont Mélange. Lithologies in the photograph include biotite-schist, migmatitic gneiss, and light-gray, equigranular gneiss.

One of the most notable features within the quarry are the sill-like granitic intrusions. The granite is comprised of biotite-muscovite-quartz-feldspar (biotite and muscovite generally about equal), medium- to coarse-grained. This unit is unsheared and unfoliated, and is very similar to the Ben Hill Granite that crops out west of the quarry.

Just north of the quarry, the Clairmont has been thrust over the Brevard Zone (BZ) rocks along the Katy Creek Fault. The BZ is a major regional zone of deformation in the Piedmont/Blue Ridge that extends from Alabama to Virginia. The BZ has been interpreted by many workers to represent various structural features, ranging from a nappe root zone, to a suture zone, to a terrain boundary. However, most agree that the BZ is a zone of intense shearing which reduced the grain size of the parent rocks forming a variety of tectonic rock types, including phyllonite, button schist, and mylonitic rocks. Generally, the BZ and associated shear fabric are subparallel to lithologic unit contacts. Preliminary age dates of mica (muscovite) growth in the BZ of Alabama by Poole (per com) suggest that the last shearing occurred ~317 Ma.

Rocks within the Clairmont show no evidence of BZ shearing and are interpreted to have been emplaced after BZ shearing culminated. Both the Clairmont and the BZ are intruded by the Ben Hill Granite. The Ben Hill cuts across button schist, phyllonite, and mylonitic rocks of the BZ, and itself is unsheared. The Ben Hill has been dated between 284 and 285Ma. Based on these contact relationships, the Clairmont Mélange must have emplaced onto the BZ between 317 and ~285 Ma.

Figure 3. Driller Mike (Tunnel Boring Machine) being assembled at the bottom of Bellwood Quarry.

14.6/0.20 Load buses and retrace route to West Marietta Street NW. Turn right onto West Marietta Street NW.

14.9/0.30 Turn left onto Marietta Boulevard NW.

15.1/0.20 Turn right onto Huff Road NW.

16.1/1.00 Turn left onto Howell Mill Road. *The Hemphill Water Treatment Plant (WTP), is a manually operated water treatment plant that is staffed 24 hours a day, 7 days per week and has a maximum capacity of 136.5 MGD. The Hemphill WTP is one of three water treatment plants in the City of Atlanta's water treatment system that provides potable water for the City of Atlanta and parts of Fulton County. Two reservoirs located at the Hemphill WTP supply water to the plant; Reservoir No. 1 and Reservoir No. 2. The reservoirs have a capacity of 180 million gallons (MG) and 345 MG, respectively, and is the sole source of raw water for the Hemphill WTP.*

16.3/0.20 Turn right onto 17th Street NW. *Atlantic Station (looking left) is a live-work-play community constructed on the former brownfield site of the Atlantic Steel Mill, which was operated from 1901 to the mid 1970's.*

17.9/1.60 Turn left onto Peachtree Street NE

18.2/0.30 Turn right onto Beverly Road NE

18.8/0.60 Turn right onto Beverly Road NE/Montgomery Ferry Drive NE

18.9/0.10 Slight right onto Maddox Drive NE

18.9/0.01 Turn left onto The Prado NE

19.3/0.40 Turn left onto Piedmont Avenue NE

19.8/0.50 Turn left onto Monroe Drive NE *As we approach Armour Drive there are excellent exposures of the Clairmont Mélange along the entrance ramp to I-85 South.*

21.0/1.20 Turn left onto Armour Drive NE.

21.4/0.40 Turn right into Buzzi Unicerm USA plant. Unload buses.

STOP 2 Buzzi Unicem USA, 348 Armour Drive NE, Atlanta, Georgia 30324

Stop Leaders: Randy L. Kath and Joseph Stika

Oblique aerial photograph of the Buzzi Unicem USA plant location.

A little bit of the history of the cement distribution terminal. It was built for Marquette Cement in 1962-1963 and was originally fed by rail cars from the Marquette Plant near Rockmart, GA. The closed plant in Rockmart is currently used as a cement distribution terminal by Cemex. The terminal at the time was unlike anything in existence and Marquette Cement presented the challenge to Dr. D. A. Polychrome of Atlanta a Professor of Structural Design at Georgia Tech. The walls are constructed of double-tee panels and the upper domed panels (Figure 4) were constructed on site and are designed to high lateral pressure. The facility has the ability to store 1,200 ton of cement and was put into service in March 1963. Over the years; modifications have been made to modernize the truck loading systems and with the currently we can load a 26.5 ton of cement in a dry bulk tanker in 5-7 minutes. The facility is currently owned and operated by Buzzi Unicem USA.

Figure 4. Marquette Cement (aka Buzzi Unicem USA) historic photograph. Note upper panels

While at Stop 1, Bellwood Quarry, we were unable to get a close look at the Clairmont Mélange due to safety considerations. At this stop, we will take a close look at the Clairmont. We will start near the north end of the railroad line for the Buzzi plant and traverse the cut toward the south.

21.4/0.00	Load busses. Turn right on Armour Drive NE.
21.5/0.10	Slight left on Ottley Drive NE.
21.8/0.30	Turn left into the Sweetwater Brewing Company.
LUNCH	
21.8/0.00	Load busses. Turn right on Ottley Drive NE.
22.1/0.30	Slight right on Amour Drive NE.
22.5/0.40	Turn right onto the ramp to I-85 S/I-75 S
22.8/0.30	Merge onto Georgia-13 S *Excellent exposures of the Clairmont in the road cut along Georgia 13 (Buford Highway Connector)*
23.1/0.30	Take the I-85 S/I-75 S exit
23.6/0.50	Merge onto I-85 S
23.9/0.30	Take exit 85 for I-75 N toward Marietta/Chattanooga
24.2/0.30	Keep right and merge onto I-75 N
41.6/17.4	Use the left 3 lanes to keep left at the fork and stay on I-75 N
42.3/0.70	Keep left to stay on I-75 N
58.1/15.8	Take Exit 285 for Red Top Mountain Road
58.5/0.40	Turn left onto Red Top Mountain Road SE
59.0/0.50	Turn right to stay on Red Top Mountain Road SE
59.2/0.20	Turn right onto US-41 N/Joe Frank Harris Parkway SE
61.3/2.10	Turn right toward Old River Road SE
61.3/0.02	Turn right toward Old River Road SE
61.4/0.10	Turn left onto Old River Road SE
64.1/2.70	Cooper Furnace. STOP 3

STOP 3: COOPER FURNACE AND OCOEE SUPERGROUP METASEDIMENTS AND YELLOW BREECHES MEMBER (?) OF THE WILHITE FORMATION; ALLATOONA DAM ABUTMENT

Stop Leaders: Randy Kath and Deana Sneyd

Google Earth overview of Cooper Furnace day use area and Allatoona Dam.

Figure 5. Cooper Furnace.

This old furnace (Figure 5) is typical of the cold-blast, charcoal furnaces operated prior to and immediately after the Civil War. By about 1880 they were replaced by hot-blast furnaces utilizing first charcoal and later coke (Figure 2). It was constructed by the renowned ironmaster, Moses Stroup (1794-1877) who was also responsible for similar furnaces at Round Mountain, Cherokee Co., Tannehill, Tuscaloosa Co., and Oxmoor, near Birmingham, Alabama. The furnace was charged from the top via a trestle, with alternating layers of charcoal, limonitic ore and limestone, fired by blasts of air from a bellows powered by an overshot water wheel. Water was apparently carried by a wooden flume connected to the stream in Hurricane Creek rather than from the Etowah River. Molten iron was tapped at the base of the furnace and fed into 'pigs' on the sand floor of a casting shed while silica and other impurities combined with limestone to form a slag of calcium silicate. The three furnaces at Tannehill, in Alabama, have been restored and No 2 was fired up during the bicentennial of 1976 (Morris & White, 1997).

According to a trail guide prepared by the Corps of Engineers, Cooper Furnace supplied pig iron for the production of nails, spikes, rails, pots, tools, cannons and other related items and was the center for the once thriving town of Etowah founded in the late 1830's by Jacob Stroup (1771-1846), father of Moses Stroup. In addition to the furnace there were spike and nail mills, a rolling mill, foundry and flour mill, as well as a hotel and homes, stretching for about a mile upstream from the iron works.

From the parking lot, follow the restricted access road east toward the right abutment of Allatoona Dam. Good exposures of metaconglomerate, quartzite, schist and phyllite are exposed in the road cut. Lithologies are characteristic of the Wilhite Formation of the Walden Creek Group of the Ocoee Series and have undergone low-grade metamorphism (Gore and Witherspoon, 2013). The Ocoee Series is an upper Precambrian clastic wedge that forms most of the Great Smoky Mountains of eastern Tennessee and western North Carolina. The Wilhite Formation has the greatest variety of rock types of the Ocoee sequence; it is the only formation that contains an appreciable amount of carbonate rock (Hanselman and others, 1974).

The carbonate rocks in the Wilhite Formation are generally in the form small discontinuous bodies completely surrounded by meta-pelitic rocks. These discontinuous carbonate bodies have been interpreted to represent olistoliths in the Yellow Breeches Member of the Wilhite Formation (LePain, 1987). There are good exposures of one of these carbonate bodies in a trail exposure from the parking lot to the main Allatoona Dam visitor's center and there are several good exposed carbonate lenses in the stilling basin for the dam, Figure 6. These carbonates are dark-gray, weakly metamorphosed, and fine-to medium-grained marble. Higgins and others (1996) interpreted these carbonates as metamorphosed Conasauga Formation that are exposed through a small structural window, just south of the Coopers Furnace Window.

Figure 6. Geologic map of the Allatoona Dam foundation by Conn (1949). Areas shown in dark blue are carbonate pods/lenses within the Wilhite Formation.

The contact with the surrounding schist and phyllite is extremely sharp which might be indicative of a fault contact. However, if this carbonate is an olistolith, the contact with the surrounding rock would expected to be sharp.

The occurrence of Conasauga Formation would be highly unlikely in this area. The Corbin Massif and Ocoee Series rocks have been thrust over the lower Cambrian (Chilhowee and Shady) part of the foreland fold and thrust belt. This well illustrated on cross section C-C'-C" by Kath and Crawford (2015), Figure 7. The nearest Conasauga Formation occurs approximately 3.7 kilometers west, in the footwall of the White Fault.

Approximately 80 pounds of this carbonate was sampled and processed in an attempt to find any non-calcareous fossils. Dr. Johnny Waters at Appalachian State University processed this sample and dissolved the limestone in acetic acid, but no fossils were found. However, Higgins and others (1996) reported the presence of tiny deformed phosphatic skeletal fragments from carbonates in the Coopers

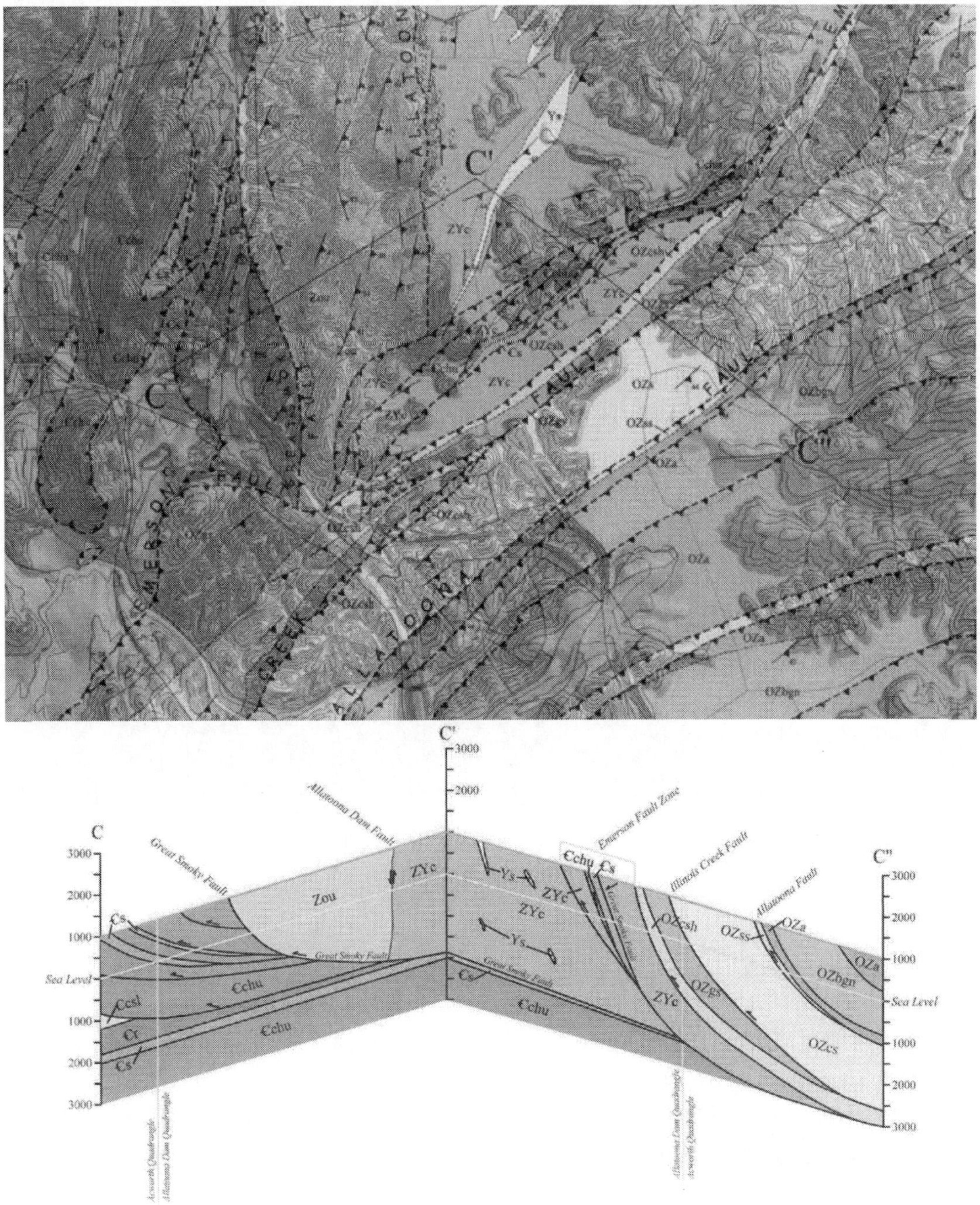

Figure 7. Geologic map and cross section showing the relationship between the Cartersville-Great Smoky, Allatoona Dam, and Emerson-Talladega Faults. Map and cross section are not to scale.

Furnace Window. The presence of these skeletal fragments suggested to Higgins and others (1996, p. 22) that the rocks were Paleozoic and they assigned them to the Lower Cambrian Shady Dolomite. In the Smoky Mountains, Unrug and Unrug (1990) found trilobite, ostracod, bryozoan, and microcrinoid fragments in the carbonates of the Wilhite. Based on this fossil assemblage, they assigned Silurian as the oldest age limit for the Walden Creek Group. A Silurian or younger age is highly unlikely for these carbonates, as there are no known Silurian rocks exposed in the Cartersville District, and the entire Allatoona Complex has been thrust over the basal Cambrian stratigraphy.

On the south side of the Etowah River, there are very good exposures of the Wilhite on the south side of the road. At the gate to the Allatoona Dam Power House, there are good exposures of metaconglomerate overlying phyllite/schist that are underlain by metagraywacke. This exposure was photographed and described by Gore and Witherspoon (2013, p. 198). Bedding is well preserved in the meta-sediments. The bedding is undulatory and dipping to the northeast. The schist/phyllite layer between the metaconglomerate and metagraywacke show some low-amplitude folds that are characteristic of flexual shear between the more competent layers. There are well developed crenulations in the phyllite/schist layers and a well-developed cleavage. The bedding-cleavage angle suggests that portions of the flexural slip folds are overturned to the north-northeast.

Allatoona Dam is located on the Allatoona Dam Fault, Figure 8. Based on detailed geologic mapping under the Allatoona Dam power house and left (south) abutment by Crawford and others (2009), the Allatoona Dam Fault is a high-angle, west-dipping normal fault that separates the Ocoee Series rocks from the Corbin Massif. Based on age relationships and geometry of the fault, we currently interpret this fault to be related to a Rodinian rift basin (graben) which preserves Ocoee Series sediments.

Figure 8. Allatoona Dam and associated geology.

64.1/0.00	Return to bus, turn west on Old River Road
67.1/3.00	Turn left onto Georgia-293 S
68.3/1.20	Turn right into New Riverside Ochre

STOP 4: NEW RIVERSIDE OCHRE- EMERSON BARITE MINE

Stop Leaders: Stanley Bearden and Randy Kath

Introduction

Production of barite from the Cartersville Mining District was first reported in 1894. New Riverside Ochre Company, Inc. (NRO) has been open-pit-mining barite in the Cartersville District since 1924. Barite deposits in the district occur stratigraphically above the ochre deposits, which occur in the lower 10-meters of the Shady. Throughout most of the district, the barite occurs as an epigenetic, stratabound and strataform, structurally deformed, Mississippi Valley Type (MVT) deposit hosted in the Shady. In this model, barium enriched low-temperature hydrothermal fluids migrated along stratigraphic and structural conduits and precipitated in open space in the Shady. Alteration is relatively minor and inconspicuous, with only minor barite replacement of the original carbonate. Most of the ore was deposited in solution cavities, solution enlarged joints, and locally, in fault breccia.

Although the MVT model is widely accepted throughout most of the district, exposures at this property (referred to throughout the text as the Emerson Mine) indicate a secondary concentration mechanism of barite and associated materials. The origin of this deposit is discussed below in the "The Emerson Mine" section.

Barite is grouped into seven general categories based on use: (1) hydrocarbon drilling fluids; (2) extenders and fillers; (3) glass and fiberglass; (4) paint; (5) chemical products; (6) high-density applications, and (7) frictional materials. All barite that is currently mined and processed by NRO is used in the extender/filler, chemical products, high-density markets, and frictional materials. Originally, barite mined in the Cartersville District was used for the manufacture of white paint pigment (lithopone), a mixture of barium sulfate and zinc sulfate.

Principal applications of barite in the filler/extender markets include paint, latex, and urethane foam manufacturing. In painting automobiles, barite is used in the primer coats to retard rusting. It also contributes to the gloss of the top coat. Processors of polyurethane foam use barite in manufacturing such products as floor mats and carpet backings and tennis balls to increase density and improve processing qualities. Other important applications include mold-release compounds in metal casting, brake systems and other frictional materials, acoustical compounds, and in high-density concrete.

Property History

The first barite prospected on this property was conducted in the late 1880's by Mr. W.R. Satterfield. After the prospecting was completed, barite mining activity was conducted by Thompson-Weinman, The Nulsen Corporation, and E.I. DuPont de Nemours & Company. The initial mining activity on this property ended in 1944, but was reactivated in 1976 when Thompson-Weinman and NRO jointly explored the property which resulted in the identification of three distinct mineral deposits: a northern deposit, central deposit, and southern deposit.

Additional exploration using wet rotary drill holes was conducted between July 1978 and July 1979. Based on this exploration, the northern and southern barite deposits were estimated to be comparable in size and tonnage, and could be mined using open pit methods. Unfortunately, the central deposit is too deeply buried to be economically mined using open pit methods.

Eventually, Thompson-Weinman and NRO divided the property with Thompson-Weinman taking the northern deposit and NRO taking the southern and central deposits. NRO mined the southern deposit between February 1996 and May 2004.

In July 2012, NRO purchased the northern deposit. A mine permit was issued in August 2012, and mining began in January 2013. Based on initial ore reserve estimates, this mine will be the 4th largest barite producer in the entire Cartersville District.

Ore Extraction

Barite is extracted from the Emerson Mine using wheel tractor elevating scrapers and track-type excavators that support the main excavation using a Manitowoc 4600 Vicon dragline (see guidebook cover). The dragline has a boom length of 140 feet, a 4 cubic yard bucket, and has a maximum excavation depth of 90 feet. Rubber tire front-end loaders with 7 cubic yard buckets load 35-ton trucks which transport crude ore matrix to stockpiles or to the washer plant. An average of three acres per year is excavated; removing approximately 500,000 cubic yards of earth per year; about half of which is crude ore matrix.

Barite Beneficiation

Crude ore and associated gangue material are processed by NRO using three different plants:

- Washer Plant
- Floatation Plant
- Magnetic Separator Plant

During this fieldtrip we will visit the washer plant; therefore, the washer plant is described below.

Ore Processing (Washer Plant)

Thirty-five ton trucks deliver payload to the washer plant at approximate 15-minute cycles. Cycle time is dictated by washing characteristics of the crude ore matrix (clay to rock content, contained ore percentage, and nature of gangue).

The function of the washer plant is to produce a slurry from which the following products are generated:

(1) jig ore
(2) scavenger jig ore
(3) hutch/screen sands
(4) Akin classifier sands
(5) dornicks
(6) gravel
(7) Akin classifier overflow

The washer plant consists of four log washers in closed circuit with screens and crushers. This circuit delivers wet classified sands and gravels to four sets of two-cell 42-inch Bendalari jigs which feed two sets of two-cell 36-inch scavenger Bendalari jigs.

Jig ore is the first identifiable and marketable product derived by the washer plant. This ore ranges between 94% and 97% weight percent $BaSO_4$. Particle size is from 3/16-inch to 1-inch. Washer plant feed rate and volume are determined by visual assessment of jig ore quality. Scavenger jig ore is the final concentrate from the last cell of the jig sets; it is too highly contaminated with iron, manganese, and silica to be blended with the jig ore and contains too great a weight percentage of barite to be blended with gravel.

In the past, the scavenger jig ore was sold as a drill-mud grade product. Since 1976, scavenger jig ore is recycled onto the crude ore matrix stockpile. $BaSO_4$ content of this ore will range from 40 to 70 weight percent. Hutch sand is a barite-enriched (30% ± 10%) minus 3/16-inch jig product. Minus 3/16-inch screen sands from up-log discharge are mixed with hutch sand. An Akin classifier removes the plus 50-mesh fraction for the log overflow stream. These sands average 10% weight percent $BaSO_4$ and are minus 3/16-inch by plus 50-mesh in particle size.

Hutch and Akin classifier sands are truck-transported to the Floatation Plant ball mill stockpile. Dornicks are plus 5 ½-inch boulders rejected by the bull screen prior to log washer feed. This material is stockpiled for future recycle. The oversize material is a minus 5 ½-inch by plus 3-inch steam rejected by the vibrating screen which segregates non-overflow log product. This material is belt-conveyed and can be hand selected upon demand for immediate barite recovery. Rejected material is stockpiled for recycle or as a source of rip-rap. Gravel is a final product from the jigs. Particle size of the gravel from 3/16-inch to 1-inch and barite content 4% ± 2%. Akin classifier overflow is approximately 3,500 gallons per minute, 15% solids slurry which is transported to a hydrocyclone for floatation recovery or to impoundment as final tailings. This stream contains an average of 2 tons per hour of plus 10 micron recoverable barite.

The Emerson Mine

Although a MVT model is called on for most of the barite deposits in the Cartersville District, the barite-concentration processes of the Emerson Mine are unique when compared to other mines in the district. This mine is characterized by thick accumulations of colluvium that are derived from the Chilhowee Group and Shady Dolostone. The colluvium locally contains well-rounded alluvial cobbles of vein-quartz that are derived from the adjacent Blue Ridge metamorphic rocks.

The Emerson Mine is one of the southernmost barite deposit in the district. Also, it is one of the mines closest to the Blue Ridge metamorphic front and lies west of the Cartersville-Great Smoky Fault, east of the Cloverleaf Fault, and north of the Emerson-Talladega Fault. Detailed geologic mapping by Kath and others (2009, 2010) of the Cartersville 7.5-minute quadrangle places the mine on the northeastern limb of a doubly plunging, west-northwest verging anticline. Bedding measured in Chilhowee quartzite dips steeply to the east. This steep dip is responsible for the high topographic relief of the mine site. The topographically highest part of the mine property, west of the open pit, is underlain by quartzite and phyllite of the Chilhowee Group. The ground surface above the mine forms a northeast facing dip slope on the Chilhowee with an average slope angle between 45 and 55 degrees. The slope angle flattens to less than 40 degrees when underlain by the Shady Dolomite.

Barite is mined at the Emerson Mine by open pit methods. The ore zones within the open pit are developed in colluvium that is derived from the upper, light-gray, Shady Dolomite. Pinnacles of the upper, light-gray Shady are exposed in mine pit (Figure 9). These pinnacles are completely surrounded by thick accumulations of colluvium that contain minor alluvial gravel. The unoxidized matrix of the colluvium is medium- to dark-brown, silty clay that contains angular fragments (sand- to boulder-size) of barite, sub-angular to angular fragments (sand- to boulder-sized) of quartzite, and well-rounded vein quartz cobbles and gravel (Figure 10). Locally within the dark-brown matrix there are zones that contain dark-red oxidized clay with similar barite, quartzite, and vein quartz material.

Figure 9. Field photograph of the Emerson Barite Mine. View looking generally southeast. Tree line on the right side of photograph is a northeast facing dip slope formed on the Chilhowee. Note pinnacles of lower Shady, light-gray dolostone.

Figure 10. Colluvium developed from the Shady Dolomite. Note the white, angular barite and rounded vein quartz cobble.

The dark-brown silty-clay material is best exposed near the bottom of the open pit. In the upper parts of the highwall, this dark-brown material has been locally oxidized during recent weathering, producing a characteristic ocher color typical of Shady residuum. This ocher-colored material is considered to be the more deeply weathered equivalent of the underlying dark-brown material based on the similar presence of abundant barite, quartzite, and rounded vein quartz.

Immediately adjacent to most of the Shady paleokarstic pinnacles there is a thin rind of light-brown to ocher-colored saprolitic residuum that contains a fabric sub-parallel to the fresh dolostone. This rind ranges in thickness from 0.1 to 1 meter and does not contain any ore-grade barite, quartzite, or vein quartz. This material is characteristic of Shady residuum developed from in-situ weathering of the light-gray dolostone seen in other mine pits throughout the District.

Overlying the Shady, the upper benches of the open pit expose at least three separate layers of colluvium derived from the Chilhowee. The uppermost, relatively youngest colluvial layer is characterized by a dark-red, silty and sandy clay matrix with abundant angular quartzite fragments. The contact between this relatively younger colluium and the underlying, relatively older colluvial (middle) layer dips around 32 degrees to the east-northeast, as shown on the left side of the upper bench in Figure 11. The middle colluvial layer is similar to the upper colluvial layer, except that it has far less angular quartzite fragments and is richer in clay matrix. The lowest (oldest) Chilhowee-sourced colluvial layer is exposed above the lower bench shown in Figure 11. This lighter-colored material is characteristic of the light-colored Chilhowee colluvium seen throughout the district.

Figure 11. Colluvium developed from the Chilhowee Group quartzite and phyllite.

As stated previously, the Emerson Mine is not characteristic of other barite mines in the Cartersville

District. Other barite mines in the district are formed mostly by in-situ weathering of the upper light-gray Shady Dolostone. The barite occurs as irregular and sub-rounded masses completely within a light- to dark-brown Shady residuum. At the Emerson Mine, the barite is angular to sub-angular and is completely contained within colluvium. In-situ weathering of the upper light-gray Shady dolostone, combined with down slope movement and gravity accumulation of the barite is responsible for this eluvial-style deposit. (Eluvial deposits consist of soils that are derived by in-situ weathering combined with gravitational movement or accumulation of soils.)

Deposition of the eluvium occurred between the paleokarstic pinnacles of the upper light-gray dolostone. During deposition, this area may have resembled a tower karst topography formed on the light-gray dolostone. Because of the proximity to metamorphic rocks, this area was protected from Paleocene and Eocene weathering; however, by late Miocene and early Pliocene, this area would have been deeply weathered. This late Miocene and early Pliocene weathering has been well documented in other ore districts throughout the world. Further, Miocene epeirogenic uplift of the southern Appalachians (Gallen, Wegmann, and Bohnenstieh, 2013) may have caused increased topographic relief and accentuated colluvium development adjacent to the more resistant ridges that were held up by quartzite of the Chilhowee. The topographic setting combined with proximity to the metamorphic front accounts for the unique nature of this barite deposit.

References Cited:

Gallen, S.F., Wegmann, K.W., Bohnenstieh, D.R., 2013, Miocene rejuvenation of topographic relief in the southern Appalachians: Geological Society of America, Today, vol. , no., pp. 4-10

Kath, R.L., Bearden, S.L., Costello, J.O., and Crawford, T.J., 2009, Bedrock Geologic Map of the Cartersville, 7.5-Minute Quadrangle, Georgia: Georgia Department of Natural Resources Open File Report 09-1, Plate 4

Kath, R.L., Bearden, S.L., Costello, J.O., and Crawford, T.J., 2010, Bedrock Geologic Map of the Cartersville, Georgia, 7.5-Minute Quadrangle: Relationships of the Chilhowee/Shady contact and structural evidence for the Emerson-Talladega and Cartersville-Great Smoky Faults: Geological Society of America, Abstracts with Programs, vol. 42, pp. 92.

68.3/0.00	Return to the bus. Retrace route to Georgia-293 South. Turn right.
69.9/1.60	Continue onto Old Allatoona Road SE
70.1/0.20	Turn right to merge onto I-75 S
70.6/0.50	Merge onto I-75 S
86.3/15.7	Take exit 267B toward Marietta
87.1/0.80	Merge onto Georgia-5 S.
88.4/1.30	Use the right 2 lanes to turn right onto North Marietta Pkwy NW
88.9/0.50	Continue onto Powder Springs Street/South Marietta Pkwy SW
89.4/0.40	Continue straight to stay on Powder Springs Street
89.7/0.30	Arrive at the Hilton Atlanta / Marietta Hotel & Conference Center

END OF FIELD TRIP